Advances in
ORGANOMETALLIC CHEMISTRY

VOLUME 14

CONTRIBUTORS TO THIS VOLUME

V. G. Albano

Kenneth P. Callahan

P. Chini

Ernst Otto Fischer

M. Frederick Hawthorne

James A. Ibers

Steven D. Ittel

M. F. Lappert

P. W. Lednor

G. Longoni

Yoshio Matsumura

Akira Nakamura

Rokuro Okawara

Sei Otsuka

V. S. Petrosyan

O. A. Reutov

Hubert Schmidbaur

Dietmar Seyferth

N. S. Yashina

Advances in Organometallic Chemistry

EDITED BY

F. G. A. STONE

DEPARTMENT OF INORGANIC CHEMISTRY
THE UNIVERSITY
BRISTOL, ENGLAND

ROBERT WEST

DEPARTMENT OF CHEMISTRY
UNIVERSITY OF WISCONSIN
MADISON, WISCONSIN

VOLUME 14

1976

ACADEMIC PRESS New York · San Francisco · London
A Subsidiary of Harcourt Brace Jovanovich, Publishers

COPYRIGHT © 1976, BY ACADEMIC PRESS, INC.
ALL RIGHTS RESERVED.
NO PART OF THIS PUBLICATION MAY BE REPRODUCED OR
TRANSMITTED IN ANY FORM OR BY ANY MEANS, ELECTRONIC
OR MECHANICAL, INCLUDING PHOTOCOPY, RECORDING, OR ANY
INFORMATION STORAGE AND RETRIEVAL SYSTEM, WITHOUT
PERMISSION IN WRITING FROM THE PUBLISHER.

ACADEMIC PRESS, INC.
111 Fifth Avenue, New York, New York 10003

United Kingdom Edition published by
ACADEMIC PRESS, INC. (LONDON) LTD.
24/28 Oval Road, London NW1

LIBRARY OF CONGRESS CATALOG CARD NUMBER: 64-16030

ISBN 0-12-031114-3

PRINTED IN THE UNITED STATES OF AMERICA

Contents

LIST OF CONTRIBUTORS		ix
PREFACE		xi

On the Way to Carbene and Carbyne Complexes
ERNST OTTO FISCHER

I.	Introduction	1
II.	Transition Metal–Carbene Complexes	3
III.	Other Syntheses of Carbene Complexes	6
IV.	Reaction Possibilities of Carbene Complexes	8
V.	Transition Metal–Carbyne Complexes	21
VI.	Reaction of Other Pentacarbonylcarbene Complexes with Boron Trihalides	24
VII.	Reaction of Pentacarbonylcarbene Complexes with Halides of Aluminum and Gallium	27
VIII.	Reaction of Lithium Benzoylpentacarbonyltungstate with Triphenyldibromophosphorane	27
IX.	Reactivity of the Carbyne Ligand	28
	References	29

Coordination of Unsaturated Molecules to Transition Metals
STEVEN D. ITTEL and JAMES A. IBERS

I.	Introduction	33
II.	Theoretical Models	35
III.	Structural Results	37
IV.	Summary	59
	References	60

Methyltin Halides and Their Molecular Complexes
V. S. PETROSYAN, N. S. YASHINA, and O. A. REUTOV

I.	Introduction	63
II.	Methods of Study	64
III.	Structures of Methyltin Halides	68
IV.	Molecular Complexes of Methyltin Halides	76
V.	Conclusion	91
	References	92

CONTENTS

Chemistry of Carbon-Functional Alkylidynetricobalt Nonacarbonyl Cluster Complexes

DIETMAR SEYFERTH

I.	Introduction: General Properties of Alkylidynetricobalt Nonacarbonyl Complexes	98
II.	Synthesis of Alkylidynetricobalt Nonacarbonyl Complexes	100
III.	Chemistry of the Tricobaltcarbon Decacarbonyl Cation	110
IV.	Highly Stable Nonacarbonyl Tricobaltcarbon-Substituted Carbonium Ions	119
V.	Decomposition Reactions and Derived Synthetic Applications of Alkylidynetricobalt Nonacarbonyl Complexes	135
VI.	Concluding Remarks	138
	References	141

Ten Years of Metallocarboranes

KENNETH P. CALLAHAN and M. FREDERICK HAWTHORNE

I.	Introduction	145
II.	Metallocarboranes: Synthetic Methods	150
III.	Twelve-Vertex Metallocarboranes	155
IV.	Thirteen-Vertex Metallocarboranes	167
V.	Fourteen-Vertex Metallocarboranes	171
VI.	Eleven-Vertex Metallocarboranes	171
VII.	Ten-Vertex Metallocarboranes	175
VIII.	Nine-Vertex Metallocarboranes	178
IX.	Oxidative Addition to B—H Bonds	180
X.	Metallocarboranes in Homogeneous Catalysis	182
	References	183

Recent Advances in Organoantimony Chemistry

ROKURO OKAWARA and YOSHIO MATSUMURA

I.	Introduction	187
II.	Hexacoordinate Mono- and Diorganoantimony Compounds	188
III.	Triorganostibine Sulfide	192
IV.	Tertiary Stibines	197
	References	202

Pentaalkyls and Alkylidene Trialkyls of the Group V. Elements

HUBERT SCHMIDBAUR

I.	Introduction	205
II.	Simple Nitrogen Ylides	207

III.	Phosphorus Ylides and Pentaalkylphosphoranes	209
IV.	Arsenic Ylides and Pentaalkylarsoranes	224
V.	Antimony Ylides and Pentaalkylstiboranes	231
VI.	Bismuth Compounds	236
VII.	Related Compounds of Vanadium, Niobium, and Tantalum	236
	References	240

Acetylene and Allene Complexes: Their Implication in Homogeneous Catalysis

SEI OTSUKA and AKIRA NAKAMURA

I.	Introduction	245
II.	Acetylene Complexes	246
III.	Allene Complexes	265
	References	279

High Nuclearity Metal Carbonyl Clusters

P. CHINI, G. LONGONI, and V. G. ALBANO

I.	Introduction	285
II.	Structural Data in the Solid State	286
III.	Structural Data in Solution	306
IV.	Syntheses	311
V.	Methods of Separation	316
VI.	Reactivity	317
VII.	Iron Derivatives	323
VIII.	Ruthenium Derivatives	324
IX.	Osmium Derivatives	325
X.	Cobalt Derivatives	325
XI.	Rhodium Derivatives	327
XII.	Iridium Derivatives	332
XIII.	Nickel Derivatives	333
XIV.	Platinum Derivatives	334
XV.	Bonding Theories	336
	References	341

Free Radicals in Organometallic Chemistry

M. F. LAPPERT and P. W. LEDNOR

I.	Introduction	345
II.	Metal-Centered Organometallic Radicals	349
III.	Other Organometallic Radicals	367
IV.	Bimolecular Homolytic Substitution (S_H2) at the Metal Center of an Organometallic Substrate	370

V. Addition or Elimination Radical Reactions	. . . . . .	381
VI. Appendix		390
References		392
SUBJECT INDEX		401
CUMULATIVE LIST OF CONTRIBUTORS		410
CUMULATIVE LIST OF TITLES		412

List of Contributors

Numbers in parentheses indicate the pages on which the authors' contributions begin.

V. G. ALBANO (285), *Istituto de Chimica Generale dell'Università, Milano, Italy*

KENNETH P. CALLAHAN (145), *Metcalf Research Laboratory, Department of Chemistry, Brown University, Providence, Rhode Island*

P. CHINI (285), *Istituto de Chimica Generale dell'Università, Milano, Italy*

ERNST OTTO FISCHER (1), *Inorganic Chemistry Laboratory, Technical University, Munich, West Germany*

M. FREDERICK HAWTHORNE (145), *Department of Chemistry, University of California, Los Angeles, California*

JAMES A. IBERS (33), *Department of Chemistry, Northwestern University, Evanston, Illinois*

STEVEN D. ITTEL (33), *Central Research and Development Department, E. I. du Pont de Nemours and Company, Wilmington, Delaware*

M. F. LAPPERT (345), *School of Molecular Sciences, University of Sussex, Brighton, England*

P. W. LEDNOR* (345), *School of Molecular Sciences, University of Sussex, Brighton, England*

G. LONGONI (285), *Istituto de Chimica Generale dell'Università, Milano, Italy*

YOSHIO MATSUMURA† (187), *Department of Applied Chemistry, Osaka University, Yamadakami, Suita, Osaka, Japan*

AKIRA NAKAMURA (245), *Department of Chemistry, Faculty of Engineering Science, Osaka University, Toyonaka, Osaka, Japan*

ROKURO OKAWARA (187), *Department of Applied Chemistry, Osaka University, Yamadakami, Suita, Osaka, Japan*

* Present address: Institut für Anorganische Chemie der Universität München, 8 München 2, Meiserstrasse 1, Germany.

† Present address: Japan Synthetic Rubber Co., Ltd., Research Laboratory, 7569 Ikuta, Tama, Kawasaki, Japan.

LIST OF CONTRIBUTORS

SEI OTSUKA (245), *Department of Chemistry, Faculty of Engineering Science, Osaka University, Toyonaka, Osaka, Japan*

V. S. PETROSYAN (63), *Chemistry Department, M. V. Lomonosov Moscow State University, Moscow, USSR*

O. A. REUTOV (63), *Chemistry Department, M.V. Lomonosov Moscow State University, Moscow, USSR*

HUBERT SCHMIDBAUR (205), *Anorganisch-chemisches Laboratorium, Technische Universität München, Munich, West Germany*

DIETMAR SEYFERTH (97), *Department of Chemistry, Massachusetts Institute of Technology, Cambridge, Massachusetts*

N. S. YASHINA (63), *Chemistry Department, M.V. Lomonosov Moscow State University, Moscow, USSR*

Preface

The first volume of *Advances in Organometallic Chemistry* was published early in 1964, and twelve other volumes have appeared since that date. The Editors have sought to produce a series of books containing specialist articles on all aspects of this field. The success of the series, as judged by the reviews of the books published in the journal literature, is due in large measure to the cooperation and help we have received from some one hundred and ten contributors. However, the demand for authoritative surveys of topics in organometallic chemistry derives mainly from the continued resilience of this area of endeavor, one measure of which is the annual appearance of over 2000 primary journal articles.

After a little over a decade of publication it seemed to the Editors that we should arrange for the appearance of a commemorative Volume containing articles by distinguished chemists which would emphasize both the wide scope of organometallic chemistry and its international character.

The number of contributors was necessarily limited by the need to keep the book to a reasonable length. This presented a problem in relation to selection of authors. Our choice is, therefore, a personal one guided to some degree by geographical distribution and the desire to balance transition metal chemistry versus main group metal chemistry.

F. G. A. STONE
R. WEST

Advances in
ORGANOMETALLIC CHEMISTRY

VOLUME 14

On the Way to Carbene and Carbyne Complexes*

ERNST OTTO FISCHER

Inorganic Chemistry Laboratory
Technical University
Munich, West Germany

   I. Introduction 1
  II. Transition Metal–Carbene Complexes 3
      A. Preparation of the First Carbene Complexes 3
      B. Bonding Concepts and Spectroscopic Findings 4
 III. Other Syntheses of Carbene Complexes 6
  IV. Reaction Possibilities of Carbene Complexes 8
      A. Addition and CO Substitution 9
      B. Transition Metal–Carbene Complex Residues as Amino-Protective Groups for Amino Acids and Peptides 11
      C. Addition-Rearrangement Reactions. 13
      D. Substitution of Hydrogen at the α-Carbon Atom . . . 13
      E. Liberation of the Carbene Ligand 14
   V. Transition Metal–Carbyne Complexes 21
      A. Preparation of the First Carbyne Complexes 21
      B. X-Ray Structural Analyses 22
  VI. Reaction of Other Pentacarbonylcarbene Complexes with Boron Trihalides 24
 VII. Reaction of Pentacarbonylcarbene Complexes with Halides of Aluminum and Gallium 27
VIII. Reaction of Lithium Benzoylpentacarbonyltungstate with Triphenyldibromophosphorane. 27
  IX. Reactivity of the Carbyne Ligand 28
      References 29

I

INTRODUCTION

In 1960 I had the honor to lecture at this University about our investigations in the area of sandwich complexes. Today I do not wish to return to the results of these earlier studies, but instead I would like to report on an area of work that has occupied us intensively for several years, namely that of carbene and, more recently, carbyne complexes.

* A Nobel lecture translated by P. Legzdins and G. O. Wiedersatz, Technical University, Munich. Copyright © The Nobel Foundation, 1974.

If one formally replaces one of the hydrogen atoms in an alkane hydrocarbon, such as ethane, by a metal atom (which may also have other ligands attached to it), one obtains an organometallic compound in which the organic entity is bonded to the metal by a σ-bond (Fig. 1a). Such compounds were first prepared more than 100 years ago by Bunsen who obtained cacodyl (tetramethyldiarsine) (*1*), as well as by Frankland who prepared various dialkylzincs (*2*). Later, Grignard succeeded in synthesizing alkylmagnesium halides by the treatment of magnesium with alkyl halides (*3*), and for this accomplishment he was awarded the Nobel Prize in 1912. In addition, the organoaluminum compounds (*4*) of Ziegler, which made possible the low-pressure polymerization of olefins such as ethylene, should be remembered. Professor Ziegler, together with Natta, were honored with the Nobel Prize in 1963.

If one next considers a system with 2 carbon atoms bonded to each other by a double bond, i.e., an alkene molecule, one recognizes a number of separate paths leading to organometallic derivatives. One path involves the replacement of a substituent by a metal atom in a manner that we have just seen and leads to σ compounds that are exemplified by vinyllithium derivatives. Alternatively, only the π-electrons of the double bond need be employed to bind the organic molecule to the metal. One, thus obtains π-complexes (*5, 6*) (Fig. 1b), the first example of which, Zeise's salt, $K[PtCl_3(C_2H_4)]$, was prepared as early as 1827 (*7*). Such metal π-complexes involving olefins are preferentially formed by transition metals; the main group elements, on the other hand, are less able to form such bonds. In this class of compound one can also include sandwich complexes (*8, 9*) in which the bond between the metal and the ligand is no longer formed by just 2 π-electrons but by a delocalized, cyclic π-electron system. A particular example is dibenzenechromium(0) (*10*) in which the chromium atom lies between two parallel and eclipsed benzene rings.

One attains the third variant when one formally cleaves the double bond and attaches one of the resulting halves of the olefin molecule to a

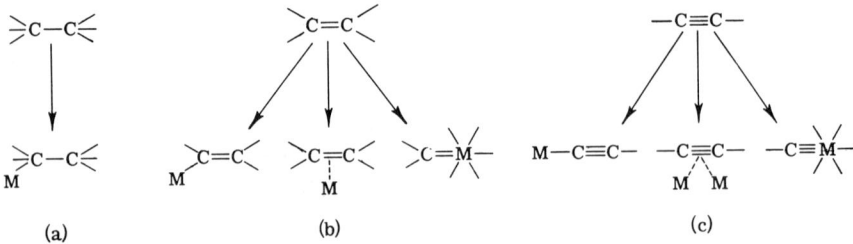

Fig. 1. Production of organometallic compounds from hydrocarbon derivatives, M = a metal or a metal-complex fragment.

transition metal component (Fig. 1b). This concept is realized in the transition metal–carbene complexes in which carbenes {:CRR'} that are short-lived in the free state are stabilized by bonding to the metal. The first part of my account will be devoted to complexes of this kind.

Finally, if one considers molecules with a carbon–carbon triple bond of the type that exists in alkynes, one realizes that there also are three possible paths to metal-containing derivatives (Fig. 1c). One may construct σ-compounds as discussed in the previous cases or one may use both π-bonds to synthesize complexes (11) in which the two metal–ligand bonds exist at a right angles to each other. Let us finally consider a cleaved triple bond in which one-half is replaced by a metal complex fragment; thus we come to the carbyne complexes about which I shall report in the second part of this review.

II

TRANSITION METAL–CARBENE COMPLEXES

A. Preparation of the First Carbene Complexes

In a short communication in 1964, Maasböl and I (12) reported for the first time stable carbene complexes. We had treated hexacarbonyltungsten with phenyl- or methyllithium in ether with the intention of adding the carbanion to a CO ligand at the carbon which, relative to the oxygen, is the more positively charged ligand site. Indeed, we did obtain the lithium acylpentacarbonyltungstates, which were converted to pentacarbonyl[hydroxy(organo)carbene]tungsten(0) complexes by subsequent acidification in aqueous solution (Scheme 1).

We quickly established that these complexes are not particularly stable. They tend to split off the carbene ligand with a simultaneous hydrogen shift thereby liberating aldehydes, a fact that Japanese investigators also discovered (13). Very recently, we learned how to prepare these hydroxycarbene complexes analytically pure (14). Previously, these complexes, without isolation, had been successfully converted to the substantially more stable methoxycarbene compounds by treatment with diazomethane (12).

We soon found a more elegant route to the latter complexes involving the direct alkylation (15) of the lithium acylcarbonylmetalates with trialkyloxonium tetrafluoroborates (16) (Scheme 1) which can be prepared

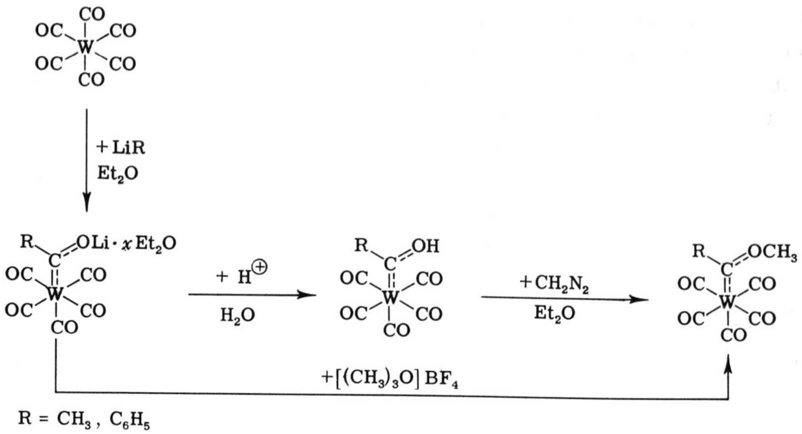

Scheme 1

R = CH₃, C₆H₅

according to the method of Meerwein *et al.* This preparative route to the methoxycarbene compounds possesses the advantage of being simple and straightforward and of leading to the desired compounds in very high yields. There arose, therefore, the possibility of synthesizing a broad spectrum of carbene complexes. Instead of phenyllithium many other organolithium reagents (*17–23*) can be employed. Moreover, hexacarbonylchromium (*17*), hexacarbonylmolybdenum (*17*), the bimetallic decacarbonyls of manganese (*24, 25*), technetium (*25*) and rhenium (*25*), and pentacarbonyliron (*26*) as well as tetracarbonylnickel (*27*) may be used instead of hexacarbonyltungsten, but the resulting carbene complexes become increasingly more labile in the indicated order. Finally, substituted metal carbonyls (*27–30*) can also be subjected to the carbanion addition and subsequent alkylation. The carbene complexes are generally quite stable, diamagnetic, soluble in organic solvents, and sublimable. Before we consider their reactions in more detail, I want to discuss briefly carbene ligand–metal bonding.

B. *Bonding Concepts and Spectroscopic Findings*

The first X-ray crystal structure determination, carried out by Mills in cooperation with us (*31*) on pentacarbonyl[methoxy(phenyl)carbene]chromium(0), confirmed our originally postulated bonding concept. According to this concept, the carbene carbon atom is sp^2 hybridized. It should therefore possess an empty p-orbital and be electron-deficient.

Substantial compensation for this strong electron deficiency is provided by a $p\pi$–$p\pi$ bond between one of the free electron pairs on the oxygen atom of the methoxy group and the unused p-orbital of the carbene carbon. To a lesser but no less certain extent, there is also a $d\pi$—$p\pi$ backbond from a filled central metal orbital of suitable symmetry to this empty p-orbital of the carbene carbon. These bonding features affect the distances of the carbene-carbon atom from the oxygen atom on the one side, and from the central chromium atom on the other side: the $C_{carbene}$—O distance, for which a value of 1.33 Å was found, lies between the values for a single (1.43 Å in diethyl ether) and a double (1.23 Å in acetone) bond. Although the average Cr—C_{CO} distance in the carbene complex amounted to 1.87 Å, the Cr—$C_{carbene}$ distance was 2.04 Å. However, according to the considerations of Cotton (*32*), a distance of 2.21 Å would have been expected for a pure chromium–carbon σ-bond. Consequently, the bond order of the Cr—$C_{carbene}$ bond is distinctly less than that of the Cr—C_{CO} bonds in the same complex, but it is greater than that of a single bond. That the phenyl group, at least in the crystal lattice, does not engage in $p\pi$—$p\pi$ bonding with the carbene carbon can be seen by the significant twisting of the plane containing the Cr, C, and O atoms relative to the one formed by the phenyl ring. At the same time, it should be recognized that the double-bond character of the $C_{carbene}$—O bond is so substantial that cis and trans isomers can easily exist relative to this bond (Fig. 2). In the case of pentacarbonyl[methoxy(phenyl)carbene]-chromium(0), only molecules of the trans type are found in the lattice, but at low temperatures ^{1}H NMR spectroscopy reveals in solution the coexistence of the cis isomers (*33, 34*).

Further important insights into the bonding relationships of the carbene complexes are made possible by a consideration of the ν_{CO} bands of vibrational spectra (*20, 35–37*). As we know, the carbonyl ligands in metal–carbonyl complexes may be considered as very weak donor systems. They donate electron density from the carbon's free electron pair to unused orbitals on the metal atom, a process that formally leads to a negative charge on the metal. This is reduced primarily by a back donation

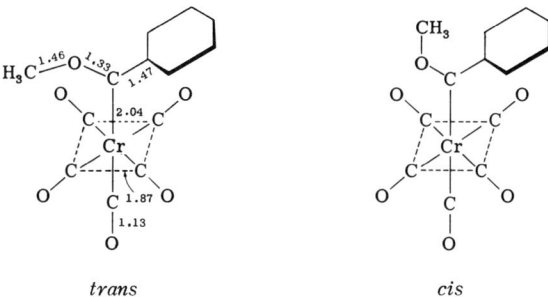

FIG. 2. The structure of pentacarbonyl [methoxy(phenyl)carbene]chromium(0); bond lengths in angstroms.

of charge density from the metal to the carbon monoxide via a $d\pi$—$p\pi$ backbond. Therefore, carbon monoxide has simultaneously an acceptor function as well as a donor function. The σ-donor/π-acceptor ratio of the CO ligand in a complex is a very sensitive probe of the electronic character of the other ligands bonded to the metal. It can be qualitatively estimated by determination of the CO-stretching frequency.

Let us compare carbon monoxide with methoxy(phenyl)carbene as a ligand in complexes of the type $(CO)_5CrL$ (where L = CO or $[C(OCH_3)-C_6H_5]$, respectively) by considering the ν_{CO} absorptions. While the totally symmetric Raman-active ν_{CO}-stretching frequency in $Cr(CO)_6$ appears at 2108 cm^{-1} (A_{1g}) (38), we find the absorption of the CO group that is situated trans to the carbene ligand to be shifted drastically to lower wave numbers and to occur at 1953 cm^{-1} (A_1) (17). This means that the carbene ligand possesses a substantially larger σ-donor/π-acceptor ratio than CO. In other words, the entire carbene ligand is positively polarized with the $Cr(CO)_5$ part being negative. Thus, the dipole moments of the complexes (ca. 5 D) are also relatively large.

In the remainder of this account I shall not consider further purely spectroscopic studies. However, special attention should be briefly given to ^{13}C NMR measurements as they represent extraordinarily valuable aids for the development of this area of chemistry. In the first study of this kind, Kreiter (39) succeeded in showing that in pentacarbonyl[methoxy-(phenyl)carbene]chromium(0) the carbene-carbon atom is very positively charged. The observed chemical shift of 351.42 ppm lies well within the range of shifts exhibited by the carbo-cations of organic chemistry. Thus, this modern investigative method confirmed once again our original concept.

With its intensely positively charged character, the carbene carbon behaves as an electrophilic center, a feature of paramount importance in the reactivity of such compounds. We shall return to this point.

III

OTHER SYNTHESES OF CARBENE COMPLEXES

Since the time that our first paper about metal carbene complexes was published in 1964, this area of research has expanded quickly. Today there are available several major review articles (40–43) dealing with the chemistry of carbene complexes. Therefore I want now only to single out particularly interesting syntheses.

In our laboratory in 1968, Öfele (44) treated 1,1-dichloro-2,3-diphenyl-2-cyclopropene with disodium pentacarbonylchromate and, with the concomitant elimination of sodium chloride, he obtained pentacarbonyl(2,3-diphenylcyclopropenylidene)chromium(O). This compound is stable up to 200°C and is notable for the fact that the carbene ligand does not contain any heteroatom. The electronic requirements of the carbene carbon here are satisfied by the three-membered cyclic π-system:

$$\begin{array}{c} H_5C_6 \\ \diagdown \\ C \\ \parallel \\ C \\ \diagup \\ H_5C_4 \end{array} \begin{array}{c} Cl \\ \diagup \\ C \\ \diagdown \\ Cl \end{array} + Na_2Cr(CO)_5 \xrightarrow[-20°C]{THF} \begin{array}{c} H_5C_6 \\ \diagdown \\ C \\ \parallel \\ C \\ \diagup \\ H_5C_6 \end{array} C=Cr(CO)_5 + 2\,NaCl \quad (1)$$

X-Ray structural analysis (45) showed that the three C—C distances of the ligand are not identical: the distance between the 2 carbon atoms bearing the phenyl substituents is shorter than the other two distances. The carbene carbon–chromium distance of 2.05 Å lies within the range of values found for our carbene complexes, thereby indicating that in this case as well an authentic carbene complex exists.

A very interesting synthetic method was published in 1969 by Richards and co-workers (46). They found that, in the reaction of alcohols with certain isocyanide complexes [such as those of platinum(II)], an addition of the alkoxy group to the carbon atom as well as of the hydrogen to the nitrogen atom of the isocyanide ligand occurs, and one thus obtains the corresponding carbene complexes:

$$\begin{array}{c} Cl \\ \diagdown \\ Cl \end{array} Pt \begin{array}{c} P(C_2H_5)_3 \\ \diagdown \\ C \equiv N \\ \diagdown \\ C_6H_5 \end{array} + C_2H_5OH \longrightarrow \begin{array}{c} Cl \\ \diagdown \\ Cl \end{array} Pt \begin{array}{c} P(C_2H_5)_3 \\ \diagdown \\ H \\ \mid \\ C=N—C_6H_5 \\ \mid \\ OC_2H_5 \end{array} \quad (2)$$

This procedure has led subsequently to many compounds of this kind. The relationship between the complex chemistry of isocyanides and of carbon monoxide comes to mind at this point.

In 1971 we succeeded, again for the first time, in transferring a carbene ligand from one metal atom to another (26, 47). For example, if one irradiates a solution of cyclopentadienyl(carbonyl)[methoxy(phenyl)carbene]-nitrosylmolybdenum(0) in the presence of an excess of pentacarbonyliron, one obtains tetracarbonyl[methoxy(phenyl)carbene]iron(0) with the simultaneous formation of cyclopentadienyl(dicarbonyl)nitrosylmolybdenum:

$$\begin{array}{c} \text{Cp} \end{array} Mo \begin{array}{c} CO \\ \diagup \\ NO \\ \diagdown \\ C=OCH_3 \\ \mid \\ C_6H_5 \end{array} + Fe(CO)_5$$

$$\downarrow h\nu, C_6H_6$$

$$(OC)_4Fe=C \begin{array}{c} OCH_3 \\ \diagdown \\ C_6H_5 \end{array} + \begin{array}{c} \text{Cp} \end{array} Mo \begin{array}{c} CO \\ \diagup \\ NO \\ \diagdown \\ CO \end{array} + \cdots \quad (3)$$

Finally, an additional synthesis of recent times (1971) comes from Lappert and his group (*48*). They treated an electron-rich olefinic system, such as 1,1′,3,3′-tetraphenyl-2,2′-biimidazolidinylidene, with a suitable complex compound. In this manner, they attained cleavage of the C=C double bond and attachment of the carbene fragment to the metal:

$$(C_2H_5)_3P\text{-PtCl}_2\text{-Pt}(P(C_2H_5)_3)\text{-Cl}_2 + \text{biimidazolidinylidene} \xrightarrow[140°C]{xylene} 2\,(C_2H_5)_3P\text{-Pt}(Cl)_2\text{=C(imidazolidine)} \quad (4)$$

This brief sketch summarizes some of the other independently discovered methods for synthesis of carbene complexes.

IV
REACTION POSSIBILITIES OF CARBENE COMPLEXES

I now wish to confine my attention to carbene complexes of "our type" and to show with recent examples the kinds of reactions we were able to produce.

We have already established that the carbene carbon is an electrophilic center and, hence, it should be very easily attacked by nucleophiles. In most reactions we believe that the first reaction step probably involves attachment of a nucleophile to the carbene carbon. In some cases, for instance with several phosphines (*49*) and tertiary amines (*50*), such addition products are isolable analytically pure under certain conditions (1 in Fig. 3). For the second step there exists the possibility that the nucleophilic agent may substitute a carbon monoxide in the complex with preservation of the carbene ligand (2 in Fig. 3). One can also very formally think of the carbene complex as an ester type of system [X=C(R′)OR with X = M(CO)$_5$ instead of X = O], because the oxygen atom as well as the metal atom in the M(CO)$_5$ residue are each missing 2 electrons for attainment of an inert gas configuration. So, it is not surprising that the

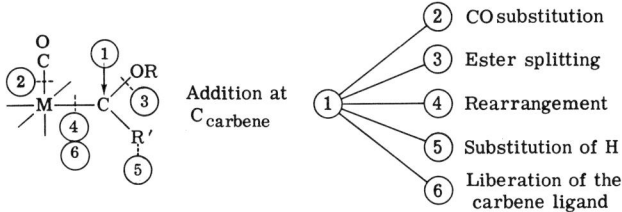

FIG. 3. Reaction possibilities of alkoxycarbene complexes.

OR group can be replaced by amino, thio, or seleno groups (3 in Fig. 3). In this way, the amino- (36, 51–54), thio- (51, 55), and seleno(organo)-carbene complexes (56) are accessible, but the synthesis of the two latter species requires a special experimental skill.

We can also observe reactions that lead to a more stable arrangement of the whole system very probably via primary addition and subsequent rearrangement (4 in Fig. 3). In addition, it can be established that because of the electron withdrawal of the $M(CO)_5$ moiety, hydrogen atoms in α-alkyl positions to the carbene carbon develop such an acidic character that their acidity corresponds to that of nitromethane (5 in Fig. 3). Finally, by cleavage of the carbene ligand from the metal complex, pathways in synthetic organic chemistry are opened (6 in Fig. 3).

A. Addition and CO Substitution

If one treats trialkylphosphines with pentacarbonyl[alkoxy(organo)-carbene] complexes of chromium(0) and tungsten(0), typically in hexane at temperatures below $-30°C$, the corresponding phosphorylide complexes (addition compounds) can be isolated analytically pure and studied (49). The formerly carbene carbon is now sp^3 hybridized and exhibits only a σ-bond to the central metal. In the case of triaryl- and mixed alkylarylphosphines, the addition–dissociation equilibrium (57) (Fig. 4) lies largely on the side of the starting materials, and so the ylide complexes can merely be detected spectroscopically. Figure 4 shows the reaction scheme for pentacarbonyl[methoxy(methyl)carbene]chromium(0) and tertiary phosphines.

Upon irradiation of solutions of these ylide complexes in hexane–toluene mixtures at $-15°C$, a CO ligand is eliminated from the cis position and thereby the cis-tetracarbonyl[alkoxy(organo)carbene]phosphine complexes are obtained (58). The phosphine that was initially added to the carbene carbon of the starting material thus takes the place of a CO

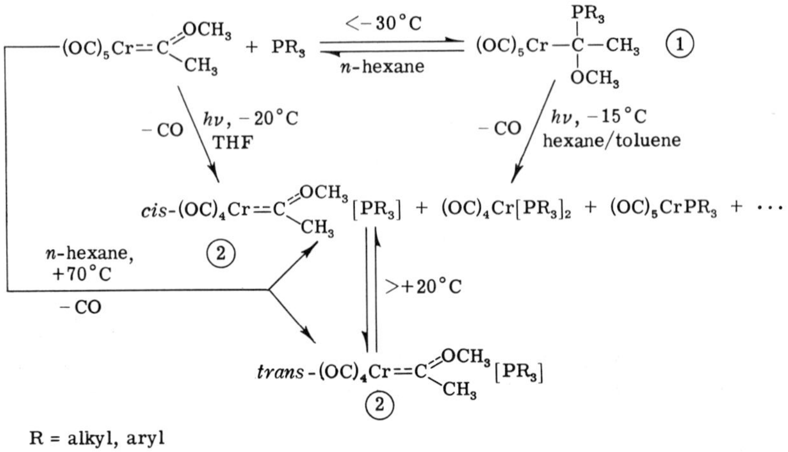

FIG. 4. Reactions of pentacarbonyl[methoxy(methyl)carbene]chromium(0) with tertiary phosphines.

ligand at the metal atom, at which site the carbene group is re-formed. In addition, but to a lesser extent, the carbene ligand is also replaced by the phosphine. One arrives at the same products if, instead of the isolated ylide complexes, one simply employs the equilibrium mixtures of pentacarbonylcarbene complexes and phosphines under slightly different conditions (at −20°C in tetrahydrofuran) (58). By contrast, if one carries out the reaction thermally (at 70°C in hexane) the pure cis-tetracarbonylcarbenephosphine complexes are no longer formed but instead mixtures of the cis and trans isomers result (59, 60). We succeeded in isolating both forms in a pure state (60).

When solutions of each of these components are warmed, isomerization occurs until an equilibrium is attained (61). We were particularly interested in the mechanism of this process, and we found that the isomerization reaction (62) follows first-order kinetics, that the reaction speed is not affected by the presence of free ligands such as phosphine or carbon monoxide, and that the isomerization velocity of tetracarbonyl[methoxy-(methyl)carbene]triethylphosphinechromium(0) is greater than that of the corresponding tricyclohexylphosphine complex. How do we now visualize the progression of this isomerization? The findings speak for an intramolecular mechanism in which none of the bonds of the six monodentate ligands to the metal is broken nor any new bonds are formed while the transitions from the cis to the trans isomer and vice-versa are occurring: Instead, a rotation of two planes, each containing three ligands, by 120° relative to each other, could take place (twist mechanism) (Fig. 5).

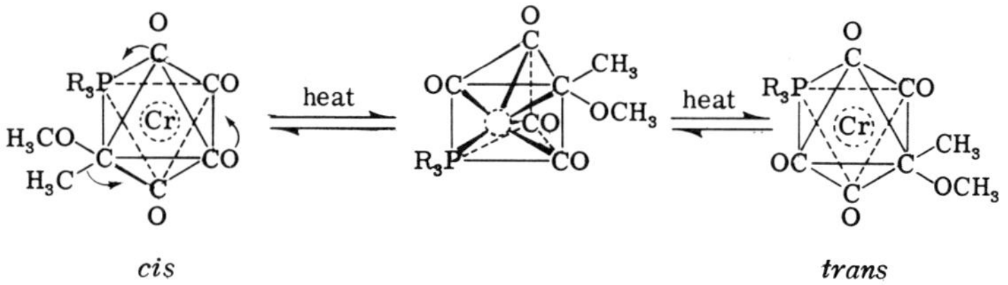

FIG. 5. Hypothesis for the isomerization of tetracarbonyl[methoxy(methyl)carbene]-phosphinechromium(0).

Since a trigonal-prismatic transition state with greater steric hindrance must be traversed, it thus becomes understandable why the compound with the very bulky tricyclohexylphosphine ligand isomerizes more slowly than does the corresponding triethylphosphine complex.

B. Transition Metal–Carbene Complex Residues as Amino-Protective Groups for Amino Acids and Peptides

If we treat alkoxycarbene complexes not with phosphines but with primary or secondary amines, we observe a new kind of reaction, reminiscent of the reactions of esters. This observation led us into peptide chemistry along a path that proved to be quite surprising to a coordination chemist. We could show that the alkoxy group of alkoxy(organo)carbene complexes can be substituted not only by mono- or dialkylamino residues but also by free amino groups of amino acid and peptide esters (63, 64). The principle of this reaction is shown in Scheme 2.

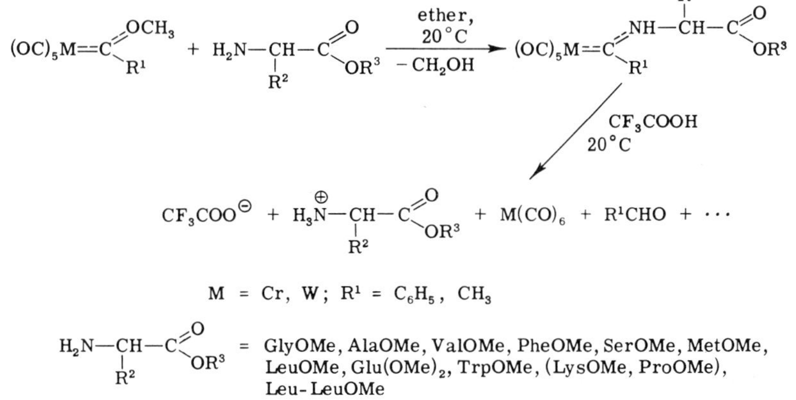

$M = Cr, W; R^1 = C_6H_5, CH_3$

$H_2N-CH(R^2)-C(=O)OR^3$ = GlyOMe, AlaOMe, ValOMe, PheOMe, SerOMe, MetOMe, LeuOMe, Glu(OMe)$_2$, TrpOMe, (LysOMe, ProOMe), Leu-LeuOMe

Scheme 2

At 20°C in ether, not only simple but also multifunctional amino acid esters react with alkoxy(organo)carbene complexes partially without protection of the third function. The organometallic residue thus shows itself to be a new and interesting protective group, particularly because it can be very easily removed again by treatment with trifluoroacetic acid. In some cases this removal can be effected under still milder conditions with acetic acid.

A further example shows that one may add more amino acids to an amino acid ester–carbene complex by employing the customary methods of peptide chemistry (64). Starting from pentacarbonyl[GlyOMe(phenyl)-carbene]chromium(0) and using the N-hydroxysuccinimide/dicyclohexyl-carbodiimide) (NHS/DCCD) method, we succeeded in synthesizing the sequence 14 to 17 of human proinsulin C-peptide (Scheme 3).

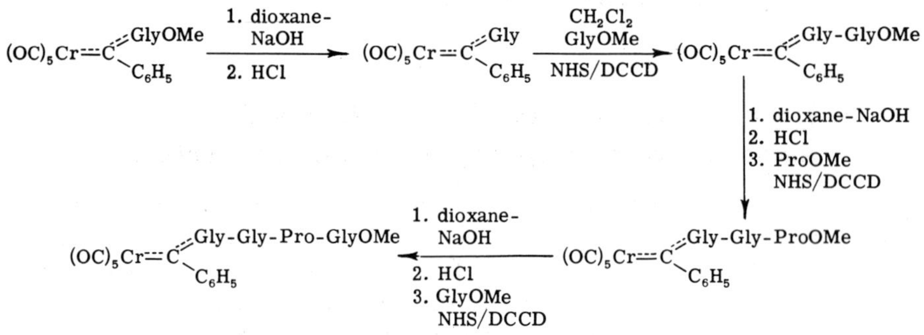

$$(OC)_5Cr{=}C\genfrac{}{}{0pt}{}{GlyOMe}{C_6H_5} \xrightarrow[\text{2. HCl}]{\text{1. dioxane-NaOH}} (OC)_5Cr{=}C\genfrac{}{}{0pt}{}{Gly}{C_6H_5} \xrightarrow[\text{NHS/DCCD}]{\text{CH}_2\text{Cl}_2,\ \text{GlyOMe}} (OC)_5Cr{=}C\genfrac{}{}{0pt}{}{Gly\text{-}GlyOMe}{C_6H_5}$$

$$\downarrow \begin{array}{l}\text{1. dioxane-NaOH}\\ \text{2. HCl}\\ \text{3. ProOMe}\\ \text{NHS/DCCD}\end{array}$$

$$(OC)_5Cr{=}C\genfrac{}{}{0pt}{}{Gly\text{-}Gly\text{-}Pro\text{-}GlyOMe}{C_6H_5} \xleftarrow[\substack{\text{2. HCl}\\ \text{3. GlyOMe}\\ \text{NHS/DCCD}}]{\text{1. dioxane-NaOH}} (OC)_5Cr{=}C\genfrac{}{}{0pt}{}{Gly\text{-}Gly\text{-}ProOMe}{C_6H_5}$$

Scheme 3

In working together with Wünsch in this area, we have concluded that the use of such carbene complexes offers a series of possibilities and advantages to the peptide chemist.

1. Amino acid as well as peptide derivatives of this kind are yellow and, hence, can be easily recognized when using methods such as chromatography.

2. This new protective group can be removed under mild conditions so that, in addition to the amino acid or peptide esters, respectively, the reaction products that originate (mostly aldehyde and hexacarbonylmetal) are volatile and hence can be easily separated.

3. Most carbene complexes of amino acid esters, as well as of some dipeptide esters, are volatile and, therefore, can be analyzed by mass spectrometry.

4. This is a method by which heavy metal atoms, such as tungsten, may be introduced into peptides and in this way free amino groups can be labeled.

C. Addition-Rearrangement Reactions

Two recent examples belong to this class of reaction for carbene complexes. Pentacarbonyl[methylthio(methyl)carbene]chromium(0) and -tungsten(0) react at low temperatures with hydrogen bromide to give pentacarbonyl[(1-bromoethyl)methylsulfide] complexes (65):

$$(OC)_5M=C\begin{smallmatrix}SCH_3\\CH_3\end{smallmatrix} + HBr \xrightarrow[T\downarrow]{pentane} (OC)_5M-S\begin{smallmatrix}CH_3\\H\\|\\C-Br\\|\\CH_3\end{smallmatrix} \quad (5)$$

M = Cr, W

In this reaction the original carbene carbon loses its bonding mode to the metal and sulfur takes over this function.

The second example shows that such a reaction need not always lead to an uncharged system. If, instead of thiocarbene complexes, aminocarbene complexes are treated with hydrogen halides, products of saltlike character are isolable (66). One finds the halogen at the metal and the hydrogen at the removed carbene ligand, and one obtains iminium halogenopentacarbonylmetalates:

$$(OC)_5M=C\begin{smallmatrix}H\\N-CH_3\\C_6H_5\end{smallmatrix} + HX \xrightarrow{ether} [(OC)_5MX]^\ominus[(CH_3)HN=CH(C_6H_5)]^\oplus \quad (6)$$

M = Cr, W; X = Cl, Br

Consequently, there is available a synthesis for cations of this kind that permits multiple variations not otherwise easily attainable.

D. Substitution of Hydrogen at the α-Carbon Atom

Kreiter (67) first demonstrated by ^{1}H NMR spectroscopic studies the acidity of hydrogen atoms that are bonded at the α-C atom of alkoxy-(alkyl)carbene complexes. In CH$_3$OD solution, in the presence of catalytic amounts of sodium methanolate, pentacarbonyl[methoxy(methyl)carbene]chromium(0) exchanges for deuterium all hydrogen atoms of the methyl group that is situated next to the carbene carbon:

$$(OC)_5M=C\begin{smallmatrix}OCH_3\\CH_3\end{smallmatrix} \xrightarrow[CH_3ONa, cat.]{+3 CH_3OD} (OC)_5M=C\begin{smallmatrix}OCH_3\\CD_3\end{smallmatrix} + 3 CH_3OH \quad (7)$$

M = Cr, Mo, W

$$\begin{smallmatrix}|\\-M=C\\|\end{smallmatrix}\begin{smallmatrix}OCH_3\\CH_2\\|\\R\end{smallmatrix} + B \rightleftharpoons \left[\begin{smallmatrix}|\\-M=C\\|\end{smallmatrix}\begin{smallmatrix}OCH_3\\CH^\ominus\\|\\R\end{smallmatrix} \quad \begin{smallmatrix}|\\-M-C\\|\end{smallmatrix}\begin{smallmatrix}\ominus OCH_3\\CH\\|\\R\end{smallmatrix}\right] + HB \quad (8)$$

Obviously, the base is able to form an anion in this situation by reversible cleavage of a proton from the position α to the carbene carbon. This reaction may also be used to introduce new groups into the carbene ligand at this position (*40, 68*).

By its nature, the α-CH acidity is closely connected with the strong positively charged character of the carbene-carbon atom. Let me, therefore, return once more to the unusual ^{13}C NMR data which are available for this atom in some characteristic chromium(0) complexes (Table I).

TABLE I

^{13}C NUCLEAR MAGNETIC RESONANCE SHIFTS FOR THE $C_{carbene}$ ATOM IN SOME CHROMIUM(0)–CARBENE COMPLEXES

Complex	δ (ppm)[a]
$(OC)_5CrC[N(CH_3)_2]C_6H_5$	270.6[b]
$(OC)_5CrC(OCH_3)C_6H_5$	351.4[c]
$(OC)_5CrC(C_6H_5)C_6H_5$	399.4[b]

[a] δ-Values are relative to internal TMS.
[b] In $(CD_3)_2CO$.
[c] In C_6D_6.

If we start with the methoxy(phenyl)carbene complex, for which 351.4 ppm was measured (*39*), and we replace the methoxy group by the better-stabilizing dimethylamino group, we find an expected decrease of the δ-value to 270.6 ppm (*66*). On the other hand, a value of 399.4 ppm is found for the diphenylcarbene complex (*69*). In this compound, with which we have been occupied just recently, there exists, therefore, a carbon atom that, even according to the standards of organic carbo-cation chemistry, is extremely positively charged. Both phenyl groups are, thus, barely able to remedy the electron deficiency at the carbene carbon. This carbene chromium complex is much more labile than the homologous tungsten compound which has been briefly described by Casey (*70*).

E. Liberation of the Carbene Ligand

1. *Reaction with Acids*

I suppose that the time is past when one, as a chemist, still was permitted to think in terms of the separate notions of "inorganic" and "organic."

Nowadays one should rather keep in view all the possibilities offered by nature.

Our carbene complexes should open the way to new organic chemistry if the carbene ligand can be successfully cleaved from the metal under not too severe reaction conditions. This expectation can be realized, for instance, by utilizing hydrogen halides in methylene chloride at $-78°C$ (71):

$$(OC)_5W{=}C\underset{C_6H_5}{\overset{OCH_3}{\diagup}} + HX \xrightarrow[-78°C]{CH_2Cl_2} (OC)_5XWH + \left\{C\underset{C_6H_5}{\overset{OCH_3}{\diagup}}\right\} \qquad (9)$$

$$X = I, Br, Cl$$

In this manner are formed pentacarbonylhalogenotungsten hydrides which, to the best of our knowledge, are new, in contrast to the corresponding anions. The neutral hydride complexes are very unstable and in aqueous solution they dissociate completely into hydronium and pentacarbonylhalogenotungstate ions. The latter can be precipitated as tetramethylammonium salts (71):

$$(OC)_5IWH + [N(CH_3)_4]Br \xrightarrow{H_2O} [(OC)_5WI]^{\ominus} [N(CH_3)_4]^{\oplus} + HBr \qquad (10)$$

We obtained from other studies hints concerning the fate of the liberated carbene ligand. Namely, if one treats tetracarbonyl[methoxy(organo)carbene]triphenylphosphinechromium(0) with benzoic or acetic acid in refluxing ether, one can isolate α-methoxyorgano esters of these acids (72):

$$cis\text{-}[(C_6H_5)_3P](CO)_4Cr{=}C\underset{R}{\overset{OCH_3}{\diagup}} + (C_6H_5)_3P + R'COOH$$

$$\Big\downarrow \text{ether} \atop \text{2-6 hr reflux}$$

$$trans\text{-}[(C_6H_5)_3P]_2Cr(CO)_4 + R'\underset{\underset{O}{\|}}{C}\text{-}O\text{-}\underset{\underset{R}{|}}{\overset{\overset{OCH_3}{|}}{C}}\text{-}H \qquad (11)$$

$$R, R' = CH_3, C_6H_5$$

This "trapping" reaction formally corresponds to an insertion of the carbene into the OH group of the carboxylic acid. The triphenylphosphine that is added to the reaction mixture merely facilitates separation of the metal-complex fragments as the slightly soluble tetracarbonylbis(triphenylphosphine)chromium(0). The reaction with hydrogen chloride proceeds in an analogous manner, but the insertion products (α-halogenoorgano(methyl)ethers) formed react at once with the phosphine present to form the corresponding phosphonium salts (72).

In this connection, the question immediately arises: What happens to the liberated carbene ligand when no suitable reaction partner is at its disposal? In this case the selected reaction conditions are of crucial importance.

2. Reaction with Pyridine

Already at the beginning of our studies involving carbene complexes, we observed that the carbene ligand can be easily released from the metal with pyridine and that the metal fragment can be isolated in the form of carbonylpyridinechromium complexes (73). In the carbene fragment of alkoxy(alkyl)carbene complexes, there ensues, with the cooperation of the base, the shift of a hydrogen atom to the original carbene-carbon atom, and enol ethers are thus formed (73, 74):

$$(OC)_5Cr=C\begin{smallmatrix}OCH_3\\ \\C\\H\ \ R^2\end{smallmatrix}R^1 \xrightarrow[\text{pyridine}]{90°C,\ 1.5\ hr} HC\begin{smallmatrix}OCH_3\\ \\C\\R^2\end{smallmatrix}R^1 + \cdots \quad (12)$$

R^1:	H	H	H	CH_3
R^2:	H	CH_3	C_2H_5	CH_3

3. Thermal Decomposition

In order to test whether the base exerts an essential influence on the subsequent reaction of the carbene ligand, we also decomposed pentacarbonyl[methoxy(methyl)carbene]chromium(0) purely thermally at 150°C in Decalin. Under these conditions we observed exclusively the formation of carbene dimers—as a mixture of cis and trans isomers (74)—thereby confirming the influence of the base:

$$(OC)_5Cr=C\begin{smallmatrix}OCH_3\\CH_3\end{smallmatrix} \xrightarrow[\text{decalin}]{150°C,\ 12\ hr} \begin{smallmatrix}H_3CO\\ \\H_3C\end{smallmatrix}C=C\begin{smallmatrix}OCH_3\\ \\CH_3\end{smallmatrix} + \begin{smallmatrix}H_3C\\ \\H_3CO\end{smallmatrix}C=C\begin{smallmatrix}OCH_3\\ \\CH_3\end{smallmatrix} + \cdots \quad (13)$$

Since the shift of a hydrogen atom is not possible with the methoxy-(phenyl)carbene ligand, it can only dimerize in reactions with bases and in thermal liberations (75).

4. Reaction with Elements of the Sixth Main Group

Naturally, reactions are especially interesting to us if the products obtained are not readily accessible by the methods of classic organic chemistry but are easily preparable with our complexes. We found one such example upon treatment of pentacarbonyl[methoxy(aryl)carbene]chromium(0) complexes with oxygen, sulfur, or selenium (*76*). In this way one obtains conveniently the corresponding methyl esters of arylcarboxylic acids and the *O*-methyl esters of arylthio- as well as arylselenocarboxylic acids; this seems to us to be synthetically useful in the last two cases:

$$(OC)_5Cr{=}C(OCH_3)(C_6H_4R) \xrightarrow{\begin{array}{c}O_2,\text{ hexane}\\68°C,\ 1\text{-}6\ \text{hr}\end{array}} O{=}C(OCH_3)(C_6H_4R)$$

$$\xrightarrow{\begin{array}{c}S_8,\text{ ether}\\34°C,\ 24\ \text{hr}\end{array}} S{=}C(OCH_3)(C_6H_4R) \quad (14)$$

$$\xrightarrow{\begin{array}{c}Se_8,\text{ dioxane}\\101°C,\ 24\ \text{hr}\end{array}} Se{=}C(OCH_3)(C_6H_4R)$$

R = H, CH_3, OCH_3, Cl

5. Reaction with Vinyl Ethers and N-Vinylpyrrolidones

Already at a very early stage of the studies on carbene complexes, we considered that these compounds only properly merit their name when they engage in reactions that are also typical for carbenes.

In organic chemistry one surely thinks at once of the construction of cyclopropane derivatives from olefins and carbenes. Indeed, it has been shown that this also is possible with our complexes and with C=C double bonds that are electron-poor and are either polarized or easily polarizable (*77–81*). As an example of this, I would like to cite the reaction of pentacarbonyl[methoxy(phenyl)carbene]chromium(0), -molybdenum(0), or -tungsten(0) with ethyl vinyl ether (*79*). One obtains the corresponding cyclopropane derivatives in this case, however, only when one removes the carbene ligand under CO pressure of 170 atm in an autoclave at 50°C:

$$(OC)_5M=C\begin{matrix}OCH_3\\C_6H_5\end{matrix} + CH_2=CH-OC_2H_5$$

$$\downarrow \text{170 atm CO, 50°C, 65 hr}$$

(a) cyclopropane with C$_6$H$_5$, OCH$_3$, H, H, OC$_2$H$_5$ + (b) cyclopropane with OCH$_3$, C$_6$H$_5$, H, H, OC$_2$H$_5$ + M(CO)$_6$ + ··· (15)

M:	Cr	Mo	W
(a)	76	80	64
(b)	24	20	36

As expected, there arise two isomers [(a) and (b) in Eq. 15] whose proportions depend on the choice of the central metal atom under otherwise similar reaction conditions. This seems to us to be a rather important hint that the reaction does not proceed via a "free" methoxy(phenyl)-carbene but that the metal atom participates in the decisive reaction step.

When we attempted the analogous reaction with N-vinyl-2-pyrrolidones as the olefin components under similar conditions, we obtained surprisingly 1-[4-methoxy-4-phenyl-3-oxo-1-butenyl]-2-pyrrolidones instead of the expected cyclopropane derivatives (*82*):

$$\text{N-vinyl-pyrrolidone} + (OC)_5Cr=C\begin{matrix}OCH_3\\C_6H_5\end{matrix} + 2\,CO \xrightarrow[150\text{ atm CO, 60 hr}]{C_6H_6,\,80°C} \text{product} + Cr(CO)_6 \quad (16)$$

R = H, CH$_3$

How can one explain the origin of these products in which one finds, aside from the former carbene ligand and the pyrrolidone, just an additional carbonyl group? It seems plausible that the carbene ligand first of all reacts with carbon monoxide to form methoxy(phenyl)ketene. This, in turn, forms with the polarized olefin a cyclobutanone derivative which by ring opening goes over to the observed product (Fig. 6).

Moreover, this hypothesis is supported by the fact that with N-(β-methylvinyl)-2-pyrrolidone the postulated four-membered ring system was isolated, in addition to the open-chained end product (*82*).

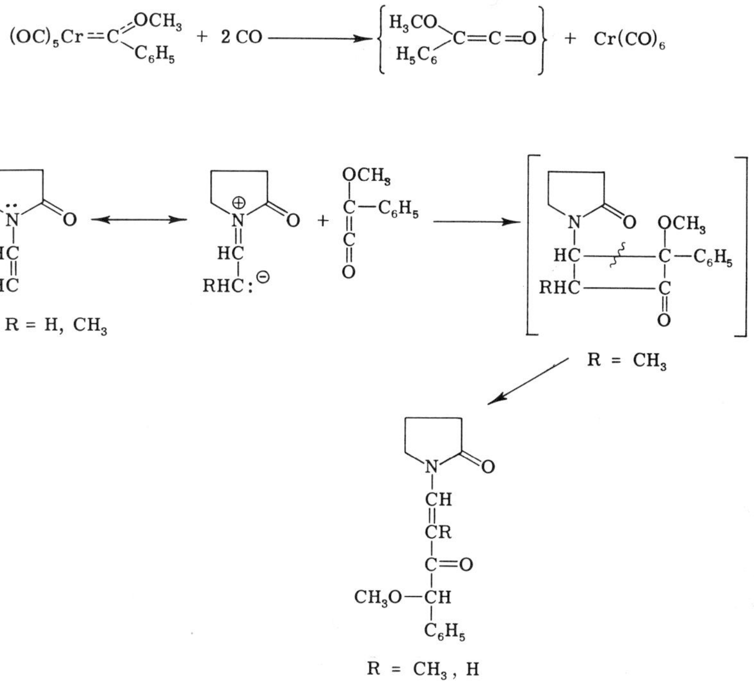

FIG. 6. Hypothesis concerning the course of the reaction during the treatment of pentacarbonyl[methoxy(phenyl)carbene]chromium(0) with N-vinyl-2-pyrrolidones under a CO pressure of 150 atm.

Our original idea to employ carbon monoxide solely for the removal of the carbene ligand led, therefore, to an unexpected result. From this observation, it is clear that the potential of the reaction of carbon monoxide with organic systems must not be neglected.

Nevertheless, to arrive at the sought-after cyclopropane derivatives, we treated the same reactants with each other thermally in benzene in the absence of carbon monoxide. In this reaction as well we did not get the desired compounds, but obtained instead, again surprisingly, the corresponding substituted α-methoxystyrenes (*83*) (Scheme 4).

A possible reaction course comes to mind. The N-vinyl-2-pyrrolidone also possesses at the oxygen a nucleophilic center which could attack the electrophilic carbene carbon and could release the carbene ligand from the metal. The intermediate product formed—irrespective of whether it is an open chain or a six-membered ring—then undergoes a heterolytic fragmentation by splitting similar to that observed by Grob (*84*) (Fig. 7).

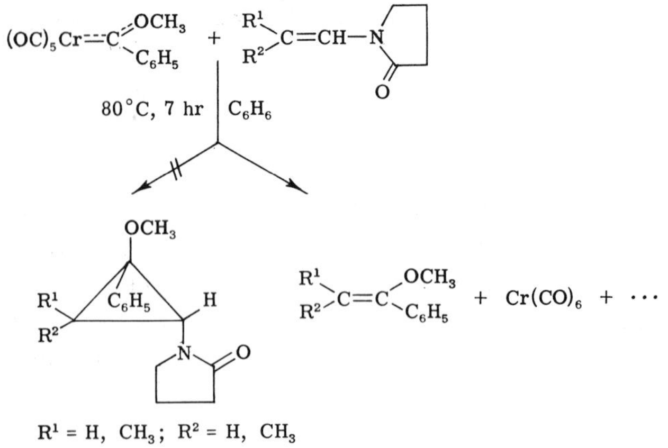

[Scheme 4 diagram]

$R^1 = H, CH_3$; $R^2 = H, CH_3$

Scheme 4

6. Reaction with Electrophilic Carbenes

As mentioned earlier, the carbene ligand in our complexes shows "nucleophilic" character with respect to the metal fragment. Therefore, we decided to combine it with an electrophilic carbene. For this purpose we treated pentacarbonyl[methoxy(phenyl)carbene]chromium(0) with phenyl(trichloromethyl)mercury (85). Compounds of this kind have been studied intensively by Seyferth et al. (86) and are known as a source of dihalogenocarbenes. The carbene complex reacted with the carbenoid compound at

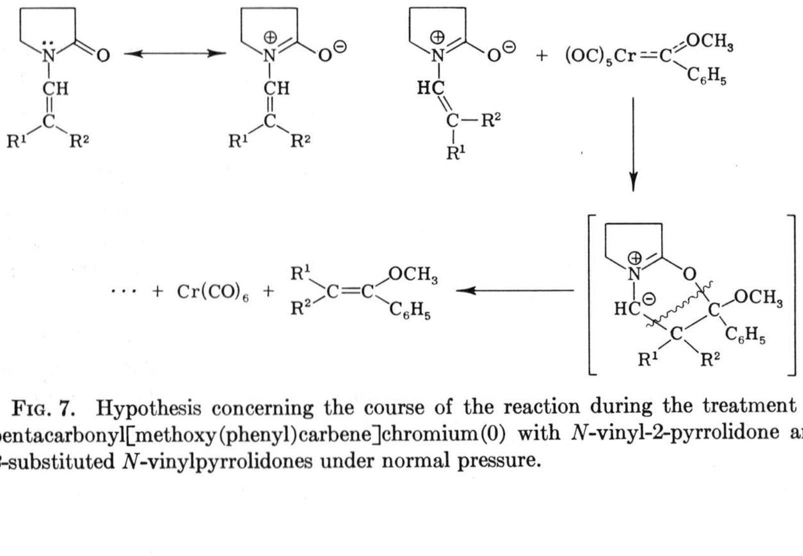

[Fig. 7 reaction scheme]

FIG. 7. Hypothesis concerning the course of the reaction during the treatment of pentacarbonyl[methoxy(phenyl)carbene]chromium(0) with N-vinyl-2-pyrrolidone and β-substituted N-vinylpyrrolidones under normal pressure.

80°C in benzene to give β,β-dichloro-α-methoxystyrene (*85*):

$$(OC)_5Cr{=}C{<}^{OCH_3}_{C_6H_5} + C_6H_5HgCCl_3 \xrightarrow[24\ hr]{C_6H_6\ 80°C,} {}^{H_3CO}_{H_5C_6}{>}C{=}C{<}^{Cl}_{Cl} + C_6H_5HgCl + Cr(CO)_6 + \cdots$$

1 : 1

$$(OC)_5Cr{=}C{<}^{OCH_3}_{C_6H_5} + C_6H_5HgCBr_3 \quad\quad (17)$$

$$\Big\downarrow {}^{80°C,}_{12\ hr}\ C_6H_6$$

$${}^{H_3CO}_{H_5C_6}{>}C{=}C{<}^{Br}_{Br} + {}^{Br}_{Br}{>}C{=}C{<}^{Br}_{Br} + C_6H_5HgBr + Cr(CO)_6 + \cdots$$

This combination reaction is very sensitive to the temperature conditions. We encountered complications when using phenyl(tribromomethyl)-mercury since mixtures of olefins arose.

I hope that I have been able to demonstrate, with this small selection of our newest research results, what a variety of reaction possibilities the chemistry of transition metal–carbene complexes display. In the following I review an area whose development we have made most recently our special task, namely that of transition metal-carbyne complexes.

V

TRANSITION METAL-CARBYNE COMPLEXES

A. *Preparation of the First Carbyne Complexes*

In order to fathom the entire range of reactions of transition metal–carbene complexes, we had undertaken years ago experiments to treat our complexes not only with nucleophilic but also with electrophilic reagents. It was our intention to exchange the methoxy group of methoxy(organo)-carbene complexes with a halogen with the aid of boron trihalides, and so to arrive at halogeno(organo)carbene complexes. Indeed, we initially observed a quick reaction but found only decomposition products. Just a short time ago, when Kreis attempted this reaction at very low temperatures, we were able indeed, to isolate definite but rather thermally labile compounds (*87*). Their composition corresponded to the sum of a tetracarbonylmetal fragment, a halogen and the carbene ligand less the methoxy group:

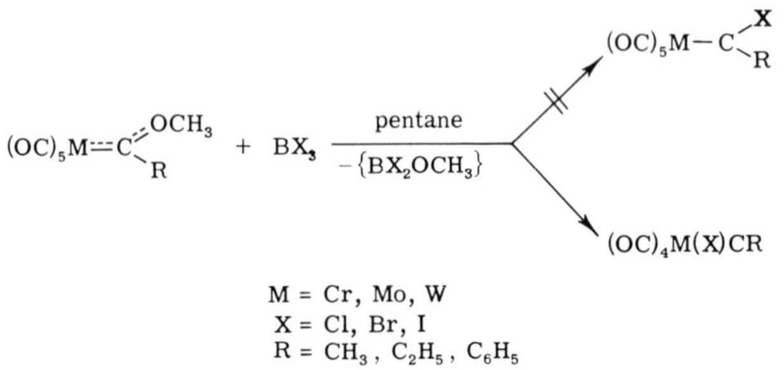

$$(OC)_5M{=}C\genfrac{}{}{0pt}{}{OCH_3}{R} + BX_3 \xrightarrow[-\{BX_2OCH_3\}]{\text{pentane}} \begin{array}{l} (OC)_5M-C\genfrac{}{}{0pt}{}{X}{R} \\[2ex] (OC)_4M(X)CR \end{array} \quad (18)$$

M = Cr, Mo, W
X = Cl, Br, I
R = CH_3, C_2H_5, C_6H_5

The IR spectra indicated the presence of disubstituted hexacarbonylmetal derivatives with two different ligands in trans positions (*trans*-$(CO)_4MR^1R^2$). Moreover, the cryoscopic molecular weight determination proved that the complexes must be monomers.

Together with further spectroscopic findings, especially from ^{13}C and ^1H NMR studies, this could only be interpreted to mean that besides the four CO ligands a halogen atom and a CR group are bonded to the metal (Fig. 8).

For this new type of compounds we would like to propose the name "carbyne complexes" for two reasons: (i) in analogy with carbene complexes, (ii) because similarly to alkyne, on the basis of the diamagnetism of these compounds, a formal metal–carbon triple bond had also to be postulated.

B. X-Ray Structural Analyses

The proposed triple bond should have a very short distance between the metal and the carbyne carbon. Only X-ray structural analyses could clarify this question as well as provide definitive confirmation of our structural proposal, and these studies were then carried out in our Institute by Huttner and his co-workers on three carbyne complexes.

The first successful X-ray study involved *trans*-(iodo)tetracarbonyl-(phenylcarbyne)tungsten(0) (*87, 88*) (Fig. 9). It confirmed in essence our concepts, and yielded an extremely short tungsten–carbon distance of

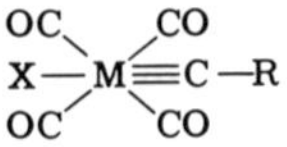

$$\begin{array}{c} OC \quad\ \ CO \\ X-M{\equiv}C-R \\ OC \quad\ \ CO \end{array}$$

FIG. 8. Structural and bonding concepts for $(CO)_4XMCR$.

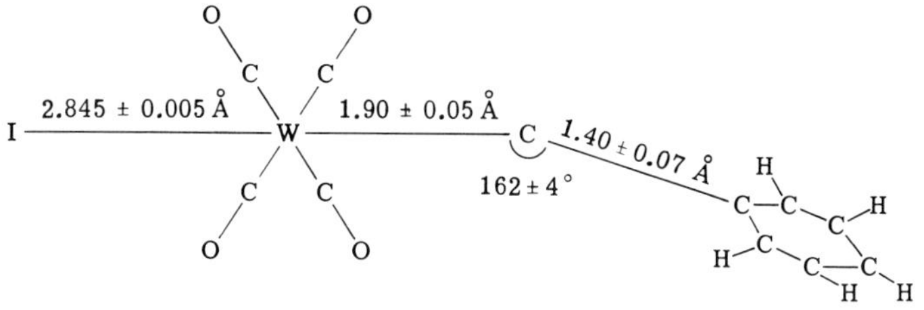

[Structure diagram showing:
I—W with 2.845 ± 0.005 Å bond, four CO ligands (W-C 1.90 ± 0.05 Å), and C—phenyl group with C—C 1.40 ± 0.07 Å, angle 162 ± 4°]

W-C$_{CO}$: 2.07 ± 0.06 Å [in W(CO)$_6$: 2.058 Å]
W-C$_{sp^2}$- single bond [in C$_5$H$_5$W(CO)$_3$C$_6$H$_5$]: 2.32 Å
W-C$_{sp}$-single bond (estimated from r_{Csp^2} = 0.74 Å and r_{Csp} = 0.69 Å): 2.27 Å

FIG. 9. Molecular structure of *trans*-(iodo)tetracarbonyl(phenylcarbyne)tungsten(0).

1.90 Å. Instead of a linear arrangement of metal, C$_{carbyne}$, and C$_{1.4(phenyl)}$ atoms, we found a distinct bend (ca. 162°). Since we cannot as yet explain whether this is attributable to electronic or lattice effects, we undertook the examination of a second complex. Figure 10 shows the result, namely, the structure of *trans*-(iodo)tetracarbonyl(methylcarbyne)chromium(0) (*88*). In this compound one finds not only the expected linear arrangement of chromium, carbon, and the methyl group, but also the shortest chromium–carbon distance (1.69 Å) known at present. This value is distinctly shorter than that for the Cr—C$_{CO}$ distance in the same complex (1.946 Å) or in hexacarbonylchromium (1.91 Å).

Of subsequent interest to us was the question of whether a third kind of ligand in the starting carbene complex could influence the orientation of the halogen in the eventual carbyne complex. Therefore, we first treated *cis*-tetracarbonyl[methoxy(methyl)carbene]trimethylphosphine-, -arsine-,

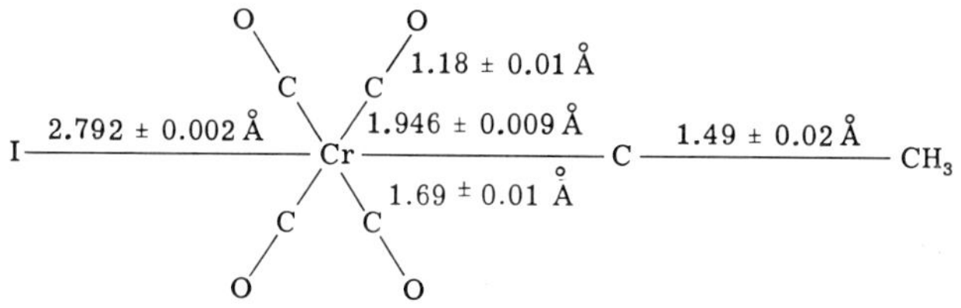

[Structure diagram showing:
I—Cr with 2.792 ± 0.002 Å bond, four CO ligands (C=O 1.18 ± 0.01 Å, Cr-C 1.946 ± 0.009 Å), Cr—C 1.69 ± 0.01 Å, C—CH$_3$ 1.49 ± 0.02 Å]

Cr-C$_{sp^2}$- single bond: 2.22 Å
Cr-C$_{sp}$- single bond: 2.17 Å
Cr-C$_{CO}$ in Cr(CO)$_6$: 1.91 Å

FIG. 10. Molecular structure of *trans*-(iodo)tetracarbonyl(methylcarbyne)chromium(0).

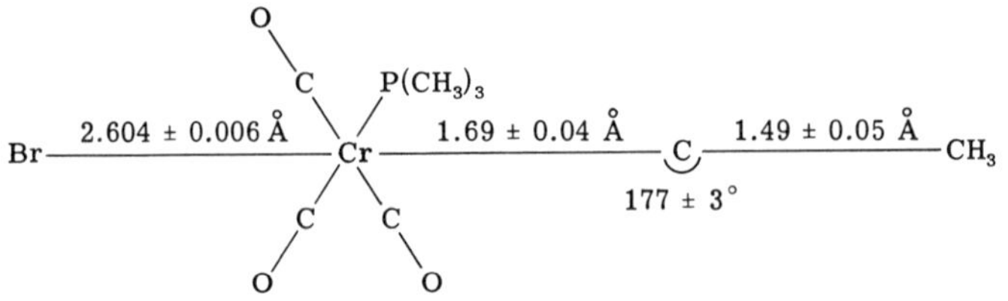

```
           O
            \\
             C      P(CH₃)₃
  2.604 ± 0.006 Å  \\ | 1.69 ± 0.04 Å         1.49 ± 0.05 Å
Br―――――――――――――――――Cr―――――――――――――C―――――――――――――CH₃
                  / \\                177 ± 3°
                 C   C
                 \\   \\
                  O    O
```

Cr–C$_{CO}$: 1.93 ± 0.04 Å
Cr–P: 2.40 ± 0.01 Å

FIG. 11. Molecular structure of *mer*-(bromo)tricarbonyl(methylcarbyne)trimethylphosphinechromium(0).

and -stibinechromium(0) likewise with boron trihalides (*89*).

$$cis\text{-}(OC)_4Cr[Y(CH_3)_3]C\genfrac{}{}{0pt}{}{OCH_3}{CH_3} + BX_3 \xrightarrow[-10°C]{pentane} (OC)_3[Y(CH_3)_3](X)Cr\equiv CCH_3 + (BX_2OCH_3) + CO$$

X = Cl, Br, I; Y = P, As, Sb   (19)

The reaction progressed just as smoothly as those previously studied, but among the products of composition $(CO)_3[Y(CH_3)_3](X)Cr\equiv CCH_3$ (X = Cl, Br, I and Y = P, As, Sb) the relative positions of the ligands could not at first be unequivocally clarified. Therefore, an X-ray structural analysis was also carried out on a representative compound of this type (*88*) (Fig. 11). For (bromo)tricarbonyl(methylcarbyne)trimethylphosphinechromium(0), it showed a meridional arrangement of the three substituents and, once again, a trans positioning of halogen and carbyne ligands. How a starting carbene complex possessing a trans configuration behaves in such reactions with boron trihalides is presently still under investigation.

VI

REACTION OF OTHER PENTACARBONYLCARBENE COMPLEXES WITH BORON TRIHALIDES

It was also of interest to us how changes in the organic residue of the carbyne ligand influence the stability and the behavior of carbyne complexes. Hence, we treated with boron tribromide a series of pentacarbonyl-[methoxy(aryl)carbene]tungsten(0) complexes which were substituted at the phenyl residue (90):

$$(OC)_5W=C(OCH_3)(C_6H_4R) + BBr_3 \xrightarrow{pentane} trans\text{-}Br(CO)_4W\equiv C\text{-}C_6H_4R + \{BBr_2OCH_3\} + CO \quad (20)$$

R = p-CH₃, p-OCH₃, p-CF₃, 2,4,6-(CH₃)₃

Also here ^{13}C NMR spectroscopy should be suitable as a probe for electronic changes. The chemical shifts of the carbyne-carbon atoms of the *trans*-(bromo)tetracarbonyl(arylcarbyne)tungsten(0) complexes obtained are presented in Table II (87, 91).

Contrary to expectation, in this series one finds for the *p*-CF₃ derivative the lowest δ-value, i.e., the strongest screening of the carbyne-carbon atom, whereas for the 2,4,6-trimethyl derivative, by comparison, the screening is distinctly weaker. For an exact interpretation of these results, it is apparent to us that further experiments (currently in progress) are necessary.

TABLE II

^{13}C NUCLEAR MAGNETIC RESONANCE SHIFTS FOR THE $C_{carbyne}$ ATOM IN SEVERAL *trans*-Br(CO)₄WC–Ar COMPLEXES[a]

Complex	δ (ppm)
≡C–C₆H₄–CF₃	266.15
≡C–C₆H₅	271.30
≡C–C₆H₄–CH₃	271.43
≡C–C₆H₄–OCH₃	273.16
≡C–C₆H₂(CH₃)₃-2,4,6	275.13

[a] CH₂Cl₂: δ-values are relative to internal TMS.

We could show further that not only methoxy(organo)carbene complexes react with boron trihalides in the above-mentioned sense. For instance, it was found that *trans*-(bromo)tetracarbonyl(phenylcarbyne)chromium(0) and -tungsten(0) are also accessible from pentacarbonyl[hydroxy(phenyl)-carbene]chromium(0) (*92*) and from pentacarbonyl[methoxycarbonyl-methylamino(phenyl)carbene]tungsten(0) (the glycine methyl ester derivative) (*93*), respectively:

$$(OC)_5Cr-C\overset{OH}{\underset{C_6H_5}{\diagup}} + BBr_3 \xrightarrow[-40°C]{\text{pentane}} Br-Cr\overset{OC}{\underset{OC}{\diagup}}\overset{CO}{\underset{CO}{\diagdown}}\equiv C-C_6H_5 + \{BBr_2OH\} + CO \quad (21)$$

$$(OC)_5W=C\overset{NHCH_2COOCH_3}{\underset{C_6H_5}{\diagup}} + BBr_3$$

$$\Big\downarrow \begin{array}{c} CH_2Cl_2 \\ -25°C \end{array} \quad (22)$$

$$\overset{OC}{\underset{OC}{\diagup}}W\overset{CO}{\underset{CO}{\diagdown}}\equiv C-C_6H_5 + CO + (BBr_2NHCH_2COOCH_3) \xrightarrow{H_2O} HBr\cdot NH_2CH_2COOH + \cdots$$
$$Br$$

I would emphasize that the reaction of the amino acid–carbene complex with boron tribromide represents a good possibility of again cleaving the "carbenyl" protective group under extremely mild conditions at $-25°C$.

That experimental results cannot always be generalized is shown by the treatment of *cis*-(bromo)tetracarbonyl[hydroxy(methyl)carbene]manganese with boron tribromide. This procedure does not lead to the analogous carbyne complex but rather to a product in which the hydrogen atom of the hydroxy group is substituted by a BBr_2 residue (*94*):

$$\begin{array}{c} Br---H \\ OC \diagup \diagdown O \\ OC-Mn=====C \\ OC \diagdown CO \quad CH_3 \end{array} + BBr_3 \xrightarrow[-20°C]{n\text{-hexane}} \begin{array}{c} Br \diagdown \diagup Br \\ B \\ OC \diagup Br \diagdown O \\ OC-Mn=====C \\ OC \diagdown CO \quad CH_3 \end{array} + HBr \quad (23)$$

Here the conditions for the formation of a carbyne complex apparently are not available because of the fixation of the OH group by formation of a bridge to the cis-situated bromine ligand.

An open question was also how pentacarbonyl[ethoxy(diethylamino)-carbene]tungsten(0) would react with boron trihalides, since, as we have learned previously, in principle both the alkoxy and the amino group are removable. The answer was given by the exclusive formation of *trans*-(bromo)tetracarbonyl(diethylaminocarbyne)tungsten(0) (*95*), a compound that is comparatively easy to handle. Its stability can be attributed to the interaction of the metal–carbon bond with the free electron pair of the nitrogen. This interpretation is supported by ^{13}C NMR findings:

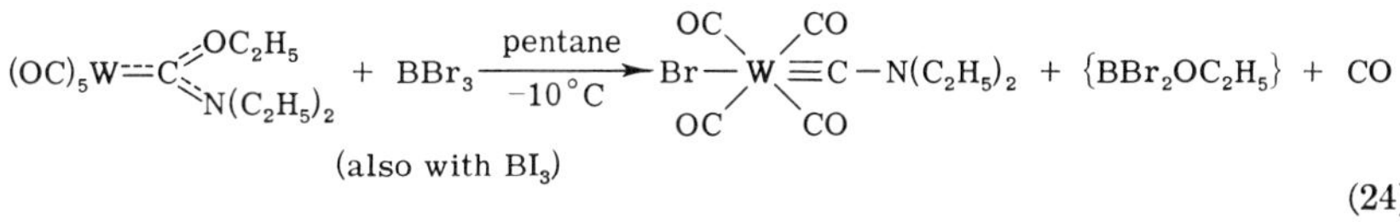

$$(OC)_5W{=}C\underset{N(C_2H_5)_2}{\overset{OC_2H_5}{\diagup}} + BBr_3 \xrightarrow[-10°C]{pentane} Br-W(CO)_4\equiv C-N(C_2H_5)_2 + \{BBr_2OC_2H_5\} + CO$$

(also with BI$_3$)

(24)

$$\underset{OC}{\overset{OC}{\diagdown}}\underset{CO}{\overset{CO}{\diagup}}Br-W\equiv C-N\underset{C_2H_5}{\overset{C_2H_5}{\diagdown}} \longleftrightarrow \underset{OC}{\overset{OC}{\diagdown}}\underset{CO}{\overset{CO}{\diagup}}Br-W=C=N\underset{C_2H_5}{\overset{C_2H_5}{\diagdown}}$$

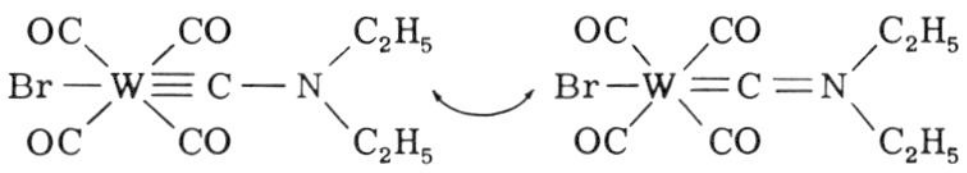

$\delta_{C(carbyne)}$ = 235.0 ppm (CD$_2$Cl$_2$; internal TMS)

VII

REACTION OF PENTACARBONYLCARBENE COMPLEXES WITH HALIDES OF ALUMINUM AND GALLIUM

In the extension of our synthetic methods, aluminum trichloride and tribromide as well as gallium trichloride may also be employed instead of boron trihalides (96):

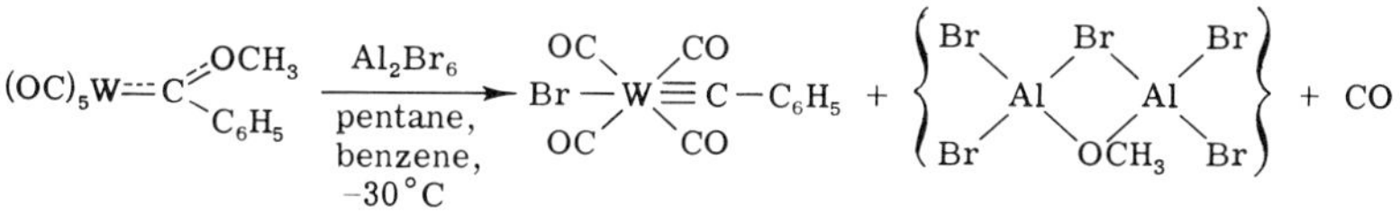

$$(OC)_5W{=}C\underset{C_6H_5}{\overset{OCH_3}{\diagup}} \xrightarrow[\substack{pentane\\ benzene,\\ -30°C}]{Al_2Br_6} Br-W(CO)_4\equiv C-C_6H_5 + \left\{\underset{Br}{\overset{Br}{\diagdown}}Al\underset{OCH_3}{\overset{Br}{\diagup}}Al\underset{Br}{\overset{Br}{\diagdown}}\right\} + CO$$

(25)

$$(OC)_5W{=}C\underset{C_6H_5}{\overset{OCH_3}{\diagup}} \xrightarrow[\substack{toluene,\\ -30°C}]{MCl_3} Cl-W(CO)_4\equiv C-C_6H_5 + \{MCl_2OCH_3\} + CO$$

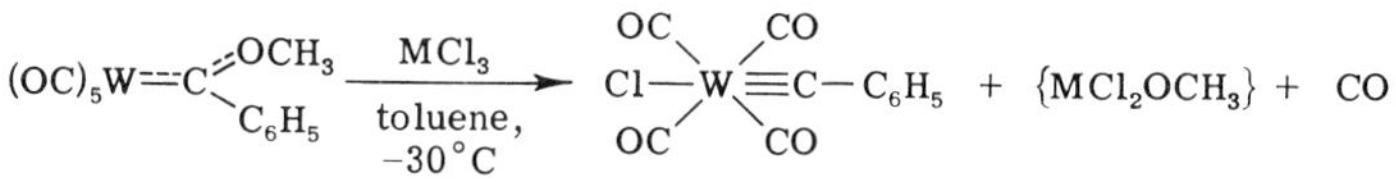

M = Al, Ga

Also in these cases we obtained carbyne complexes in good yields.

VIII

REACTION OF LITHIUM BENZOYLPENTACARBONYLTUNGSTATE WITH TRIPHENYLDIBROMOPHOSPHORANE

In principle, a new synthetic route resulted from the treatment of lithium benzoylpentacarbonyltungstate with triphenyldibromophosphorane at low temperature in ether (97):

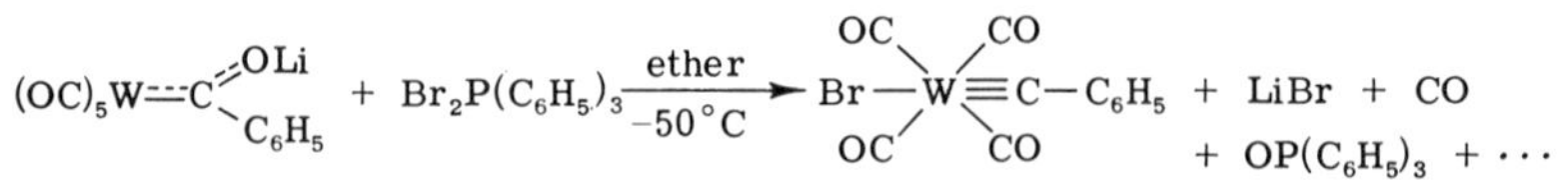

$$(OC)_5W{=}C\overset{OLi}{\underset{C_6H_5}{}} + Br_2P(C_6H_5)_3 \xrightarrow[-50°C]{\text{ether}} Br{-}W(CO)_4{\equiv}C{-}C_6H_5 + LiBr + CO \quad (26)$$
$$+ OP(C_6H_5)_3 + \cdots$$

One may surely suppose that the first step involves the creation of a $C_{carbene}$—O—P linkage by formation of lithium bromide. The resultant intermediate product could stabilize itself by attack of a second bromine atom at the metal, elimination of a CO ligand, and liberation of the thermodynamically favored triphenyloxophosphorane, thereby forming the carbyne complex.

IX

REACTIVITY OF THE CARBYNE LIGAND

Also with carbyne complexes we do not want to confine ourselves only to preparing and spectroscopically examining new variants of this type of compound, and thus we have already begun to study their reaction behavior. Initially we looked for a possibility to compare such a metal–carbon triple bond with a carbon–carbon triple bond. In this connection, it seemed appropriate to react dimethylamine with *trans*-(halogeno)tetracarbonyl-(phenylethynylcarbyne)tungsten(0) (98); the latter compounds is accessible from pentacarbonyl[ethoxy(phenylethynyl)carbene]tungsten(0) (21) and boron trihalides. We found that, at −40°C in ether, only addition to the "organic" triple bond took place, whereas the carbyne–metal bond remained unchanged (99) (Scheme 5).

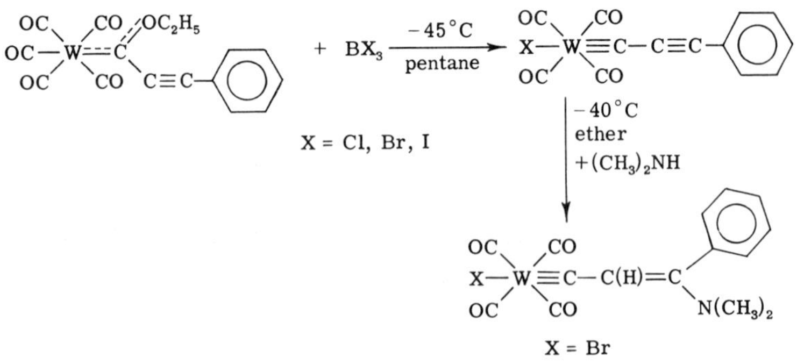

Scheme 5

Simultaneously, we have been occupied with the question as to how the carbyne ligand behaves when it is split off from the metal. Analogous to the carbene complexes, in the absence of a suitable reaction partner, one also observes dimerization, in this case to alkynes (*100*). The conditions for the removal are very mild. In nonpolar solvents, diphenylacetylene or dimethylacetylene are accessible in this way at 30°C. One arrives at the same result when the solid methylcarbyne complex is heated to 50°C:

$$\text{Br}-\underset{\underset{\text{OC}}{|}}{\overset{\overset{\text{OC}}{|}}{\text{Cr}}}\equiv\text{C}-\text{C}_6\text{H}_5 \xrightarrow[+30°C,\ 1.5\ hr]{\text{hexane,}} \text{C}_6\text{H}_5-\text{C}\equiv\text{C}-\text{C}_6\text{H}_5 + \cdots$$

(27)

$$\text{Br}-\underset{\underset{\text{OC}}{|}}{\overset{\overset{\text{OC}}{|}}{\text{Cr}}}\equiv\text{C}-\text{CH}_3 \xrightarrow[2.\ +50°C,\ 3\ hr]{1.\ n\text{-heptane,}\ +30°C,\ 2\ hr} \text{CH}_3-\text{C}\equiv\text{C}-\text{CH}_3 + \cdots$$

Thus, the way appears to be open to make carbyne complexes for use in the syntheses of organic compounds. Since to our knowledge there is no specific and selective "carbyne source" available for preparative purposes, presumably there arises here a wide field of interesting possibilities of application, especially because of the mild conditions required to transfer the carbyne ligand.

It is hoped that this report has shown that there are still many exciting possibilities open in the field of organometallic chemistry.

ACKNOWLEDGMENTS

The chemistry which I have had the honor to report is in large part the work of my co-workers. I would like once more to thank very heartily Miss Dipl.-Chem. K. Weiss as well as Dr. K.H. Dötz, Dr. H. Fischer, Dr. F.R. Kreissl, Dr. S. Riedmüller, Dr. K. Schmid, Dr. A. de Renzi, Dipl.-Chem. B. Dorrer, Dipl.-Chem. W. Kalbfus, Dipl.-Chem. H.-J. Kalder, Dipl.-Chem. G. Kreis, Dipl.-Chem. E.W. Meineke, Dipl.-Chem. D. Plabst, Dipl.-Chem. K. Richter, Dipl.-Chem. U. Schubert, Dipl.-Chem. A. Schwanzer, Dipl.-Chem. T. Selmayr, Dipl.-Chem. S. Walz, and Cand. Chem. W. Held for their collaboration. At the same time, that sentiment applies also to the colleagues of our Institute and their co-workers: Dr. J. Müller for the evaluation of the mass spectra, Dr. C.G. Kreiter for the ^{13}C NMR studies, and Dr. G. Huttner in collaboration with Dipl.-Chem. W. Gartzke as well as Dipl.-Chem. H. Lorenz for the X-ray structural analyses.

REFERENCES

1. Bunsen, R., *Liebigs. Ann. Chem.* **42**, 41 (1842).
2. Frankland, E., *Liebigs Ann. Chem.* **21**, 171, 213 (1849); **95**, 36 (1855).
3. Grignard, V., *C. R. Acad. Sci.* **130**, 1322 (1900).

4. Ziegler, K., *Angew. Chem.* **76,** 545 (1964).
5. Fischer, E. O., and Werner, H., "Metal-π-Complexes," Vol. I: Complexes with Di- and Oligo-olefinic Ligands. Elsevier, Amsterdam, 1966.
6. Herberhold, M., "Metal-π-Complexes," Vol. II: Complexes with Monoolefinic Ligands. Elsevier, Amsterdam, 1972.
7. Zeise, W. C., *Poggendorfs Ann.* **9,** 632 (1827).
8. (a) Fischer, E. O., and Fritz, H. P., *Advan. Inorg. Chem. Radiochem.* **1,** 55 (1959); *Angew. Chem.* **73,** 353 (1961); (b) Fischer, E. O., *Angew. Chem.* **67,** 475 (1955).
9. Wilkinson, G., and Cotton, F. A., *Progr. Inorg. Chem.* **1,** 1 (1959).
10. Fischer, E. O., and Hafner, W., *Z. Naturforsch. B* **10,** 665 (1955).
11. Hübel, W., *in* "Organic Syntheses via Metal Carbonyls" (I. Wender and P. Pino, eds.), Wiley (Interscience), New York, 1967.
12. Fischer, E. O., and Maasböl, A., *Angew. Chem.* **76,** 645 (1964); *Angew. Chem., Int. Ed. Engl.* **3,** 580 (1964).
13. Ryang, M., Rhee, I., and Tsutsumi, S., *Bull. Chem. Soc. Jap.* **37,** 341 (1964).
14. Fischer, E. O., Kreis, G., and Kreissl, F. R., *J. Organometal. Chem.* **56,** C37 (1973).
15. Aumann, R., and Fischer, E. O., *Chem. Ber.* **101,** 954 (1968).
16. Meerwein, H., Hinz, G., Hofmann, P., Kroning, E., and Pfeil, E., *J. Prakt. Chem.* [2], **147,** 257 (1937); Meerwein, H., Bettenberg, E. Gold, H., Pfeil, E., and Willfang, G., *ibid.* [2] **154,** 83 (1940).
17. Fischer, E. O., and Maasböl, A., *Chem. Ber.* **100,** 2445 (1967).
18. Fischer, E. O., and Kollmeier, H. J., *Angew. Chem.* **82,** 325 (1970); *Angew. Chem., Int. Ed. Engl.* **9,** 309 (1970).
19. Fischer, E. O., Winkler, E., Kreiter, C. G., Huttner, G., and Krieg, B., *Angew. Chem.* **83,** 1021 (1971); *Angew. Chem., Int. Ed. Engl.* **10,** 922 (1971).
20. Fischer, E. O., Kreiter, C. G., Kollmeier, H. J., Müller, J., and Fischer, R. D., *J. Organometal. Chem.* **28,** 237 (1971).
21. Fischer, E. O., and Kreissl, F. R., *J. Organometal. Chem.* **35,** C47 (1972).
22. Fischer, E. O., Kreissl, F. R., Kreiter, C. G., and Meineke, E. W., *Chem. Ber.* **105,** 2558 (1972).
23. Wilson, J. W., and Fischer, E. O., *J. Organometal. Chem.* **57,** C63 (1973).
24. Fischer, E. O., and Offhaus, E., *Chem. Ber.* **102,** 2449 (1969).
25. Fischer, E. O., Offhaus, E., Müller, J., and Nöthe, D., *Chem. Ber.* **105,** 3027 (1972).
26. Fischer, E. O., Beck, H.-J., Kreiter, C. G., Lynch, J., Müller, J., and Winkler, E., *Chem. Ber.* **105,** 162 (1972).
27. Fischer, E. O., Kreissl, F. R., Winkler, E., and Kreiter, C. G., *Chem. Ber.* **105,** 588 (1972).
28. Fischer, E. O., and Riedel, E., *Chem. Ber.* **101,** 156 (1968).
29. Fischer, E. O., and Aumann, R., *Chem. Ber.* **102,** 1495 (1969).
30. Fischer, E. O., and Beck, H.-J., *Chem. Ber.* **104,** 3101 (1971).
31. Mills, O. S., and Redhouse, A. D., *J. Chem. Soc. A* 642 (1968).
32. Cotton, F. A., and Richardson, D. C., *Inorg. Chem.* **5,** 1851 (1966).
33. Moser, E., and Fischer, E. O., *J. Organometal. Chem.* **13,** 209 (1968).
34. Kreiter, C. G., and Fischer, E. O., *Angew. Chem.* **81,** 780 (1969); *Angew. Chem., Int. Ed. Engl.* **8,** 761 (1969).
35. Kreiter, C. G., and Fischer, E. O., *Chem. Ber.* **103,** 1561 (1970).
36. Connor, J. A., and Fischer, E. O., *J. Chem. Soc. A* 578 (1969).
37. Fischer, E. O., and Kollmeier, H. J., *Chem. Ber.* **104,** 1339 (1971).
38. Hawkins, N. J., Mattraw, H. C., Sabol, W. W., and Carpenter, D. R., *J. Chem. Phys.* **23,** 2422 (1955).

39. Kreiter, C. G., and Formáček, V., *Angew. Chem.* **84**, 155 (1972); *Angew. Chem., Int. Ed. Engl.* **11**, 141 (1972).
40. Fischer, E. O., *Pure Appl. Chem.* **24**, 407 (1970); **30**, 353 (1972).
41. Cardin, D. J., Cetinkaya, B., and Lappert, M. F., *Chem. Rev.* **72**, 545 (1972).
42. Cotton, F. A., and Lukehart, C. M., *Progr. Inorg. Chem.* **16**, 487 (1972).
43. Cardin, D. J., Cetinkaya, B., Doyle, M. J., and Lappert, M. F., *Chem. Soc. Rev.* **2**, 99 (1973).
44. Öfele, K., *Angew. Chem.* **80**, 1032 (1968); *Angew. Chem., Int. Ed. Engl.* **7**, 950 (1968).
45. Huttner, G., Schelle, S., and Mills, O. S., *Angew. Chem.* **81**, 536 (1969); *Angew. Chem., Int. Ed. Engl.* **8**, 515 (1969).
46. Badley, E. M., Chatt, J., Richards, R. L., and Sim, G. A., *Chem. Commun.* 1322 (1969).
47. Fischer, E. O., and Beck, H.-J., *Angew. Chem.* **82**, 44 (1970); *Angew. Chem., Int. Ed. Engl.* **9**, 72 (1970).
48. Cardin, D. J., Cetinkaya, B., Lappert, M. F., Manojlovič-Muir, Lj., and Muir, K. W., *Chem. Commun.* 400 (1971).
49. Kreissl, F. R., Fischer, E. O., Kreiter, C. G., and Fischer, H. *Chem. Ber.* **106**, 1262 (1973).
50. Kreissl, F. R., and Fischer, E. O., *Chem. Ber.* **107**, 183 (1974).
51. Klabunde, U., and Fischer, E. O., *J. Amer. Chem. Soc.* **89**, 7141 (1967).
52. Fischer, E. O., Heckl, B., and Werner, H., *J. Organometal. Chem.* **28**, 359 (1971).
53. Fischer, E. O., and Leupold, M., *Chem. Ber.* **105**, 599 (1972).
54. Fischer, E. O., and Fontana, S., *J. Organometal. Chem.* **40**, 367 (1972).
55. Fischer, E. O., Leupold, M., Kreiter, C. G., and Müller, J., *Chem. Ber.* **105**, 150 (1972).
56. Fischer, E. O., Kreis, G., Kreissl, F. R., Kreiter, C. G., and Müller, J., *Chem. Ber.* **106**, 3910 (1973).
57. Fischer, H., Fischer, E. O., Kreiter, C. G., and Werner, H., *Chem. Ber.* **107**, 2459 (1974).
58. Fischer, H., Fischer, E. O., and Kreissl, F. R., *J. Organometal. Chem.* **64**, C41 (1974).
59. Werner, H., and Rascher, H., *Inorg. Chim. Acta* **2**, 181 (1968).
60. Fischer, E. O., and Fischer, H., *Chem. Ber.* **107**, 657 (1974).
61. Fischer, H., and Fischer, E. O., *Chem. Ber.* **107**, 673 (1974).
62. Fischer, H., Fischer, E. O., and Werner, H., *J. Organometal. Chem.* **73**, 331 (1974).
63. Weiss, K., and Fischer, E. O., *Chem. Ber.* **106**, 1277 (1973).
64. Weiss, K., and Fischer, E. O., unpublished.
65. Kreis, G., and Fischer, E. O., *Chem. Ber.* **106**, 2310 (1973).
66. Fischer, E. O., Schmid, K. R., Kalbfus, W., and Kreiter, C. G., *Chem. Ber.* **106**, 3893 (1973).
67. Kreiter, C. G., *Angew. Chem.* **80**, 402 (1968); *Angew. Chem., Int. Ed. Engl.* **7**, 390 (1968).
68. Casey, C. P., Boggs, R. A., and Anderson, R. L., *J. Amer. Chem. Soc.* **94**, 8947 (1972).
69. Fischer, E. O., Held, W., Riedmüller, S., and Köhler, F., unpublished.
70. Casey, C. P., and Burkhardt, T. J., *J. Amer. Chem. Soc.* **95**, 5833 (1973).
71. Fischer, E. O., Walz, S., and Kreis, G., unpublished.
72. Schubert, U., and Fischer, E. O., *Chem. Ber.* **106**, 3882 (1973).
73. Fischer, E. O., and Maasböl, A., *J. Organometal. Chem.* **12**, P15 (1968).

74. Fischer, E. O., and Plabst, D., *Chem. Ber.* **107,** 3326 (1974).
75. Fischer, E. O., Heckl, B., Dötz, K. H., Müller, J., and Werner, H., *J. Organometal. Chem.* **16,** P29 (1969).
76. Fischer, E. O., and Riedmüller, S., *Chem. Ber.* **107,** 915 (1974).
77. Fischer, E. O., and Dötz, K. H., *Chem. Ber.* **103,** 1273 (1970).
78. Dotz, K. H., and Fischer, E. O., *Chem. Ber.* **105,** 1356 (1972).
79. Fischer, E. O., and Dötz, K. H., *Chem. Ber.* **105,** 3966 (1972).
80. Cooke, M. D., and Fischer, E. O., *J. Organometal. Chem.* **56,** 279 (1973).
81. Cf. Fischer, E. O., Weiss, K., and Burger, K., *Chem. Ber.* **106,** 1581 (1973).
82. Dorrer, B., and Fischer, E. O., *Chem. Ber.* **107,** 2683 (1974).
83. Fischer, E. O., and Dorrer, B., *Chem. Ber.* **107,** 374 (1974).
84. Grob, C. A., *Angew. Chem.* **81,** 543 (1969); *Angew. Chem., Int. Ed. Engl.* **8,** 535 (1969).
85. De Renzi, A., and Fischer, E. O., *Inorg. Chim. Acta* **8,** 185 (1974).
86. Seyferth, D., Burlitch, M., Minasz, R. J., Yick-Pui Mui, J., Simmons, H. D., Jr., Treiber, A. J. H., and Dowd, S. R., *J. Amer. Chem. Soc.* **87,** 4259 (1965).
87. Fischer, E. O., Kreis, G., Kreiter, C. G., Müller, J., Huttner, G., and Lorenz, H., *Angew. Chem.* **85,** 618 (1973); *Angew. Chem., Int. Ed. Engl.* **12,** 564 (1973).
88. Huttner, G., Lorenz, H., and Gartzke, W., *Angew. Chem.* **86,** 667 (1974); *Angew. Chem., Int. Ed. Engl.* **13,** 609 (1974).
89. Fischer, E. O., and Richter, K., unpublished.
90. Fischer, E. O., and Schwanzer, A., unpublished.
91. Fischer, E. O., Schwanzer, A., and Kreiter, C. G., unpublished.
92. Fischer, E. O., and Kreis, G., unpublished.
93. Fischer, E. O., and Weiss, K., unpublished.
94. Fischer, E. O., and Meineke, E. W., unpublished.
95. Fischer, E. O., Kreis, G., Kreissl, F. R., Kalbfus, W., and Winkler, E., *J. Organometal. Chem.* **65,** G53 (1974).
96. Fischer, E. O., and Walz, S., unpublished.
97. Fischer, H., and Fischer, E. O., *J. Organometal. Chem.* **69,** C1 (1974).
98. Kreis, G., Thesis, Technical University, Munich, 1974.
99. Fischer, E. O., Kalder, H. J., and Köhler, F. H., *J. Organometal. Chem.* **81,** C23 (1974).
100. Fischer, E. O., and Plabst, D., unpublished.

Coordination of Unsaturated Molecules to Transition Metals

STEVEN D. ITTEL

Central Research & Development Department
E. I. du Pont de Nemours and Company
Wilmington, Delaware

JAMES A. IBERS

Department of Chemistry
Northwestern University
Evanston, Illinois

I. Introduction	33
II. Theoretical Models	35
III. Structural Results	37
A. Geometry of the Coordinated Olefin	38
B. Geometry of the Complex	53
C. Complexes of Nonolefinic Unsaturated Molecules	55
IV. Summary	59
References	60

I
INTRODUCTION

The interaction of unsaturated molecules, for example olefins and acetylenes, with transition metals is of paramount importance for a variety of chemical processes. Included among such processes are stereospecific polymerization of olefin monomers, the production of alcohols and aldehydes in the hydroformylation reaction, hydrogenation reactions, cyclopropanation, isomerizations, hydrocyanation, and many other reactions.

Many of these processes take place under both homogeneous and heterogeneous conditions. Because of their apparently greater simplicity, homogeneous reactions have served increasingly as models for heterogeneous reactions. Many of these proceed catalytically in solution, and the direct methods available to characterize such reactions are necessarily limited to kinetic and spectroscopic investigations. As a result, unequivocal information on the nature of the all-important interaction of the unsaturated molecule with the transition metal is necessarily limited. A very useful modeling approach to gain more information on the metrical and bonding aspects of this interaction is the isolation and structural characterization of related, more stable complexes. Although such an approach must necessarily be used with caution, since, for example, one might succeed in isolating the least soluble rather than the most important species or since the Ir analogue of a Rh catalyst may not be a perfect one, the approach has yielded considerable information recently on the nature of the interaction between unsaturated molecules and transition metals.

In this review we will concentrate on the metrical aspects of this interaction as derived from a considerable body of structural data. We will not attempt a comprehensive review; rather, we use selected examples and discuss the results within the framework of current theories of bonding in these systems. These examples will be limited to nonring molecules interacting through one double bond, that is, olefins, acetylenes, diazenes, ketones, and imines. Consequently, the emphasis in this review differs from that of other recent reviews (*40, 46, 55, 58, 72*) on metal-olefin and related complexes.

In order to provide an overview of the interaction of unsaturated molecules with transition metals, the following remarks may prove helpful. If we take a simple olefin as an example, then the major characteristics of its interaction with a transition metal are that (*a*) the C atoms are essentially equidistant from the metal center, and (*b*) the formerly planar olefin becomes nonplanar with the substituents on the carbon atoms bending away from the metal. The orientation of the C=C bond with respect to the coordination geometry surrounding the metal shows some variation but generally is that shown in Fig. 1. (In Fig. 1 and throughout this review we consider the olefin to be a monodentate ligand.) Figure 1 illustrates the observation that in square-planar (SP) complexes the C=C bond is approximately perpendicular to the plane, in trigonal complexes (TR) it is

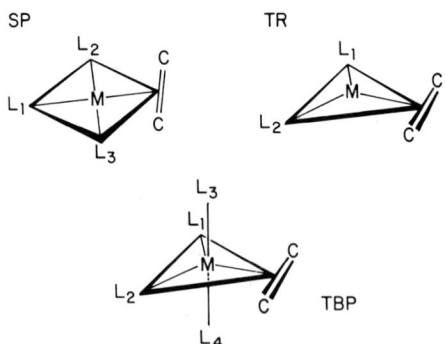

FIG. 1. Four-coordinate square-pyramidal (SP), three-coordinate trigonal (TR), and five-coordinate trigonal-bipyramidal (TBP) olefin complexes.

approximately in the plane, and in trigonal-bipyramidal complexes (TBP) the olefin occupies an equatorial site with the C=C bond approximately in the equatorial plane.

We now turn to a discussion of theoretical models for and structural information on this interaction between an unsaturated molecule and a transition metal.

II
THEORETICAL MODELS

The most widely accepted model for the patterns of bonding illustrated in Fig. 1 is that proposed by Dewar (*23*) and by Chatt and Duncanson (*8*). This model, illustrated in Fig. 2, involves formation of a σ-bond by donation of π-electrons to the metal atom and formation of a π-bond by back donation from the d-orbitals of the metal to the olefin π^*-orbital. Although this model was designed with square-planar olefin complexes in mind, it has proved to be of great utility in the description of bonding of olefins in other coordination geometries as well as the description of the bonding of other unsaturated molecules to metals (*56*). The model for an olefin will be described in detail. As the π^*-orbital of the olefin becomes appreciably populated, the geometry of the coordinated molecule approaches that of its first excited state (*59*). The order of the multiple bond is reduced, and the molecular geometry is changed; the olefin becomes nonplanar as the substituent groups bend away from the metal atom. The addition of substituent groups that lower the energy of the π^*-orbitals facilitates π-backbonding, and there are classes of molecules for which π-backbonding makes the major contribution to the stability of the complex. The metal in this instance acts as a base toward the olefin. Thus, it is not surprising that the metals interact strongly with strong π-acids, such as tetracyanoethylene (TCNE) (*69*) and quinones (*33*).

The relative importance of σ and π contributions to the overall bonding is unclear, but several different combinations of relative strengths lead to "limiting case" models. When there are 2 electrons in the forward σ-bond and 2 electrons in the π-backbond, there are 2 bonding electrons for each metal–carbon bond. This is mathematically equivalent to 2σ-bonds and a metallocyclopropane structure (*72*). This model does not necessitate strict sp^3 hybridization at the carbon atoms. Molecular orbital calculations for cyclopropane (*15*) indicate that the C—C bonds have higher carbon atom p character than do the C—H bonds. Thus, the metallocyclopropane model allows π interactions with substituent groups on the olefin (*68*).

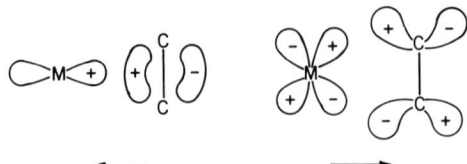

FIG. 2. The Dewar-Chatt-Duncanson model for olefin bonding showing the σ forwardbonds (left) and π backbonds (right).

The Walsh model (73), which has been used to explain the conjugation of cyclopropane with π-bonding substituents, can be used to gain an insight into the π interaction in olefin complexes. There is close similarity between the orbitals involved in classic π-bonding (see Fig. 2) and the highest occupied molecular orbital of the proper symmetry to interact with π substituents (Fig. 3) in the Walsh model. Examination of known cyclopropane structures (49) demonstrates that π-withdrawing substituents cause a shortening of the bond across the ring from the substituent. Calculations have shown (41) that removal of π-electron density from the orbital weakens the two adjacent bonds while strengthening the opposite bond. This model has important implications for the interactions of olefins with transition metals despite the fact that the many extra d orbitals available complicate the situation.

Finally, when the σ forward bond makes a negligible contribution to the bonding and 2 electrons have been π-backbonded, the metal is oxidized and the olefin functions as a bidentate dicarbanion with two σ-bonds to the metal. This model is consistent with the conventional practice of assigning a negative charge to coordinated alkyl groups. Although some justification for this model will be presented, it has not proved to be as useful as either the Dewar-Chatt-Duncanson or the Walsh models.

Recently, there have been several detailed calculations on the bonding

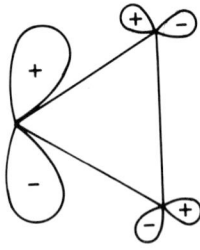

FIG. 3. The highest occupied molecular orbital capable of π-bonding in the Walsh model (73) for cyclopropane.

of unsaturated molecules to transition metals. The results for Zeise's anion,[1] $[PtCl_3(C_2H_4)]^{-1}$, are more complex than the simple σ-π type interaction (67). When the results are interpreted in terms of the Dewar-Chatt-Duncanson model, the σ-bonding effects are found to be considerably more important than the π-backbonding effects. When a similar calculation is performed for a complex of the ethylene-like molecule, oxygen, in the complex $Pt(PH_3)_2(O_2)$ (57), it is found that π back donation is now equal in importance to σ forward bonding, but it takes place by a reorganization of electrons throughout the complex as it is formed. These two calculations demonstrate, as expected, that as the electron density on the metal increases and the electronegativity of the unsaturated molecule increases, π-backbonding becomes more important.

III
STRUCTURAL RESULTS

There have been too many crystallographic studies of transition metal–olefin complexes to present a comprehensive survey in this limited space. Therefore, only representative structures of major classes of compounds will be discussed, drawing on pertinent structural determinations as they are needed. Many of the important features of olefin bonding can be illustrated in the d^{10} system where most of the complexes are approximately trigonal-planar (Fig. 4).

Before proceeding with a discussion of bond lengths and related details, it is essential to consider the inherent accuracies of structure determinations. In the usual X-ray diffraction experiment the scattering power of

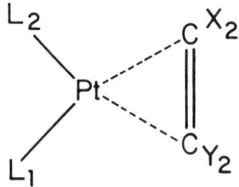

FIG. 4. A d^{10} trigonal-planar complex. L_1 is considered to be trans to the X substituents and L_2 trans to the Y substituents.

---

[1] For purposes of clarity, we depart from the usual IUPAC nomenclature rules and place the unsaturated molecule at the end of the formula.

an atom increases with atomic number. Clearly the location of the carbon atoms of an olefin can be accomplished with greater precision the greater the contribution these atoms make to the total scattering power of the molecule. Thus greater precision is possible if the transition metal is lighter (first-row versus third-row) and if the other ligands are simpler [methyl isocyanide versus tri(cyclohexyl)phosphine]. In most X-ray diffraction determinations the final parameters are obtained by a nonlinear least-squares fit of calculated structure amplitudes to observed structure amplitudes derived from the intensity data. Standard deviations on the derived atomic parameters are obtained by standard mathematical techniques in this process. In doing so it is assumed that all errors rest with the observations and none with the theoretical model. Despite the fact that this assumption is not strictly valid, it is generally found that, for example, by comparing presumed equivalent bond distances within the molecule with their errors obtained from the least-squares procedure, the estimated standard deviations are reasonable. But because of inherent limitations in the overall process, there is considerable evidence, based, for example, on comparisons of crystal structures performed on the same substance in different laboratories, that these standard deviations should be doubled or trebled before applying the usual statistical significance tests. There is no doubt that considerable overinterpretation of the significance of small differences has occurred when these facts have not been kept in mind.

A. Geometry of the Coordinated Olefin

1. C═C Bond Lengths and Spectroscopic Manifestations

The aspect of coordination geometry most readily explained by theoretical models is the lengthening of the olefin bond. An examination of the data listed in Table I reveals that the bond length of a coordinated olefin is significantly longer than that of a free olefin [C_2H_4, 1.337(2); $C_2(CN)_4$, 1.34(2); C_2F_4, 1.31(4) Å] but it is rather insensitive to the nature of the substituents on the olefin or to the other ligands trans to the olefin. This insensitivity has prompted the hypothesis (2) that in a series of complexes where the trans ligands are held constant, the separation of the olefinic carbon atoms will remain constant for all simple olefins. This hypothesis, although lacking in a theoretical basis, does have limited validity. Yet a clear-cut failure is found in the structure of $(C_5H_5)Rh(C_2H_4)(C_2F_4)$ (XXIV, Table I) where the C═C distances are 1.405(7) and 1.358(9) Å for C_2F_4 and C_2H_4, respectively. These distances are significantly different,

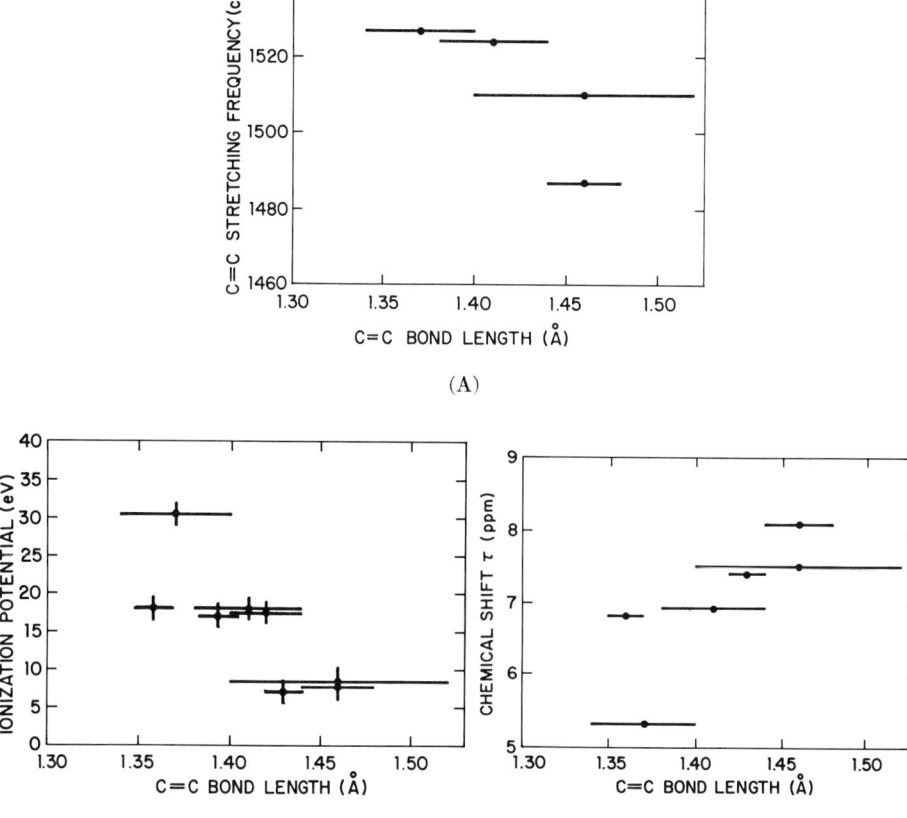

FIG. 5. (A) The C=C stretching frequency, (B) the metal ionization potential, and (C) the chemical shift (τ) plotted versus the C=C bond length for coordination complexes of ethylene.

even if one trebles the estimated standard deviations. It is, therefore, clear that there are some significant variations in C=C distances.

For the detection of subtle differences within a given series of closely related compounds, spectroscopic studies are generally more useful than diffraction studies. An inverse correlation between the C=C bond lengths and the infrared stretching frequencies, $\nu_{C=C}$, in a series of ethylene complexes has been observed (63) (Fig. 5A). There is at present some discussion about the nature of the infrared band observed around 1500 cm^{-1} [(70) and references therein]. It is found that the normal mode contains

TABLE I

STRUCTURAL PARAMETERS FOR SELECTED TRANSITION METAL COMPLEXES OF UNSATURATED MOLECULES[a]

No.	Compound[b]	Structure[c]	C—C (Å)	M—C[d] (Å)	M—L[d,e] (Å)	α^f (deg)	$\beta^{d,f}$ (deg)	$\gamma^{d,f}$ (deg)	$\delta^{d,f}$ (deg)	$\phi^{d,f}$ (deg)	$\theta^{d,f}$ (deg)	Ref.
					Olefin complexes							
I	Ni(PPh₃)₂(H₂C=CH₂)	TR	1.43(1)	1.99(1)	2.152(5)	—	—	—	—	—	5.0	9
II	Ni[P(O-o-Tol)₃]₂(H₂C=CH₂)	TR	1.46(2)	2.02(2)	2.095(2)	—	—	—	—	—	6.6(11)	36
III	Ni[P(O-o-Tol)₃]₂(H₂C=CHCN)	TR	1.46(2) {H, CN}	2.016(10), 1.911(12)	2.121(4), 2.096(4)	—	—	—	103.2	—	3.9(1)	36
IV	Ni[P(p-Tol)]₃(PhHC=CPhH)	TR	1.471(19)	2.019(13)	2.181(4)	—	—	162.7(11)	95.5(11), 101.8(11)	71.6(9)	18.5(9)	43
V	Ni(DCPE)(MeC=CMe₂)	TR	1.421(3)	1.981(2)	2.156(6)	54.5	62.8(6)	139.0, 152.0	—	73.5	16.5	6
VI	Ni(t-BuN=C)₂[(CN)₂C=C(CN)₂]	TR	1.476(5)	1.954(4)	1.866(5)	56.8(5)	61.6(5)	148.9(4), 143.2(4)	107.0(4)	82.2(2)	23.9(2)	69
VII	Ni(PCy₃)₂(H₂C=CH₂)	TR	1.401(14)	2.014(11)	2.196(2)	—	—	—	—	—	—	48
VIII	Ni(BCH)₂	TR	1.40	2.063	—	—	—	—	—	—	29
IX	Ni(CDT)	TR	1.372(5)	2.024(2)	—	—	72(2)	169.5(5)	—	—	32(1)	5
X	Ni(COD)₂	TET	1.39(1)	2.12(1)	—	—	—	—	—	—	—	26
XI	Ni(DUR)(COD)	TET	1.325(13)	2.10(1)	—	—	—	—	—	—	—	33
XII	Ni(TDPME)(F₂C=CF₂)	TET	1.37(3)	1.86(2)	2.25(1)	84	48	—	—	—	—	7
XIII	Pt(PPh₃)₂(H₂C=CH₂)	TR	1.43(1)	2.11(1)	2.268(2)	—	—	—	—	—	1.3	10
XIV	Pt(PPh₃)₂[(CN)HC=C(CN)H]	TR	1.53(4)	2.11(2)	2.287(5)	—	—	—	—	—	5	61
XV	Pt(PPh₃)₂[(CN)₂C=C(CN)₂]	TR	1.49(5)	2.11(3)	2.290(9)	66.1	63.8	141.3	—	—	8.3	4
XVI	Pt(PPh₃)₂(Cl₂C=CCl₂)	TR	1.62(3)	2.04(3)	2.285(8)	81.3(22)	49.4(30)	132(2)	—	78.1(15)	12.3(15)	31

Unsaturated Molecule–Metal Complexes

	Complex												
XVII	Pt(PPh$_3$)$_2$(Cl$_2$C=C(CN)$_2$)	TR	1.42(3)	CN 2.10(2)	2.260(6)	62.0(26)	69.9(33)	131(3)	107(3)	88.2(13)	1.9(13)	41
				Cl 2.00(2)	2.339(6)		48.4(23)	153(3)	115(3)	—	—	
XVIII	Pt(PPh$_3$)$_2$[(PNP)HC=C(PNP)H]	TR	1.416(15)	2.13(2)	2.280(6)	38	85(2)	150(1)	99(1)	—	8.7(7)	2
									111(1)			
XIX	Pt(PPh$_3$)$_2$[(CF$_3$)FC=C(CF$_3$)F]	TR	1.429(14)	2.038(12)	2.312(3)	80(1)	40(1) CF$_3$ 120(1)	—	—	10.8(7)	1
							F 132(1)					
XX	K[PtCl$_3$(H$_2$C=CH$_2$)]	SP	1.37(3)	2.127(19)	2.327(5)	34.7	72.7	—	—	5.8(12)	87.4	45
			1.354(15)	2.139(10)	2.33					5.9	—	38
			1.36(6)	2.17(5)	—					13(4)		3
XXI	Pt(NH$_2$C$_8$H$_{11}$)Cl$_2$(MeHC=CMeH)	SP						150(5)				
XXII	Rh(ACAC)(H$_2$C=CH$_2$)	SP C$_2$H$_4$	1.42(2)	2.19(1)	2.027(8)	—	—	—	—	—	87.8	28
	(F$_2$C=CF$_2$)	C$_2$F$_4$	1.40(2)	2.01(1)	2.047(8)						84.2	
XXIII	Rh(ACAC)(H$_2$C=CH$_2$)	SP	1.41(3)	2.14(2)	2.03(1)	—	—	—	—	—	87.4	28
XXIV	Rh(Cp)(H$_2$C=CH$_2$)(F$_2$C=CF$_2$)	—C$_2$H$_4$	1.358(9)	2.167(2)	2.171(6)	42.4	69.1	154.4	102.8	2.6	—	37
		C$_2$F$_4$	1.405(7)	2.024(2)	2.244(7)	74.3	52.8	131.4	114.3	1.1	—	
XXV	Ir(PPh$_3$)$_2$H(CO)[(CN)HC=C(CN)H]	TBP	1.431(20)	2.110(9)	2.317(3)	—	—	136(1)	112(1)	89.8(9)	0.2(9)	53
XXVI	Ir(PPh$_3$)$_2$Br(CO)[(CN)$_2$C=C(CN)$_2$]	TBP	1.506(15)	2.148(11)	2.400(3)	70.4(13)	54.8	136	112(3)	89.5(6)	0.7(5)	54
XXVII	Ir(AsPh$_3$)$_2$Cl(CO)[(CN)$_2$C=C(CN)$_2$]	TBP	1.447(23)	2.107(18)	2.480(2)	68.5	56.0	—	—	—	3.6	27
XXVIII	Ir(PPh$_3$)$_2$(C$_6$HN$_4$)(CO)[(CN)$_2$C=C(CN)$_2$]	TBP	1.526(12)	2.166(15)	2.392(7)	67.4(12)	56(2)	—	110(2)	86.0(5)	5.4(4)	64
XXIX	Ir[PMe$_2$Ph$_2$](Me)(COD)	TBP EQ	1.386(28)	2.191	2.323(9)	—	—	—	—	—	—	11
		AX	1.362(27)	2.224	—							
XXX	Ir(DPPE)(Me)(COD)	TBP EQ	1.459(21)	2.139	2.308(3)	—	—	—	—	90	0	12
		AX	1.374(21)	2.215	—							
XXXI	Fe(CO)$_4$[(HO$_2$C)HC=CH(CO$_2$H)]	TBP 1	1.40(5)	2.09(3)	1.84(4)	—	—	154(3)	103(3)	—	17	62
		TBP 2	1.30(4)	2.06(3)	1.85(3)			153(3)	103(3)	—	17	
		TBP 3	1.40(4)	2.03(3)	1.81(4)			143(3)	108(3)	—	0	
XXXII	Fe(CO)$_4$[H$_2$C=CH(CN)]	TBP	1.40	2.10	1.77	—	—	—	—	—	—	50
XXXIII	Fe(CO)$_4$(H$_2$C=CH$_2$)	TBP	1.46(6)	—	—	—	—	—	—	—	—	22

(*Continued*)

TABLE I Continued

No.	Compound[b]	Structure[c]	C—C (Å)	M—C[d] (Å)	M—L[d,e] (Å)	α^{f} (deg)	$\beta^{d,f}$ (deg)	$\gamma^{d,f}$ (deg)	$\delta^{d,f}$ (deg)	$\varphi^{d,f}$ (deg)	$\theta^{d,f}$ (deg)	Ref.
				Acetylene complexes								
XXXIV	Ni(t-BuN≡C)₂(PhC≡CPh)	TR	1.284(16)	1.899(19)	1.832(28)	31.4(14)	—	8(4)	172(4)	87.8(7)	2.6(7)	24
XXXV	Pt(PPh₃)₂(PhC≡CPh)	TR	1.32(9)	2.03	2.28	40	—	—	—	—	14	32
XXXVI	Pt(PPh₃)₂[(CF₃)C≡C(CF₃)]	TR	1.255(9)	2.028(5)	2.281(2)	39.9(5)	—	—	—	—	3.7(4)	20
XXXVII	Ir(PPh₃)₂(C₆HN₂)(CO)[(CN)C≡C(CN)]	TBP	1.29(2)	2.09(1)	—	40(1)	—	9(1)	—	87(1)	—	47
XXXVIII	[Pt(Me)(PMe₂Ph)₂(MeC≡CMe)]PF₆	SP	1.22(3)	2.28(3)	2.11(3)	12(4)	—	—	—	—	86.5(15)	19
XXXIX	Pt(C₇H₇NH₂)Cl₂(t-BuC≡C-t-Bu)	SP	1.24(2)	—	—	15	—	0	180	—	90	21
XL	Pt(TPB)(Me)(CF₃C≡CCF₃)	TBP	1.292(12)	2.018(6)	—	34.4(4)	—	0	180	90	1.8(2)	18
				Heteroatom complexes								
XLI	Ni(t-BuN≡C)₂(PhN=NPh)	TR	1.385(5)	1.898(4)	1.841(5)	—	—	153.2(4)	103.4(3)	88.8(3)	1.2(3)	25
XLII	Ni(P-p-Tol₃)₂(PhN=NPh)	TR	1.371(6)	1.930(5)	2.198(3)	—	—	156.5(3)	101.7(3)	83.1(3)	7.6(3)	42
XLIII	Ni(PPh₃)₂[(CF₃)₂C=O]	TR	1.32(2) C / 1.87(1) O	1.89(2) / 1.87(1)	2.249(7) / 2.175(6)	—	48	—	—	—	6.9	16
XLIV	Ni(t-BuN≡C)₂[t-BuN=C=C(CN)₂]	TR	1.245(4) C / N	1.855(4) / 1.843(3)	1.876(4) / 1.819(5)	35.8(2) / 41.3(3)	—	11.9(10)	179.6(8) / 168.5(2)	82.5(3)	7.8(3)	75
XLV	Ni(C₆H₅NO)[(C₆H₄O)HC=NHMe]	TR	1.43(2) C / N	1.917(14) / 1.867(10)	—	—	77 / 55	—	—	—	4.4	52
XLVI	Pt(PPh₃)₂[(CF₃)C=N[N=C(CF₃)₂]]	TR	1.44 C / N	2.02 / 2.11	—	—	—	127	117	—	—	13

[a] The complexes in this table have been selected from the many structures available to illustrate pertinent details of coordination. Cyclic olefins and olefins with substituents which would complicate the interpretation have not been included except when necessary.

[b] The compounds have been named with the ligands ordered as in Fig. 1, with the olefin last. Abbreviation are Ph = phenyl, Tol = tolyl, Me = methyl, *t*-Bu = *tert*-butyl, Cy = cyclohexyl, DCPE = 1,2-bis(dicyclohexylphosphino)ethane, TDPME = tris(diphenylphosphinomethyl)ethane, BCH = bicycloheptene, CDT = cyclododecatriene, COD = 1,5-cyclooctadiene, DUR = duraquinone, PNP = *para*-nitrophenyl, ACAC = 2,4-pentanedionato, Cp = cyclopentadienyl, DPPE = 1,2-bis(diphenylphosphino)ethane, TPB = hydrotris(1-pyrazolyl)borato.

[c] Abbreviations: TR = trigonal planar, TET = tetrahedral, SP = square planar, TBP = trigonal bipyramidal.

[d] These values represent averages except in special cases where individual values are given to illustrate differences.

[e] Values given are for the trans ligands.

[f] These angles are defined in the text. Note that α is defined differently for double- and triple-bonded molecules.

a significant portion of CH_2 scissoring deformation. A lower-energy, strongly polarized Raman line may contain more C=C stretching character, but this band is found to be rather insensitive to the nature of the complex. The variation of the 1500 cm^{-1} band, which may result from differences in the mixing of the C=C stretching and the CH_2 scissoring motions, is thus a more sensitive probe of the nature of the complex.

The ground-state ionization potentials of the metals involved in the ethylene complexes are found to correlate directly with the C=C stretching frequencies or inversely with the C=C bond lengths (Fig. 5B). These ionization potentials of the free metals are an indication of the ability of the complex to modify the π-electron density to the olefin, although they are undoubtedly affected by the variety of remaining coordinated ligands. This modified electron density on the olefin is also reflected in the chemical shift (τ) of the olefinic protons in the NMR spectra (Fig. 5C). Larger values of τ indicate a greater electron density on the olefin π-orbitals resulting in increased magnetic shielding. Although these vibrational and NMR effects would be expected to hold for substituted olefins, it is difficult to predict the additional perturbations on the spectra caused by substituents.

It is evident that the electron density on the olefin and the resultant changes in bond length, nuclear shielding, and vibrational modes are related to the ionization potential of the metal atoms. It is not clear whether this effect is caused by inhibition of the σ forward donation or enhancement of the π back donation as the ionization potential is lowered. When electronegative substituents are placed on the olefin to "activate" it, there is also the question of the extent to which the backbonded electron density remains localized on the double bond or is delocalized to the substituent groups. Any means of probing the electron density on the metal or olefin would be useful in ascertaining the relative importance of the σ and π effects.

Electron spectroscopy for chemical analysis (ESCA) has been employed to measure the electron density centered on the metal in a series of platinum complexes (14). The binding energies of the metal electrons were interpreted on the assumption that within a series of complexes the binding energies of a given level are dependent only on the net electronic charge transferred from the metal to the ligand. On this basis it is possible to assign a degree of oxidation to the metal. In the platinum series, oxidation states of 0.0 and 2.0 were assigned to $Pt(PPh_3)_4$ and $Pt(PPh_3)_2Cl_2$, respectively. It was then found that in the complex $Pt(PPh_3)_2L$, 0.7, 0.8, and 1.8 electrons were transferred to L when L was diphenylacetylene, ethylene, and dioxygen, respectively. These results indicate that there has been a net transfer of electron density to the unsaturated molecules. Thus, as expected, the high electronegativity of dioxygen, which is bonded to the metal in a sideways π fashion, results in a complex in which almost 2 electrons have been removed from the metal center. The degree of platinum oxidation in the olefin and acetylene complexes is not as great but it is, nonetheless, significant.

Another probe of the electronic properties of a metal center is the stretching frequency of a carbonyl or isocyanide (*42*) ligand in the complex of interest. The limitations of this technique are the constraints on the nature of the complexes studied and the uncertainty of the relation between the stretching frequencies and the actual electron density on the metal, but these are more than offset by the greater ease of measurement and the greater resolution of this technique when compared with ESCA measurements. Table II presents the isocyanide stretching frequencies for a variety of complexes, Ni(*t*-BuN≡C)₂L (*35, 42, 44, 60*), listed in order of increasing frequency. The complexes at the top of the list are considered to contain nickel(0), whereas those at the bottom contain nickel(II). In the intermediate complexes, varying degrees of electron transfer from the nickel atom have occurred. Although ethylene does not appear on the list, some substituted olefins do. If one again adopts the procedure of assigning degrees of electron transfer, then $\nu_{N\equiv C}$ around 2000 cm⁻¹ indicates no elec-

TABLE II

Isocyanide Stretching Frequencies for Complexes Ni (*t*-BuN≡C)₂ (Un)[a]

Un	$\nu_{N\equiv C}$ (cm⁻¹)	
(*t*-BuN≡C)₂	2000	—
(PEt₃)₂	2025	1970
trans-PhHC=CPhH	2100	2057
CH₂=CH(CN)	2138	2102
PhC≡CPh	2138	2110
trans-(CN)HC=CH(CN)	2162	2138
PhN=NPh	2168	2140
(CF₃)C≡C(CF₃)	2169	2134
(CF₃)FC=CF₂	2170	2130
(CF₃)₂C=NH	2182	2150
trans-(CF₃)FC=CF(CF₃)	2185	2152
(CN)₂C=C(CN)₂	2194	2179
O=O	2196	2178
(CF₃)₂C=O	2199	2183
(CF₃)₂C=C(CN)₂	2210	2190
—(CF₂)₄—	2230	2219

[a] Un = an unsaturated molecule.

tron transfer and $\nu_{N\equiv C}$ above 2200 represents the 2+ oxidation state. It is then found that π-bonded molecules accept between 1 and 2 electrons, depending on the substituent groups. It is again evident that π-bonding can make a substantial contribution to the overall bonding.

α-Halogen substituents, which may be considered to be σ-acceptors and π-donors, apparently result in less backbonding when compared with the cyano group, which is a weak σ- and a strong π-acceptor. This is in contrast to the β-halogen substituent, CF_3, which seems to result in greater removal of electron density from the metal than does the cyano group. The classic activation of the double bonds by introduction of electronegative substituent groups can also be accomplished by introducing nitrogen or oxygen into the double bond. When similarly substituted, the electron-withdrawing powers of the various 2-atom bridges increase as $C=C < C\equiv C < N=N$, and $C=C < C=N < C=O$. Substituting one cyano group onto a bridging carbon atom makes it approximately as effective in electron-withdrawing power as a bridging nitrogen atom, and two cyano groups make it approximately as effective as a bridging oxygen atom. Thus *intrinsic* activation by modification of the bridging double bond has the same effect as classic *extrinsic* activation by substituent groups (*43*).

A probe of the electron density on the double bond is discussed in Section III,C,2.

2. *Metal–Carbon Bond Lengths*

The metal olefin bond lengths presented in Table I exhibit two interesting trends. The first is that the shortest M—C bond lengths are associated with olefins bearing halogen substituents. Although it is well known that cyano groups make olefins much better π-acids, halogen substituents result in shorter metal–carbon bond lengths. This effect has to be rationalized on the basis of the σ-π bonding scheme. The recent calculations on the bonding for $[PtCl_3(C_2H_4)]^-$ show that σ forward bonding is the major contributor (*67*). Whereas in a zero-valent platinum complex π backbonding would increase in importance because of the higher electron density on the metal, we would still expect σ forward bonding to make an important contribution. Introduction of cyano groups lowers both the π- and π^*-olefin orbital energies. This results in reduced σ forward bonding and greatly increased π-backbonding. There is a net increase in the strength of the metal olefin bond, and the bond becomes more polar or ionic. Introduction of halogen substituents, specifically fluorine, does not change the level of the π-orbital, but the energy of the π^*-orbital increases greatly (*70*). This should have the effect of reducing the overall bonding by reducing the π-backbonding contribution. Although this has been observed for mono-

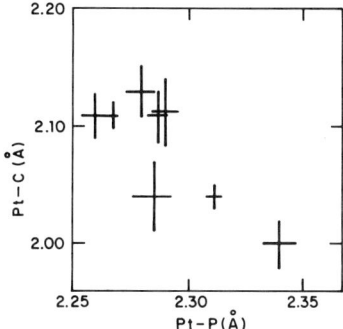

Fig. 6. Plot of Pt—C bond distances versus *trans*-Pt—P bond distances for a series of olefin(L) complexes, Pt(PPh$_3$)$_2$(L).

and disubstitution (*70*), it is not observed for the tetrasubstituted olefins studied crystallographically. Two explanations have been given for this seemingly anomalous result. The first is that there must be an extensive reorganization of molecular orbitals owing to the change in geometry of the olefin, and this rearrangement renders use of uncoordinated molecular orbitals inappropriate (*70*); the resultant molecule has a high degree of cyclopropane character. The second explanation is based on a three-center bonding molecular orbital (*2, 51*) in which the π-donating properties of halogens increase the electron density in that orbital. This would result in a more covalent bond. Whereas both of these explanations are useful, it may be convenient to attribute the enhanced bonding to a purely inductive enhancement of the π-backbonding based on the electronegativity of the substituents.

A second trend cited in metal–olefin bond lengths is their inverse correlation with trans metal–ligand bond lengths (*2*). This phenomenon has been observed in several structures in which the 2 olefinic carbon atoms are at different distances from the metal. A vivid illustration is found in the platinum complex Pt(PPh$_3$)$_2$[Cl$_2$C=C(CN)$_2$] (XVII, Table I). The effect was attributed to the strong σ-donation capability of the CCl$_2$ group. A plot of Pt—C distances versus Pt—P distances for the series of complexes Pt(PPh$_3$)$_2$(olefin) (Fig. 6) reveals a poor correlation. It is found that several complexes in Table I show a direct relationship between metal–olefin and metal–ligand bond distances. For example, in Ni[P(O-*o*-Tol)$_3$]$_2$[H$_2$C=CH(CN)] (III) the cyano end of the olefin is significantly closer to the metal, and the *trans*-phosphite ligand is closer to the metal than is the *cis*-phosphite. This can be rationalized on the basis of the stronger π-accepting capability of the cyano end of the olefin. The metal–

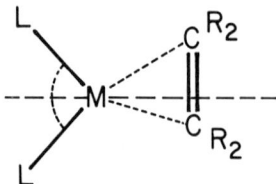

FIG. 7. A trigonal complex displaying displacement of the olefin along the double bond.

carbon bond is shortened by enhanced π-backbonding, and the *trans*-phosphite can then form a stronger and thus shorter σ-bond to the metal. Again we see a difference between the activation of an olefin by π-withdrawing or by π-donating substituents.

This discussion of the metal–carbon distances should include another closely related effect: sliding of the olefin along the carbon–carbon bond. This effect, pictured in Fig. 7, involves a displacement of the centroid of the olefin double bond in the coordination plane relative to the bisector of the ligand–metal–ligand angle. This effect is apparently not a steric one, as both Ni[P(O-o-Tol)$_3$]$_2$[H$_2$C=CH(CN)] (III) and Pt(PPh$_3$)$_2$-[Cl$_2$C=C(CN)$_2$] (XVII) display a marked shift to bring the cyano end of the olefin closer to the bisector, even though in the platinum complex the chloro end is closer to the metal. The same effect is observed in Fe(CO)$_4$[H$_2$C=CH(CN)] (XXXII). The shift of the cyanoolefins is perhaps caused by interaction of the π^*-orbital associated with the cyano groups with the d orbitals of the metal.

3. *Nonplanarity of the Bound Olefin*

We next discuss the nonplanarity of the bound olefin. As we indicated in Section I, this nonplanarity manifests itself in the bending back of substituent groups away from the metal. There are a number of ways to describe this bending back. The measure easiest to picture but least informative is the olefin double bond to substituent angle, which would be 120° for an idealized sp^2 hybridized carbon atom, and 109.5° for an idealized sp^3 hybridized carbon atom. This measure will be used in the description of metal–acetylene bonds. A more informative measure of the bending back are the α and β angles defined in Fig. 8, which have come into common usage (*69*). The angle α is the angle between the normals to the planes defined by the substituent groups; β and β' are the angles between the olefin bond and the plane normals. As bending back of the substituents occurs, α increases from 0° and the β angles decrease from 90°. The sum

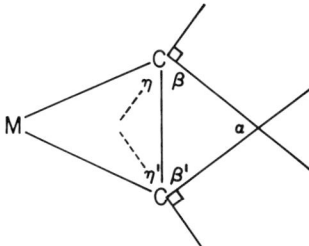

FIG. 8. Illustration of angles α, β, β', η, and η'.

of the three angles need not be 180° unless the bending back of the substituent planes takes place so that the two plane normals and the C=C bond all remain in the same plane. It is generally found that the sum of the angles is near 180°. Inspection of the values of α and β in Table I reveals that bending back is least for hydrogen atoms, increases for cyano groups and alkyl groups, and is greatest for halogen substituents. This parallels the trend observed for metal–carbon distances (Section III, A, 2).

If one fashions a geometrical model based on the "metallocyclopropane" model for bonding, the substituent atoms on the olefin should exhibit tetrahedral geometry. They cannot do this because of the severe steric and electronic nontetrahedral restraints imposed by a three-membered ring. A geometrical model proposed (37) to account for the simplest of steric considerations is constructed by bisecting the M—C=C angle, η (the dotted line in Fig. 8). The substituent atoms are required to lie on a plane perpendicular to the M—C=C plane and containing the bisector. This construction results in the new "tetrahedral" values of α and β defined by

$$\alpha_T = \frac{\eta}{2} + \frac{\eta'}{2}, \qquad \beta_T = 90 - \frac{\eta}{2}, \qquad \beta_T' = 90 - \frac{\eta'}{2}$$

The only distance criterion in assessing the geometries is fixed by the ratio of the M—C to C=C distances.

There are several possible tests for this assumed tetrahedral geometry. The geometry of ethylene oxide has been determined (71). The observed values of α and β of 47.8° and 68.6° compared with the calculated values of 59.2° and 60.4° indicate that the model predicts excessive bending back. In the molecule ethylene sulfide, the ring shape more closely resembles an olefin complex. The observed values of α and β of 56.6° and 61.7° (17) are closer to the calculated values of 65.8° and 57.1°, but again the model predicts excessive bending back.

When the measured values of α and β are compared with the respective values of α_T and β_T calculated for the various complexes listed in Table I,

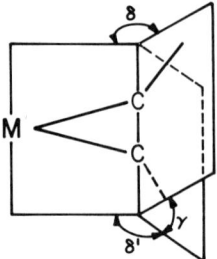

Fig. 9. Illustration of angles γ, δ, and δ'.

it is found that the only complexes in which the bending back is greater than predicted are those of halogen-substituted olefins. Thus, again we observe that halogen substituents cause greater changes in the olefin geometry upon coordination than do other substituents.

In complexes of olefins where the positions of all four substituents cannot be determined, another measure of the bending back must be used. The measure commonly employed involves the torsional angles about the carbon–carbon bond (42). This is particularly useful in the case of *trans*-disubstituted olefins. Figure 9 illustrates the angles for a *trans*-olefin. There is an M—C=C—R torsional angle, δ, for each substituent and an R—C=C—R torsional angle, γ, for each pair of trans substituents. The angles δ and δ' in Fig. 9 increase from 90° as bending back occurs and γ decreases from 180°. The sum of the three angles must necessarily equal 360°. For complexes with no trans substituents, *cis*-disubstituted olefins, or monosubstituted olefins, only the δ angles can be measured. As might be expected, there is a good correlation between δ and α, which was defined in Fig. 8, for tetrasubstituted olefins as shown in Fig. 10. Thus δ, or the related γ, is a good measure of the bending back of substituent groups.

In complexes of symmetrical *trans*-olefins where the positions of all four substituents have been determined, it is possible for one set of trans substituents to be bent back more than the other set. This twist of the CR_2 groups about the C=C bond would not be obvious from an inspection of the α and β angles. This is true in compound XIX (Table I) where the CF_3 groups are bent back more than the F atoms; $\gamma = 132(1)°$ and $120(1)°$ for F and CF_3, respectively. It is difficult to ascribe this effect solely to either steric or electronic causes. It seems reasonable that certain groups bend back more for electronic reasons, but in this instance it is the bulkier group that is bent back more. It is not possible to predict on electronic grounds whether F or CF_3 should bend back more.

In the Ni complex of tetramethylethylene (V, Table I), where there is no difference in the size of the four substituents, the two observed values

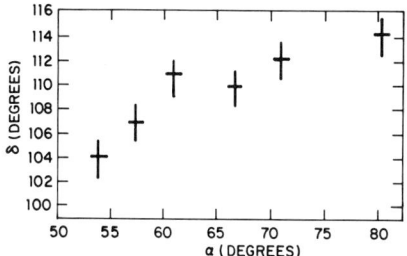

FIG. 10. Plot of δ versus α for olefin complexes.

of γ are 139° and 152°. This twist of the two $C(CH_3)_2$ groups about the olefin C=C bond has the effect of orienting the π-orbitals of the carbon atoms slightly toward the Ni,P,P plane. The opposite is found in the octafluorobutene structure (XIX, Table I) and more importantly in the tetrachloroethylene structure (XVI). Perhaps the twist is steric in nature, having little significance in the description of the bonding, but this remains to be determined.

4. *Twist of the Olefin out of the Coordination Plane*

A twist of the olefin out of the coordination plane is common in trigonal-planar d^{10} structures. Two independent measures of this twist have been used somewhat interchangeably although they are not necessarily equivalent. The more common measure is the angle θ between the normals n_1 to the L—M—L plane and n_2 to the C—M—C plane, illustrated in Fig. 11. A second measure is the angle φ between the C=C vector and the normal n_1 to the L—M—L plane. These two angles are complementary only when the rotation of the olefin takes place about the line D, bisecting both ≮LML and ≮CMC (Fig. 11A). Frequently this is not the only distortion

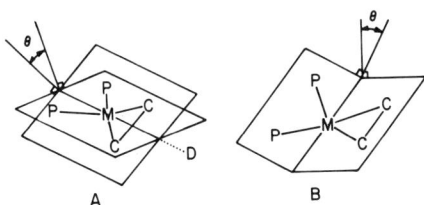

FIG. 11. Illustration of the two possible modes of twisting of the olefin out of the P—M—P plane. The angles are, of course, measured between the normals to the planes.

of the coordination plane, but there is also a trigonal pyramidal distortion with the metal moving out of the ligand plane (Fig. 11B). The most marked example of this pyramidal distortion is found in the nickel TCNE complex VI (Table I). The interplanar angle θ of 23.9(2)° is one of the largest observed for a simple olefin complex, but the vector-plane normal angle φ of 82.2(2)° is not exceptional. The 16.1° deviation from complementarity results from the pyramidal distortion reflected in the deviations of the 5 atoms from a least-squares plane. The nickel atom is above and the rest of the atoms average 0.25 Å below the plane. This extreme pyramidal distortion has not been observed in any other complex; it may be attributable to intermolecular packing forces.

The more commonly observed distortion, pure rotation of the olefin about axis D, results in complementary angles. There have been several conflicting theoretical calculations dealing with this distortion, predicting a potential minimum for both planar (*74*) or slightly twisted (*39*) complexes. These calculations are based on C_{2v} symmetry and have not considered the reduction of symmetry to C_2 often brought about by the packing of the bulky groups on the trans ligands. It is now well-established that in solution olefins exhibit a very low barrier to rotation about the metal–olefin bond in a variety of complexes. Hence a slight rotation about the axis D requires little energy and probably occurs frequently in order to lessen nonbonded contacts.

5. Orientation of π Substituents

Another aspect of the geometry of the bound olefin that has been barely studied is the orientation of substituents capable of π interactions with the olefin. The most studied π substituent is the cyano group, but its linearity precludes discussion of the nature of the interaction. The structures of two complexes of diphenylethylenes (IV and XVIII, Table I) have been determined. On the basis of electronic effects one would expect a phenyl ring either to be coplanar with the olefin double bond for better conjugation or to be perpendicular to the metal–olefin plane for greatest π overlap (in the cyclopropane model). The limited evidence favors the second orientation. However, structural studies of olefins with substituent groups such as -COH, -COOR, or -NO$_2$ would be useful for the further definition of the orientation of π substituents.

6. "Pointing" of the Trans Ligands

One final effect which has been noted (*5*) is the variation of the direction in which the trans ligands "point." This effect is not peculiar to olefin

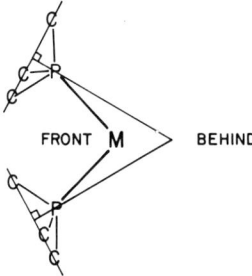

FIG. 12. Illustration of the "pointing" of the *trans*-phosphorus ligands.

complexes but is observed in all types of complexes involving phosphorus and other ligands. There are two manifestations of this effect that can be measured. It is very commonly noted that, whereas the three carbon–phosphorus–carbon angles are equal, the three metal–phosphorus–carbon angles for phosphine ligands are not equal; therefore, the ligand is not pointing directly at the metal atom. Another indication of this effect is found in the angle between the normals to the planes described by the 3 α atoms of each phosphorus ligand. As Fig. 12 indicates, the normals may intersect in front of or behind the metal atom, resulting in an interplanar angle greater or less than the phosphorus–metal–phosphorus angle, respectively. It is found that in complexes containing bidentate phosphorus ligands, such as V (Table I), the phosphorus atoms point behind the metal owing to the structural constraints of the chelate ring. It is interesting that for the remainder of the applicable complexes listed in Table I the phosphorus ligands generally point in front of the metal atom, although there are also distortions above and below the coordination plane. These effects most probably arise from steric causes, with the phosphorus ligand turning toward the greater nonbonded interaction. Nonetheless, these distortions should be considered in discussions of olefin coordination.

B. Geometry of the Complex

The foregoing discussion dealt primarily with d^{10} complexes, and, although deviations from the coordination plane were discussed, the overall geometry of the complexes was not dealt with in detail. A recent calculation (*66*) predicts that tris(ethylene)nickel(0) will be a trigonal complex with all 3 ethylene molecules lying in the coordination plane. The structures of nickel (bicycloheptene)$_3$ (VIII, Table I) and $Pt(C_2H_4)_2(C_2F_4)$ (*34*) are consistent with the calculation. This same geometry is found when one

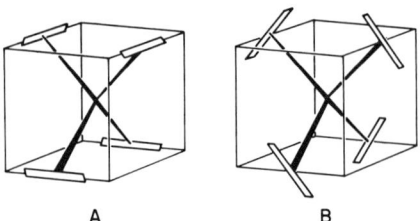

FIG. 13. The so-called cubic and dodecahedral modifications of tetrahedral four-coordinate olefin complexes.

olefin is replaced by a phosphorus ligand; thus, complex VII (Table I) has both ethylene molecules in the coordination plane. The structures of many complexes $Ni(PR_3)_2$(olefin) are known, and in each the olefin is found in the coordination plane, so that a planar or nearly planar geometry results. In many of these complexes there is little steric interaction between the two substituent ligands and the olefin. In others there is pronounced steric interaction and yet the complexes remain essentially planar.

Four-coordinate complexes can assume two geometries—square-planar or tetrahedral. It is found that d^{10} complexes are usually tetrahedral. A tetrakisolefin complex can assume one of two modifications of this geometry, either "cubical" or "dodecahedral," as shown in Fig. 13. It has been predicted that the dodecahedral conformation would be more stable (66), but the known d^{10} four-coordinate complexes (X and XI, Table I) involve cyclic olefins that limit the geometries to cubical. The only structure (XII) of a tetrahedral complex in which the olefin is free to assume its preferred geometry involves a tridentate phosphorus ligand. Details on this structure are limited, but it appears that the olefin lies close to a plane defined by 1 phosphorus atom and the bisector of the P—Ni—P angle between the other 2 phosphorus atoms. If so, the tetrafluoroethylene has assumed the dodecahedral geometry. This conclusion is tenuous, and other complexes of this type should be investigated.

Four-coordinate complexes involving d^8 metals are found to have a square-planar geometry. An olefin could potentially assume a geometry parallel or perpendicular to the coordination plane; it is found in all cases that the olefin is perpendicular to the plane. This can be rationalized partially on steric grounds; molecular models indicate impossibly close nonbonded interactions between an olefin in the equatorial plane and the ligands cis to the olefin. Of course, these arguments do not eliminate modest variations from the perpendicular position. In addition, calculations (66) have provided evidence on electronic grounds that the perpendicular conformation is more stable than the parallel conformation.

Five-coordinate complexes are generally either trigonal bipyramidal or square-pyramidal, with the former geometry favored by d^8 metals and the latter by d^6 metals [(30) and references therein]. The known five-coordinate monoolefin complexes are trigonal bipyramidal, presumably because d^8, rather than d^6, systems have been extensively investigated. In these trigonal bipyramidal monoolefin complexes, the olefin is equatorial and lies in the trigonal plane. The nonbonded interactions between the olefin and the axial ligands would be equivalent to the interactions between the olefin and the cis ligands 90° away, noted above for square-planar complexes. The nonbonded contacts in the equatorial plane are much less because the ligands are 120° away. Thus, again the observed geometry of these trigonal bipyramidal complexes can be rationalized on steric grounds alone. A recent calculation on d^{10} five-coordinate complexes (66) has shown that a π-acceptor ligand in an equatorial position will prefer a configuration in which its π-acceptor orbital is in the plane. Although there are no known five-coordinate d^{10} olefin complexes, the results are applicable to d^8 complexes also. Thus both steric and electronic effects again predict the same conformation.

Several five-coordinate complexes involving the cyclic diolefin, 1,5-cyclooctadiene (COD), have been investigated (XXIX and XXX, Table I). The olefin coordinates in an equatorial–axial manner, thus allowing the equatorial double bond to lie in the trigonal plane. If the olefin had coordinated in an equatorial–equatorial fashion, the two double bonds would necessarily be perpendicular to the trigonal plane. This preferred orientation may also be influenced by the approximately 90° bite of COD. In keeping with other five-coordinate complexes, the axial M—C distances are greater than the corresponding equatorial distances. The stronger equatorial interaction also results in a longer C=C distance.

C. Complexes of Nonolefinic Unsaturated Molecules

1. *Acetylene Complexes*

A modest number of acetylene complexes has been investigated structurally. Many of the features of olefin complexes are also observed in acetylene complexes, the major difference being the change in geometry of the coordinated acetylene.

The acetylene molecule approaches the geometry of a *cis*-olefin with the C≡C—R angle deviating from 180° by the angle α. From Table I one finds that, for a variety of complexes, α ranges from 12° to 40°. There again seems to be no correlation between C≡C distances and the bending back of substituent groups, but the number of structures is limited. It should be noted that metal–carbon distances are about 0.07 Å shorter in acetylene complexes than in related olefin complexes. Much of this difference could be attributed to the 0.04 Å change in the carbon single-bond radius on going from sp^2 to sp hybridization. Yet the carbon–carbon distance between the bridge and ring carbon atoms of coordinated diphenylacetylene and *trans*-stilbene are not appreciably different. Thus, if one equates bond length with bond strength, then one concludes that acetylenes interact more strongly with metal complexes than do similarly substituted olefins. This conclusion is in accord with theoretical predictions (56). When the usual description of metal–olefin bonding is extended to metal–acetylene systems, the same orbitals are used in the model. But acetylenes possess an additional set of π- and π^*-orbitals orthogonal to the metal–olefin plane. A calculation indicates that interaction of this π^*-orbital with additional metal d orbitals could be significant. Thus the metal–acetylene bond could be strengthened beyond that of a metal–olefin bond.

In general it is found that the carbon–carbon bond of acetylenes is lengthened less upon coordination than that of olefins. This might be interpreted as an indication that acetylenes interact to a lesser degree. Yet, note that the correlation between bond length and bond strength is not linear. Thus the increases in going from a triple bond to a double bond and then to a single bond are 0.13 and 0.21 Å, respectively. A smaller lengthening of an acetylene molecule relative to an olefin molecule can still indicate a comparable change in bond order.

As was noted in the discussion of olefin complexes, a twisting of the C≡C bond is usually observed in acetylene complexes. This twisting is manifested in a nonzero R—C≡C—R torsion angle. Angles γ and δ, defined the same way as for olefin complexes, should ideally be 0° and 180°, respectively. Values of γ up to 9° have been observed, and the two δ angles are not necessarily equal (Table I). As for olefin complexes, the differences in the δ's can be attributed to nonbonded interactions, both intra- and intermolecular.

The variations in γ for various diphenylacetylene complexes can be attributed to minimization of nonbonded contacts between the two phenyl rings. The contacts between *ortho*-hydrogen atoms would be rather close in a strictly planar molecule, and conjugation between the phenyl rings and the acetylene bridge is interfered with if the rings twist too much. The remaining method for relief of the hydrogen atom contacts is twisting about the acetylene bond. Although this argument seems plausible, a twist of 9° is observed in a dicyanoacetylene complex (XXXVII, Table I), where steric crowding is minimal, and 0° (by symmetry) is observed in a bis(*tert*-butyl)acetylene complex (XXXIX), where nonbonded contacts might become important.

2. *Diazene Complexes*

The structures of two complexes (XLI and XLII, Table I) of the diphenyl-substituted diazene, azobenzene, have been determined. Azobenzene is found to be capable of stronger π-backbonding than the isoelectronic *trans*-diphenylethylene. This effect is manifested in several structural aspects. After the difference in nitrogen and carbon atomic radii is considered, it is found that the N=N bond is longer and the M—N bond is shorter than the respective bonds in the olefin complex. Angle γ is 6.2(13)° less in the azobenzene complex, indicating a greater bending back of the phenyl rings.

Molecular orbital calculations (*43*) show that the highest occupied molecular orbitals of azobenzene and *trans*-stilbene are at approximately the same energy, but the lowest unoccupied molecular orbital of azobenzene is much lower than that of *trans*-stilbene. Thus azobenzene should be capable of better π-backbonding. The higher electronegativity of the N=N bridge compared with that of a C=C bridge would also argue for better π-backbonding for a diazene.

In the Ni(0) complexes of azobenzenes, the $n \to \pi^*$ transition observed in the visible spectrum provides an additional probe of the multiple bond (*42, 44*). It is found that the transition shifts to higher energy as the electron density on the double bond is increased. When electron density is increased by putting more electron-donating ligands on the nickel atom, the nickel–azobenzene bond is strengthened. If, however, the electron density on the N=N bond is increased by putting electron-donating substituents on the phenyl rings of the azobenzene, the complex is destabilized.

3. *Ketone Complexes*

Several complexes of hexafluoroacetone with d^{10} metals have been prepared. Based on ^{19}F NMR data, these complexes were thought to involve a ketonic C=O group involved in a π interaction with the metal. This was confirmed by a structural determination of the nickel(0) bisphosphine complex (XLIII, Table I). In this structure the CF_3 groups are bent back about as far as in typical halogenated olefin complexes. Interestingly, the oxygen atom and the carbon atom (with a larger atomic radius), are approximately equidistant from the nickel atom, whereas the phosphine ligand trans to the carbon atom shows a significant trans effect. This suggests that a CF_3-substituted carbon atom may be a better π-acceptor than an oxygen atom, consistent with the spectroscopic results of Section III, A,1.

4. *Imine Complexes*

Few π-bonded imine complexes have been prepared and investigated structurally. The only complexes of nonbridging π-bonded C=N investigated structurally are a substituted ketenimine complex (XLIV), an iminium complex (XLV), and a complex of the azine $[(CF_3)_2C=N-]_2$ (XLVI) (see Table I).

The ketenimine complex is unusual in that angle γ for the complex is 11.9(10)° and the average δ is 174°. These angles are indicative of a coordinated triple bond, rather than the expected double bond. This seemingly anomalous result can be rationalized on the basis of a large contribution of resonance form B to the structure of the ketenimine. This

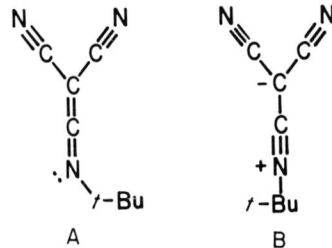

form would be enhanced by the highly electron-withdrawing cyano groups. The $-C\equiv N-C(CN)_2$ portion of the complex is very nearly planar, and this should enhance electron withdrawal by the two cyano groups. The deviations from linearity, α, at either end of the C—N bond are similar to those observed for acetylene complexes, thus lending additional support to the triple-bond model of the ketenimine.

The π-iminium complex (N-methylsalicylaldiminato)-(N-methylsalicylaldiminium)nickel(0) (XLV) is the first example of a nonchelating salicylaldimine complex. The nitrogen atom of the C=N double bond is made more electron-withdrawing by introduction of the positive charge. Thus, the greater bending back of the nitrogen substituents relative to the carbon substituents (β = 55° and 77°, respectively) is reasonable.

The azine complex (XLVI), formed by the reaction of bis(trifluoromethyl)diazomethane with Pt(0) is a rather simple imine complex. The feature of note in this complex is angle γ that represents one of the greatest observed deviations from planarity (γ = 127°).

IV
SUMMARY

The broad geometrical features of the interaction between unsaturated molecules and transition metals are now well-defined as a result of a large number of structural studies. In a general way these results can be rationalized by current, crude bonding models. But if the overall understanding of the bonding of unsaturated molecules to transition metals is to be improved, additional experimental and theoretical work needs to be done.

On the experimental side, a number of factors must be considered. The diffraction results on a given compound could be improved considerably if the experiment were done at low temperatures so that the smearing effects of thermal motion could be minimized. As yet, very few such studies have been performed. Now that more powerful neutron sources have been developed, there is a very important need for extensive neutron diffraction experiments on these types of complexes. The scattering of neutrons does not follow a simple pattern with atomic number and, generally speaking the location of both the C or N atoms *and* the H atoms of an unsaturated molecule in the presence of a transition metal could be made with much greater precision using neutrons. This is particularly true now that ligand systems involving fewer atoms are available (e.g., t-BuNC with 19 atoms versus PPh_3 with 34 atoms).

But even the usual diffraction studies at room temperature could yield valuable information through a systematic approach. One needs systematic studies of the same unsaturated molecule bound to an ML_x system in which first M and then L is varied. Ideally one needs far more studies of systems in which two or more different olefins are bound simultaneously to the same ML_x system. Clearly such systematic studies are dependent on new syntheses.

Although improvement of the theoretical models, especially in the direction of *a priori* calculations, presents formidable computational problems, there is the general trend that if reliable, interesting, and perhaps tantalizing observations are available, these will serve as an incentive for theoreticians to proceed.

Finally there is a desperate need for greater correlations of these structural and theoretical studies of the nature of the metal-unsaturated molecule interaction with other experimentally derived quantities. We have in mind correlations with spectroscopically derived quantities, such as stretching frequencies and NMR shielding parameters. But we also have in mind the most important problem of the correlation of the metrical details of the bonding with the reaction chemistry. If the discovery and utilization of new catalyst systems, so essential today in view of shifting patterns of feedstocks and of energy considerations, is to be anything but empirical, then an understanding of the relation between the metal-unsaturated molecule interaction and the reaction chemistry is of paramount importance.

REFERENCES

1. J. M. Baraban and J. A. McGinnety, *Inorg. Chem.* **13**, 2864 (1974).
2. J. M. Baraban and J. A. McGinnety, *J. Amer. Chem. Soc.* **97**, 4232 (1975).
3. E. Benedetti, P. Corradini, and C. Pedone, *J. Organometal. Chem.* **18**, 203 (1969).
4. G. Bombieri, E. Forsellini, C. Panattoni, R. Graziani, and G. Bandoli, *J. Chem. Soc. A* 1313 (1970).
5. D. J. Brauer and C. Krüger, *J. Organometal. Chem.* **44**, 397 (1972).
6. D. J. Brauer and C. Krüger, *J. Organometal. Chem.* **77**, 423 (1974).
7. J. Browning and B. R. Penfold, *Chem. Commun.* 198 (1973).
8. J. Chatt and L. A. Duncanson, *J. Chem. Soc.* 2939 (1953).
9. P. T. Cheng, C. D. Cook, C. H. Koo, S. C. Nyburg, and M. T. Shiomi, *Acta Crystallogr., Sect B* **27**, 1904 (1971).
10. P. T. Cheng and S. C. Nyburg, *Can. J. Chem.* **50**, 912 (1972).
11. M. R. Churchill and S. A. Bezman, *Inorg. Chem.* **11**, 2243 (1972).
12. M. R. Churchill and S. A. Bezman, *Inorg. Chem.* **12**, 260 (1973).
13. J. Clemens, R. E. Davis, M. Green, J. D. Oliver, and F. G. A. Stone, *Chem. Commun.*, 1095 (1971).
14. C. D. Cook, K. Y. Wan, U. Gelius, K. Hamrin, G. Johansson, E. Olsson, H. Siegbahn, C. Nordling, and K. Sieghban, *J. Amer. Chem. Soc.* **93**, 1904 (1971).
15. C. A. Coulson and T. H. Goodwin, *J. Chem. Soc.* 3161 (1963).
16. R. Countryman and B. R. Penfold, *Chem. Commun.* 1598 (1971); *J. Cryst. Mol. Struct.* **2**, 281 (1972).
17. G. L. Cunningham, A. W. Boyd, R. J. Meyers, W. D. Gwinn, W. I. LeVan, *J. Chem. Phys.* **19**, 676 (1951).
18. B. W. Davies and N. C. Payne, *Inorg. Chem.* **13**, 1843 (1974).
19. B. W. Davies and N. C. Payne, *Can. J. Chem.* **51**, 3477 (1973).
20. B. W. Davies and N. C. Payne, *Inorg. Chem.* **13**, 1848 (1974).
21. G. R. Davies, W. Hewertson, R. H. B. Mais, P. G. Owston, and C. G. Patel, *J. Chem. Soc. A* 1873 (1970).
22. M. I. Davis and C. S. Speed, *J. Organometal. Chem.* **21**, 401 (1970).
23. M. J. S. Dewar, *Bull. Soc. Chim. Fr.* **18**, C 71 (1951).
24. R. S. Dickson and J. A. Ibers, *J. Organometal. Chem.* **36**, 191 (1972).
25. R. S. Dickson and J. A. Ibers, *J. Amer. Chem. Soc.* **94**, 2988 (1972).
26. H. Dierks and H. Dietrich, *Z. Kristallogr.* **122**, 1 (1965).
27. J. B. R. Dunn, R. Jacobs, and C. J. Fritchie, *J. Chem. Soc., Dalton, Trans.* 2007 (1972).
28. J. A. Evans and D. R. Russell, *Chem. Commun.* 197 (1971).
29. K. Fischer, K. Jonas, P. Misbach, R. Stabba, and G. Wilke, *Angew. Chem. Int. Ed.* **12**, 943 (1973).
30. B. A. Frenz and J. A. Ibers, *M.T.P. Int. Rev. Sci., Phys. Chem. Ser.* **1**, 11, 33 (1972).
31. J. N. Francis, A. McAdam, and J. A. Ibers, *J. Organometal. Chem.* **29**, 131 (1971).
32. J. O. Glanville, J. M. Stewart, and S. O. Grim, *J. Organometal. Chem.* **7**, 9 (1967).
33. M. D. Glick and L. F. Dahl, *J. Organometal. Chem.* **3**, 200 (1965).
34. M. Green, J. A. K. Howard, J. L. Spencer, F. G. A. Stone, private communication.

35. M. Green, S. K. Shakshooki, and F. G. A. Stone, *J. Chem. Soc. A* 2828 (1971).
36. L. J. Guggenberger, *Inorg. Chem.* **12**, 499 (1973).
37. L. J. Guggenberger and R. Cramer, *J. Amer. Chem. Soc.* **94**, 3779 (1972).
38. W. C. Hamilton, K. A. Klanderman, and R. Spratley, *Acta Crystallogr., A* **25**, S172 (1969).
39. F. R. Hartley, *Angew. Chem., Int. Engl. Ed.* **11**, 596 (1972).
40. F. R. Hartley, *Chem. Rev.* **69**, 799 (1969).
41. R. Hoffmann, *Tetrahedron Lett.* **33**, 2907 (1970).
42. S. D. Ittel and J. A. Ibers, *J. Organometal. Chem.* **57**, 389 (1973).
43. S. D. Ittel and J. A. Ibers, *J. Organometal. Chem.* **74**, 121 (1974).
44. S. D. Ittel, unpublished results.
45. J. A. J. Jarvis, B. T. Kilbourn, and P. G. Owston, *Acta Crystallogr., Sect. B* **27**, 366 (1971).
46. M. M. T. Khan and A. E. Martell, "Homogeneous Catalysis by Metal Complexes," Vols. I and II. Academic Press, New York, 1974.
47. R. M. Kirchner and J. A. Ibers, *J. Amer. Chem. Soc.* **95**, 1095 (1973).
48. C. Krüger and Y. H. Tsay, *J. Organometal. Chem.* **34**, 387 (1972).
49. J. W. Lauher and J. A. Ibers, *J. Amer. Chem. Soc.*, **97**, 561 (1975).
50. A. R. Luxmoore and M. Truter, *Acta Crystallogr.* **15**, 1117 (1962).
51. A. McAdam, J. N. Francis, and J. A. Ibers, *J. Organometal. Chem.* **29**, 149 (1971).
52. M. Matsumoto, K. Nakatsu, K. Tani, A. Nakamura, and S. Otsuka, *J. Amer. Chem. Soc.* **96**, 6777 (1974).
53. K. W. Muir and J. A. Ibers, *J. Organometal. Chem.* **18**, 175 (1969).
54. L. M. Muir, K. W. Muir, and J. A. Ibers, *Discuss. Faraday Soc.* **47**, 84 (1969).
55. J. H. Nelson and H. B. Jonassen, *Coord. Chem. Rev.* **6**, 27 (1971).
56. J. H. Nelson, K. S. Wheelock, L. C. Cusachs, and H. B. Jonassen, *J. Amer. Chem. Soc.* **91**, 7005 (1969).
57. J. G. Norman, *J. Amer. Chem. Soc.* **96**, 3327 (1974).
58. M. Orchin, *Advan. Catal.* **16**, 1 (1966).
59. L. E. Orgel, "Introduction to Transition-Metal Chemistry," p. 137. Meuthen, London, 1960.
60. S. Otsuka, A. Nakamura, Y. Tatsuno, *J. Amer. Chem. Soc.* **91**, 6994 (1969).
61. C. Panattoni, R. Graziani, C. Bandoli, D. A. Clemente, U. Belluco, *J. Chem. Soc., B* 371 (1970).
62. C. Pedone and A. Sirigu, *Inorg. Chem.* **7**, 2614 (1968).
63. H. W. Quinn and J. H. Tsai, *Advan. Inorg. Chem. Radiochem.* **12**, 217 (1969).
64. J. S. Ricci and J. A. Ibers, *J. Amer. Chem. Soc.* **93**, 2391 (1971).
65. G. B. Robertson and P. O. Whimp, *J. Organometal. Chem.* **32**, C69 (1971).
66. N. Rösch and R. Hoffmann, *Inorg. Chem.* **13**, 2656 (1974).
67. N. Rösch, R. P. Messmer, and K. H. Johnson, *J. Amer. Chem. Soc.* **96**, 3855 (1974).
68. P. von R. Scheyer and V. Bass, *J. Amer. Chem. Soc.* **91**, 5880 (1969).
69. J. K. Stalick and J. A. Ibers, *J. Amer. Chem. Soc.* **92**, 5333 (1970).
70. C. A. Tolman, *J. Amer. Chem. Soc.* **96**, 2780 (1974).
71. T. E. Turner and J. A. Howe, *J. Chem. Phys.* **24**, 924 (1956).
72. R. Ugo, *Coord. Chem. Rev.* **3**, 319 (1968).
73. A. Walsh, *Trans. Faraday Soc.* **45**, 179 (1949).
74. K. S. Wheelock, J. H. Nelson, L. C. Cusachs, and H. B. Jonassen, *J. Amer. Chem. Soc.* **92**, 5110 (1970).
75. D. J. Yarrow, J. A. Ibers, Y. Tatsuno, and S. Otsuka, *J. Amer. Chem. Soc.* **95**, 8590 (1973).

Methyltin Halides and Their Molecular Complexes

V. S. PETROSYAN, N. S. YASHINA, and O. A. REUTOV

Chemistry Department
M. V. Lomonosov Moscow State University
Moscow, USSR

I. Introduction 63
II. Methods of Study 64
    A. Infrared and Raman Spectroscopy 64
    B. Nuclear Magnetic Resonance 65
    C. Nuclear Quadrupole Resonance 66
    D. Mössbauer Spectroscopy 66
    E. Electron and X-Ray Diffraction 67
    F. Other Methods 67
III. Structures of Methyltin Halides 68
    A. Trimethyltin Halides 68
    B. Dimethyltin Dihalides 71
    C. Methyltin Trihalides 72
IV. Molecular Complexes of Methyltin Halides 76
    A. Complexes of Trimethyltin Halides 77
    B. Complexes of Dimethyltin Dihalides 84
    C. Complexes of Methyltin Trihalides 90
V. Conclusion 91
    References 92

I
INTRODUCTION

Of the various organotin compounds which have been studied (*118*), the most intensive research has been carried out on organotin halides. The synthetic and mechanistic significance of these species was surveyed by Clark and Puddephatt (*28*), whose review covers the literature up to 1968. The simplest representatives, methyltin halides, are, in turn, the most important for structural studies forming the subject of more than a hundred papers during the last 6 years. The present review aims at a detailed, up-to-date discussion of methyltin halides and their molecular complexes with organic ligands. The halides and their complexes are very promising for the study of the chemistry of organometallic and complex compounds, throwing light on the problem of solvent effects on the kinetics and mechanisms of organometallic reactions (*110, 117*).

II

METHODS OF STUDY

To study the electronic structures and stereochemistries of methyltin halides and their molecular complexes, infrared and Raman spectroscopy, nuclear magnetic resonance, nuclear quadrupole resonance, Mössbauer spectroscopy, diffraction, and other techniques are usually employed today. Consequently, the application of these methods to the objects of interest, and the information to be gained will be discussed in this section. The conclusions made will help us to understand the behavior of specified compounds.

A. Infrared and Raman Spectroscopy

The most important structural information may be obtained by analyzing intensities and frequencies associated with tin–carbon, tin–halogen, and tin–ligand stretching vibrations. Let us first discuss intensities. A C_3Sn site in Me_3SnX gives rise to vibrations of two types, $\nu_s(Sn$—$C)$ and $\nu_{as}(Sn$—$C)$. If $\nu_s(Sn$—$C)$ is absent or its intensity is low compared with that of $\nu_{as}(Sn$—$C)$, the geometry of the site can be interpreted as planar or close to planar. Similarly, the absence of $\nu_s(Sn$—$C)$ or the fact that it is markedly less intense than is $\nu_{as}(Sn$—$C)$ shows that a C_2Sn site in Me_2SnX_2 is linear or close to linear. Intensity analyses of tin–halogen or tin–ligand stretching vibrations permit geometries of the sites to be assigned.

The frequency variation occurring on going from one molecule to another or to the molecular complex depends on force constants in noncomplexed molecules and reflects, in the final analysis, the ionic nature of the bonds. When no spectra can be obtained in solution or in the gas phase, the intermolecular contributions should be taken into account, especially in interpreting the Sn—X patterns. It should also be remembered that Sn—X frequencies are markedly dependent on the mass of the halogen involved and increase with atomic number. Nevertheless, the frequencies, which are a function of the nature and the number of halogens and vary over a wide range, are a good reflection of the bond ionicities and of the coordination number.

The Sn—C bond frequencies are rather less sensitive to electronic effects in methyltin halides and depend weakly on either the number or the nature of the halogens. The increases associated with complex formation are also insignificant. Vibrational spectroscopy is a good tool not only for studying the spatial arrangement and ionicities of Sn—X bonds but also for measuring stability constants of the complexes (*96, 15, 91*). The method is based on estimating the intensity due to the uncomplexed ligand as a function of the concentrations of the methyltin halide and of the ligand.

B. Nuclear Magnetic Resonance

The application of NMR to a study of the structure and complexation of tin compounds was reported by us at the 11th European Conference on Molecular Spectroscopy (*114*). In the present review we would like to emphasize that a study of electronic and spatial structures of methyltin halides requires, in the first place, a study of spin-spin coupling between ^{119}Sn and ^{1}H and ^{13}C in methyl groups.

In 1961, Burke and Lauterbur (*21*) showed that the constants $J(^{119}\text{Sn}-\text{C}-^{1}\text{H})$ in methyltin halides depend significantly on both the number of halogens and the solvent used. Later work (*66, 148*), dealing with solutions of methyltin halide in water, postulated a linear dependence of $J(^{119}\text{Sn}-\text{C}-^{1}\text{H})$ on the s contribution of sp^n hybrid tin orbitals in Sn—C bonds. The dependence found seemed very attractive since it would allow, *inter alia*, the assignment of the geometries of molecular methyltin halide complexes in solution (*16, 137, 61, 24, 4, 71*) by quantitatively estimating the s contribution in Sn—C bonds. However, in 1965 Verdonck and van der Kelen (*150*) studied the proton magnetic resonance (PMR) spectra of the ethyltin halides showing that the dependence postulated (*66, 148*) should be viewed with great caution and that the variations in constants $J(^{119}\text{Sn}-\text{C}-^{1}\text{H})$ diverged from a simple correlation involving the s contribution. Similar conclusions were made by McFarlane (*97*), who showed that $J(^{119}\text{Sn}-\text{C}-^{1}\text{H})$ did vary as $J(^{119}\text{Sn}-^{13}\text{C})$ but that the straight line did not pass through the origin, and by Lorberth and Vahrenkamp (*87*) who studied the spectral characteristics of methyltin and ethyltin halides in detail.

In 1972 we found (*109*) that $J(^{199}\text{Hg}-\text{C}-^{1}\text{H})$ increased from 98.0 to 104.5 Hz, whereas $J(^{199}\text{Hg}-\text{C}-\text{C}-^{1}\text{H})$ (127.5 Hz) was unaffected, on going from diethylmercury solutions in inert solvents to solutions in solvating solvents. These findings and the fact (*46*) that $J(^{199}\text{Hg}-\text{C}-^{1}\text{H})$ and $J(^{199}\text{Hg}-\text{C}-\text{C}-^{1}\text{H})$ behave similarly to $J(^{199}\text{Hg}-^{13}\text{C})$ and $J(^{199}\text{Hg}-\text{C}-^{13}\text{C})$ suggested (*109, 114*) that these constants, analogous to other heteronuclear constants, depended on the relative content of s electrons in the respective sites (Hg—C, Hg—C—H, Hg—C—C, Hg—C—C—H) rather than on just the s contribution of sp^n hybrid orbitals of the metal. That this conclusion applies to methyltin halides and their molecular complexes is substantiated by a thorough study (to be discussed in the following) of concentration and temperature dependences of $J(^{119}\text{Sn}-\text{C}-^{1}\text{H})$ measured for the chlorides and bromides in various solvents (116).

Constants $J(^{119}\text{Sn}-^{13}\text{C})$ have been studied less fully and will be dealt with when discussing the structures of individual compounds of interest. A review on ^{119}Sn chemical shifts has appeared recently (134).

C. Nuclear Quadrupole Resonance

In 1973, Bryukhova, Semin, and the present writers (112) and, later, van der Kelen and his co-workers (149) showed that nuclear quadrupole resonance (NQR) spectroscopy, especially of ^{35}Cl, ^{81}Br, and ^{127}I is highly applicable to the study of the structure and complex formation of methyltin halides in the solid phase. The most important NQR characteristics are resonance frequency, multiplicity of signals, and relative intensities. The frequency shifts are due to the alteration of the electric field gradient on the halogen atoms, that is, to alteration of the tin–halogen bond ionicities (130). Consequently, NQR frequencies of ^{35}Cl, ^{81}Br, and ^{127}I may provide information on the tin–halogen bond nature as a function of spatial and electronic factors. At the same time, if two or more halogens are present in a molecule, the resonance frequencies may differ and thus reflect (88, 89) a difference in the electronic environment (chemical nonequivalence) or a difference in positions of the nuclei in the crystal lattice (crystallographic nonequivalence). Crystallographic splittings $[(\Delta\nu/\nu_{\text{av}}) \cdot 100\%]$ are, as a rule, of about 2–3%, whereas the chemical nonequivalence may be significantly greater (130). Analysis of the splittings and relative intensities often allows one to depict reliably the spatial arrangement in solid methyltin halides and their molecular complexes (112).

D. Mössbauer Spectroscopy

More than 10 years ago, Gol'danskii (53) had shown how promising ^{119}Sn Mössbauer spectroscopy is for the study of tin compounds. The data obtained up to 1971 for various organotin compounds have been surveyed by Smith (132), Zuckerman (154), Parish (101), and Bancroft and Platt (1). Mössbauer spectroscopy of monoorganotin(IV) derivatives has been reviewed very recently (7).

These data, as analyzed in the light of the recent results (119, 51, 6, 83, 33, 17, 120, 84, 58, 113, 128, 63, 70, 107, 108, 121, 2, 3), demonstrate that the principal Mössbauer parameters, isomer shift (IS) and quadrupole splitting (QS), are highly informative of chemical structure and coordination. For example, IS is a measure of the total electron density at a tin atom (*53*) and depends mainly on the population of the valence shell, that is, it reflects those alterations of the structure affecting the *s*-electron density (*82*). The alterations are chiefly due to alterations of tin–atom or tin–ligand bond ionicities (*102*), although a contribution to the *s*-electron density may sometimes be provided by the shielding associated with *d* orbitals (*53*). Regardless of the mechanism that governs the IS behavior, the IS pattern in any methyltin halide complexed with monodentate electron-donor ligands reflects charge transfer, i.e., the relative donor ability of the ligands (*113*).

Recent data (*113, 3*) show that QS values depend, first of all, on the tin coordination number in the compounds and the stereochemistry. They also depend on the nature of the ligand in isostructural molecules. This picture, illustrated in original papers (*64, 104*) and surveyed in reviews (*1, 101*), gives a clue to assigning spatial arrangements of ligands, L, in complexes of the type $RSnX_3 \cdot 2L$ (*113*), $R_2SnX_2 \cdot 2L$ (*113*), and $R_3SnX \cdot L$ (*3*) as well as to constructing donor ability series for ligands in the compounds.

E. Electron and X-Ray Diffraction

Diffraction methods, which are beyond doubt the most informative approach, are at the same time the most cumbersome. It should also be stressed that, although this approach may allow the complete structure of a molecule to be given in the gas or solid phase, in practice nonvolatility, instability, and difficulties inherent in crystal growth, of the samples may interfere significantly. Where, however, these difficulties are overcome the results obtained are profitable whatever the effort invested.

Recently, Ho and Zuckerman have published an extensive review (*65*) on structural organotin chemistry and discussed the experimental evidence accumulated up to 1971. The number of papers devoted to organotin structures has increased (*13, 19, 20, 23, 59, 60, 74, 125, 127*) lately, the increase being markedly favored by progress in electronic and X-ray hardware. The bond lengths and angles obtained are very important for correlations with the spectral evidence (*57*). Such correlations may make other spectral evidence quite reliable even in the absence of the respective diffraction data.

F. Other Methods

Other approaches include UV spectroscopy (*151, 73, 92, 93, 133*), photoelectron spectroscopy (*106, 55, 18, 11*), conductometry (*78, 79, 140, 141,*

47, 24), thermochemistry (*14, 15*), and dipole moment measurements (*144, 67, 56*). The works cited have demonstrated (*54, 122*), however, that these methods are markedly less applicable.

III
STRUCTURES OF METHYLTIN HALIDES

In this section, structures in the gas phase, in solution, and in the solid state are discussed. Let us first give a classification of the solvents used for studying structures in solutions. These may be divided into (i) poorly solvating and nonionizing solvents (cyclohexane, CCl$_4$, benzene, chloroform, dichloromethane, nitromethane, nitrobenzene, and others), (ii) strongly solvating and nonionizing solvents [pyridine, acetone, tetrahydrofuran, dioxan, N,N-dimethylformamide (DMF), dimethyl sulfoxide (DMSO), hexametapol (HMPT), and others], and (iii) strongly solvating and ionizing solvents such as water, alcohols, amines, and others.

Methyltin halides dissolved in solvents of type (i) do not enter into a noticeable donor–acceptor interaction with the solvent; the tin coordination number is, therefore, unaffected. At the same time, conductivity techniques (*78, 79, 140, 141, 47, 24*) show that methyltin halides do not ionize in solvents such as nitromethane and nitrobenzene whose dielectric permeabilities are high (35.9 and 34.8, respectively). In this section the halide structures are discussed only in solvents of type (i). When they are dissolved in solvents of type (ii), the halides enter into complex formation; these systems are dealt with under complexation of methyltin halides (see Section IV).

Organotin cations formed through dissolving the halides in solvents of type (iii) are not discussed in this review. Their recent study has been rather limited, but they have been reviewed by Tobias (*142*).

A. Trimethyltin Halides

1. Me_3SnF

In the gas phase, no structure is known for Me$_3$SnF. In the solid state, IR spectroscopy suggested (*100, 80*) a structure of the Me$_3$Sn$^+$F$^-$ type. X-Ray studies (*26, 152*) demonstrate, however, that orthorhombic Me$_3$SnF crystals have a covalent structure in which fluorine atoms alternate with

Me₃Sn groups in an infinite chain, the tin being five-coordinate and approximately trigonal-bipyramidal. There are two sorts of fluorines: Sn—F, 2.15 Å and Sn···F, 2.45 Å. It is believed (*26, 49*) that the electron density distribution data may be interpreted in terms of either planar or pyramidal structures for the SnMe₃ fragment, but a weak band found at 515 cm⁻¹ is thought to indicate that the structure is pyramidal. The planarity, however, has been proved unambiguously in studies (*44, 136, 83*) of Sn—C stretching vibrations in the IR and Raman spectra. The Mössbauer parameters [δ, 1.28 mm/sec; Δ, 3.86 mm/sec (*62, 34, 103*)] agree with the five-coordinate structure in terms of correlation (*62*) or point charge (*104*) approaches. The Debye-Weller factor as a function of temperature (*135*) also suggests that Me₃SnF is a strongly bonded polymer.

The compound is insoluble in inert organic solvents, hence no structure in solution has been obtained. As for the PMR parameters found in methanol solution (*87*), the peculiarities of methanol (see above) suggest that the parameters [$\delta(^1H)$, 0.45 ppm; $J(^1H—C—^{119}Sn)$, 69.0 Hz] bear no relation to the structure of Me₃SnF.

2. *Me₃SnX (X = Cl, Br, I)*

Electron diffraction showed (*131*) that in the gas phase the compounds are distorted tetrahedra with the Sn—C and Sn—X bond lengths roughly corresponding to the sums of the covalent radii: Me₃SnCl (Sn—C, 2.11 Å; Sn—Cl, 2.36 Å), Me₃SnBr (Sn—C, 2.17 Å; Sn—Br, 2.49 Å), Me₃SnI (Sn—I, 2.72 Å). No X-ray studies have been published. Infrared (Table I), NQR (*112, 149*), and Mössbauer spectroscopy data (Table II) demonstrate that the solid compounds are associates containing five-coordinate tin, the associates being decomposed on melting or on dissolving in inert solvents. The increase in ν(Sn—X) across the series solid–melt–solution suggests that the covalence of the bond increases across the same series, i.e., with a decrease in association. The fact that ν(Sn—Cl) varies more than ν(Sn—Br) shows (*30*) that Me₃SnCl is associated more strongly than is Me₃SnBr, whereas the constancy of ν(Sn—I) suggests that there is no significant association in Me₃SnI. That association decreases in the series Me₃SnF ≫ Me₃SnCl > Me₃SnBr > Me₃SnI agrees also with the Mössbauer parameters (Table II) and especially with the quadrupole splittings.

When trimethyltin chloride is dissolved in inert solvents, ν(Sn—Cl) remains essentially invariant: CCl₄, 336 cm⁻¹ (*8, 9*); benzene, 331 cm⁻¹ (*8, 9*); cyclohexane, 331 cm⁻¹ (*32*); carbon disulfide, 331 cm⁻¹ (*80*). Compound Me₃SnCl may, therefore, be assumed to be a tetrahedral monomer in these solvents.

A similar conclusion was arrived at in a study of the concentration de-

TABLE I

TIN–HALOGEN AND TIN–CARBON STRETCHING FREQUENCIES OF Me$_3$SnX IN THE SOLID AND LIQUID STATES AND IN SOLUTION[a]

Vibration	Me$_3$SnF Solid IR	Me$_3$SnF Solid R	Me$_3$SnCl Solid IR	Me$_3$SnCl Solid R	Me$_3$SnCl Liquid IR	Me$_3$SnCl Liquid R	Me$_3$SnCl Solution in CCl$_4$ IR	Me$_3$SnCl Solution in CCl$_4$ R	Me$_3$SnBr Solid IR	Me$_3$SnBr Liquid R	Me$_3$SnBr Solution in cyclohexane IR	Me$_3$SnBr Solution in cyclohexane R	Me$_3$SnI Solid IR	Me$_3$SnI Solid R	Me$_3$SnI Solution in cyclohexane IR
ν(Sn—X)	335vs	—	—	288	325	315	336	331	—	199	219	234	229	—	177	189
ν_s(Sn—C)	—	521s	514w	513	514	514	513m	514	512m	512	512	511m	512	509	511	—
ν_{as}(Sn—C)	555s	559w	543vs	546	545	545	543s	544	541vs	545	543	539s	541	540	538	—

[a] Values are expressed in cm^{-1}.

TABLE II

MÖSSBAUER DATA FOR TRIMETHYLTIN HALIDES

Parameter	Me₃SnF	Me₃SnCl	Me₃SnBr	Me₃SnI
δ(mm/sec)	1.28	1.42	1.49	1.48
Δ(mm/sec)	3.86	3.41	3.25	3.05

pendences of the ^{119}Sn chemical shifts (*69, 134, 143*) and $J(^1\text{H}-\text{C}-^{119}\text{Sn})$ constants (*116*) in nonpolar solvents. The NMR data (*45, 25, 126, 87, 39, 98, 146, 147*) listed in Table III for trimethyltin halides in CCl₄ show that the most informative are the $J(^1\text{H}-\text{C}-^{119}\text{Sn})$ values which increase slightly with the electronegativity of the halogen. Their behavior, which reflects the growth of the *s* contribution in the Sn—C—H site, fits well with the Bent rehybridization theory (*10*). On the other hand, the fact that the $J(^1\text{H}-\text{C}-^{119}\text{Sn})$ values embrace a rather small range, whereas the $J(^1\text{H}-^{13}\text{C})$ couplings are constant across the series Me₃SnCl > Me₃SnBr > Me₃SnI, points out that rehybridization of tin orbitals is rather insignificant in this series.

B. Dimethyltin Dihalides

Electron diffraction techniques showed (*131, 50*) that in the gas phase Me₂SnX₂ (X = Cl, Br, I) are somewhat distorted tetrahedra: Me₂SnCl₂ (∠CSnCl, 109.8°; ∠ClSnCl, 107.5°; Sn—C, 2.11 Å; Sn—Cl, 2.33 Å), Me₂SnBr₂ (Sn—C, 2.17 Å; Sn—Br, 2.48 Å), Me₂SnI₂ (Sn—I, 2.69 Å).

X-Ray analysis of crystalline Me₂SnF₂ showed that (*129*) in the solid state the compound is an associate containing six-coordinate tin. The structure is a two-dimensional infinite net in which every tin atom is

TABLE III

NUCLEAR MAGNETIC RESONANCE PARAMETERS FOR SOLUTIONS OF Me₃SnX IN CCl₄

Me₃SnX	δ(CH₃) (ppm)	δ(^{119}Sn) (ppm)	$J(^{119}\text{Sn}-\text{C}-^1\text{H})$ (Hz)	$J(^{119}\text{Sn}-^{13}\text{C})$ (Hz)	$J(^{13}\text{C}-^1\text{H})$ (Hz)
Me₃SnCl	0.61	158.0	58.1	386	131.6
Me₃SnBr	0.73	128.0	57.8	372	131.8
Me₃SnI	0.88	38.6	57.2	—	132.1

bonded to 4 other tin atoms by fluorines situated symmetrically in between. Methyl groups lie under and above the resulting plane and, thus, complete the octahedral structure. The Sn—C distance of 2.08 ± 0.01 Å is the shortest known in organotin compounds and was ascribed (*129*) to ionicity of the $(SnF_2)_\infty$ site. Infrared spectra of the compound have shown, however, that ν_s(Sn—C) cannot be detected, whereas ν(Sn—F) is observed. Thus, covalence of the octahedral polymer with *trans*-methyl groups is beyond doubt (Table IV), (*86, 22, 51, 83*). The structure fits well with the Mössbauer data (Table V) processed via a correlation (*62*) or a point-charge method leading to Δ values for the *trans*-R_2SnL_4 structures (*48*). The IR data of Table IV and the Mössbauer data of Table V show that the association falls across the series $Me_2SnF_2 > Me_2SnCl_2 > Me_2SnBr_2 > Me_2SnI_2$ (*105*).

A sharp difference between the Me_2SnF_2 structure and the structures of the other dihalides is verified by an X-ray study of Me_2SnCl_2 (*41*). The tin environment was shown to be intermediate between a tetrahedron and an octahedron, owing to association of adjacent molecules through Sn—Cl···Sn bridges. The structure consists of molecular chains, with tin and chlorine atoms being coplanar in each of the chains. Methyl groups lie under and above the plane. The chain has a zigzag shape. The bond lengths and angles are listed in Table VI.

The distorted octahedron of Me_2SnCl_2 agrees also with the Mössbauer quadrupole splitting found for the compound (Table V), lying between the tetrahedral (2.3 mm/sec) and octahedral (4.1 mm/sec) values.

Nuclear quadrupole resonance ^{79}Br and ^{81}Br spectra of solid Me_2SnBr_2 display (*112*) weakly split doublets which suggest that σ-electron density is equal in both the bromines accurately to within the crystallographic splitting. The associate assumed on the basis of Mössbauer data (*105*) is thus not a five-coordinate species, because otherwise a doublet of considerable splitting, with one of the components corresponding to a bridge bromine and the other to a terminal bromine, would have arisen.

The NMR parameters in Table VII, especially the $J(^1H—C—^{119}Sn)$ couplings, imply that the tin rehybridization in the halides is significant and dependent on the halogen. As in the trimethyltin halides, the *s*-electron content in the Sn—C—H site increases across the series I < Br < Cl.

C. Methyltin Trihalides

Electron diffraction showed (*131*) that, in the gas phase the trihalides have a slightly distorted tetrahedral structure: $MeSnCl_3$ (Sn—C, 2.19 Å; Sn—Cl, 2.32 Å), $MeSnBr_3$ (Sn—Br, 2.45 Å), $MeSnI_3$ (Sn—I, 2.68 Å). No

TABLE IV
TIN–HALOGEN AND TIN–CARBON STRETCHING FREQUENCIES OF Me_2SnX_2 IN THE SOLID AND LIQUID STATES AND IN SOLUTION.[a]

Vibration	Me_2SnF_2 Solid IR	Me_2SnF_2 Solid R	Me_2SnCl_2 Solid IR	Me_2SnCl_2 Solid R	Me_2SnCl_2 Liquid IR	Me_2SnCl_2 Solution in benzene IR	Me_2SnCl_2 Solution in cyclohexane IR	Me_2SnBr_2 Solid IR	Me_2SnBr_2 Solution in benzene IR	Me_2SnBr_2 Solution in cyclohexane IR	Me_2SnI_2 Solid IR	Me_2SnI_2 Solid R	Me_2SnI_2 Solution in cyclohexane IR
$\nu_s(Sn{-}X)$	360	—	307	344	320	350	356	—	—	240	—	182	186
$\nu_{as}(Sn{-}X)$	360	—	332	344	320	350	361	—	—	250	—	197	204
$\nu_s(Sn{-}C)$	—	536	515	531	515	521	524	514	518	—	511	513	—
$\nu_{as}(Sn{-}C)$	598	—	567	566	566	559	560	563	554	—	547	544	—

[a] Values are expressed in cm^{-1}.

TABLE V

MÖSSBAUER DATA FOR DIMETHYLTIN DIHALIDES

Parameter	Me$_2$SnF$_2$	Me$_2$SnCl$_2$	Me$_2$SnBr$_2$
δ(mm/sec)	1.24	1.52	1.59
Δ(mm/sec)	4.11	3.62	3.41

crystal structures have been reported (we are studying them now in collaboration with L. A. Aslanov), but a natural assumption is that the greater Lewis acidity of the trihalides (compared with Me$_2$SnX$_2$) will cause the molecules to associate in the solid phase. For MeSnF$_3$, this assumption agrees with the IR spectrum (Table VIII) and with the temperature dependence of the Mössbauer spectrum which contains a well-resolved doublet at 298°K. The resulting assignment (83) leads to a polymeric structure containing six-coordinate tin atoms, two bridging fluorines, and one terminal fluorine. It also agrees with the relative intensities of the IR signals ν(Sn—F)$_t$ and ν(Sn—F)$_b$.

With MeSnCl$_3$ and MeSnBr$_3$, the six-coordinate association in the solid state is verified by ^{35}Cl, ^{79}Br, and ^{81}Br NQR data (112). The spectra contain strongly split doublets, with the intensity ratio being 2:1 for the low- and high-frequency components. Consequently, there are two bridging halogens and one terminal halogen. That the association is stronger in MeSnCl$_3$ than in MeSnBr$_3$ is suggested by the temperature dependence of the NQR spectra and by a comparison of the Mössbauer spectra of the compounds with those of the molecular complexes containing various electron-donor ligands (113). The isomer shifts demonstrate that MeSnCl$_3$ (δ, 1.32 mm/sec) is a stronger Lewis acid than is MeSnBr$_3$ (δ, 1.44 mm/sec).

As with Me$_2$SnX$_2$, the $J(^1$H—C—^{119}Sn) values (116) and the δ(^{119}Sn) chemical shifts (134) found for MeSnX$_3$ dissolved in solvents such as benzene or dichloromethane depend neither on concentration nor on tempera-

TABLE VI

INTERATOMIC DISTANCES AND ANGLES IN Me$_2$SnCl$_2$

Bond	Length (Å)	Fragment	Angle
Sn—C	2.21	Cl—Sn—Cl	93°0'
Sn—Cl	2.40	C—Sn—C	123°30'
Sn···Cl	3.54	C—Sn—Cl	109°0'

TABLE VII

NUCLEAR MAGNETIC RESONANCE PARAMETERS FOR SOLUTIONS OF
Me_2SnX_2 IN CCl_4

Me_2SnX_2	$\delta(CH_3)$ (ppm)	$\delta(^{119}Sn)$ (ppm)	$J(^{119}Sn-C-^1H)$ (Hz)	$J(^{13}C-^1H)$ (Hz)
Me_2SnCl_2	1.15	140	69.0	136.2
Me_2SnBr_2	1.33	70	66.3	136.6
Me_2SnI_2	1.63	−157	62.4	136.8

ture. This may be interpreted by assuming that there is no significant association in the systems.

Molten $MeSnCl_3$ has $\delta(^{119}Sn)$ of 6.0 ppm (42), markedly different from the value found for the benzene solution (Table IX). Consequently, there is some association in the melt, which increases on going to the solid state. Nuclear magnetic resonance parameters, especially the $J(^1H-C-^{119}Sn)$ values, found for solutions of $MeSnX_3$ in CCl_4 show that the s-electron content in the Sn—C—H site is higher than in Me_3SnX or Me_2SnX_2, and depends markedly on the nature of the halogen, in agreement with the Bent theory (10).

On summarizing the structures of methyltin halides (35, 36), we may say that, in the series $Me_3SnX < Me_2SnX_2 < MeSnX_3$, on the one hand, the increase in the NMR constants $J(^1H-C-^{119}Sn)$ reflect an increase in the s-electron content in the Sn—C—H site and, on the other hand, the increases in the IR $\nu(Sn-Hal)$ frequencies and the NQR $\nu(Hal)$ frequen-

TABLE VIII

TIN–CARBON AND TIN–HALOGEN STRETCHING FREQUENCIES OF $MeSnX_3$ IN THE SOLID AND LIQUID STATES AND IN SOLUTION[a]

Vibration	$MeSnF_3$ Solid		$MeSnCl_3$ Solid	Liquid	Solution in cyclohexane	$MeSnBr_3$ Solution in cyclohexane		$MeSnI_3$ Solution in cyclohexane	
$\nu(Sn-C)$	548vs	535s	542	546	551	539m		527w	
$\nu(Sn-X)$	646vs	629s 425vs	384	360s	382vs	264vs	235m	207vs	174w

[a] Values are expressed per centimeter.

TABLE IX

NUCLEAR MAGNETIC RESONANCE PARAMETERS FOR SOLUTIONS OF MeSnX$_3$ IN CCl$_4$.

MeSnX$_3$	δ(CH$_3$) (ppm)	δ(^{119}Sn) (ppm)	J(^{119}Sn—C—^{1}H) (Hz)	J(^{119}Sn—^{13}C) (Hz)	J(^{13}C—^{1}H) [Hz(in C$_6$H$_6$)]
MeSnCl$_3$	1.69	21	100.0	—	141.2
MeSnBr$_3$	1.85	−165	88.6	−640 (in C$_6$H$_6$)	141.2
MeSnI$_3$	2.32	−600	73.4	—	141.2

cies both reflect a decrease in the Sn—X bond polarity. These experimental data draw attention to a quantum chemistry calculation (56) of electric dipole moments and σ and π charges in Me$_n$SnCl$_{4-n}$. The calculation has shown that the Sn—Cl bond polarity decreases across the series Me$_3$SnCl < Me$_2$SnCl$_2$ < MeSnCl$_3$ (Table X) and that the Sn—C polarity falls in the same direction. Sign inversion of the dipole occurs in MeSnCl$_3$ which is why the MeSnCl$_3$ dipole moment is greater than that of Me$_3$SnCl.

Thus, the methods discussed in the foregoing allow all the necessary information on the structures of methyltin halides in the gas phase (electron diffraction), in the crystal state (X-ray techniques, NQR, Mössbauer spectroscopy, IR spectroscopy), and in solution (NMR and IR methods) to be obtained.

IV
MOLECULAR COMPLEXES OF METHYLTIN HALIDES

The data reported in the literature on molecular complexes of trimethyltin halides demonstrate that these have a 1:1 composition with mono-

TABLE X

DIPOLE MOMENTS AND BOND POLARITIES IN Me$_n$SnCl$_{4-n}$ (n = 1–3)

Me$_n$SnCl$_{4-n}$	μ(D)		Polarity (%)		
	Calculated	Observed	Sn—Cl	Sn—C	C—H
Me$_3$SnCl	3.46	3.46–3.52	38.68	10.66	3.30
Me$_2$SnCl$_2$	4.04	4.14–4.21	33.70	4.48	3.64
MeSnCl$_3$	3.63	3.62–3.77	28.07	−2.77	4.05

dentate ligands and a 2:1 composition with bidentate ligands. On the other hand, dimethyltin dihalides form, as a rule, 1:2 complexes with monodentate ligands and 1:1 complexes with bidentate ligands. Under certain conditions, however, 1:1 complexes with monodentate ligands may be isolated for Me_2SnX_2 and $MeSnX_3$; their formation in solution may also be possible. In the following discussion, we denote the number of organotin molecules by the first figure and the number of coordinated donor centers by the second figure, regardless of whether the ligand is mono- or bidentate. Molecular complexes of methyltin halides may exist in various stereoisomeric forms whose types are listed in Table XI together with the notation to be employed below.

It is also noteworthy that there are many complexes whose isolation in the individual state is impossible. However, methods such as NQR or Mössbauer spectroscopy allow one to study the complexes in frozen solutions.

As for the spectral parameters [e.g., $\nu(Sn-X)$ or $J(^1H-C-^{119}Sn)$] of methyltin halides dissolved in electron-donor solvents, they should be tested in each individual case to see whether they correspond to nondissociated complexes or to an equilibrium between the complex and the uncomplexed halide.

Together with stereochemical problems, the ways in which the Sn—C and Sn—X bonds are affected on going from a methyltin halide to its molecular complex are of interest. Such information is quite helpful in interpreting the reactivities of methyltin halides in various solvents (110, 117).

More than twenty molecular complexes of methyltin halides with electron-donor solvents (111) were reported in 1973 and their structures were studied by NQR (112), Mössbauer (113), and NMR (116) techniques. These and other data will be discussed under individual types of the complexes.

A. Complexes of Trimethyltin Halides

In 1963, Gielen and Nasielski (52) studied the NMR spectra of trimethyltin bromide in various solvents and showed that $J(^1H-C-^{119}Sn)$ increased across the series CCl_4 < MeCOOH < dioxan < acetone < MeOH < pyridine < water < DMSO < DMF. Simultaneously, Beattie and McQuillan (8) found that $\nu(Sn-Cl)$ in the IR spectra of trimethyltin chloride is also solvent-dependent and increases as follows: CCl_4 < benzene < acetonitrile < pyridine. In both cases the data were interpreted as indicating formation of 1:1 complexes. Finally, Hulme (68) found by

TABLE XI

THE BASIC TYPES OF STRUCTURES FOR MOLECULAR COMPLEXES OF METHYLTIN HALIDES

$Me_3SnX \cdot L$	$Me_2SnX_2 \cdot L$	$Me_2SnX_2 \cdot 2L$	$MeSnX_3 \cdot L$	$MeSnX_3 \cdot 2L$
| ```
Me   X
 \  |
  Sn—Me
 /  |
Me   L
 (Ia)
``` | ```
Me   X
 \  |
  Sn—L
 /  |
Me   X
 (IIa)
``` | ```
L  Me   X
 \  |  /
    Sn
 /  |  \
L  Me   X
 (IIIa)
``` | ```
Me   X
 \  |
  Sn—L
 /  |
X    X
 (IVa)
``` | ```
L  Me   X
 \  |  /
    Sn
 /  |  \
L   X   X
 (Va)
``` |
| ```
Me  Me
 \  |
  Sn—X
 /  |
Me   L
 (Ib)
``` | ```
Me   X
 \  |
  Sn—X
 /  |
Me   L
 (IIb)
``` | ```
X  Me   L
 \  |  /
    Sn
 /  |  \
L  Me   X
 (IIIb)
``` | ```
X    X
 \  |
  Sn—Me
 /  |
X    L
 (IVb)
``` | ```
L  Me   X
 \  |  /
    Sn
 /  |  \
X   X   L
 (Vb)
``` |
| ```
Me   X
  | /
Me—Sn
  | \
Me   L
 (Ic)
``` | ```
X   Me
 \  |
  Sn—L
 /  |
X   Me
 (IIc)
``` | ```
L  Me  Me
 \  |  /
    Sn
 /  |  \
L   X   X
 (IIIc)
``` | ```
X   Me
 \  |
  Sn—X
 /  |
X    L
 (IVc)
``` | ```
X  Me   X
 \  |  /
    Sn
 /  |  \
L   L   X
 (Vc)
``` |
| ```
Me  Me
  | /
Me—Sn
  | \
X    L
 (Id)
``` | ```
Me  Me
 \  |
  Sn—L
 /  |
X    X
 (IId)
``` | ```
L  Me  Me
 \  |  /
    Sn
 /  |  \
X   X   L
 (IIId)
``` | ```
X    X
 \  |
  Sn—L
 /  |
X   Me
 (IVd)
``` | |
| | ```
X   Me
 \  |
  Sn—Me
 /  |
X    L
 (IIe)
``` | ```
X  Me  Me
 \  |  /
    Sn
 /  |  \
L   L   X
 (IIIe)
``` | | |

X-ray methods that the trimethyltin chloride–pyridine complex has a 1:1 composition. It is a molecular adduct of trigonal bipyramidal structure, with three methyls lying on the equatorial plane and with the pyridine and a chlorine atom at the axial positions. The Sn—Cl distance found in the complex exceeds that in the initial halide.
```

TABLE XII

Tin-Carbon and Tin-Halogen Stretching Frequencies for Me₃SnX·L Complexes

Compound	ν_{as}(Sn—C) (cm⁻¹)	ν_s(Sn—C) (cm⁻¹)	ν(Sn—X) (cm⁻¹)
Me₃SnCl[a]	543s	513m	336vs
Me₃SnCl·Py	546s	512vw	250
Me₃SnCl·PyO	557vs	—	232s
Me₃SnCl·bipy	554w	—	246m
Me₃SnBr[a]	539s	511m	234s
Me₃SnBr·Py	542s	509vw	—
Me₃SnBr·PyO	550s	—	—
Me₃SnBr·bipy	551w	—	158s
Me₃SnBr·DMSO	550s	—	—
Me₃SnBr·Bz₂SO	545s	—	—
Me₃SnBr·Ph₃PO	542s	—	—
Me₃SnBr·Ph₃AsO	543s	—	—
Me₃SnI[a]	536m	508w	189s
Me₃SnI·Py	541s	504vw	—
Me₃SnI·bipy	567w	538w	—

[a] Stretching frequencies are given for solutions in inert solvents.

These pioneer works were followed by numerous studies on complexes of trimethyltin halides, the data from which are discussed in the following.

1. *Complexes with Pyridine and N-Oxopyridines*

The preceding results for the trimethyltin chloride–pyridine complex (*68*) fit well with IR (*8, 9*) and Mössbauer data (*1, 101*). The ν_s(Sn—C): ν_{as}(Sn—C) intensities ratio (Table XII) suggests a planar arrangement of the methyl groups, whereas ν(Sn—Cl) suggests that the bond polarity is markedly higher than in uncomplexed Me₃SnCl.

The chloride, bromide, and iodide complexes probably possess similar structures since all three have identical Sn—C stretching frequency ranges (Table XII). An NQR ⁸¹Br spectrum of Me₃SnBr·Py points to a noticeable crystallographic nonequivalence in the crystal cell and to a higher Sn—Br bond ionicity compared with Me₃SnBr.

The PMR spectra of Me₃SnX·Py complexes dissolved in CHCl₃ implied that the compounds dissociated completely (*145*) in solution. We showed (*116*), however, in a study of the concentration and temperature dependence of $J(^1\text{H}—\text{C}—^{119}\text{Sn})$ in Me₃SnX–pyridine–CH₂Cl₂ mixtures, that the equilibrium

$$Me_3SnX + Py \rightleftharpoons Me_3SnX \cdot Py$$

depends on the pyridine-to-CH_2Cl_2 ratio and on the sample temperature, and is totally shifted toward the complex at higher pyridine concentrations and lower temperatures. The limiting $J(^1H—C—^{119}Sn)$ values observed (Table XIII) in the system reflect the s-electron density content in the Sn—C—H site and allow one to deduce a series for the electron-donor ability of coordinating solvents. Pyridine occurs between weak and strong electron donors.

Complexes of the type

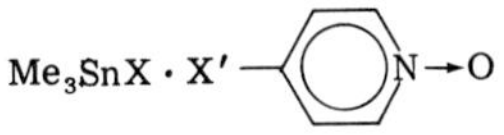

$$Me_3SnX \cdot X'-\!\!\!\left\langle\bigcirc\right\rangle\!\!\!N \rightarrow O$$

(X = Cl, Br; X' = H, CH_3, OCH_3, Cl, NO_2) were studied both in the solid state and in solution (73, 76). Their $\nu(N \rightarrow O)$ frequencies are lower than in the free ligands, implying coordination at the oxygen. The $\nu(Sn—X)$ frequencies (Table XII) show that complex formation raises the bond polarity.

TABLE XIII

THE $J(^{119}Sn—C—^1H)$ LIMITING VALUES FOR $Me_3SnX \cdot L$ COMPLEXES

	$J(^{119}Sn—C—^1H)$ (Hz)	
L in $Me_3SnX \cdot L^a$	Me_3SnCl	Me_3SnBr
MeCN	66.7	65.1
THF	64.5	—
Acetone	66.1	66.0
DMTAA	65.7	—
Pyridine	68.0	67.6
DMAA	68.8	—
TMED	69.7	68.5
DMF	70.0	69.6
DMSO	70.1	69.6
HMPT	71.8	71.3

[a] Abbreviations: THF, tetrahydrofuran; DMF, N,N-dimethylformamide; DMSO, dimethylsulfoxide; HMPT, hexametapol; DMTAA, dimethylthioacetamide; DMAA, dimethylacetamide; TMED, tetramethylethylenediamine.

Spectrophotometric data for the stability constants of complexes of Me$_3$SnCl with substituted pyridine N-oxides in acetonitrile show (*76*) that these constants correlate linearly with the σ parameters of the γ substituents.

No ν_s(Sn—C) was found in Me$_3$SnBr·PyO. It was, therefore, interpreted (*38*) in terms of planarity of the C$_3$Sn site, namely, structure Ia (Table XI).

2. *Complexes with Phosphine or Arsine Oxides*

Conductivities of the complexes in absolute ethanol were shown (*38*) to be very low when compared with tetramethylammonium bromide as a reference. This finding was interpreted to indicate that molecular adducts exist in solution. The absence of ν_s(Sn—C) from the IR spectra and the J(^{1}H—C—^{119}Sn) magnitude (ca. 70 Hz) were interpreted in terms of planarity of the C$_3$Sn site and a noticeable rehybridization of the tin orbitals.

3. *Complexes with Sulfoxides*

Trimethyltin chloride and bromide form DMSO complexes of 1:1 composition, melting at 49°C (*73*) and 63°C (*111*) respectively. The decrease in ν(S=O) ($\Delta\nu$ is 95 cm^{-1} for Me$_3$SnCl·DMSO, and 45 cm^{-1} for Me$_3$SnBr·Bz$_2$SO) may show (*38, 77*) that the coordination is via the oxygen, whereas the absence of ν_s(Sn—C) from the complexes of Me$_3$SnBr (Table XII) suggests a structure of type Ia.

As the NQR ^{81}Br frequency is shifted on going from Me$_3$SnBr to its complex with DMSO (*112*), a strong charge transfer and an increase in the Sn—Br bond polarity may be assumed.

Studies of the concentration and temperature dependences of mixtures of DMSO and Me$_3$SnX (X = Cl, Br) in CH$_2$Cl$_2$ gave limiting values for the J(^{1}H—C—^{119}Sn) constants (Table XIII) and the Me$_3$SnX·DMSO stability constants (Table XIV). These results suggest that the values of 62.0 and 63.0 Hz, obtained earlier for CHCl$_3$ solutions of Me$_3$SnBr·DMSO (*38*) and Me$_3$SnBr·Bz$_2$SO (*77*), respectively, are nothing but the equilibrium values and do not reflect structural specificity of the complexes formed. The J(^{1}H—C—^{119}Sn) values found for Me$_3$SnX solutions in DMSO (*4*) exceed our values (Table XIII) by 1.8 and 2.4 Hz, but the couplings reported (*4*) for Me$_3$SnX solutions in CDCl$_3$ or for other methyltin halides are also 1.5 to 3.8 Hz greater than the respective couplings reported elsewhere (*66, 87, 116, 144*).

TABLE XIV
STABILITY CONSTANTS FOR $Me_3SnX \cdot L$ COMPLEXES

Complex[a]	Solvent	Method	$t(°C)$	K_p (moles/ liter)	ΔH^0 (kcal/ mole)
$Me_3SnCl \cdot$ acetone	Acetone	HNDMR $^1H-\{^{119}Sn\}$	+37	1.1	4.2
	Acetone	NMR(^{119}Sn)	+20	7.0	—
	Isooctane	Calorimetry	+26	0.9	−6.0
	CCl_4	PMR	+26	0.4	−5.7
	CH_2Cl_2	PMR	−30	0.8	—
$Me_3SnBr \cdot$ acetone	Acetone	NMR(^{119}Sn)	+20	3.0	—
	CH_2Cl_2	PMR	−30	0.7	—
$Me_3SnCl \cdot MeCN$	CCl_4	PMR	+26	0.5	−48
	MeCN	NMR(^{119}Sn)	+20	2.7	—
	MeCN	HNDMR $^1H-\{^{119}Sn\}$	+20	2.1	4.1
$Me_3SnBr \cdot MeCN$	MeCN	NMR(^{119}Sn)	+20	3.5	—
$Me_3SnCl \cdot DMTAA$	CCl_4	PMR	+26	0.5	−5.9
	CH_2Cl_2	PMR	−70	1.5	—
$Me_3SnCl \cdot Py$	CCl_4	IR	+27	1.8	−6.5
	CH_2Cl_2	PMR	−30	36.0	—
$Me_3SnBr \cdot Py$	CH_2Cl_2	PMR	−30	28.0	—
$Me_3SnCl \cdot DMAA$	CCl_4	IR	+27	3.7	−7.9
	CCl_4	PMR	+34	3.1	−7.9
$Me_3SnCl \cdot DMSO$	CCl_4	Calorimetry	+26	9.1	−8.2
	CCl_4	IR	+27	8.3	—
	CCl_4	PMR	+34	6.2	—
	CH_2Cl_2	PMR	+28	2.8	—
$Me_3SnBr \cdot DMSO$	CH_2Cl_2	PMR	+28	3.6	—
$Me_3SnCl \cdot HMPT$	Isooctane	PMR	+26	384	−10.1
	CH_2Cl_2	PMR	+28	102	—
$Me_3SnBr \cdot HMPT$	CH_2Cl_2	PMR	+28	99	—
$Me_3SnCl \cdot$ dioxan	Dioxan	HNDMR $^1H-\{^{119}Sn\}$	+20	2.1	5.1
$Me_3SnCl \cdot DMF$	CH_2Cl_2	PMR	−30	2.5	—
$Me_3SnBr \cdot DMF$	CH_2Cl_2	PMR	−30	3.1	—

[a] Abbreviations: DMSO, dimethylsulfoxide; HMPT, hexametapol; DMF, N,N-dimethylformamide; DMAA, dimethylacetamide; DMTAA, dimethylthioacetamide; HNDMR, heteronuclear double magnetic resonance.

The data obtained, when analyzed as a whole, show that Me₃SnBr·DMSO may be the strongest complex among molecular complexes of trimethyltin halides with monodentate ligands.

4. *Other Complexes*

The molecular complexes of methyltin halides with various electron-donating solvents have been given a great deal of attention by a number of workers.

Drago *et al.* (*96, 14–16*) employed calorimetry and IR and PMR methods to measure the stability constants and heats of formation of Me₃SnCl complexes with dimethyl sulfoxide, dimethylacetamide, DMF, acetonitrile, pyridine, and HMPT (see Table XIV).

Okavara and co-workers (*91*), who studied solvent effects on the PMR spectra, and on the ν_s(Sn—C):ν_{as}(Sn—C) intensities ratio in the IR spectra, of Me₃SnCl in various solvents concluded that the coordination ability of the solvents may be represented by the series DMSO ~ DMF ~ DMAA > pyridine > acetone > dioxan > MeCN > PhCN.

We isolated (*111*) molecular complexes of Me₃SnBr, studied their structure by NQR in the solid state (*112*), and showed that complexes of type Ia are formed with monodentate electron-donor solvents, whereas complexes with bidentate solvents may have structure Ib. The latter assumption agrees with the fact that the limiting J(¹H—C—¹¹⁹Sn) values measured in dioxan and DME at different temperatures are different; they may reflect the relative percentages of Ia and Ib complexes in solution.

The ⁸¹Br NQR frequencies of the complexes lead (*112*) to the following series of electron-donor ability with respect to Me₃SnBr: DEE < dioxan < acetone < DME < THF < DMF < HMPT < pyridine < DMSO < TMED. This series fits, on the whole, with the one based on NMR data obtained for Me₃SnBr in the same solvents (*116*). The strongest electron donor is HMPT, which gives a molecular complex even with Me₃SnCF₃ (*114, 115*), in contrast to the commonly held belief that organotin compounds containing four C—Sn bonds cannot form molecular complexes (*4*).

An X-ray study showed (*20*) that Me₃SnCl·Ph₃P=CH—COCH₃ has a Ia type structure, with the oxygen atom in the second axial position.

In conclusion, trimethyltin halides give Ia complexes with all monodentate ligands. All three Sn—C bonds are coplanar and contain more *s*-electron density than does the uncomplexed Me₃SnX. The donor atom of the coordinating ligand, and the halogen, occupy axial positions in the structure, with the Sn—X bond in the complex being more polar than in the free halide. The *s*-electron content in the Sn—C bond and the Sn—X bond polarity increase with the electron-donor ability of the ligand.

B. Complexes of Dimethyltin Dihalides

The complexes of dimethyltin dihalides were readily studied since they may be obtained easily, owing to the high acceptor ability of the starting halides, and may be purified by simple recrystallization or sublimation.

1. *Complexes with Sulfoxides*

Many sulfoxide complexes, mainly with Me_2SnCl_2, have been reported. The simplest, $Me_2SnCl_2 \cdot 2Me_2SO$, was described for the first time by Langer and Blut (*81*) and studied in great detail by various physical methods. Infrared spectroscopy led to the conclusion (*27*) that the two methyls (as well as the two chlorines) were trans-arranged (structure IIIb in Table XI), but more sophisticated IR (*84, 85*) and Raman (*123, 124*) studies of Me_2SnCl_2 complexes with sulfoxides have shown that the Tanaka (*137*) IIIa structure is correct. The trans-arrangement of the methyl groups agrees also with the Mössbauer quadrupole splitting data (4.40 mm/sec) (*40, 84, 113*) and with an X-ray study (*153*) that revealed a slightly distorted octahedron with the atom–atom distances listed in Table XV.

The $J(^1H\text{—}C\text{—}^{119}Sn)$ values (113.0–117.1 Hz) (*75, 91, 116*) obtained for the complex in excess DMSO suggest that the *trans*-methyl configuration is retained in solution and reflect a strong increase in the *s*-electron content in the Sn—C—H site.

As for $J(^1H\text{—}C\text{—}^{119}Sn)$, equal to 86.0 Hz for the solution in $CHCl_3$ (*38*), it reflects nothing but an equilibrium between the complexed and uncomplexed Me_2SnCl_2 molecules.

The absence of $\nu_s(Sn\text{—}C)$ from the IR spectrum of $Me_2SnBr_2 \cdot 2DMSO$, and a high value (430 cm^{-1}) for $\nu(Sn\text{—}O)$ suggested structure IIIb for the complex (*137*). This agrees with the ^{81}Br NQR singlet (*112*) which,

TABLE XV

INTERATOMIC DISTANCES AND ANGLES IN $Me_2SnCl_2 \cdot 2DMSO$

Bond	Length (Å)	Fragment	Angle
Sn—Cl$_1$	2.53	Cl$_1$—Sn—Cl$_2$	96°
Sn—Cl$_2$	2.48	Cl$_1$—Sn—O$_1$	92°
Sn—O$_1$	2.32	Cl$_1$—Sn—O$_2$	175°
Sn—O$_2$	2.38	Cl$_1$—Sn—C$_1$	90°
Sn—C$_1$	2.08	Cl$_1$—Sn—C$_2$	94°
Sn—C$_2$	2.07	Cl$_2$—Sn—O$_1$	172°
		C$_1$—Sn—C$_2$	170°

among other things, shows the greatest charge transfer among Me$_2$SnBr$_2$ complexes with electron-donor solvents such as pyridine, DMF, and HMPT. The sulfoxide complex also has the highest (113) quadrupole splitting (4.49 mm/sec). The $J({}^1\text{H}-\text{C}-{}^{119}\text{Sn})$ of 115.5 Hz found for the solution in excess DMSO (116) shows that the *trans*-dimethyl arrangement is retained in solution.

Complexes of Me$_2$SnCl$_2$ with Et$_2$SO, Pr$_2$SO, Bu$_2$SO, Ph$_2$SO (84) Bz$_2$SO, and BzMeSO (77) were characterized by their PMR (77), IR (77, 84), and Mössbauer (84) parameters. Many 1:1 and 1:2 complexes were isolated. The decrease in $\nu(\text{S}=\text{O})$ shows coordination via oxygen in all cases. A strong $\nu(\text{Sn}-\text{C})$ and a complex multiplet in the Sn—Cl and Sn—O region found in the 1:2 complexes led to a structure of type IIIa. A *trans*-dimethyl arrangement in the solid state fits well with the Mössbauer quadrupole splittings (3.94–4.19 mm/sec). The $J({}^1\text{H}-\text{C}-{}^{119}\text{Sn})$ values for solutions of Me$_2$SnCl$_2\cdot$2Et$_2$SO, Me$_2$SnCl$_2\cdot$2BzMeSO, and Me$_2$SnCl$_2\cdot$Bz$_2$SO in CH$_2$Cl$_2$ are 88.5, 85.5, and 80.0 Hz, respectively. In our opinion they reflect the complexed–uncomplexed equilibrium rather than the geometry of the complexes in solution.

The fact that Bz$_2$SO forms a 1:1 complex, whereas Me$_2$SO and MeBzSO form both 1:1 and 1:2 complexes may imply an important role for steric factors on coordination. The 1:1 complexes have two intense $\nu(\text{Sn}-\text{C})$ vibrations and a multiplet in the Sn—Cl region; therefore, they were ascribed (84) a IIa structure.

2. *Complexes with Pyridine, Phosphine, and Arsine Oxides*

Pyridine, phosphine, and arsine oxide complexes have also been studied extensively. In 1966, Clark and Wilkins (29) studied the IR spectra of Me$_2$SnCl$_2$ complexes with pyridine and triphenylphosphine oxides (Table XVI) and assumed a structure of the IIIa type (see Table XI). Later, however, it was shown (30, 73, 38, 84) that the Sn—C multiplicity in Me$_2$SnCl$_2\cdot$PyO does not significantly differ from that in Me$_2$SnCl$_2\cdot$2DMSO. These facts, when collated with the absence of $\nu_s(\text{Sn}-\text{C})$, the singlet of $\nu(\text{Sn}-\text{O})$, and the magnitude of Δ (3.96 mm/sec), lead to the IIIb structure. This assignment was verified by an X-ray study (12) which revealed a slightly distorted octahedron with the distances and angles as listed in Table XVII.

The IR spectrum of Me$_2$SnCl$_2\cdot$PyO (Table XVI) contains two intense $\nu(\text{Sn}-\text{C})$ vibrations and a $\nu(\text{Sn}-\text{Cl})$ vibration. The complex was, therefore, assigned (84) a IIa structure, in agreement with the rule of predominantly axial arrangement of electron-donor groups in a trigonal bipyramid (99). In Me$_2$SnCl$_2$ complexes with substituted pyridine N-oxides, a linear

TABLE XVI

Tin–Carbon, Tin–Halogen and Tin–Ligand Stretching Frequencies of Dimethyltin Dichloride Complexes

Complex	ν_{as}(Sn—C) (cm^{-1})	ν_s(Sn—C) (cm^{-1})	ν(Sn—Cl) (cm^{-1})			ν(Sn—O) (cm^{-1})	
Me$_2$SnCl$_2$·2PyO	574m	—	233s	228s	223sh	324vs	
	541s	—	229vs	234sh		—	
Me$_2$SnCl$_2$·PyO	570ms	508ms		252ms		324vs	320sh
Me$_2$SnCl$_2$·2Ph$_3$PO	577m	—	261s	246s		323m	305s
	575ms	507w	248s	242sh		264sh	258s
Me$_2$SnCl$_2$·2Ph$_3$AsO	574	—		245–220s		375m	

correlation was found (76) between ν(Sn—O) and the σ constants of the γ substituents. The J(^{1}H—C—^{119}Sn) values found (51) for the complexes in solution are insensitive to the geometry but are very sensitive, unlike Me$_3$SnBr complexes, to the nature of X in

$$X-\underset{}{\bigcirc}N\rightarrow O$$

No ν_s(Sn—C) was observed in Me$_2$SnCl$_2$·2Ph$_3$PO or in Me$_2$SnCl$_2$·2Ph$_3$AsO (29, 38, 84) and a Δ value of 4.23 mm/sec was found for the former (84); both facts indicate a *trans*-dimethyl arrangement. A multiplicity of bands found in the Sn—O frequency region suggests a IIIa structure.

TABLE XVII

Interatomic Distances and Angles in Me$_2$SnCl$_2$·2PyO

Bond	Length (Å)	Fragment	Angle
Sn—Cl	2.58	Cl—Sn—C$_1$	89.5°
Sn—O	2.25	Cl—Sn—O	89.5°
Sn—C$_1$	2.22	Cl$_1$—Sn—O	95.6°
O—N	1.37	Sn—O—N	117°

3. Pyridine Complexes

In the IR spectra of pyridine complexes (*29, 27, 139, 85*), there is only one ν_{as}(Sn—C) vibration between 550 and 564 cm^{-1}. This fact, when collated with the Δ values of 3.98 to 4.05 mm/sec (*34, 113*) suggests the *trans*-dimethyl arrangement. The fact that the Sn—C frequencies pattern in Me$_2$SnCl$_2$·2Py closely resembles that found for Me$_2$SnCl$_2$·bipy (*43*) was interpreted in terms of a IIIa structure for both the complexes, and for Me$_2$SnBr$_2$·2Py, which was shown to be isomorphous with Me$_2$SnCl$_2$·2Py in an X-ray study. The presence of ν_s(Sn—C) in Me$_2$SnI$_2$·2Py was assigned to a distortion of the octahedral structure. These assumptions fit well with the ^{81}Br NQR spectrum of Me$_2$SnBr$_2$·2Py which contains just one resonance signal (*112*).

Proton magnetic resonance spectroscopy was of no avail in the study of complex formation of Me$_2$SnCl$_2$ with pyridine in solution since insoluble species are formed with any ratio of reactants. An IR spectrum of the Me$_2$SnCl$_2$·γ-picoline complex contains (*61*) a strong singlet for ν_{as}(Sn—C), at 560 cm^{-1}, and a broad Sn—Cl multiplet at 233 cm^{-1}, suggesting the IIIa structure. The J(^{1}H—C—^{119}Sn) value of 102.0 Hz found for the solution in CHCl$_3$ (*145*), and corrected for the possibility of dissociation, may confirm that the configuration is retained in solution.

4. Complexes with DMF and Other Monodentate Ligands

In 1968, Okawara *et al.* (*92*) isolated the 1:1 and 1:2 Me$_2$SnX$_2$ (X = Cl, Br) complexes with DMF and the 1:1 complexes with aromatic carbonyl compounds. In the IR spectra of these complexes (*92*), there is a shift of ν(CO) toward lower frequencies, suggesting coordination via the carbonyl oxygen. For Me$_2$SnX$_2$·2DMF, containing weak ν_s(Sn—C) vibrations in their IR spectra, a structure of the IIIa type (see Table XI) was proposed, although angle CSnC was admittedly below 180°. The geometry was believed to be due to the lower coordinative ability of DMF compared with, for example, DMSO. Indeed, ν(Sn—O) values are lower, and ν(Sn—Cl) values higher, in the DMF complexes compared with Me$_2$SnX$_2$·2DMSO. A IIb structure corrected via the Muetterties rule (*99*) was proposed for Me$_2$SnX$_2$·DMF since there are two intense ν(Sn—C) vibrations and two intense ν(Sn—Cl) vibrations that imply nonlinearity of the CSnC and ClSnCl skeletons.

Complexes of type Me$_2$SnX$_2$·L (L is a para-substituted carbonyl compound) were also assigned a IIb structure on the basis of similar IR data. The application of IR and UV techniques to solutions of the complexes in

1,2-dichloroethane gave (92) stability constants lower than the constants for the respective bipyridyl complexes discussed in the following.

In 1968, Tanaka and Kamitani (138) synthesized, and obtained IR spectra of the 1:1 and 1:2 Me_2SnCl_2 complexes with dimethylselenoxide. Complex formation shifts $\nu(Se—O)$ toward lower frequencies, suggesting probable coordination via the oxygen. The magnitude (60–100 cm^{-1}) of the shifts observed is, however, markedly lower than that (150–200 cm^{-1}) in similar complexes with DMSO. The presence of only one $\nu(Sn—C)$ vibration and of two well-resolved $\nu(Sn—X)$ vibrations suggests the IIIa structure. Complex $Me_2SnCl_2 \cdot Me_2SeO$ shows two $\nu(Sn—C)$ bands and one $\nu(Sn—Cl)$ band, indicating a IIa structure.

5. Complexes with Bi- and Polydentate Ligands

Complexes of Me_2SnX_2 with bipyridyl and o-phenanthroline have been studied extensively (47, 31, 29, 32, 37, 85). In nitrobenzene, they have very low conductivities (47) showing that no ionization occurs in solution. The presence of only one $\nu(Sn—C)$ vibration (ca. 570 cm^{-1}) in the IR spectra of the complexes (Table XVIII) and the Mössbauer quadrupole splittings [4.03 mm/sec for $Me_2SnCl_2 \cdot$ phen; 4.09 mm/sec for $Me_2SnCl_2 \cdot$ bipy, (34)] imply the trans-dimethyl arrangement. The presence of two $\nu(Sn—N)$ frequencies and a broad Sn—Cl multiplet suggest a IIIa structure. Complex formation lowers the $\nu(Sn—Cl)$ frequencies by 80–100 cm^{-1}, so that bond polarity rises considerably. Spectrophotometry gave (90) the stability constant for $Me_2SnCl_2 \cdot$ bipy in acetonitrile (log K = 3.36 ± 0.04 liters/mole) and thermodynamical parameters of its formation (ΔH = 17.25 kcal/mole, ΔG = 4.50 kcal/mole, and ΔS = 24.8 eu).

The $\nu(CO)$ vibration found for the 1:1 Me_2SnCl_2 complex with N,N-dimethylpicolinamide (94) is lower, whereas the ring-bending frequencies

TABLE XVIII

TIN–CARBON AND TIN–HALOGEN STRETCHING FREQUENCIES OF $Me_2SnX_2 \cdot L$ COMPLEXES[a]

Vibration	bipy			phen				
	Cl	Br	I	Cl		Br		I
$\nu(Sn—C)$	575s	571w	569m	578	—	572m	551w	560m 554sh
$\nu(Sn—X)$	244s	151s	145m	247s	239sh	157m	149s	139m 126s

[a] Values are expressed in cm^{-1}.

are higher, than in the free ligand. This was explained by assuming that both the carbonyl oxygen and the ring nitrogen participate in coordination. The presence of two strong ν(Sn—Cl) vibrations and of one ν_s(Sn—C) and one ν_{as}(Sn—C) vibration suggests a IIIa structure with the CSnC backbone slightly nonlinear. The fact that ν(C=S) is unaffected in the corresponding complex with N,N-dimethylthiopicolinamide was explained on the basis of the lower donor ability of a C=S group compared with that of C=O.

In complexes of Me_2SnCl_2 with N,N-dimethylnicotinamide and the isonicotinamide, ν(C=O) falls but the ring-bending frequencies rise. The structures were assumed (94) to be polymeric with ligand bridges, since chelate structures are impossible in this case.

Recently, complexation of organotin halides with polydentate ligands has been intensively studied in various countries (71, 125, 94, 95, 72, 107, 108). An X-ray study made in 1973 by Randaccio (125) showed that the molecular complex of bis(salicylaldehyde)ethylenediimine with Me_2SnCl_2 is, in the solid state, a polymer. The tin atom, *trans*-dimethyl groups, and *trans*-chlorines lie on a plane of an octahedron, whose axial positions are occupied by the oxygen atoms of the ligand (a IIIb structure). A similar conclusion was arrived at in an IR and Mössbauer study (5) of R_2SnX_2 complexes ($R = CH_3, C_4H_9, C_6H_5$; $X = Cl, Br$) with bis(acetylacetone)-ethylenediimine. The same explanation probably holds for the NQR spectra of Me_2SnX_2 complexes with dioxan (112), although a IIIc structure is possible for the complexes with DME. The PMR spectra of solutions of Me_2SnCl_2 complexes with various picolinealdimines showed (94) that the imine nitrogen also takes part in the coordination since the imine proton–tin couplings were observed. The presence of a strong ν(Sn—C) vibration (575 cm^{-1}) and two intense ν(Sn—Cl) vibrations (240 and 253 cm^{-1}) suggest the IIIa structure. The $J(^1H—C—^{119}Sn)$ value of 111 to 115 Hz shows linearity of the C—Sn—C skeleton. The concentration dependence of the couplings resulted in a determination of the stability constants of the complexes, and thermodynamic parameters for the complexation were obtained calorimetrically. The interaction of organotin halides with tri- and tetradentate Schiff bases more often than not leads to cleavage of the Sn—X bonds, resulting in ionic, rather than molecular complexes (71, 72, 95). These will not be discussed here.

6. *Complexes with Electron-Donor Solvents*

In order to study the interaction of Me_2SnX_2 with electron-donor solvents, we have recorded and analyzed the ^{81}Br, NQR spectra (112) of Me_2SnBr_2 complexes with dioxan, THF, acetone, DME, pyridine, HMPT,

DMF, and DMSO and found that the donor ability of the solvents increases across the series given. The singlet pattern obtained suggested structures of types IIIa or IIIb which agreed with the Δ values of 3.85 to 4.40 mm/sec found from the Mössbauer spectra (*113*). A good correlation was found between ν (^{81}Br) and Δ. Both of these quantities vary slightly (within 5 to 15% and 8 to 13%, respectively) in Me$_2$SnBr$_2$ complexes with weak electron donors such as THF, acetone, dioxan. In complexes with stronger donors (DMF, DMSO, HMPT), however, the variations are rather high (40–45% and 25–30%, respectively).

It is interesting to note that the NQR spectra of the 1:2 Me$_2$SnCl$_2$ complexes with acetone, and the 1:1 complexes with DME, differ sharply from the respective complexes of Me$_2$SnBr$_2$ in that they are clear-cut doublets. This is readily explained in terms of a IIIc structure for the former complexes.

A study of the concentration and temperature dependences of J(^{1}H—C—^{119}Sn) in Me$_2$SnCl$_2$/CH$_2$Cl$_2$/electron-donor solvent systems showed (*116*) that regularities found in the solid state also operate, on the whole, in solution. A collation of NQR (*112*), Mössbauer (*113*), and PMR (*116*) data suggests that in the Me$_2$SnCl$_2$ complexes the *s*-electron content in the Sn—C—H site, and the Sn—Cl bond polarity are greater than in the free halides. As for the stereochemistries of the 1:2 complexes, they are regular or slightly distorted octahedra of the IIIa or IIIb types, depending on the solvent applied.

C. Complexes of Methyltin Trihalides

In 1963, Beattie and McQuillan (*8*) studied the IR spectra of MeSnCl$_3$ complexes with pyridine, bipyridyl, and *o*-phenanthroline, the compositions being 1:2, 1:1, and 1:1, respectively. They obtained a rather complicated pattern of Sn—Cl vibrations and could make no structural assignment. Later it was shown (*31, 81, 32*) that IR spectroscopy is hardly applicable to the study of complexes of this type. In 1967, Wardell (*151*) studied the UV spectra of diethyl ether solutions of MeSnCl$_3$ complexes with nitroanilines and measured the equilibrium constants. The data were interpreted in terms of 1:1 complexes containing a five-coordinate tin (with 2,5- or 2,4-diaminonitrobenzene) or a six-coordinate tin (with 3,4-diaminonitrobenzene).

In 1974, Barbieri and co-workers (*107*) synthesized the 1:1 complex of MeSnCl$_3$ with N,N-ethylenebis(salicylideneiminato)nickel and studied its Mössbauer, IR, and electronic spectra in the solid state, and its electronic and PMR spectra in solution. The data led to a structure of the Va or Vc type (Table XI).

In 1973, we described eight complexes (*111*) of the type MeSnX$_3 \cdot$ 2L (X = Cl, Br; L = Py, DMF, DMSO, HMPT) as well as MeSnCl$_3 \cdot$ dioxan and 2MeSnBr$_3 \cdot$ dioxan. We studied (*113*) the Mössbauer spectra of these complexes and of frozen solutions of MeSnX$_3$ in DEE, DME, THF, acetone, and TMED. The isomer shifts were found to be very sensitive to complex formation, unlike the cases of Me$_2$SnX$_2$ and Me$_3$SnX, and led to a series for the electron-donor ability of the solvents toward MeSnX$_3$. The quadrupole splitting pattern found for MeSnX$_3$ complexes is also very different. Unlike Me$_2$SnX$_2$, in which the splitting increases with electron-donor activity of the ligands, in MeSnX$_3$ it is greater for the complexes with weaker donors. These facts agree with the present-day quadrupole splitting theories (see Section II, D) and with the NQR spectra of the complexes (*112*). The pattern of $\nu(^{35}\text{Cl})$ and $\nu(^{81}\text{Br})$ for complexes of MeSnX$_3$ (X = Cl, Br) with dioxan shows that the structure is of type Va. In this structure the total cis- and trans-influence of the coordinated ligands on the Sn—X (equatorial) bond ionicity is higher than is the total cis-influence on the Sn—X (axial) bond ionicity. These and other data lead to the conclusion that the trans-influence in octahedral complexes of methyltin halides is stronger than is the cis-influence. The NQR spectra (*112*) and the magnitude (2.18 mm/sec) of Δ obtained for 2MeSnBr$_3 \cdot$ dioxan (*113*) suggest that the structure is a IVd trigonal bipyramide (see Table XI).

Concentration and temperature dependences studied for $J(^1\text{H}—\text{C}—^{119}\text{Sn})$ in MeSnX$_3$/CH$_2$Cl$_2$/electron-donor solvent systems showed (*116*) that an even stronger redistribution of s and p electrons over the Sn—C and Sn—X bonds occurs in MeSnX$_3$ complexes, whereas $J(^1\text{H}—\text{C}—^{119}\text{Sn})$ equal to 146.0 Hz in MeSnX$_3 \cdot$ 2HMPT is the greatest among all the geminal proton–^{119}Sn couplings observed. It is interesting that in solution electron-donor ability series are identical for all methyltin halides: acetone < pyridine < DMF < DMSO < HMPT.

V
CONCLUSION

The data reported in the literature and analyzed in this review demonstrate that when we deal with the twelve methyltin halides, which have been studied for more than 125 years and have constituted the subject of hundreds of papers, we are still far from a complete understanding of the structures and, even more so, the properties of these rather elementary but, nevertheless, extremely important compounds.

A promising fact, however, is that the number of researches devoted to the subject has risen steeply during recent years, as is clear from the references cited. Unfortunately, the number of electron diffraction and X-ray studies, which give the most unambiguous information on the structures in the gas and solid states, is still rather low. As for solution studies, the use of IR and PMR methods has given a satisfactory picture of the electronic structures and stereochemistries of the molecules and their complexes. In this field as well, however, the number of thorough studies is not high and it is possible that the main problems still await attack.

ACKNOWLEDGMENT

The authors are greatly indebted to Dr. A. V. Grib who kindly translated this review.

REFERENCES

1. Bancroft, G. M., and Platt, R. H., *Advan. Inorg. Chem. Radiochem.* **15,** 59 (1972).
2. Bancroft, G. M., Das, V. G. K., and Butler, K. D., *J. Chem. Soc., Dalton Trans.* 2355 (1974).
3. Bancroft, G. M., Das, V. G. K., Sham, T. K., and Clark, M. G., *Chem. Commun.* 236 (1974).
4. Barbieri, G., and Taddei, F., *J. Chem. Soc., Perkin Trans. II* 1327 (1972).
5. Barbieri, R., Cefalú, R., Chandra, S. C., and Herber, R. H., *J. Organometal. Chem.* **32,** 97 (1971).
6. Barbieri, R., and Herber, R. H., *J. Organometal. Chem.* **42,** 65 (1972).
7. Barbieri, R., Pellerito, L., Bertazzi, N., and Stocco, G. C., *Inorg. Chim. Acta* **11,** 173 (1974).
8. Beattie, I. R., and McQuillan, G. P., *J. Chem. Soc.* 1519 (1963).
9. Beattie, I. R., and Ozin, G. A., *J. Chem. Soc. A* 370 (1970).
10. Bent, H. A., *Chem. Rev.* **61,** 275 (1961).
11. Bischof, P. K., Dewar, M. J. S., Goodman, D. W., and Jones, T. B., *J. Organometal. Chem.* **82,** 89 (1974).
12. Blom, E. A., Penfold, B. R., and Robinson, W. T., *J. Chem. Soc. A* 913 (1969).
13. Bokii, N. G., Struchkov, Yu. T., and Prokof'yev, A. K., *Zh. Strukt. Khim.* **13,** 665 (1972).
14. Bolles, T. F., and Drago, R. S., *J. Amer. Chem. Soc.* **87,** 5015 (1965).
15. Bolles, T. F., and Drago, R. S., *J. Amer. Chem. Soc.* **88,** 3921 (1966).
16. Bolles, T. F., and Drago, R. S., *J. Amer. Chem. Soc.* **88,** 5730 (1966).
17. Brown, J. M., Chapman, A. C., Harper, R., Mowthorpe, D. J., Davies, A. G., and Smith, P. J., *J. Chem. Soc., Dalton Trans.* 338 (1972).
18. Brown, R. S., Eaton, D. F., Hosomi, A., Traylor, T. G., and Wright, J. M., *J. Organometal. Chem.* **66,** 249 (1974).
19. Buckle, J., Harrison, P. G., and Das, M. K., *Inorg. Chim. Acta* **6,** 17 (1972).

20. Buckle, J., Harrison, P. G., King, T. J., and Richards, J. A., *Chem. Commun.* 1104 (1972).
21. Burke, J. J., and Lauterbur, P. C., *J. Amer. Chem. Soc.* **83**, 326 (1961).
22. Butcher, F. K., Gerrard, W., Mooney, E. F., Rees, R. G., Willis, H. A., Anderson, A., and Gebbe, H. A., *J. Organometal. Chem.* **1**, 431 (1964).
23. Calligaris, M., Randaccio, L., Barbieri, R., and Pellerito, L., *J. Organometal. Chem.* **76**, C56 (1974).
24. Cefalú, R., Pellerito, L., and Barbieri, R., *J. Organometal. Chem.* **32**, 107 (1971).
25. Chan, S. O., and Reeves, L. W., *Inorg. Chem.* **12**, 1704 (1973).
26. Clark, H. C., O'Brien, R. J. O., and Trotter, J., *J. Chem. Soc.* 2332 (1964).
27. Clark, H. C., and Goel, R. C., *J. Organometal. Chem.* **7**, 263 (1967).
28. Clark, H. C., and Puddephatt, R. J., *in* "Organometallic Compounds of Group IV Elements" (A. G. McDiarmid, ed.), Vol. 2, Part II. Dekker, New York, 1972.
29. Clark, J. P., and Wilkins, C. J., *J. Chem. Soc. A* 871 (1966).
30. Clark, J. P., Langford, V. M., and Wilkins, C. J., *J. Chem. Soc. A* 792 (1967).
31. Clark, R. J. H., and Williams, C. S., *Spectrochim. Acta* **21**, 1861 (1965).
32. Clark, R. J. H., Davies, A. G., and Puddephatt, R. J., *J. Chem. Soc. A* 1828 (1968).
33. Clark, M. G., Maddock, A. G., and Platt, R. H., *J. Chem. Soc., Dalton Trans.* 281 (1972).
34. Cordey-Hayes, M., Peacock, R. D., and Vucelić, M., *J. Inorg. Nucl. Chem.* **29**, 1177 (1967).
35. Crasnyanskii, M. E., Litinskii, A. O., and Schifrovich, E. J., *Teor. Exp. Khim.* **10**, 536 (1974).
36. Crasnyanskii, M. E., Lysenko, Yu. A., Litinskii, A. O., and Schifrovich, E. J., *Zh. Strukt. Khim.* **15**, 711 (1974).
37. Das, V. G. K., *Inorg. Nucl. Chem. Lett.* **9**, 155 (1973).
38. Das, V. G. K., and Kitching, W., *J. Organometal. Chem.* **13**, 523 (1968).
39. Davies, A. G., Harrison, P. G., Kennedy, J. D., Mitchell, T. N., Puddephatt, R. J., and McFarlane, W., *J. Chem. Soc. C* 1136 (1969).
40. Davies, A. G., Smith, L., and Smith, P. J., *J. Organometal. Chem.* **23**, 135 (1970).
41. Davies, A. G., Milledge, H. J., Puxley, D. C., and Smith, P. J., *J. Chem. Soc. A* 2862 (1970).
42. Davies, A. G., Smith, L., and Smith, P. J., *J. Organometal. Chem.* **39**, 279 (1972).
43. Dey, K., *J. Inorg. Nucl. Chem.* **32**, 3125 (1970).
44. Edgel, W. F., and Ward, C. H., *J. Mol. Spectrosc.* **8**, 343 (1962).
45. Fedorov, L. A., and Fedin, E. I., *Izv. Akad. Nauk SSSR, Ser. Khim.* 787 (1971).
46. Fedorov, L. A., Stumbrevichute, Z. A., Prokof'yev, A. K., and Fedin, E. I., *Dokl. Akad. Nauk SSSR* **209**, 134 (1973).
47. Fergüsson, J. E., Roper, W. R., and Wilkins, C. J., *J. Chem. Soc.* 3716 (1965).
48. Fitzsimmons, B. W., Seely, N. J., and Smith, A. W., *Chem. Commun.* 390 (1968).
49. Forder, R. A., and Sheldrik, G. M., *J. Organometal. Chem.* **21**, 115 (1970).
50. Fujii, H., and Kimura, M., *Bull. Chem. Soc. Jap.* **44**, 2643 (1971).
51. Gassend, R., Delmas, M., Maire, J.-C., Richard, Y., and More, C., *J. Organometal. Chem.* **42**, C29 (1972).
52. Gielen, M., and Nasielski, J., *J. Organometal. Chem.* **1**, 173 (1963).
53. Gol'danksii, V. I., "Mössbauer Effect and Its Application in Chemistry" (in Russian), Moscow, 1963.
54. Gol'dstein, I. P., Zemlyanskii, N. N., Perepyolkova, T. I., Mel'nichenko, L. S., Gur'yanova, E. N., and Kocheshkov, K. A., *Dokl. Akad. Nauk SSSR* **217**, 849 (1974).

55. Grutsch, P. A., Zeller, M. V., and Fehlner, T. P., *Inorg. Chem.* **12,** 1431 (1973).
56. Gupta, R., and Majee, B., *J. Organometal. Chem.* **33,** 169 (1971).
57. Gupta, R., and Majee, B., *J. Organometal. Chem.* **36,** 71 (1972).
58. Gupta, R., and Majee, B., *J. Organometal. Chem.* **49,** 203 (1973).
59. Harrison, P. G., King, T. J., and Richards, J. A., *J. Chem. Soc., Dalton Trans.* 1723 (1974).
60. Harrison, P. G., and King, T. J., *J. Chem. Soc., Dalton Trans.* 2298 (1974).
61. Hendricker, D. G., *Inorg. Nucl. Chem. Lett.* **5,** 115 (1969).
62. Herber, R. H., Stöckler, H. A., and Reichle, W. T., *J. Chem. Phys.* **42,** 2447 (1965).
63. Herber, R. H., *J. Inorg. Nucl. Chem.* **35,** 67 (1973).
64. Hill, J. C., Drago, R. S., and Herber, R. H., *J. Amer. Chem. Soc.* **91,** 1644 (1969).
65. Ho, B. I. K., and Zuckerman, J. J., *J. Organometal. Chem.* **49,** 1 (1973).
66. Holmes, J. R., and Kaesz, H. D., *J. Amer. Chem. Soc.* **83,** 3903 (1961).
67. Huang, H. H., Hui, K. M., and Chiu, K. K., *J. Organometal. Chem.* **11,** 515 (1968).
68. Hulme, R., *J. Chem. Soc.* 1524 (1963).
69. Hunter, B. K., and Reeves, L. W., *Can. J. Chem.* **46,** 1399 (1968).
70. Jaura, K. L., and Verma, V. K., *J. Inorg. Nucl. Chem.* **35,** 2361 (1973).
71. Kawakami, K., and Tanaka, T., *J. Organometal. Chem.* **49,** 409 (1973).
72. Kawakami, K., Miya-Uchi, M., and Tanaka, T., *J. Organometal. Chem.* **70,** 67 (1974).
73. Kawasaki, Y., Hori, M., and Uenaka, K., *Bull. Chem. Soc. Jap.* **40,** 2463 (1967).
74. King, T. J., and Harrison, P. G., *Chem. Commun.* 815 (1972).
75. Kitching, W., *Tetrahedron Lett.* 3689 (1966).
76. Kitching, W., and Das, V. G. K., *Aust. J. Chem.* **21,** 2401 (1968).
77. Kitching, W., Moore, C. J., and Doddrell, D., *Aust. J. Chem.* **22,** 1149 (1969).
78. Kraus, C. A., and Callis, C. C., *J. Amer. Chem. Soc.* **45,** 2624 (1923).
79. Kraus, C. A., and Greer, W. N., *J. Amer. Chem. Soc.* **45,** 2946, 3078 (1923).
80. Kriegsmann, V. H., and Pischtschan, S., *Z. Anorg. Allg. Chem.* **308,** 212 (1961).
81. Langer, H. D., and Blut, A. H., *J. Organometal. Chem.* **5,** 288 (1966).
82. Lees, J. K., and Flinn, P. A., *J. Chem. Phys.* **48,** 882 (1968).
83. Levchuk, L. E., Sams, J. R., and Aubke, F., *Inorg. Chem.* **11,** 43 (1972).
84. Liengme, B. V., Randall, R. S., and Sams, J. R. *Can. J. Chem.* **50,** 3212 (1972).
85. Limouzin, Y., and Maire, J. C., *J. Organometal. Chem.* **82,** 99 (1974).
86. Lippincott, E. R., Mercier, R., and Tobin, M. C., *J. Phys. Chem.* **57,** 939 (1953).
87. Lorberth, J., and Vahrenkamp, H., *J. Organometal. Chem.* **11,** 111 (1968).
88. Maksyutin, Yu. K., Khrapov, V. V., Mel'nichenko, L. S., Semin, G. K., Zemlyanskii, N. N., and Kocheshkov, K. A., *Izv. Akad. Nauk SSSR, Ser. Khim.* 602 (1972).
89. Maksyutin, Yu. K., Guryanova, E. N., Kravchenko, E. A., and Semin, G. K., *J. Chem. Soc., Chem. Commun.* 492 (1973).
90. Matsubayashi, G., Kawasaki, Y., Tanaka, T., and Okawara, R., *J. Inorg. Nucl. Chem.* **28,** 2937 (1966).
91. Matsubayashi, G., Kawasaki, Y., Tanaka, T., and Okawara, R., *Bull. Chem. Soc. Jap.* **40,** 1566 (1967).
92. Matsubayshi, G., Tanaka, T., and Okawara, R., *J. Inorg. Nucl. Chem.* **30,** 1831 (1968).
93. Matsubayashi, G., Nishii, N., and Tanaka, T., *Bull. Chem. Soc. Jap.* **42,** 2369 (1969).
94. Matsubayashi, G., Hiroshima, M., and Tanaka, T., *J. Inorg. Nucl. Chem.* **33,** 3787 (1971).

95. Matsubayashi, G., Okunaka, M., and Tanaka, T., *J. Organometal. Chem.* **56**, 215 (1973).
96. Matwiyoff, N. A., and Drago, R. S., *Inorg. Chem.* **3**, 337 (1964).
97. McFarlane, W., *J. Chem. Soc. A* 528 (1967).
98. Mitchell, T. N., *J. Organometal. Chem.* **59**, 189 (1973).
99. Muetterties, E. L., Mahler, W., Packer, K. J., and Schmutzler, R., *Inorg. Chem.* **3**, 1298 (1964).
100. Okawara, R., Webster, D. E., and Rochow, E. G., *J. Amer. Chem. Soc.* **82**, 3287 (1960).
101. Parish, R. V., *Progr. Inorg. Chem.* **15**, 101 (1972).
102. Parish, R. V., and Johnson, C. E., *Chem. Phys. Lett.* **6**, 239 (1970).
103. Parish, R. V., and Platt, R. H., *Chem. Commun.* 1118 (1968).
104. Parish, R. V., and Platt, R. H., *J. Chem. Soc. A* 2145 (1969).
105. Parish, R. V., and Platt, R. H., *Inorg. Chim. Acta* **4**, 65 (1970).
106. Parshall, G. W., *Inorg. Chem.* **11**, 373 (1972).
107. Pellerito, L., Cefalú, R., Gianguzza, A., and Barbieri, R., *J. Organometal. Chem.* **70**, 303 (1974).
108. Pellerito, L., Cefalú, R., Silvestri, A., Di Bianca, F., Barbieri, R., Haupt, H.-J., Preut, H., and Huber, F., *J. Organometal. Chem.* **78**, 101 (1974).
109. Petrosyan, V. S., Yashina, N. S., and Reutov, O. A., *Izv. Akad. Nauk SSSR, Ser. Khim.* 1018 (1972).
110. Petrosyan, V. S., and Reutov, O. A., *J. Organometal. Chem.* **52**, 307 (1973).
111. Petrosyan, V. S., Yashina, N. S., and Reutov, O. A., *J. Organometal. Chem.* **52**, 315 (1973).
112. Petrosyan, V. S., Yashina, N. S., Reutov, O. A., Bruchova, E. V., and Semin, G. K., *J. Organometal. Chem.* **52**, 321 (1973).
113. Petrosyan, V. S., Yashina, N. S., Sacharov, S. G., Reutov, O. A., Rochev, V. Ya., and Gol'danskii, V. I., *J. Organometal. Chem.* **52**, 333 (1973).
114. Petrosyan, V. S., and Reutov, O. A., *Pure Appl. Chem.* **37**, 147 (1974).
115. Petrosyan, V. S., Permin, A. B., and Reutov, O. A., *Izv. Akad. Nauk SSSR, Ser. Khim.* 1305 (1974).
116. Petrosyan, V. S., Yashina, N. S., Bakhmutov, V. I., Permin, A. B., and Reutov, O. A., *J. Organometal. Chem.* **72**, 71 (1974).
117. Petrosyan, V. S., and Permin, A. B., *Int. Conf. Organometal. Chem., 7th, Abstr. Papers, 1975* 250.
118. Poller, R. C., "The Chemistry of Organotin Compounds." Logos Press, London, 1970.
119. Poller, R. C., and Ruddick, J. N. R., *J. Organometal. Chem.* **39**, 121 (1972).
120. Poller, R. C., and Ruddick, J. N. R., *J. Chem. Soc., Dalton Trans.* 555 (1972).
121. Potts, D., Sharma, H. D., Carty, A. J., and Walker, A., *Inorg. Chem.* **13**, 1205 (1974).
122. Pudovik, A. N., Muratova, A. A., Medvedeva, M. D., and Yamaliyeva, L. N., *Zh. Obsch. Khim.* **42**, 2402 (1972).
123. Ramos, V. B., and Tobias, R. S., *Inorg. Chem.* **11**, 2451 (1972).
124. Ramos, V. B., and Tobias, R. S., *Spectrochim. Acta* **30A**, 181 (1974).
125. Randaccio, L., *J. Organometal. Chem.* **55**, C58 (1973).
126. Redl, G., and Winokur, M., *J. Organometal. Chem.* **26**, C36 (1971).
127. Ronova, I. A., Sinitsina, N. A., Struchkov, Yu. T., Okhlobystin, O. Yu., and Prokof'yev, A. K., *Zh. Strukt. Khim.* **13**, 15 (1972).
128. Ruddick, J. N. R., and Sams, J. R., *J. Organometal. Chem.* **60**, 233 (1973).

129. Schlemper, E. O., and Hamilton, W. C., *Inorg. Chem.* **5,** 995 (1966).
130. Semin, G. K., Babushkina, T. A., and Yacobson, G. G., "Application of Nuclear Quadrupole Resonance in Chemistry" (in Russian), Moscow, 1972.
131. Skinner, H. A., and Sutton, L. E., *Trans. Faraday Soc.,* **40,** 164 (1944).
132. Smith, P. J., *Organometal. Chem. Rev.,* A **5,** 373 (1970).
133. Smith, P. J., and Dodd, D., *J. Organometal. Chem.* **32,** 195 (1971).
134. Smith, P. J., and Smith, L., *Inorg. Chim. Acta Rev.* **7,** 11 (1973).
135. Stöckler, H. A., and Sano, H., *Chem. Commun.* 954 (1969).
136. Taimsalu, P., and Wood, J. L., *Spectrochim. Acta* **20,** 1043, 1357 (1964).
137. Tanaka, T., *Inorg. Chim. Acta* **1,** 217 (1967).
138. Tanaka, T., and Kamitani, T., *Inorg. Chim. Acta* **2,** 175 (1968).
139. Tanaka, T., Matsumura, Y., Okawara, R., Musha, Y., and Kinumaki, S., *Bull. Chem. Soc. Jap.* **41,** 1497 (1968).
140. Thomas, A. B., and Rochow, E. G., *J. Amer. Chem. Soc.* **79,** 1843 (1957).
141. Thomas, A. B., and Rochow, E. G., *J. Inorg. Nucl. Chem.* **4,** 205 (1957).
142. Tobias, R. S., *Organometal. Chem. Rev.* **1,** 93 (1966).
143. Torocheshnikov, V. M., Tupciauskas, A. P., Sergeev, N. M., and Ustynyuk, Yu. A., *J. Organometal. Chem.* **35,** C25 (1972).
144. Van den Berghe, E. V., and Van der Kelen, G. P., *J. Organometal. Chem.* **6,** 515 (1966).
145. Van den Berghe, E. V., and van der Kelen, G. P., *J. Organometal. Chem.* **11,** 479 (1968).
146. Van den Berghe, E. V., and van der Kelen, G. P., *J. Organometal. Chem.* **59,** 175 (1973).
147. Van den Berghe, E. V., and van der Kelen, G. P., *J. Organometal. Chem.* **72,** 65 (1974).
148. Van der Kelen, G. P., *Nature (London)* **193,** 1069 (1962).
149. Van den Vondel, D. F., Willemen, H., and van der Kelen, G. P., *J. Organometal. Chem.* **63,** 205 (1973).
150. Verdonck, L., and van der Kelen, G. P., *Ber. Bunsenges. Phys. Chem.* **69,** 478 (1965).
151. Wardell, J. L., *J. Organometal. Chem.* **9,** 89 (1967).
152. Yasuda, K., Kawasaki, Y., Kosai, N., and Tanaka, T., *Bull. Chem. Soc. Jap.* **38,** 126 (1965).
153. Isaacs, N. W., and Kennard, C. H. L., *J. Chem. Soc. A* 1257 (1970).
154. Zuckerman, J. J., *Advan. Organometal. Chem.* **9,** 21 (1970).

Chemistry of Carbon-Functional Alkylidynetricobalt Nonacarbonyl Cluster Complexes*

DIETMAR SEYFERTH

Department of Chemistry
Massachusetts Institute of Technology
Cambridge, Massachusetts

OR

I'll call it "Fred" 'cause alkylidynetricobalt nonacarbonyl's too damn long for me. [1]

I. Introduction: General Properties of Alkylidynetricobalt Nonacarbonyl Complexes 98
II. Synthesis of Alkylidynetricobalt Nonacarbonyl Complexes . . 100
III. Chemistry of the Tricobaltcarbon Decacarbonyl Cation . . 110
IV. Highly Stable Nonacarbonyl Tricobaltcarbon-Substituted Carbonium Ions 119

* This review is based in large part on research carried out at the Massachusetts Institute of Technology by Ralph J. Spohn, John E. Hallgren, Anthony T. Wehman, Gary H. Williams, Paul L. K. Hung, C. Scott Eschbach, Mara Ozolins Nestle, Cynthia L. Nivert, and Ying-Ming Cheng.

[1] Claiming that it took longer to say "alkylidynetricobalt nonacarbonyl" than to make one, R. J. Spohn, who began our research program in this area, informally christened this compound class "Fred." Thus $ClCCo_3(CO)_9$ is chlorofred, $(OC)_9Co_3CCO_2H$, fredoic acid, $(OC)_9Co_3CC(O)CH_3$, acetofredone, etc.

V. Decomposition Reactions and Derived Synthetic Applications of
   Alkylidynetricobalt Nonacarbonyl Complexes 135
VI. Concluding Remarks 138
   References 141

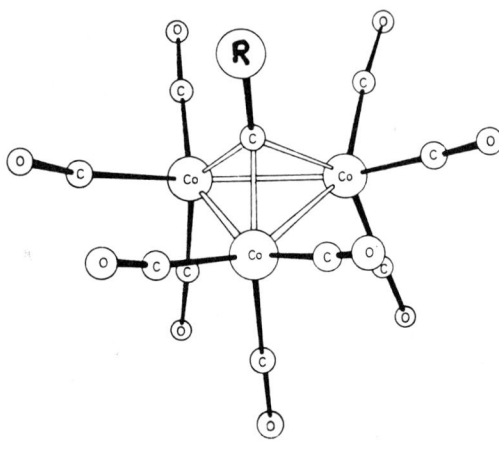

I

INTRODUCTION: GENERAL PROPERTIES OF ALKYLIDYNETRICOBALT NONACARBONYL COMPLEXES

In 1971 we reviewed the chemistry of alkylidynetricobalt nonacarbonyl[2] complexes (1). We have continued our research in this area and have made substantial progress in the development of the organic chemistry of this interesting class of organometallic compounds. In this article we review principally the chemistry of carbon-functional alkylidynetricobalt nonacarbonyl complexes.

In view of the unusual structure of these complexes, it will be useful to begin with a general review of their properties, structure, and formation. The serendipitous preparation of the first member of this class of compounds was reported in 1958 by workers at the Bureau of Mines laboratories at Bruceton, Pennsylvania (2):

$$(OC)_3Co \cdots \overset{\overset{\displaystyle H}{|}}{\underset{\underset{\displaystyle Co(CO)_3}{|}}{C}} \cdots C-H \xrightarrow[H_2O/EtOH]{H_2SO_4;} CH_3CCo_3(CO)_9 \quad (1)$$

[2] Many of the other workers in this field call these complexes methinyltricobalt enneacarbonyls. However, we are not Greek scholars and prefer the more prosaic Latin numerical prefixes.

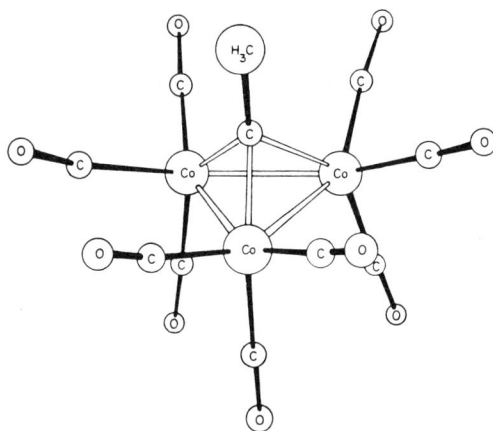

FIG. 1. Structure of $CH_3CCo_3(CO)_9$. [From P. W. Sutton and L. F. Dahl, *J. Amer. Chem. Soc.* **89**, 261 (1967).]

After initial confusion concerning the structure of this compound (*2*), chemical studies by other workers (*3a*, *3b*) and, quite conclusively, an X-ray crystal structure determination by Sutton and Dahl, (*4*) established the presence of a tetrahedral CCo_3 cluster in this compound. This structure is shown in Fig. 1. The apical carbon atom is coordinated symmetrically, apparently via σ-bonds, to 3 cobalt atoms. Each cobalt atom may be regarded as achieving a closed-shell configuration by σ-bonding to 2 other cobalt atoms, the apical carbon atom, and three terminal carbon monoxide ligands. The C—CH_3 distance is 1.53 Å, the normal C_{sp^3}—C_{sp^3} distance. The Co—C(apical)—Co bond angles average 81.1°, the H_3C—C—Co angles average 131.3° (*4*). The arrangement of the carbon monoxide ligands in this complex merits special notice for it has important consequences with regard to the chemistry of this class of compounds. Six of the nine CO ligands are disposed upward in the general direction of the apical carbon atom and its substituent. As a result, any reactions at the apical carbon atom or at its substituent group will be subject to substantial steric hindrance. Not only is the backside of the apical carbon atom well shielded, but attack from other directions also will be hindered by steric interference of these six CO groups. Thus, steric hindrance plays an important role in the chemistry of alkylidynetricobalt nonacarbonyls.

The alkylidynetricobalt nonacarbonyl complexes all are highly colored, with colors ranging from red to purple to brown to black, depending on the apical substituent. Their thermal stability also depends on the apical substituent; some survive heating to 100°–185° and many may be sublimed in vacuum at 50°–80°C. Most, but not all, are air-stable (in contrast to their pyrophoric parent, dicobalt octacarbonyl) in the solid and in solution.

Most of the complexes that we have worked with are crystalline solids, usually (but not always) with well-defined melting and/or decomposition points. However, when the apical carbon atom bears a long carbon chain substituent, the melting points of these complexes are lowered and oils, rather than crystalline solids, can result. Most of the $RCCo_3(CO)_9$ complexes are soluble in the usual organic solvents. In view of their color, stability, and solubility properties, thin-layer and column chromatography are well suited for their detection, purification, and isolation. Their combustion analysis presents no difficulties. The general chemical reactivity of the alkylidynetricobalt nonacarbonyls includes instability toward attack by oxidizing agents and many bases and nucleophiles. Thus, the action of ceric ammonium nitrate or potassium permanganate on an $RCCo_3(CO)_9$ compound results in evolution of carbon monoxide and the formation of inorganic cobalt salts and the carboxylic acid derived from the apical carbon, RCO_2H (5). Many alkylidynetricobalt nonacarbonyl complexes, however, are stable toward protonic and Lewis acids. This property provides the basis for much of the chemistry which we have developed since 1970.

II

SYNTHESIS OF ALKYLIDYNETRICOBALT NONACARBONYL COMPLEXES

Procedures are available for the preparation of $RCCo_3(CO)_9$ complexes with a wide variety of substituents, R, at the apical carbon atom. The original preparation of $CH_3CCo_3(CO)_9$ via acetylenedicobalt hexacarbonyl [Eq. (1)] can be extended to dicobalt hexacarbonyl complexes of terminal acetylenes, $(RC_2H)Co_2(CO)_6$ (*1*), but the effect of R on the success or failure of this reaction has not been explored with a systematic variation of R. In any case, the $(RC_2H)Co_2(CO)_6 \rightarrow RCH_2CCo_3(CO)_9$ conversion is not always successful. Furthermore, it is limited to the synthesis of complexes with a CH_2 group attached to the apical carbon atom. Complexes of the type $(OC)_9Co_3CCH_2R$ have been prepared by this route with R = H, CH_3, CF_3, MeO_2C, C_2H_5, $HO(CH_2)_2$, $n\text{-}C_3H_7$, Me_2CH, Ph_2CH, Me_3C, C_6H_5, $p\text{-}BrC_6H_4$, C_6F_5 (*6*). Some of these products were formed directly in $RC{\equiv}CH/Co_2(CO)_8$ reactions (e.g., R = Ph, CF_3); presumably the acetylene in these cases is a strong enough acid to induce conversion of the initially formed acetylenedicobalt hexacarbonyl to the alkylidynetricobalt nonacarbonyl.

Another more general preparative route to such complexes was developed by other research groups (*3b*, *7*, *8*):

$$RCX_3 + Co_2(CO)_8 \longrightarrow RCCo_3(CO)_9 + CoX_2 \quad (2)$$

This procedure may be used to prepare a great diversity of alkylidynetricobalt nonacarbonyl complexes, including $RCCo_3(CO)_9$ compounds where R = H, halogen, alkyl, and aryl. Some dihalomethyl compounds also may be used as starting materials, e.g., $PhCHCl_2$, CH_3OCHCl_2, and $[Me_2N=CHCl]Cl$, which react with $Co_2(CO)_8$ to give $PhCCo_3(CO)_9$, $CH_3OCCo_3(CO)_9$, and $Me_2NCCo_3(CO)_9$, respectively (*9*). Of particular interest are those reactions that introduce reactive organic functional groups at the apical carbon atom:

$$CCl_3CO_2R + Co_2(CO)_8 \longrightarrow (OC)_9Co_3CCO_2R \quad (3b, 7) \quad (3)$$

$$CCl_3\text{—}\underset{\underset{O}{\|}}{C}\text{—}R + Co_2(CO)_8 \longrightarrow (OC)_9Co_3C\underset{\underset{O}{\|}}{C}R \quad (9) \quad (4)$$

$$CCl_3CONR_2 + Co_2(CO)_8 \longrightarrow (OC)_9Co_3CCONR_2 \quad (9) \quad (5)$$

$$CCl_3P(O)(OR)_2 + Co_2(CO)_8 \longrightarrow (OC)_9Co_3CP(O)(OR)_2 \quad (9) \quad (6)$$

$$CCl_3CR\!=\!CHR' + Co_2(CO)_8 \longrightarrow (OC)_9Co_3CCR\!=\!CHR' \quad (10) \quad (7)$$

The mechanism of the reaction of organic di- and trihalo compounds with dicobalt octacarbonyl has not been elucidated yet. It seems certain that the active reagent is the $Co(CO)_4^-$ anion. This species can be used instead of dicobalt octacarbonyl, as can $Co_4(CO)_{12}$. Some thoughts concerning the mechanism of $RCX_3/Co_2(CO)_8$ reactions were given in our previous review (*1*). If the atom transfer/radical mechanism which we suggested is, indeed, operative, it is not surprising that the $RCCo_3(CO)_9$ product yields are, in general, only low to moderate, as shown in Table I, which presents some yield data from our own work. In the meantime, it has become clear that some attempted $RCX_3/Co_2(CO)_8$ reactions are unsuccessful not because of difficulties with this reaction per se, but because of the instability of the product to the reaction and/or work-up conditions. For instance, $CCl_3CR'\!=\!CHR'/Co_2(CO)_8$ reactions must be carried out under rigorously anhydrous conditions and hydrolytic work-up must be avoided or else reduced products, $(OC)_9Co_3CCHRCH_2R'$, are obtained instead of the desired vinylic products (*10*). Thus we were surprised that our first attempted preparation of $Me_3SiCH\!=\!CHCCo_3(CO)_9$ by reaction of $Me_3SiCH\!=\!CHCBr_3$ with $Co_2(CO)_8$ gave $Me_3SiCH_2CH_2CCo_3(CO)_9$ as the sole product. Maintenance of scrupulously anhydrous conditions throughout reaction and product isolation steps gave the desired product. Attempted preparation of α-hydroxy derivatives of type $(OC)_9Co_3CC(OH)RR'$ has been

TABLE I

Preparation of Alkylidynetricobalt Nonacarbonyl Complexes by Reactions of Polyhalides with Dicobalt Octacarbonyl[a]

Starting halide	R in $RCCo_3(CO)_9$ produced	% Yield
CCl_4	Cl	46
CBr_4	Br	43
$CHBr_3$	H	34
CH_3CCl_3	CH_3	43
$C_6H_5CCl_3$	C_6H_5	29
$C_6H_5CHCl_2$	C_6H_5	34
Me_3SiCCl_3	Me_3Si	41
$PhMe_2SiCCl_3$	$PhMe_2Si$	50
$(EtO)_2P(O)CCl_3$	$(EtO)_2P(O)$	42
CCl_3CO_2Et	EtO_2C	53
$CCl_3CO_2SiMe_3$	Me_3SiO_2C	38
CCl_3CONEt_2	$Et_2NC(O)$	19
CCl_3COCH_3	$CH_3C(O)$	40
$CCl_3COC_4H_9$-n	n-$C_4H_9C(O)$	49
CCl_3COCMe_3	$Me_3CC(O)$	24
CCl_3COPh	$PhC(O)$	33
CCl_3CH_2OH	CH_2OH	0.8
$CBr_3CH_2OSiMe_3$	CH_2OH (after hydrolysis)	5
CH_3OCHCl_2	CH_3O	27
$[ClCH=NMe_2]Cl$	Me_2N	26
$CH_2=CHCCl_3$[b]	$CH_2=CH$	45
$CH_2=C(CH_3)CCl_3$[b]	$CH_2=C(CH_3)$	59
$Me_3SiCH=CHCBr_3$[b]	$Me_3SiCH=CH$	28
$CH_3C(O)CH=CHCCl_3$[b]	$CH_3C(O)CH=CH$	19

[a] From Seyferth et al. (9).
[b] Seyferth et al. (10).

quite unsuccessful to date, giving either none of the desired product at all (11) or only very low yields (<1%) (9) when trichloroethanol derivatives of type $CCl_3C(OH)RR'$ were allowed to react with dicobalt octacarbonyl. The major products of such reactions were the alkyl derivatives, $(OC)_9Co_3CCHRR'$, but the yields of these also were low (<10%) (11). Conversion of the starting alcohols to their trimethylsiloxy ethers prior to the reaction with dicobalt octacarbonyl resulted in an increase in the $(OC)_9Co_3CC(OH)RR'$ yield to 4–5% (9). To cite one more example, the aldehyde $(OC)_9Co_3CCH=O$ was not obtained when chloral was allowed to react with dicobalt octacarbonyl, but it could be isolated in 3% yield when the ethylene glycol-derived acetal of chloral or bromal was used in place of the aldehyde (9). In all of these reactions we believe that it is the formation of cobalt tetracarbonyl hydride, $HCo(CO)_4$, a strong protonic acid, during the reaction or during work-up, that results in diversion of the desired product; in fact, we have shown that $HCo(CO)_4$ or $Co_2(CO)_8$-strong acid systems will convert $(OC)_9Co_3CCH=CH_2$ to $(OC)_9Co_3CCH_2CH_3$ (10). One may also postulate reasonable reaction courses for $HCo(CO)_4$-induced destruction of $(OC)_9Co_3CC(OH)RR'$ and $(OC)_9Co_3CCH=O$, especially with after-the-fact knowledge of our later research results in the area which we review here.

Interconversions of one alkylidynetricobalt nonacarbonyl complex to another also are possible but are rather limited in scope. First, as already mentioned, many strong bases and nucleophiles ultimately destroy the $RCCo_3(CO)_9$ cluster. In a halogen derivative, $XCCo_3(CO)_9$, any nucleophilic attack in which the halogen atom is displaced must occur by way of nucleophilic attack at halogen or by an electron transfer mechanism, since, as already mentioned, the backside of the apical carbon atom is very effectively shielded. A usually nonproductive alternative which very likely results in ultimate destruction of the cluster is nucleophilic attack at the carbon atoms of coordinated CO ligands. Bromo- and chloromethylidynetricobalt nonacarbonyl are arylated in moderate yield by the action of an aryl Grignard reagent in considerable excess (Table II) (12). Primary and secondary alkyl Grignard reagents, on the other hand, destroy rather than alkylate these cluster halides. It was suggested that either course of reaction possibly begins with RMgX attach at coordinated carbon monoxide (12), but, in our opinion, reaction via an electron-transfer mechanism seems a reasonable alternative. Organolithium reagents react with $ClCCo_3(CO)_9$ to form a radical anion of some stability, $[ClCCo_3(CO)_9]^{\bar{\cdot}}$, which is detectable by ESR (13) [also formed by the action of an alkali metal on the chloro compound (14)], but subsequent events do not lead to alkylation of the cluster, but rather to its destruction.

A very useful reaction for the conversion of $HCCo_3(CO)_9$ to $ArCCo_3(CO)_9$ compounds has been developed in these laboratories (5):

$$Ar_2Hg + HCCo_3(CO)_9 \xrightarrow[\text{CO atmosphere, 2-8 hr}]{\text{benzene or THF, reflux}} ArCCo_3(CO)_9 + ArH + Hg \quad (8)$$

Arylmercuric halides may be used in place of diarylmercurials. Excellent yields, often 90–98%, were obtained in many cases. Table III lists some examples. Oxidative degradation studies of selected products showed that the aryl carbon atom originally attached to mercury was the one that became bonded to the apical carbon atom of the cluster. This reaction was much less useful in the preparation of alkyl derivatives, because of very long reaction times (1–2 weeks) giving at best moderate yields (Table III). The reaction of α-haloalkylmercurials with $HCCo_3(CO)_9$ occurred

TABLE II
Grignard Synthesis of Substituted Benzylidynetricobalt Nonacarbonyl Complexes from Bromomethylidynetricobalt Nonacarbonyl[a]

Ar in ArMgBr[b]	Product	% Yield
2-CH₃-C₆H₄-	2-CH₃-C₆H₄-CCo₃(CO)₉	40
3-CH₃-C₆H₄-	3-CH₃-C₆H₄-CCo₃(CO)₉	70
4-CH₃-C₆H₄-	4-CH₃-C₆H₄-CCo₃(CO)₉	60
2-CH₃O-C₆H₄-	2-CH₃O-C₆H₄-CCo₃(CO)₉	30
3-CH₃O-C₆H₄-	3-CH₃O-C₆H₄-CCo₃(CO)₉	35
4-CH₃O-C₆H₄-	4-CH₃O-C₆H₄-CCo₃(CO)₉	40
3-CH₃O-4-CH₃-C₆H₃-	3-CH₃O-4-CH₃-C₆H₃-CCo₃(CO)₉	25
2-CH₃O-4-CH₃-C₆H₃-	2-CH₃O-4-CH₃-C₆H₃-CCo₃(CO)₉	20
3-CH₃O-4-Cl-C₆H₃-	3-CH₃O-4-Cl-C₆H₃-CCo₃(CO)₉	10
4-Me₂N-C₆H₄-	4-Me₂N-C₆H₄-CCo₃(CO)₉	23
4-Me₂N-3-CH₃-C₆H₃-	4-Me₂N-3-CH₃-C₆H₃-CCo₃(CO)₉	25
4-Me₂N-3-Cl-C₆H₃-	4-Me₂N-3-Cl-C₆H₃-CCo₃(CO)₉	21
1-naphthyl	1-naphthyl-CCo₃(CO)₉	30

[a] From Dolby and Robinson (12).
[b] More than 9 moles of ArMgBr per mole of BrCCo₃(CO)₉ is required.

Alkylidynetricobalt Nonacarbonyl Complexes

TABLE III
PREPARATION OF $RCCo_3(CO)_9$ COMPLEXES BY REACTIONS OF ORGANOMERCURIALS WITH $HCCo_3(CO)_9$[a]

Organomercurial[b]	R in $RCCo_3(CO)_9$ produced	% Yield
$(C_6H_5)_2Hg$	C_6H_5	93
$(p\text{-}CH_3OC_6H_4)_2Hg$	$p\text{-}CH_3OC_6H_4$	64
$(p\text{-}CH_3C_6H_4)_2Hg$	$p\text{-}CH_3C_6H_4$	92
$(m\text{-}CH_3C_6H_4)_2Hg$	$m\text{-}CH_3C_6H_4$	96
$(o\text{-}CH_3C_6H_4)_2Hg$	$o\text{-}CH_3C_6H_4$	49
$(p\text{-}ClC_6H_4)_2Hg$	$p\text{-}ClC_6H_4$	83
$(m\text{-}ClC_6H_4)_2Hg$	$m\text{-}ClC_6H_4$	93
$(o\text{-}ClC_6H_4)_2Hg$	$o\text{-}ClC_6H_4$	57
$(m\text{-}FC_6H_4)_2Hg$	$m\text{-}FC_6H_4$	85
$(p\text{-}BrC_6H_4)_2Hg$	$p\text{-}BrC_6H_4$	86
$(p\text{-}IC_6H_4)_2Hg$	$p\text{-}IC_6H_4$	51
$(C_6F_5)_2Hg$	C_6H_5	69
$(C_6H_5CH_2)_2Hg$	$C_6H_5CH_2$	75
C_6H_5HgBr	C_6H_5	58
$p\text{-}FC_6H_4HgBr$	$p\text{-}FC_6H_4$	57
$p\text{-}H_2NC_6H_4HgCl$	$p\text{-}H_2NC_6H_4$	26
$p\text{-}ClC_6H_4HgBr$[c]	$p\text{-}ClC_6H_4$	54
$\alpha\text{-}C_{10}H_7HgBr$[c]	$\alpha\text{-}C_{10}H_7$	11
ferrocenyl-HgCl[c]	methylferrocenyl	14
$(\text{C}_6\text{H}_5\text{-Cr(CO)}_3)_2Hg$[d]	$p\text{-tolyl-Cr(CO)}_3$	37
$CH_3OCH_2CH_2HgCl$	$CH_3OCH_2CH_2$	32
$(n\text{-}C_5H_{11})_2Hg$	$n\text{-}C_5H_{11}$	32

[a] From Seyferth et al. (5).
[b] Reactions were carried out under an atmosphere of carbon monoxide unless otherwise specified.
[c] Reaction carried out under an atmosphere of dry nitrogen.
[d] D. Seyferth and C. S. Eschbach, *J. Organometal. Chem.* **94,** C5 (1975).

with concomitant reduction of the carbon–halogen bond but proved to be an unexpectedly effective synthesis of alkyl-substituted cluster complexes (5):

$$\text{RCH(X)—HgX} + \text{HCCo}_3(\text{CO})_9 \xrightarrow{\text{C}_6\text{H}_6} \text{RCH}_2\text{CCo}_3(\text{CO})_9 \quad (9)$$

(R = H, X = Br: 59% product yield)
(R = Me$_3$Si, X = Br: 70% product yield)

$$(\text{RCHI})_2\text{Hg} + \text{HCCo}_3(\text{CO})_9 \xrightarrow{\text{C}_6\text{H}_6} \text{RCH}_2\text{CCo}_3(\text{CO})_9 \quad (10)$$

(R = H, 77% product yield)
(R = CH$_3$, 88% product yield)

This reaction also was useful in the preparation of specific deuterated derivatives:

$$\text{Hg(CH}_2\text{Br})_2 + \text{DCCo}_3(\text{CO})_9 \longrightarrow \text{CH}_2\text{DCCo}_3(\text{CO})_9 \quad (11)$$

$$\text{Hg(CD}_2\text{I})_2 + \text{HCCo}_3(\text{CO})_9 \longrightarrow \text{CHD}_2\text{CCo}_3(\text{CO})_9 \quad (12)$$

Reactions of organomercurials with HCCo$_3$(CO)$_9$ appear to be subject to steric hindrance. As seen in Table III, ortho-substituted arylmercury compounds usually give lower product yields than those with substituents in meta or para positions. The presence of two ortho substituents larger than fluorine causes such reactions to fail completely. Thus, no organocobalt cluster product was obtained in reactions of HCCo$_3$(CO)$_9$ with bis(pentachlorophenyl)mercury and dimesitylmercury. Unfortunately, nothing is known at this time concerning the mechanism of these alkylation and arylation reactions, and the reduction process accompanying the reactions of α-haloalkylmercurials with HCCo$_3$(CO)$_9$ is at present still a complete mystery.

Another synthesis of aryl-substituted methylidynetricobalt nonacarbonyls was developed by Dolby and Robinson (15) who found that chloromethylidynetricobalt nonacarbonyl alkylates aromatic compounds in a Friedel-Crafts-type reaction. High product yields were obtained when equimolar amounts of ClCCo$_3$(CO)$_9$ and aluminum chloride and an excess of the arene were stirred at 60°–70°C for 2 hours. When the arene was a solid, the reaction was carried out in dichloromethane solution. Both ortho and para substitution was encountered, the size of the substituent(s) already present on the benzene ring appearing to determine the position of substitution (Table IV). Noteworthy is that milder temperature conditions affected the position of substitution; thus, reaction of chlorobenzene with chloromethylidynetricobalt nonacarbonyl in dichloromethane at 42° gave

TABLE IV

FRIEDEL-CRAFTS SYNTHESIS OF SUBSTITUTED BENZYLIDYNETRICOBALT
NONACARBONYL COMPLEXES FROM CHLOROMETHYLIDYNETRICOBALT NONACARBONYL[a]

Arene	Solvent	Reaction temp. (°C)	Ar in $ArCCo_3(CO)_9$ produced	% Yield
C_6H_5F	None	70	$o\text{-}FC_6H_4$	90
C_6H_5F	CH_2Cl_2	42	$o\text{-}FC_6H_4$	80
C_6H_5Cl	None	70	$o\text{-}ClC_6H_4$	90
C_6H_5Cl	CH_2Cl_2	42	$p\text{-}ClC_6H_4$	85
C_6H_5Br	None	70	$p\text{-}BrC_6H_4$	90
C_6H_5Br	CH_2Cl_2	42	$p\text{-}BrC_6H_4$	85
C_6H_5I	CH_2Cl_2	42	$p\text{-}IC_6H_4$	90
$CH_3C_6H_5$	None	70	$o\text{-}$ and $p\text{-}CH_3C_6H_4$ ($o > p$)	40
$CH_3C_6H_5$	CH_2Cl_2	42	$o\text{-}$ and $p\text{-}CH_3C_6H_4$ ($p > o$)	60
C_6H_6	None	70	C_6H_5	50
$C_6H_5\text{—}C_6H_5$	CH_2Cl_2	42	$p\text{-}C_6H_5C_6H_4$	75
$C_{10}H_8$	CH_2Cl_2	42	$\beta\text{-}C_{10}H_7$	80
$o\text{-}Me_2C_6H_4$	CH_2Cl_2	42	$3,4\text{-}Me_2C_6H_3$	80
$m\text{-}Me_2C_6H_4$	CH_2Cl_2	42	$2,4\text{-}Me_2C_6H_3$	10
$p\text{-}Me_2C_6H_4$	CH_2Cl_2	42	$2,5\text{-}Me_2C_6H_3$	10
$o\text{-}Cl_2C_6H_4$	CH_2Cl_2	42	$3,4\text{-}Cl_2C_6H_3$	22
$m\text{-}Cl_2C_6H_4$	CH_2Cl_2	42	$2,4\text{-}Cl_2C_6H_3$	15

[a] From Dolby and Robinson (15).

the p-Cl-substituted product rather than the ortho isomer obtained at 70°. Deactivated arenes, such as nitrobenzene, and hindered and polysubstituted arenes did not react with $ClCCo_3(CO)_9/AlCl_3$. Although the Friedel-Crafts procedure appears to be useful for the preparation of various substituted benzylidynetricobalt nonacarbonyl complexes, one is at the mercy of directive effects and so the arylmercurial/$HCCo_3(CO)_9$ reaction has a much broader scope of application. An intriguing feature of the aluminum chloride-induced reactions of $ClCCo_3(CO)_9$ is that they may involve the unusual apical carbonium ion (I) or at least a species with a great deal of positive charge at the apical carbon atom. We will encounter this species again later.

The addition of $H\text{—}CCo_3(CO)_9$ to an olefinic $C=C$ bond is a newer method of introducing organic functionality at the apical carbon atom of a $CCo_3(CO)_9$ cluster and was reported first from these laboratories in 1971 (1). Allyl acetate and allyl ethyl ether were found to react with $HCCo_3(CO)_9$ to give $(OC)_9Co_3C(CH_2)_3O_2CCH_3$ and $(OC)_9Co_3C(CH_2)_3\text{-}OC_2H_5$, respectively, when the olefin was used as solvent. This addition occurred only in the presence of about 20 mole % of a free radical initiator, azoisobutyronitrile (16):

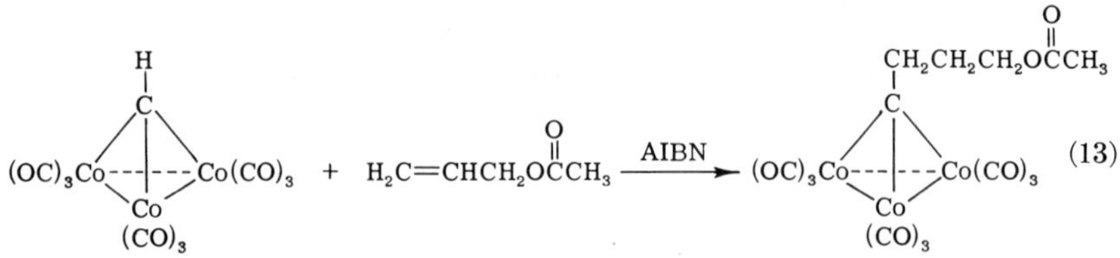

$$(OC)_3Co\text{---}|\text{---}Co(CO)_3 \overset{H}{\underset{Co(CO)_3}{\overset{|}{C}}} + H_2C=CHCH_2O\overset{O}{\overset{||}{C}}CH_3 \xrightarrow{AIBN} (OC)_3Co\text{---}|\text{---}Co(CO)_3 \overset{CH_2CH_2CH_2O\overset{O}{\overset{||}{C}}CH_3}{\underset{Co(CO)_3}{\overset{|}{C}}}$$ (13)

The implication that the apical radical (II) is an intermediate in these reactions is of some interest and suggests that other radical reactions of such organocobalt cluster complexes may be found in the future. Results reported by Japanese workers (17, 18) confirmed that such reactions are free-radical processes and provided more examples of this addition reaction. Unfortunately, the synthetic potential of such C—H addition seems limited in view of the low product yields generally obtained (Table V). Some of the more reactive functional groups were reduced in such reactions. Thus the reactions of $HCCo_3(CO)_9$ with allyl bromide and with vinyl acetate gave $CH_2=CHCH_2CCo_3(CO)_9$ and $C_2H_5CCo_3(CO)_9$, respectively, instead of the expected functionally substituted cluster complexes.

(I) (II)

These then are the more generally applicable routes by which such $RCCo_3(CO)_9$ cluster complexes may deliberately be prepared. At times such complexes are unexpected main products or by-products in reactions involving cobalt carbonyl complexes. For instance, synthesis of succinic acid from acetylene, carbon monoxide, and water in the presence of a cobalt carbonyl catalyst is accompanied by formation of organocobalt carbonyl by-products, including the organo-functional cluster complex $(OC)_9Co_3CH_2CH_2CO_2H$ (19a). Related to this transformation is the acid-induced conversion of acetylenedicobalt hexacarbonyl carbonylation products to unsaturated acid derivatives of the $(OC)_9Co_3C$ cluster (19a, b):

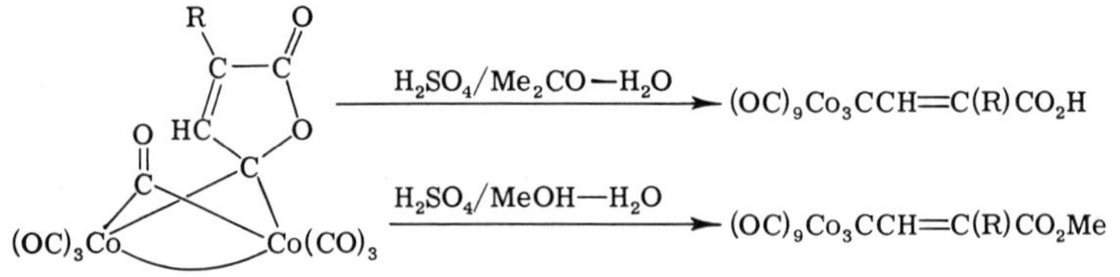

$$\xrightarrow{H_2SO_4/Me_2CO-H_2O} (OC)_9Co_3CCH=C(R)CO_2H$$ (14a)

$$\xrightarrow{H_2SO_4/MeOH-H_2O} (OC)_9Co_3CCH=C(R)CO_2Me$$ (14b)

TABLE V

REACTIONS OF METHYLIDYNETRICOBALT NONACARBONYL WITH OLEFINS[a,b]

Olefin	Product	% Yield
$CH_2=CH$ (at 130°C)	$C_2H_5CCo_3(CO)_9$	20
$CH_3CH=CH_2$ (at 130°C)	$C_3H_7CCo_3(CO)_9$ (5:1 n/iso)	10
norbornene	norbornene-$CCo_3(CO)_9$	11
cyclooctene	cyclooctene-$CCo_3(CO)_9$	4
1,5-cyclooctadiene	cyclooctenyl-$CCo_3(CO)_9$	26
1,3,5,7-cyclooctatetraene	cyclooctatrienyl-$CCo_3(CO)_9$	45
1,3-cyclohexadiene	cyclohexenyl-$CCo_3(CO)_9$	11
1,4-cyclohexadiene	cyclohexenyl-$CCo_3(CO)_9$	25
$CH_2=CH-CH=CH_2$	$CH_3CH=CHCH_2CCo_3(CO)_9$	32
$CH_2=CH-CH=CHCH_3$	$CH_3\underset{\underset{CH=CHCH_3}{\mid}}{C}HCCo_3(CO)_9$	11
$CH_2=CHCO_2CH_3$	$CH_3O_2C\underset{\underset{CH_3}{\mid}}{C}HCCo_3(CO)_9$	19

[a] From Sakamoto et al. (17).
[b] Reaction for 5 hours at 100°C.

Other reactions that one may cite include the following:

$$(OC)_4CoCF_2CF_2Co(CO)_4 \xrightarrow{100°C} CF_3CCo_3(CO)_9 \quad (20) \tag{15}$$

$$Co_2(CO)_8 + C_6H_5SH \xrightarrow{\text{hexane, room temp.}} C_6H_5CCo_3(CO)_9 \quad (21) \tag{16}$$

$$Co_2(CO)_8 + R_3N \cdot BY_3 \xrightarrow[{[Y = H\ (22a);\ Y = Cl, Br\ (22b)]}]{C_6H_6,\ 60°-65°C} (OC)_9Co_3COBY_2 \cdot NR_3 \tag{17a}$$

$$Co_2(CO)_8 + Et_3N \cdot AlBr_3 \xrightarrow[(22b)]{C_6H_6,\ 60°C} (OC)_9Co_3COAlBr_2 \cdot NEt_3 \quad (17b)$$

$$Co_2(CO)_8 + AlBr_3 \xrightarrow[(22b)]{C_6H_6,\ 60°C} (OC)_9Co_3COAlBr_2 \cdot AlBr_3 \quad (17c)$$

$$Me_3SiCo(CO)_4 \xrightarrow{THF} Me_3SiOCCo_3(CO)_9 \quad (23) \quad (18)$$

$$Co_2(CO)_8 + M \xrightarrow{Et_2O} (OC)_9Co_3CO^-M^+ \quad (M = Li,\ Na,\ K) \quad (24) \quad (19)$$

Finally, the formation of a "supercluster" by the reaction of dicobalt octacarbonyl with hexachlorocyclopropane in THF solution may be noted (25):

$$Co_2(CO)_8 + \underset{\substack{C \\ Cl_2}}{\overset{Cl_2C-CCl_2}{\diagup\!\!\!\!\diagdown}} \longrightarrow (OC)_9Co_3C-C\equiv C-C\underset{(CO)_3}{\overset{(CO)_3}{\underset{Co}{\overset{Co}{\diagdown\!\!\!/}}}}C-CCo_3(CO)_9 \quad (20)$$

A single-crystal X-ray diffraction study of its benzene solvate was required to establish its identity.

III

CHEMISTRY OF THE TRICOBALTCARBON DECACARBONYL CATION

One of the major goals in our study of the chemistry of the alkylidynetricobalt nonacarbonyl complexes was to obtain experimental information concerning the electronic effects of the nonacarbonyl tricobaltcarbon substituent. In order to obtain such information, it was necessary to have available some $RCCo_3(CO)_9$ complexes in which the substituent R contained a conventional, well-studied organic functional group. At the time we began these studies, the only such derivatives in the literature were the esters, $(OC)_9Co_3CCO_2R$ (R = Me, Et). These could be prepared by direct reaction of the appropriate ester of trichloroacetic acid with dicobalt octacarbonyl (3b, 7) or by the methanolysis of bromomethylidynetricobalt nonacarbonyl at 50° (7, 26). These, we thought, should be a good starting point, since directly, or by way of the free acid and its various derivatives, they could be converted to many other organic functional groups. The first conversion of $(OC)_9Co_3CCO_2Et$ which we attempted was its hydrolysis to the acid, $(OC)_9Co_3CCO_2H$ (27). Base-catalyzed hydrolysis using sodium carbonate in aqueous THF resulted in destruction of the complex within 30 minutes. Our initial attempts to effect acid hydrolysis of the ethyl ester also were unsuccessful; the use of catalytic, stoichiometric, or above stoichiometric amounts of mineral acids in aqueous THF was without effect, and the ethyl ester was recovered unchanged. Why, we asked ourselves, is this cluster-substituted ester so resistant to acid hydrolysis? One could envision that both electronic effects and steric effects are responsible. We have already noted in our initial discussion of the structure of $RCCo_3(CO)_9$ complexes that steric hindrance is expected to play an important role in their chemistry, and our further efforts to effect the hydrolysis of $(OC)_9Co_3CCO_2Et$ were based on the assumption that we were dealing with a problem of steric hindrance.

Some observations by Hammett *et al.* (28) on solutions of carboxylic acids in concentrated sulfuric acid had led to a procedure which served in the esterification of hindered weak acids and in the hydrolysis of their esters (28, 29). It had been found that most carboxylic acids dissolved in concentrated sulfuric acid to give the monoprotonated species. However, some highly hindered carboxylic acids reacted further to give the much less hindered acylium ion:

$$RC(=O)OH + H_2SO_4 \rightleftharpoons [RC(OH)_2]^+ + HSO_4^- \quad (21)$$

$$[RC(OH)_2]^+ + H_2SO_4 \rightleftharpoons RC\overset{+}{=}O + HSO_4^- + H_3O^+ \quad (22)$$

When a sulfuric acid solution containing such a reactive acylium ion was poured onto ice, the carboxylic acid was produced; when it was treated with an alcohol, the ester was formed. The magnitude of the steric factor is critical; in contrast to 2,4,6-trimethylbenzoic acid which formed the acylium ion in concentrated sulfuric acid, 2,4,5-trimethylbenzoic acid did not. We decided to try the Hammett procedure for the hydrolysis of $(OC)_9Co_3CCO_2Et$, although it was not at all certain that this compound would survive treatment with such strong acid. However, the ester dissolved in concentrated sulfuric acid without evolution of carbon monoxide, giving a yellow-brown solution which was stable at room temperature when maintained under a dry nitrogen atmosphere. When such a solution was poured onto cracked ice, the desired $(OC)_9Co_3CCO_2H$ was obtained in virtually quantitative yield. When such a sulfuric acid solution of the acid or of the ethyl ester was poured into a large excess of another alcohol, the ester of that alcohol was obtained, generally in high yield (Table VI).

TABLE VI

PREPARATION OF ESTERS OF $(OC)_9Co_3CCO_2H$ BY THE SULFURIC ACID PROCEDURE[a]

ROH	R in $(OC)_9Co_3CCO_2R$	% Yield
CH_3OH	CH_3	95
$(CH_3)_2CHOH$	$(CH_3)_2CH$	96
$(CH_3)_3COH$	$(CH_3)_3C$	39
$CH_2=CHCH_2OH$	$CH_2=CHCH_2$	99
$HOCH_2CH_2OH$	$HOCH_2CH_2$	85

[a] From Seyferth et al. (27).

This procedure, however, was not applicable to the preparation of aryl esters.

Of greater interest to us than the availability of the desired acid was the fact that $(OC)_9Co_3CCO_2R$ complexes dissolved in concentrated sulfuric acid, apparently to give the tricobaltcarbon decacarbonyl cation, $(OC)_9Co_3CCO^+$, by the reaction sequence shown in Eqs. (21) and (22). This novel acylium ion would be expected to be more reactive than any other derivative of the acid, e.g., the acid halides. Sulfuric acid, of course, was not the solvent of choice in which to study the chemistry of the $(OC)_9Co_3CCO^+$ ion since it is not compatible with so many of the nucleophiles with which one might want the acylium ion to react. Maximum development of the chemistry of this species required its generation in an inert organic medium, and well-known techniques from carbonium ion chemistry made this possible. Solution of either carboxymethylidynetricobalt nonacarbonyl or one of its esters in a minimum volume of propionic anhydride followed by addition of an excess of 65% aqueous hexafluorophosphoric acid resulted in the immediate precipitation of a black solid. This material was moisture sensitive but stable to oxygen and could be handled using Schlenk tube techniques. It was insoluble in and unaffected by ethers, chlorinated hydrocarbons, acetone, and many arenes; it dissolved in nitromethane without reaction. An analytically pure sample was obtained simply by washing it with dichloromethane. The analysis, as well as all of the chemical conversions of this solid, was compatible with its formulation as the salt $(OC)_9Co_3CCO^+ PF_6^-$. In most of its reactions which we studied the solid salt was allowed to react directly with the nucleophilic substrate or it was used in the form of a dichloromethane slurry. In a few cases, nitromethane solutions of the salt were used.

The chemical reactions of $(OC)_9Co_3CCO^+PF_6^-$ are summarized in Fig. 2. Individual examples are given in Table VII. As expected this reagent

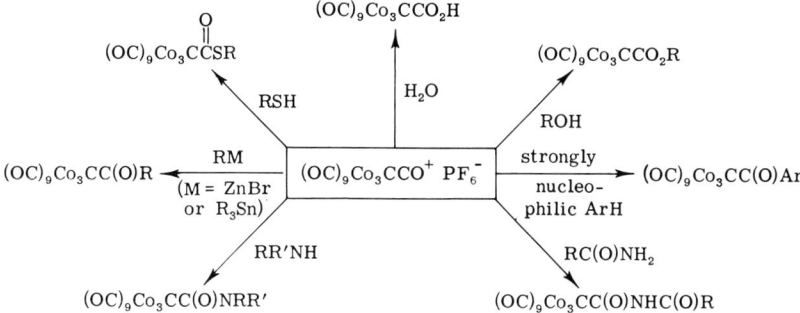

FIG. 2. A general survey of the reactions of $(OC)_9Co_3CCO^+PF_6^-$.

acylates primary, secondary, and tertiary alcohols; it also reacts with phenol, a reaction that could not be accomplished in sulfuric acid medium. Acylation of ammonia, primary and secondary amines gave amide derivatives of the $(OC)_9Co_3C$ cluster which were not very stable thermally. Even less stable were the imides obtained by acylation of formamide and acetamide with $(OC)_9Co_3CCO^+PF_6^-$; these decomposed slowly even at 0°C. The reaction of $(OC)_9Co_3CCO^+PF_6^-$ with diphenylamine gave the expected $(OC)_9Co_3CC(O)NPh_2$ in 42% yield, but a second product also was formed and identified as $(OC)_9Co_3CC(O)C_6H_4(NHPh)$-$p$, i.e., as the Friedel-Crafts acylation product. The lower nucleophilicity of and the greater steric hindrance to attack at nitrogen in the case of diphenylamine (compared with N-methylaniline) no doubt are responsible for the attack at the aromatic nucleus. Acylation of the amino function of ethyl glycinate was effected by reaction of the cluster acylium hexafluorophosphate with $EtO_2CCH_2NH_3^+Cl^-$ in the presence of pyridine. A tripeptide hydrochloride was acylated in similar manner:

$$Cl^-H_3\overset{+}{N}CH_2\overset{\underset{\Vert}{O}}{C}NHCHCNHCH_2COEt + (OC)_9Co_3CCO^+PF_6^- + 2C_5H_5N$$
$$\underset{CH_2Ph}{}$$

$$\longrightarrow (OC)_9Co_3CCNHCH_2CNHCHCNHCH_2COEt$$
$$\underset{CH_2Ph}{}$$

$$+ C_5H_5NH^+PF_6^- + C_5H_5NH^+Cl^- \quad (23)$$

This reaction suggests that the $(OC)_9Co_3CCO$ group can be introduced into more complex molecules of biological interest via acylation of N—H, O—H,

TABLE VII: Reactions of $(OC)_9Co_3CCO^+PF_6^-$ [a]

Nucleophile	R in $(OC)_9Co_2Cr$ product	% Yield
CH_3OH	CO_2CH_3	82
C_2H_5OH	$CO_2C_2H_5$	96
$(CH_3)_2CHOH$	$CO_2CH(CH_3)_2$	89
$(CH_3)_3COH$	$CO_2C(CH_3)_3$	84
$HOCH_2CH_2OH$	$CO_2CH_2CH_2O_2CCCo_3(CO)_9$	35
$HOCH_2C{\equiv}CCH_2OH$	$CO_2CH_2C{\equiv}CCH_2O_2CCCo_3(CO)_9$	52
C_6H_5OH	$CO_2C_6H_5$	72
C_2H_5SH	$C(O)SC_2H_5$	80
C_6H_5SH	$C(O)SC_6H_5$	80
NH_3 (g)	$C(O)NH_2$	89
CH_3NH_2 (g)	$C(O)NHCH_3$	96
$(CH_3)_2NH$ (g)	$C(O)NCH_3)_2$	84
$(C_2H_5)_2NH$	$C(O)N(C_2H_5)_2$	93
morpholine (O(CH₂CH₂)₂NH)	$C(O)N$(morpholino)	52
$C_6H_5NH_2$	$C(O)NHC_6H_5$	73
$C_6H_5NHCH_3$	$C(O)N(CH_3)(C_6H_5)$	66
$(C_6H_5)_2NH$	$C(O)N(C_6H_5)_2$	42
$HC(O)NH_2$	$C(O)C_6H_4NHC_6H_5$	20
$HC(O)NH_2$	$C(O)NHC(O)H$	42
$CH_3C(O)NH_2$	$C(O)NHC(O)CH_3$	65
$Cl^-\ H_3\overset{+}{N}CH_2CO_2Et$	$C(O)NHCH_2CO_2Et$	63
$Cl^-\ H_3\overset{+}{N}CH_2C(O)NHCH(CH_2Ph)C(O)NHCH_2CO_2Et$	$C(O)NHCH_2C(O)NHCH(CH_2Ph)C(O)NHCH_2CO_2Et$	58
pyrrole	$C(O)$-pyrrolyl	80
indole	$C(O)$-indolyl	70
$(CH_3)_2NC_6H_5$	$C(O)C_6H_4N(CH_3)_2$	69
ferrocene	$C(O)$-ferrocenyl	31
$(CH_3)_4Sn$	$C(O)CH_3$	58
$(C_2H_5)_4Sn$	$C(O)C_2H_5$	9
C_2H_5ZnBr	$C(O)C_2H_5$	54

[a] From Seyferth et al. (27).

or S—H functions. Applications in polypeptide X-ray crystallography and electron microscopy are conceivable.

Attempted alkylation of $(OC)_9Co_3CCO^+PF_6^-$ with organolithium or organomagnesium reagents was not successful. However, mild alkylating agents did serve in the preparation of ketones from this salt. Ethylzinc halides reacted rapidly in THF with the PF_6^- salt to give $(OC)_9Co_3CC(O)Et$ in variable (20–55%) yield. Tetramethyl- and tetraethyltin also were found to react with the cluster acylium ion, but much less rapidly, to give the expected ketones. The arylation of $(OC)_9Co_3CCO^+PF_6^-$ with nucleophilic aromatic substrates was of special interest since it provided clear evidence that the cluster acylium ion was relatively poorly reactive as an electrophile and that this was due to electronic factors. The acylation of diphenylamine has been mentioned already. Only highly nucleophilic aromatic substrates were found to undergo such reactions: N,N-dimethylaniline, pyrrole, indole, and ferrocene, but not anisole or benzofuran. The fact that anisole, which is fairly high on the Friedel-Crafts reactivity scale, does not react must be due to an electronic factor, since N,N-dimethylaniline does react.

At the time the chemistry of $(OC)_9Co_3CCO^+PF_6^-$ was being developed (27, 30), another route to this novel acylium ion was found in these laboratories (31, 32). This discovery was a result of our intention to adapt the Friedel-Crafts synthesis of benzylidynetricobalt nonacarbonyl complexes of Dolby and Robinson (15) to the preparation of alkyl derivatives of methylidynetricobalt nonacarbonyl, whose general preparation was not well in hand. The reaction chosen for investigation was the aluminum chloride-induced interaction of tetraalkyltin compounds with $ClCCo_3(CO)_9$. In analogy to a known ketone synthesis (33),

$$R_4Sn + R'COCl \xrightarrow{AlCl_3} R'COR + R_3SnCl \quad (24)$$

this reaction might have been expected to give the alkyl-substituted cluster, $RCCo_3(CO)_9$. However, when this reaction was carried out in dichloromethane solution with tetramethyltin, the product obtained (in 26% yield) was the acetyl derivative, $CH_3C(O)CCo_3(CO)_9$, rather than the methyl compound, $CH_3CCo_3(CO)_9$.

Since we had found tetramethyltin to react with $(OC)_9Co_3CCO^+PF_6^-$ to give this acetyl derivative, it was a reasonable conclusion that the action of aluminum chloride on chloromethylidynetricobalt nonacarbonyl under our reaction conditions had led to formation of an acylium ion salt, perhaps $(OC)_9Co_3CCO^+AlCl_4^-$. Further investigation of this system indicated that this actually was what was occurring (34). The chemistry of the formation of the acylium ion in this system must be somewhat complicated.

When $ClCCo_3(CO)_9$ and a two- to threefold excess of aluminum chloride are mixed under nitrogen in dichloromethane, an initially purple solution is formed as most of the solids dissolve. (It is such solutions that can alkylate aromatic substrates.) When such a solution is stirred under nitrogen for 20 to 30 minutes, the color changes from purple to an intense yellow-brown. At this point, thin-layer chromatography shows that the $ClCCo_3(CO)_9$ has been consumed completely; the reagent formed is completely dissolved or almost so. Such reagent solutions are stable at room temperature for at least 2 days and probably longer. Treatment of such yellow-brown solutions with any of the nucleophiles known to react with $(OC)_9Co_3CCO^+PF_6^-$, at room temperature, results in formation of the expected acylated product, generally in good yield, in what appear to be nearly instantaneous reactions as indicated by the observed color changes. Table VIII shows some of the reactions that were carried out with the $(OC)_9Co_3CCO^+AlCl_4^-$ reagent. Acylation of alcohols, phenol, thiols, ammonia, and amines could be accomplished; reactions with tetraalkyltins gave ketones; highly nucleophilic aromatic compounds were acylated. Especially noteworthy is the reaction with triethylsilane which gave the aldehyde $(OC)_9Co_3CCH=O$. By contrast, $(OC)_9Co_3CCO^+PF_6^-$ reacted with triethylsilane in dichloromethane to give a mixture of $CH_3CCo_3(CO)_9$ (30%) and $HCCo_3(CO)_9$ (43%). However, when 3 molar equivalents of $AlCl_3$ was added to a slurry of $(OC)_9Co_3CCO^+PF_6^-$ in dichloromethane, the brown, nearly homogeneous solution which formed reacted with an excess of triethylsilane to give the cluster aldehyde in 74% yield. We suggest that aluminum chloride effects release of the aldehyde from the intermediate in the first step in the reduction of the acylium ion, $(OC)_9Co_3$-$CCHOSiEt_3^+$, via Cl^- attack at silicon. In the absence of aluminum chloride, further reduction to give the methyl derivative is possible.

The conversion of chloromethylidynetricobalt nonacarbonyl to the acylium ion by the action of aluminum chloride is a remarkable process. The reaction does not require external carbon monoxide; it proceeds perfectly satisfactorily under a nitrogen atmosphere. The CO function at the apical carbon atom of the products obtained thus was derived from carbon monoxide ligands on cobalt in $ClCCo_3(CO)_9$. This transfer of CO to the apical carbon atom is very efficient. The yields given in Table VIII are based on the amount of $ClCCo_3(CO)_9$ charged. If one recalculates the yield of $(OC)_9Co_3CCO_2CH_3$ in the first entry in Table VIII based on the available $(OC)_9Co_3CCO^+$ from the $ClCCo_3(CO)_9$ used [assuming the destruction of an amount of $ClCCo_3(CO)_9$ equivalent to the theoretically required amount of CO for the transfer to carbon], it is 92%.

No improvement in yield was achieved in a reagent preparation carried out under an atmosphere of carbon monoxide rather than nitrogen. The mechanism of this conversion of $ClCCo_3(CO)_9$ to the $(OC)_9Co_3CCO^+$ ion is of interest, especially since this reagent is preparatively useful. The aluminum chloride is essential in order to obtain good product yields.

TABLE VIII

Synthetic Applications of the $(OC)_9Co_3CCO^+AlCl_4^-$ Reagent[a]

Nucleophile	R in $(OC)_9Co_3CR$ product	% Yield[b]
CH_3OH	CO_2CH_3	83
C_2H_5OH	$CO_2C_2H_5$	78
Me_2CHOH	CO_2CHMe_2	78
Me_3COH	CO_2CMe_3	77
$CH_2{=}CHCH_2OH$	$CO_2CH_2CH{=}CH_2$	66[c]
$HC{\equiv}CCH_2OH$	$CO_2CH_2C{\equiv}CH$	62[c]
CCl_3CH_2OH	$CO_2CH_2CCl_3$	62[c]
C_6H_5OH	$CO_2C_6H_5$	66
$p\text{-}CH_3OC_6H_4OH$	$CO_2C_6H_4OCH_3\text{-}p$	39[c]
Me_3CSH	$C(O)SCMe_3$	51
PhSH	C(O)SPh	58
NH_3 (g)	$C(O)NH_2$	64
$PhNH_2$	C(O)NHPh	76
Me_2NH	$C(O)NMe_2$	55
Et_2NH	$C(O)NEt_2$	75
▷NH (aziridine)	C(O)N◁	60[c]
morpholine (O⌒NH)	C(O)N-morpholinyl	64[c]
$(CH_3)_4Sn$	$C(O)CH_3$	61
$(C_2H_5)_4Sn$	$C(O)C_2H_5$	66
$Me_2N\text{-}C_6H_4\text{-}$	$C(O)\text{-}C_6H_4\text{-}NMe_2$	69
ferrocenyl	C(O)-ferrocenyl	41
Et_3SiH	C(O)H	63

[a] Seyferth and Williams (34).
[b] Yields are based on the quantity of $ClCCo_3(CO)_9$ charged.
[c] Unpublished work, C. L. Nivert.

Chloromethylidynetricobalt nonacarbonyl does react with alcohols, but only very slowly. Thus its reaction with an excess of ethanol at room temperature for 15 days gave $(OC)_9Co_3CCO_2Et$, but only in 6% yield. [The bromo cluster is more reactive, giving $(OC)_9Co_3CCO_2Me$ in 59% yield on reaction with methanol at 55° (7, 26).] To obtain the high yields shown in

Table VIII, at least 2 (and preferably 3) moles of $AlCl_3$ per mole of $ClCCo_3(CO)_9$ are required. The acylating agent is not formed instantaneously; only when the purple to yellow-brown color change has taken place is the acylium ion species present in high yield. When a mixture of 2 molar equivalents of $AlCl_3$ and 1 of $ClCCo_3(CO)_9$ in dichloromethane was quenched with methanol immediately after mixing, no ester was formed and a high recovery of the chloro cluster was obtained.

At the present time, both the exact constitution of the final acylating agent and the mechanism of its formation remain unclear. The picture which we favor is admittedly incomplete and very possibly subject to change. We suggest that initially a $(OC)_9Co_3CCl \cdot AlCl_3$ complex is formed in which substantial polarization of the C—Cl bond has occurred and which is capable of Friedel-Crafts substitution on aromatic systems. Subsequent complexation of a second molecule of $AlCl_3$ at a carbon monoxide ligand [a known process in metal carbonyl chemistry (35, 36)] provides the activation for CO migration from cobalt to the electron-deficient apical carbon atom. This is not unreasonable since it is known that in binuclear metal carbonyls the bridging carbonyl ligands are stronger Lewis basic sites than are the terminal carbon monoxides (35, 36). In fact, aluminum alkyls have been reported to promote a terminal to bridging carbon monoxide ligand shift in a binuclear ruthenium complex (37). Such a cobalt-to-carbon CO transfer in our system, occurring either intra- or intermolecularly, would leave coordinatively unsaturated cobalt atoms in cluster acylium ions which would require efficient CO transfer from other molecules in order to obtain the $(OC)_9Co_3CCO^+$ species in high yield. We are still working in this area in the hope of achieving a better understanding of this process.

We note that systems that give the cluster acylium ion more rapidly are available. Thus, $BrCCo_3(CO)_9$ is converted to the yellow-brown acylium ion solution in dichloromethane by an excess of aluminum chloride at a faster rate than is $ClCCo_3(CO)_9$. This transformation of $ClCCo_3(CO)_9$ can be accelerated by using a larger excess of aluminum chloride (10 molar equivalents of $AlCl_3$ rather than 2 or 3) or by carrying out the $ClCCo_3(CO)_9/3AlCl_3$ reaction in the presence of iodomethane. In the latter case, it appears that the rate acceleration is due to the formation of the more reactive aluminum iodide rather than of $ICCo_3(CO)_9$. The preparative utility of these systems has not yet been assessed in detail, but in all cases the yields of $(OC)_9Co_3CCO_2R$ on treatment of the solutions with methanol or ethanol were good.

Robinson and co-workers have claimed, without providing any experimental details, that boron trifluoride also converts $BrCCo_3(CO)_9$ to the cluster acylium ion under a carbon monoxide atmosphere (38). They also report that the $ClCCo_3(CO)_9/AlCl_3$ reaction can be carried out *in situ* in the presence of the nucleophile (e.g., H_2O, EtOH, PhOH) (*38*).

IV
HIGHLY STABLE NONACARBONYL TRICOBALTCARBON-SUBSTITUTED CARBONIUM IONS

Our discovery of the easily formed, very stable, and preparatively very useful $(OC)_9Co_3CCO^+$ ion suggested to us that carbonium ions of type $(OC)_9Co_3CCRR'^+$ might provide another fruitful area of study. The electronic effects of the $(OC)_9Co_3C$ cluster were not at all clear and one could entertain the possibility that this organometallic substituent might stabilize an adjacent positive charge. A suitable carbonium ion precursor was needed in order to carry out such an investigation and at that time no such $(OC)_9Co_3C$ derivative was available in useful amounts. Alcohols of type $(OC)_9Co_3CC(OH)RR'$, as mentioned already, were available only in yields of 5% or less, and halides of type $(OC)_9Co_3CC(X)RR'$ were (and still are) unknown. The first problem, then, was to develop a more useful, high-yield alcohol synthesis.

The reduction of a ketone to an alcohol is a well-known organic reaction, and since ketones of type $(OC)_9Co_3CC(O)R$ and the aldehyde $(OC)_9Co_3CCHO$ could be prepared in good yield, we chose to examine this route

TABLE IX

REDUCTION OF $RC(O)CCo_3(CO)_9$ TO $RCH_2CCo_3(CO)_9$ BY Et_3SiH/CF_3CO_2H[a]

R in $RC(O)CCo_3(CO)_9$ (mmole)	Et_3SiH (mmole)	CF_3CO_2H (mmole)	$RCH_2CCo_3(CO)_9$ (% yield)
CH_3 (2.9)	7	6	$C_2H_5CCo_3(CO)_9$ (90)
C_2H_5 (3)	7	6	$n-C_3H_7CCo_3(CO)_9$ (92)
$n-C_3H_7$ (3)	7	6	$n-C_4H_9CCo_3(CO)_9$ (87)
$n-C_4H_9$ (3)	7	6	$n-C_5H_{11}CCo_3(CO)_9$ (80)
$n-C_6H_{13}$ (3)	8	8	$n-C_7H_{15}CCo_3(CO)_9$ (85)
cyclo-C_6H_{11} (3)	8	9	cyclo-$C_6H_{11}CH_2CCo_3(CO)_9$ (75)
$(CH_3)_2CH$ (2)	5	6	$(CH_3)_2CHCH_2CCo_3(CO)_9$ (81)
C_6H_5 (2)	5	5	$C_6H_5CH_2CCo_3(CO)_9$ (82)
$p-CH_3C_6H_4$ (2)	5.2	6	$p-CH_3C_6H_4CH_2CCo_3(CO)_9$ (78)
$p-BrC_6H_4$ (2.9)	7	6	$p-BrC_6H_4CH_2CCo_3(CO)_9$ (67)

[a] From Seyferth *et al.* (*39*).

(*39*). The initial results were not encouraging. Attempted conversion of formylmethylidynetricobalt nonacarbonyl to the primary alcohol was unsuccessful. Treatment of this aldehyde with sodium borohydride in THF at reflux gave a mixture of $CH_3CCo_3(CO)_9$ and $HCCo_3(CO)_9$, whereas reaction with lithium aluminum hydride or sodium borohydride in benzene gave only $HCCo_3(CO)_9$. In view of the stability of some of these cluster complexes to strong acids, we turned our attention to an acidic reducing system, triethylsilane/trifluoroacetic acid, that had been developed by Russian workers [see Kursanov *et al.* (*40*) for a recent review]. This reduction proceeds by initial protonation of the carbonyl compound, followed by reduction, via hydride transfer from the silicon hydride, of the protonated species. If one of the carbonyl substituents can stabilize an adjacent positive charge (e.g., an aryl group), the reduction proceeds past the alcohol stage to give a hydrocarbon:

$$\underset{R}{\overset{R'}{>}}C=O \underset{}{\overset{H^+}{\rightleftharpoons}} \underset{R}{\overset{R'}{>}}\overset{+}{C}{-}OH \xrightarrow{Et_3SiH} R\underset{H}{\overset{R'}{\underset{|}{-}}}\overset{|}{C}-OH \underset{}{\overset{H^+}{\rightleftharpoons}} R\underset{H}{\overset{R'}{\underset{|}{-}}}\overset{|}{C}-\overset{+}{O}H_2 \quad (25)$$

$$R\underset{H}{\overset{R'}{\underset{|}{-}}}\overset{|}{C}-\overset{+}{O}H_2 \longrightarrow \underset{R}{\overset{R'}{>}}\overset{+}{C} \xrightarrow{Et_3SiH} \underset{R}{\overset{R'}{>}}CH_2 \quad (26)$$

A number of cluster-substituted ketones and the aldehyde all reacted with the Et_3SiH/CF_3CO_2H to give the alkyl derivatives rather than the expected alcohols:

$$(OC)_9Co_3CC(O)R \xrightarrow{Et_3SiH/CF_3CO_2H} (OC)_9Co_3CCH_2R \quad (27)$$

The examples studied are given in Table IX. The product yields were excellent (75–92%) and, in fact, this reaction is the best available procedure for the preparation of such alkyl derivatives. The R in Eq. (27) may be primary or secondary alkyl or aryl; when R becomes more bulky (e.g., Me_3C, $(OC)_9Co_3C$) the reduction fails. Although this reaction did not result in the alcohols we required for our further studies, the results were of considerable interest to us since they provided good indication that the $(OC)_9Co_3C$-substituted carbonium ions would be rather stable species. As mentioned, the complete reduction observed occurs only when at least one of the substituents on the ketone carbonyl function is capable of stabilizing an adjacent positive charge. In those cluster complexes in Table IX where R = H or alkyl, it must be the $(OC)_9Co_3C$ substituent that is providing such stabilization. However, in order to pursue this question, we still required $(OC)_9Co_3CC(OH)RR'$ compounds in useful amounts.

Hydrosilylation of ketones and aldehydes converts these to silyl ethers whose hydrolysis gives alcohols. Phenylsilanes were found to add to benzophenone, but the reaction conditions (reaction temperatures of 220°–270°C) were not attractive (41). Catalyzed hydrosilylations proceeded under much milder conditions, e.g., zinc chloride (42), $H_2PtCl_6 \cdot 6H_2O$ (43), $(Ph_3P)_3$-RhCl (44). The hydrosilylation of our cluster-substituted ketones was found to occur under surprisingly mild conditions. It was sufficient to heat equimolar quantities of triethylsilane and the $(OC)_9Co_3CC(O)R$ compounds in benzene at reflux, under a carbon monoxide atmosphere for about 8 hours. The crude silyl ether was converted to the alcohol by solution in concentrated sulfuric acid and subsequent hydrolysis by pouring into an ice–water mixture:

$$(OC)_9Co_3C-\underset{\underset{O}{\|}}{C}R + Et_3SiH \longrightarrow (OC)_9Co_3C\underset{\underset{R}{|}}{C}HOSiEt_3 \quad (28)$$

$$(OC)_9Co_3C\underset{\underset{R}{|}}{C}HOSiEt_3 \xrightarrow{H_2SO_4} (OC)_9Co_3C\overset{+}{C}\overset{R}{\underset{H}{\diagdown}} \quad (29)$$

$$(OC)_9Co_3C\overset{+}{C}\overset{R}{\underset{H}{\diagdown}} \xrightarrow{H_2O} (OC)_9Co_3C\underset{\underset{R}{|}}{C}H-OH \quad (30)$$

It was found important to carry out the hydrosilylation under an atmosphere of carbon monoxide in order to obtain the good yields given in Table X for a number of these reactions. The question remains why these hydrosilylation reactions proceed so readily under mild conditions in the absence of a catalyst. One possibility which we considered was that the cluster ketone provided its own catalyst by way of minor decomposition to give mononuclear cobalt carbonyl intermediates which could be the actual catalytic species. However, $RCCo_3(CO)_9$ were not found to catalyze the hydrosilylation of cyclohexanone. A second possibility is that the C=O bond in these $(OC)_9Co_3CC(O)R$ compounds is exceptionally reactive. Table XI lists the ketonic C=O stretching frequencies of some of the $(OC)_9Co_3$-CC(O)R compounds that we have prepared. These are found to be in the range of 1560 to 1645 cm^{-1}, at much lower frequency than in dialkyl ketones (1725–1705 cm^{-1}) or aryl ketones (1660–1700 cm^{-1}), and this implies greater carbon–oxygen bond polarization:

$$(OC)_9Co_3C-\underset{\underset{O}{\|}}{C}-R \longleftrightarrow (OC)_9Co_3C-\underset{\underset{O^-}{|}}{\overset{+}{C}}-R \quad (31)$$

TABLE X

REDUCTION OF $RC(O)CCo_3(CO)_9$ TO $RCH(OH)CCo_3(CO)_9$[a]

R in $RC(O)CCo_3(CO)_9$ (mmole)	Et_3SiH (mmole)	$RCH(OH)CCo_3(CO)_9$ (% yield)
CH_3 (16.0)	17.5	$CH_3CH(OH)CCo_3(CO)_9$ (84)
(19.7)	20.8	(90)
C_2H_5 (3)	3.2	$C_2H_5CH(OH)CCo_3(CO)_9$ (73)
n-C_3H_7 (3)	3.2	n-$C_3H_7CH(OH)CCo_3(CO)_9$ (75)
n-C_4H_9 (3)	3.4	n-$C_4H_9CH(OH)CCo_3(CO)_9$ (81)
n-C_6H_{13} (3)	3.4	n-$C_6H_{13}CH(OH)CCo_3(CO)_9$ (81)
$(CH_3)_2CH$ (3)	3.0	$(CH_3)_2CHCH(OH)CCo_3(CO)_9$ (80)
cyclo-C_6H_{11} (3)	3.4	cyclo-$C_6H_{11}CH(OH)CCo_3(CO)_9$ (52)
C_6H_5 (64.0)	87.5	$C_6H_5CH(OH)CCo_3(CO)_9$ (87)
p-$CH_3C_6H_4$ (3)	3.4	p-$CH_3C_6H_4CH(OH)CCo_3(CO)_9$ (70)
p-BrC_6H_4 (3)	3.4	p-$BrC_6H_4CH(OH)CCo_3(CO)_9$ (84)
H (10.6)	31.4	$HOCH_2CCo_3(CO)_9$ (46)

[a] From Seyferth et al. (39).

This should facilitate attack by the silicon hydride, $Et_3Si^{\delta+}$—$H^{\delta-}$, if a polar mechanism is operative.

With the $(OC)_9Co_3CC(OH)RR'$ alcohols available, experiments whose goal was the generation of cluster-substituted carbonium ions were now possible. Our initial experiments were carried out with $(OC)_9Co_3C$-$CH(OH)R$, where R = H, CH_3 and C_6H_5 (45). The procedure used in the conversion of $(OC)_9Co_3CCO_2Et$ to $(OC)_9Co_3CCO^+PF_6^-$ was found to be applicable to cluster-carbonium ion synthesis. Thus, treatment of $(OC)_9Co_3CCH(OH)CH_3$ in propionic anhydride solution with a small excess of 65% aqueous hexafluorophosphoric acid under nitrogen resulted in precipitation of a black solid whose analysis after several washes with dichloromethane indicated the constitution $(OC)_9Co_3CCHCH_3^+PF_6^-$. Similar reactions gave $(OC)_9Co_3CCH_2^+PF_6^-$ and $(OC)_9Co_3CCHPh^+PF_6^-$. These salts were quite stable in the absence of air and moisture. Reactions at carbon were observed with alcohols, a thiol, aniline, and N,N-dimethylaniline with all three salts, e.g.,

$$(OC)_9Co_3CCH_2^+PF_6^- + CH_3OH \longrightarrow (OC)_9Co_3CCH_2OCH_3 + H^+PF_6^- \quad (32)$$

An exception was the reaction of $(OC)_9Co_3CCHCH_3^+PF_6^-$ with N,N-dimethylaniline, in which the basic substrate abstracted a proton to give the vinyl-substituted cluster. The results of these reactions are summarized in Table XII. The $(OC)_9Co_3CCHR^+PF_6^-$ salts, like the cluster-substituted acylium ion, are rather weakly electrophilic. Although they alkylate N,N-

TABLE XI

FREQUENCIES OF KETONIC CARBONYL VIBRATIONS IN $(OC)_9Co_3CC(O)R$ COMPOUNDS[a]

R in $(OC)_9Co_3CC(O)R$	ν (C=O) (cm^{-1}, in CCl$_4$)
H[b]	1625[c]
CH$_3$	1640
C$_2$H$_5$	1645
n-C$_3$H$_7$	1640
Me$_2$CH	1635
Me$_3$C	1618
n-C$_4$H$_9$	1635
Ph	1610
p-CH$_3$C$_6$H$_4$	1611
p-BrC$_6$H$_4$	1610
p-Me$_2$NC$_6$H$_4$[d]	1582
p-PhHNC$_6$H$_4$[d]	1586
2-pyrrolyl[d]	1555
3-indolyl[d]	1573
ferrocenyl[d]	1590

[a] From Seyferth et al. (9). [c] In CHCl$_3$.
[b] Williams (32). [d] Seyferth (27).

dimethylaniline when R = H or Ph, neither of these salts will react with anisole.

It seems clear that cluster-substituted carbonium ions are easily generated, are very stable thermally, and are of sufficient electrophilic reactivity to be useful in the synthesis of many new functional cluster compounds. Before we consider further their exceptional stability and the mode of their stabilization, it is of interest to mention some chemical consequences of the high stability of a positive charge generated α to the apical carbon atom of the $(OC)_9Co_3C$ cluster. For instance, one might expect the addition of the positive part of an ionic or polar reagent, X$^+$Y$^-$, to the

TABLE XII

ORGANOCOBALT CLUSTER DERIVATIVES PREPARED FROM THE
$(OC)_9Co_3CCHR^+PF_6^-$ SALTS[a]

R in $(OC)_9Co_3CCHR^+PF_6^-$	Reactant	Product (% yield)
H	MeOH	$(OC)_9Co_3CCH_2OMe$ (85)
H	EtOH	$(OC)_9Co_3CCH_2OEt$ (76)
H	$PhNH_2$	$(OC)_9Co_3CCH_2NHPh$ (67)
H	$C_6H_5NMe_2$	$(OC)_9Co_3CCH_2C_6H_4NMe_2\text{-}p$ (49)
Me	MeOH	$(OC)_9Co_3CCH(Me)OMe$ (85)
Me	EtOH	$(OC)_9Co_3CCH(Me)OEt$ (82)
Me	PhSH	$(OC)_9Co_3CCH(Me)SPh$ (42)
Me	$PhNH_2$	$(OC)_9Co_3CCH(Me)NHPh$ (73)
Me	$C_6H_5NMe_2$	$(OC)_9Co_3CCH=CH_2$ (68)
Ph	MeOH	$(OC)_9Co_3CCH(Ph)OMe$ (59)
Ph	EtOH	$(OC)_9Co_3CCH(Ph)OEt$ (75)
Ph	PhSH	$(OC)_9Co_3CCH(Ph)SPh$ (38)
Ph	$PhNH_2$	$(OC)_9Co_3CCH(Ph)NHPh$ (59)
Ph	$C_6H_5NMe_2$	$(OC)_9Co_3CCH(Ph)C_6H_4NMe_2\text{-}p$ (54)

[a] From Seyferth et al. (45).

terminal carbon atom of a C=C substituent attached to the apical carbon of the $(OC)_9Co_3C$ cluster to be a rather favorable process, so that the direction of addition would be

$$(OC)_9Co_3CC=C\diagup\diagdown + X^+Y^- \longrightarrow (OC)_9Co_3C-\overset{+}{\underset{|}{C}}-C-X, Y^- \longrightarrow (OC)_9Co_3C-\underset{|}{\overset{|}{C}}-\underset{Y}{\overset{X}{C}}-X \quad (33)$$

All of the reactions of such vinylic derivatives that we have studied proceed in this manner (46). This reaction is of special utility in the generation of tertiary cluster-substituted carbonium ions:

$$(OC)_9Co_3CC=CH_2 + H^+PF_6^- \xrightarrow{(EtCO)_2O} (OC)_9Co_3CC(CH_3)_2^+PF_6^- \quad (34)$$
$$\underset{CH_3}{|}$$

The carbonium ion salt obtained in this manner reacted with methanol to give the methyl ether, $(OC)_9Co_3CC(CH_3)_2OCH_3$, in 86% yield and with aniline to produce $(OC)_9Co_3CC(CH_3)_2NHC_6H_5$ in 49% yield. Attempted purification of the methyl ether by chromatography on pH4 silicic acid resulted in formation of the alcohol, $(OC)_9Co_3CC(CH_3)_2OH$, which is a further indication of the easy accessibility of the tertiary carbonium ion. Reduction of isopropenyl-substituted cluster, by way of the carbonium ion, to $(OC)_9Co_3CCH(CH_3)_2$ could be effected with zinc amalgam in trifluoroacetic or concentrated hydrochloric acid. Similar protonation of $(OC)_9Co_3CCH=CH_2$ with $HPF_6/(EtCO)_2O$ gave the $(OC)_9Co_3CHCH_3{}^+$-$PF_6{}^-$ salt which we had prepared previously from the alcohol. In the case of $(OC)_9Co_3CCH=CHSiMe_3$, protonation resulted in desilylation:

$$Me_3SiCH=CHCCo_3(CO)_9 + H_2SO_4 \longrightarrow Me_3SiCH_2-\overset{+}{C}H-CCo_3(CO)_9, HSO_4{}^- \longrightarrow$$

$$CH_2=CHCCo_3(CO)_9 + Me_3SiOSO_2OH \xrightarrow{H_2SO_4} CH_3-\overset{+}{C}H-CCo_3(CO)_9, HSO_4{}^-$$

$$\downarrow H_2O$$

$$Me_3SiOSiMe_3 + (OC)_9Co_3CCHCH_3 \quad (35)$$
$$\qquad\qquad\qquad\qquad\qquad\quad |$$
$$\qquad\qquad\qquad\qquad\qquad\;\; OH$$

Since olefin elimination of this type from a species with a positive charge β to a silyl group is well known *(47)*, this result was not unexpected.

Electrophilic Friedel-Crafts acylation of $(OC)_9Co_3CCH=CH_2$ with acetyl chloride/aluminum chloride under a carbon monoxide atmosphere gave three products (all of which had the acetyl group attached to the terminal carbon atom of the vinyl group, as expected): $(OC)_9Co_3CCH_2$-$CH_2C(O)CH_3$ (13%), *trans*-$(OC)_9Co_3CCH=CHC(O)CH_3$ (12%), and $(OC)_9Co_3CCH(OH)CH_2C(O)CH_3$ (6%). All three are derivable from the initially formed cation, $(OC)_9Co_3CCHCH_2C(O)CH_3{}^+$, through reduction, proton loss, and hydrolysis, respectively. Similar addition of $CH_3CO^+AlCl_4{}^-$ to the isopropenyl-substituted cluster gave $(OC)_9Co_3C(CH_3)HCH_2C(O)$-$CH_3$. The oxymercuration of allylidynetricobalt nonacarbonyl also proceeded in the direction expected:

$$CH_2=CHCCo_3(CO)_9 + CF_3CO_2Hg^+ \longrightarrow CF_3CO_2HgCH_2\overset{+}{C}HCCo_3(CO)_9$$

$$\downarrow CH_3OH$$

$$CF_3CO_2HgCH_2CHCCo_3(CO)_9 \quad (36)$$
$$\qquad\qquad\qquad\qquad\quad |$$
$$\qquad\qquad\qquad\qquad\; OCH_3$$

The radical-initiated addition of $CBrCl_3$ or CBr_4 to olefins followed by base-induced dehydrobromination of the adducts can be used to prepare a wide variety of $RR'C=C(R'')CX_3$-type compounds whose reaction with dicobalt octacarbonyl will give the respective vinylic cluster derivatives. The reactions of the latter with appropriate reagents will extend further the number and types of organo-functional cluster complexes. Also, their catalytic hydrogenation or their reduction with systems that generate $HCo(CO)_4$ can make available as well simple or functional alkyl derivatives that are difficult or impossible to prepare by other routes. It may be noted that two examples of the catalytic hydrogenation of unsaturated cluster complexes have been reported:

$$(OC)_9Co_3CCH=CHCO_2H \xrightarrow[{[Co_2(CO)_8], 110°C\ (19)}]{H_2/CO\ (260\ atm)} (OC)_9Co_3CCH_2CH_2CO_2H \quad (37)$$

$$(OC)_9Co_3CC=CH_2 \atop {|\atop CH_3} \xrightarrow[{5\%\ Pd/C,\ 50°C\ (48)}]{H_2\ (40\ psi)} (OC)_9Co_3CCH(CH_3)_2 \quad (38)$$

(The 2-methylallylidynetricobalt nonacarbonyl used in Eq. (38) was prepared by the novel reaction of dimethylketene with dicobalt octacarbonyl.)

The stabilization of a positive charge on a carbon atom α to a $(OC)_9Co_3C$ cluster appears to play an important role in the chemistry of aryl-substituted cluster complexes. During an investigation of the chemical transformations of benzylidynetricobalt nonacarbonyl complexes, we discovered that $C_6H_5CCo_3(CO)_9$ and its o- and m-methyl and chloro derivatives react with acetyl chloride/aluminum chloride to give p-acetylated cluster complexes in good yield (Scheme 1). Similar Friedel-Crafts acylation was observed with the benzoyl chloride/aluminum chloride reagent, but not in the case of the Cl-substituted cluster complexes (49). Formylation of $C_6H_5CCo_3(CO)_9$ in the para position in low yield could be achieved with $CH_3OCHCl_2/AlCl_3$. The acetylation reactions, in particular, proceeded rapidly in high yield and under rather mild conditions. In contrast to the

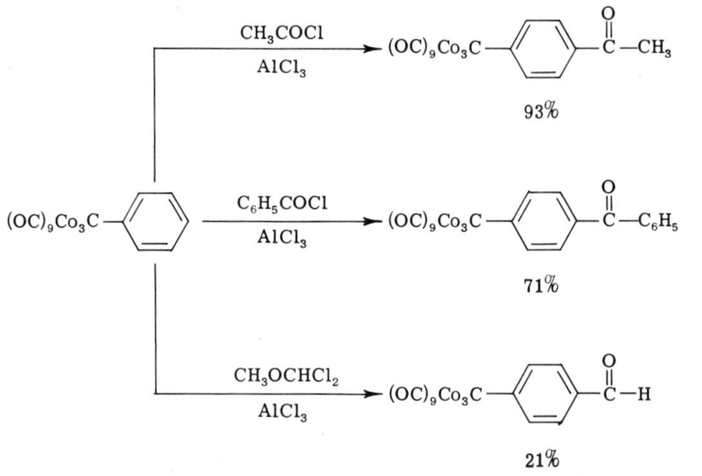

Scheme 1

high reactivity of $C_6H_5CCo_3(CO)_9$, the benzyl-substituted complex, $C_6H_5CH_2CCo_3(CO)_9$, did not appear to react with $CH_3COCl/AlCl_3$. These reactions all were complicated by the ability of the Lewis acid catalyst to coordinate to carbon monoxide ligands of the benzylidynetricobalt nonacarbonyls. Therefore, the reaction conditions used were of critical importance to the success or failure of the reaction studied (*49b*).

These reactions represent another route to organo-functional cluster complexes whose scope of application, however, is rather limited. Nevertheless, we were struck by the facility of these reactions in those cases where they proceeded in high yield, in particular, the acetylation of $C_6H_5CCo_3(CO)_9$ and *o*- and *p*-$CH_3C_6H_4CCo_3(CO)_9$. In order to obtain a more quantitative estimate of their reactivity, we carried out competition experiments in which a mixture of $C_6H_5CCo_3(CO)_9$ and another aromatic substrate was allowed to react with a deficiency of the $CH_3COCl/AlCl_3$ reagent in dichloromethane. The high reactivity of benzylidynetricobalt nonacarbonyl was immediately apparent when it was found that in its competition reaction with anisole, a rather nucleophilic benzene derivative, no acetylation product of the latter was formed. In a competition experiment with N,N-dimethylaniline, a more potent nucleophile, the acetylation product yields established that $k[\text{PhCCo}_3(\text{CO})_9]/k(C_6H_5NMe_2)$ for reaction with $CH_3COCl/AlCl_3$ under the conditions used is 1.3. A similar competition experiment with ferrocene, known as a *supernucleophile* in the Friedel-Crafts reaction (*50*), showed $C_6H_5CCo_3(CO)_9$ approximately as reactive as this organoiron complex $\{k[\text{PhCCo}_3(\text{CO})_9]/k(\text{ferrocene})$ values of 0.9–1.2, depending on reagent ratios and reaction times, assuming also in the calculations that ferrocene has ten reactive positions to only one in benzylidynetricobalt nonacarbonyl$\}$ (*49*). Thus benzylidynetricobalt nonacarbonyl is one of the most, if not the most, reactive monosubstituted benzenes in the Friedel-Crafts reaction.

This observation may be rationalized in terms of the possibilities for the stabilization of the charged intermediates in such reactions. The attack of the $CH_3COCl/AlCl_3$ reagent (as $CH_3CO^+AlCl_4^-$) at the para position of $C_6H_5CCo_3(CO)_9$ would lead to an ionic intermediate in which the positive charge can be displaced to the ring carbon which is bonded to the apical carbon atom of the cluster, as shown in structure III. Thus, further delocalization into the cluster would be possible. The net stabilization gained must be considerable, in view of the high reactivity of benzylidynetricobalt nonacarbonyl. Ortho substitution in these systems would lead to a cluster-stabilized intermediate (IV). However, the steric effect of the six carbon monoxide ligands disposed in the general direction of the apical carbon atom and its substituent would be expected to hinder or possibly completely prevent attack at an ortho position. In any case, meta substitution is not expected. As a result of these electronic and steric factors, reactions of $ArCCo_3(CO)_9$ complexes with the $RCOCl/AlCl_3$ reagents is limited to para-substitution, irrespective of electronic effects due to other substituent groups on the benzene ring. Stabilization of the type shown in structure III is, of course, not possible in the case of $C_6H_5CH_2CCo_3(CO)_9$, which explains its lack of reactivity toward $CH_3COCl/AlCl_3$.

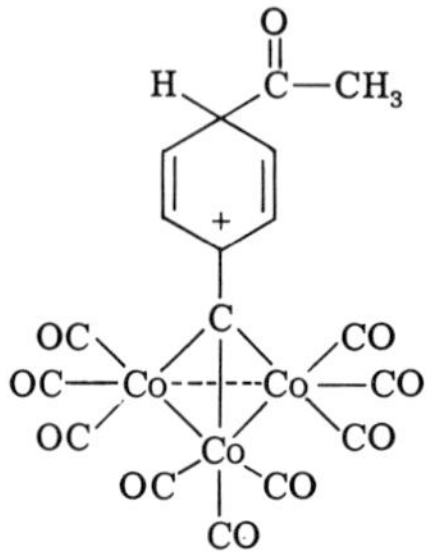

(III) 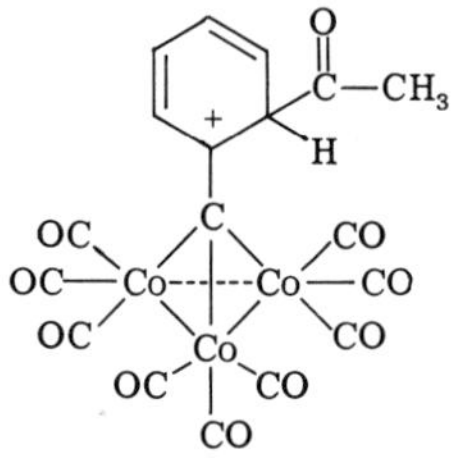 (IV)

The carbonyl stretching frequency of p-$CH_3C(O)C_6H_4CCo_3(CO)_9$ in dilute carbon tetrachloride solution was found to be 1685 cm^{-1}. Fuson et al. (51) had found that a good linear correlation existed between Hammett σ constants and ν (C=O) of substituted acetophenones in dilute CCl_4. From the position of the ν (C=O) of p-$CH_3C(O)C_6H_4CCo_3(CO)_9$ on their ν (C=O)/σ plot, we could estimate the σ constant for the $(OC)_9Co_3C$ substituent to be about -0.35. That this substituent is an electron donor with respect to a p-acetyl group in an attached benzene ring is not unexpected since charge delocalization into the cluster via complex V should be possible.

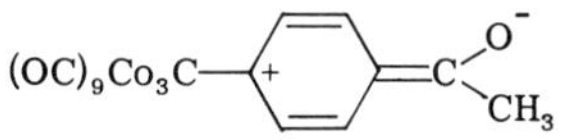

(V)

Other observations, which we and others have made during the course of investigations of the chemistry of alkylidynetricobalt nonacarbonyl complexes, may have been consequences of the high stability of a positive charge generated at a carbon atom α to the $(OC)_9Co_3C$ cluster. For instance, unexpected reduction of certain functional groups during organocobalt cluster preparations when these would be expected to appear on the α-carbon atom of the resulting complex could be a result of the generation of acidic conditions during work-up or even during the reaction itself. This might result in formation of a cluster-substituted carbonium ion whose ultimate fate (reduction, hydrolysis, destruction of the cluster) will be dictated by the further reaction and work-up conditions. In all such considerations, it must be kept in mind that $HCo(CO)_4$ could be formed during reaction or work-up and that this compound, in aqueous medium, is as strong an acid as HCl (52).

The fact that all attempts thus far to prepare alkylidynetricobalt nonacarbonyl complexes with an α-halogen substituent have been unsuccessful may also be a result of the ease of formation of cluster-substituted carbonium ions. Even very mild procedures, e.g., the action of lithium chloride on $(OC)_9Co_3CCH(CH_3)OSO_2C_6H_4CH_3$-$p$ in THF, resulted in decomposition of the cluster (39). Possibly the cluster carbonium ion is generated and is destroyed by way of halide ion attack at cobalt, which, as we show in the following discussion, must bear a substantial portion of the positive charge. Another reaction suggesting that the carbon atom at which C—O bond heterolysis in the alcohol had occurred was not the only site of electrophilic reactivity in these carbonium ions was that with triphenylphosphine. One might have expected this nucleophile to form stable phosphonium salts by reaction with the carbonium ion hexafluorophosphates at carbon, but its addition to a slurry of such a salt in dichloromethane resulted in vigorous gas evolution and complete decomposition.

The carbonium ion salts that we had prepared and studied could possibly be pictured as structure VI, but this representation does not seem adequate in view of their ease of formation, high stability, and rather low reactivity. These properties suggested that extensive delocalization of the positive charge originally generated at the carbon atom had occurred. Such delocalization, however, could not be accommodated by conventional inductive and resonance effects. To obtain more information on the question of structure and bonding of these novel organometallic carbonium ions, we began a study of their NMR spectra (53). This investigation is not yet completed, but the preliminary results provide interesting and useful information.

(VIa) R = H
(VIb) R = CH_3
(VIc) R = C_6H_5

Table XIII summarizes proton chemical shift data for the three carbonium ion salts (VIa, b, and c) and for the alcohols from which they were prepared by treatment with concentrated sulfuric acid or trifluoroacetic acid. It can be seen that the hydrogen atoms on the carbon α to the $(OC)_9Co_3C$ cluster become less shielded on going from the alcohols to the carbonium ions, but the shifts, $\Delta\delta$, are not at all large, compared to those observed in other less stabilized systems [e.g., Me_2CHOH to Me_2CH^+, $\Delta\delta = 8.5$ ppm (54)].

The ^{13}C NMR spectra of compounds VIa, b, and c were more informative. The chemical shifts of carbon resonances in ^{13}C NMR spectra have been used as a measure of the electron densities of the carbon atoms being studied (55), and so a comparison of the carbinyl carbon resonances of compounds VIa, b, and c with those of the alcohols from which they were

TABLE XIII

Proton Nuclear Magnetic Resonance Spectra of $(OC)_9Co_3CCH(R)OH$ and $(OC)_9Co_3CCHR^+X^{-a,b}$

Alcohol, δ (ppm)		Cation, δ (solvent) (ppm)	$\Delta\delta$ (ppm)
$(OC)_9Co_3CCH_2OH$		$(OC)_9Co_3CCH_2^+$	
CH_2	5.21	5.8 (H_2SO_4)	-0.6
		5.7 (CF_3CO_2H)	-0.5
$(OC)_9Co_3CCH(CH_3)OH$		$(OC)_9Co_3CCHCH_3^+$	
CH	5.4	6.7 (H_2SO_4)	-1.3
		6.9 (CF_3CO_2H)	-1.5
		6.9 (PF_6^- salt in CH_3NO_2)	-1.5
CH_3	1.8	2.4 (H_2SO_4)	-0.6
		2.5 (CF_3CO_2H)	-0.7
		2.4 (PF_6^- salt in CH_3NO_2)	-0.6
$(OC)_9Co_3CCH(C_6H_5)OH$		$(OC)_9Co_3CCHC_6H_5^+$	
CH	6.2	7.7 (H_2SO_4)	-1.5
		8.2 (CF_3CO_2H)	-2.0
C_6H_5	7.4	7.2 (H_2SO_4)	$+0.2$
		7.6 (CF_3CO_2H)	-0.2

[a] From Seyferth et al. (53).

[b] Alcohol spectra were obtained in chloroform-d and are referenced to internal tetramethylsilane. Cation spectra are referenced to tetramethylsilane contained in a capillary inside the NMR tube. The methyl group signal for the methyl-substituted carbonium ions (in nitromethane, sulfuric acid, and trifluoroacetic acid) was a doublet ($J = 7$ Hz). The methyne proton for these same carbonium ions appears as a quartet ($J = 7$ Hz). For the phenyl-substituted and unsubstituted carbonium ions, the methyne proton and methylene proton signals (respectively) were singlets.

derived should give an indication of the degree of charge delocalization in these carbonium ions. The data obtained are given in Table XIV. Noteworthy is how small the changes in the carbinyl carbon atom chemical shifts are on going from the alcohol to the carbonium ion. By way of comparison, $\Delta\delta$ for the carbinyl carbon atom when Me_2CHOH is converted to $Me_2CH^+SbF_6^-$ is 255.3 ppm (*55*).

A comparison with similar data for an alcohol/carbonium ion set where the latter is highly stabilized would be of interest. Ferrocene chemistry provides such examples. The extraordinarily high stability of ferrocenylmethyl carbonium ions is well documented and for some years has been the subject of much discussion and some controversy (*56*). We measured the ^{13}C NMR spectra of ferrocenylcarbinol, 1-ferrocenylethanol and ferrocenylphenylcarbinol and of their derived carbonium ions (VIIa, b, and c) obtained by dissolving the alcohols in concentrated sulfuric acid (*57*). The $\Delta\delta$ values for each are shown with the structures. They are slightly larger than those observed for the analogous cluster-substituted carbonium ions and so the latter are certainly of comparable stability to the ferrocenylmethyl carbonium ions and perhaps even more stable.

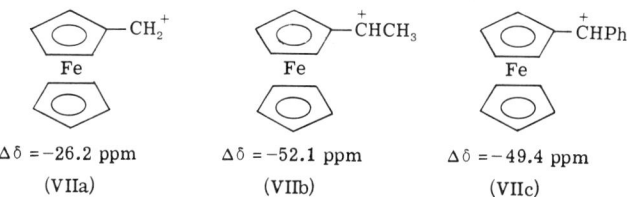

We note upon inspection of the data in Table XIV that the chemical shifts in these carbonium ions (VIa, b, and c) follow the normal order observed when substituents on a fully substituted carbon atom are varied from R = H (most shielded) to CH_3 to C_6H_5 (least shielded), rather than for substituents on an electron-deficient trivalent carbon atom, where the CH_3-substituted carbon atom is less shielded than the analogous phenyl-substituted carbon atom (*58*). This is a rather clear-cut indication that the carbinyl carbon atoms in complexes VIa, b, and c are nearly fully bonded and only slightly electron-deficient. The slight increase in shielding of the CH_3 carbon when VIb is formed from the alcohol and of the C-1 carbon of the phenyl group of VIc provides further confirmation of this.

The data in Table XIV suggest that the positive charge in complexes VIa, b, and c has been delocalized to a large extent into the cluster substituent. The observed slight increase in shielding of the carbon atoms of the carbon monoxide ligands when the alcohols are converted to the carbonium ions speak in favor of this view. If the cobalt atoms are more

TABLE XIV
Carbon-13 Nuclear Magnetic Resonance Spectra for $(OC)_9Co_3CCH(R)OH$ and $(OC)_9Co_3CCHR^+HSO_4^{-\ a,b}$

	Alcohol (ppm)	Cation (ppm)	Δ (ppm)
	$(OC)_9Co_3CCH_2OH$	$(OC)_9Co_3CCH_2^+HSO_4^-$	
Carbinyl carbon	77.6	91.1	−13.5
C≡O	200.6	192.7	+7.9
	$(OC)_9Co_3CCH(CH_3)OH$	$(OC)_9Co_3CCHCH_3^+HSO_4^-$	
Carbinyl carbon	82.5	119.9	−37.4
CH_3	28.5	26.2	+2.3
C≡O	200.5	193.2	+7.2
	$(OC)_9Co_3CCH(C_6H_5)OH$	$(OC)_9Co_3CCH(C_6H_5)^+HSO_4^-$	
Carbinyl carbon	88.8	124.5	−35.7
Ph—C_1	146.2	135.4	+10.8
Ph—C_2	128.8	129.9	−1.1
Ph—C_3	126.8	129.3	−2.5
Ph—C_4	128.6	132.6	−4.0
C≡O	200.0	192.4	+7.6

[a] From Seyferth et al. (53).
[b] Alcohol spectra are referenced to internal tetramethylsilane. Cation spectra are referenced to external tetramethylsilane through the ^{19}F lock signal. All cation spectra were obtained in concentrated sulfuric acid as solvent. All spectra were proton decoupled.

electron-deficient in VIa, b, and c than in the corresponding alcohols, then the carbon monoxide ligands would be expected to be bonded more tightly in the cations. The consequent movement of the CO carbon atoms to a position closer to the cobalt atoms might be expected to result in increased shielding due to the diamagnetic anisotropic shielding effect of the cobalt atom.

In our early efforts (53) we had been unable to see a ^{13}C signal of the apical carbon atom in the alcohols and the carbonium ions. In more recent work, we have located the ^{13}C signals due to the apical carbon atom in the carbonium ions in the region δ_C 255–275 ppm [VIa, 286.2 ppm; VIb, 273.5 ppm; VIc, 267.0 ppm; $(OC)_9Co_3CCMe_2^+$, 257.8 ppm; $(OC)_9Co_3CC_9H_{19}$-n^+, 273.3 ppm (59)]. Most of the alcohols from which these carbonium ions were derived were not soluble enough to enable the apical carbon atom resonance to be seen. The long-chain alcohol, $(OC)_9Co_3CCH(OH)C_9H_{19}$-$n$, however, was sufficiently soluble in CH_2Cl_2, and for its apical carbon atom $\delta_C = 306.0$ ppm. The ^{13}C NMR spectrum of a neat sample of the ester $(OC)_9Co_3CCO_2Et$ above its melting point of 46° showed $\delta_C = 258.4$ ppm for its apical carbon atom. These ^{13}C NMR shifts are far downfield from those of most carbon atoms in organic and organometallic compounds with the exception of alkyl-substituted carbonium ions (60), some transition metal carbene complexes [e.g., $(OC)_5CrC^*(OMe)Ph$, $\delta_{C^*} = 351.4$ ppm (61) and $(OC)_5WC^*(OEt)C\equiv CPh$, $\delta_{C^*} = 286.1$ ppm (62)] and carbyne complexes [e.g., $CH_3C^*\equiv W(CO)_4Cl$, $\delta_{C^*} = 288.8$ ppm (63) and $C_6H_5C\equiv C-C^*\equiv W(CO)_4Br$, $\delta_{C^*} = 230.6$ ppm (62)]. In our previous review on $RCCo_3(CO)_9$ complexes (1) we suggested that these compounds could, in principle, be regarded as adducts of a carbyne intermediate, $R\dot{C}$:, and the $Co_3(CO)_9$ unit, i.e., as triply bridged carbyne complexes. The similarity of the ^{13}C NMR chemical shifts of the apical carbon atom in $RCCo_3(CO)_9$ complexes and those of the carbyne carbon atoms in the nonbridged carbyne complexes of type $RC\equiv M(CO)_4X$, which were first reported by Fischer in 1973 (63), is striking and very likely not coincidental.

How then is the positive charge in the $(OC)_9Co_3C$-substituted carbonium ions delocalized into the cluster substituent? In our first preliminary report on this new class of carbonium ions (45), we suggested that their structure presents an especially favorable opportunity for σ-π hyperconjugation. It is this type of bonding, lateral overlap of a filled σ-bonding orbital of a metal–carbon bond and a vacant p orbital on an electron-deficient carbonium ion center β to the metal (VIII), that is believed responsible for the high

(VIII)

stability of carbonium ion centers β to both main group and transition metal atoms (64). In our cluster-substituted carbonium ions, such stabilization could make substantial contribution to their stability. It should, however, be recognized that the RC—C—Co bond angle in the neutral clusters of about 131° is too large to allow very effective lateral overlap of the type shown in complex VIII. To achieve better overlap, the substituent at the apical carbon atom in the carbonium ion might bend down toward the cobalt atom, as shown in compound IX. With 3 cobalt atoms giving equally good opportunity for such σ-π overlap, a very stable species, with a high concentration of positive charge at the cobalt atoms would be expected. There also is a good possibility that carbonium ions of type IX would be fluxional species. An alternate way of regarding these carbonium ions, (X) suggests itself when one considers that the ^{13}C NMR data for

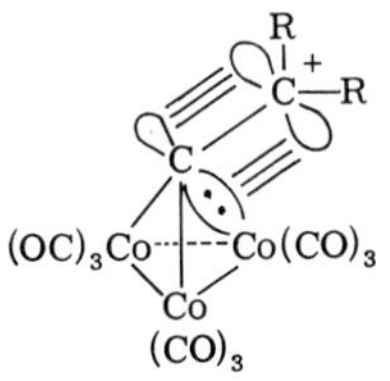

(IX) 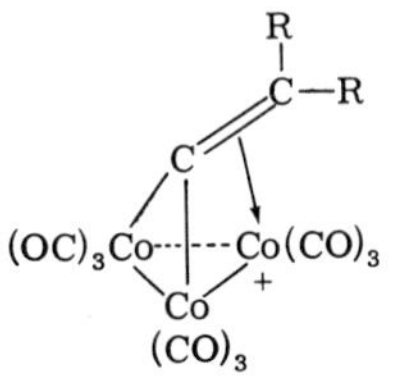   (X)

the α-carbon atoms in complexes VIa, b, and c show trends typical of olefinic carbon atoms (32). A structure of this type (XI) was suggested for the protonation product of the unsaturated osmium cluster (XII) by Deeming et al. (65). The proton NMR spectrum of complex XI showed two high field singlets of relative intensity 1:2 in $CDCl_3$—CF_3CO_2H at $-10°$ and this required an unsymmetrical structure.

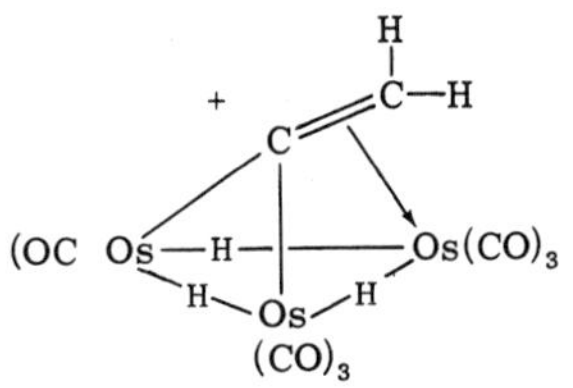

(XI) 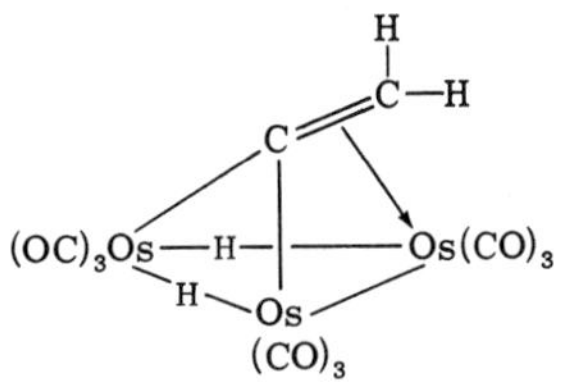   (XII)

The α-carbon atom resonances in the ^{13}C NMR spectra of the carbonium ions VIa, b, and c fall within the range reported for the sp^2 carbon atoms of olefins complexed to transition metal centers (66), but those of the apical carbon atom in these ions are far downfield from this range. In the case of the n-C_9H_{19}-substituted alcohol and carbonium ion cluster systems, $\Delta\delta_C$ on going from the alcohol to the carbonium ion is only 33 ppm and the δ_C values for all of the apical carbon atoms in the carbonium ion systems which we have studied are still within the range of δ_C values for apical carbon atoms in neutral $RCCo_3(CO)_9$ complexes. Thus far we have no spectroscopic evidence against the symmetrical arrangement shown in structure VI. However, intuition suggests that an unsymmetrical arrangement (IX or X) is more reasonable since it would optimize bonding in these carbonium ions. This question is receiving further attention.

V

DECOMPOSITION REACTIONS AND DERIVED SYNTHETIC APPLICATIONS OF ALKYLIDYNETRICOBALT NONACARBONYL COMPLEXES

One might hope that some of the transformations described in this review could be useful in synthetic chemistry outside the area of cluster chemistry. This brings up the question of how best to release the apical carbon atom and its substituent from the complex. Since the alkylidynetricobalt nonacarbonyls can be viewed as complexes of a "carbyne," $R\dot{C}$:, with the $Co_3(CO)_9$ moiety, it is of interest to see if their controlled decomposition could give an interceptable carbyne fragment. Fischer's newly discovered carbyne complexes of type $RC\equiv M(CO)_4X$ (X = halogen, M = Cr, Mo, W) decompose to give an acetylene, $RC\equiv CR$, as the organic product (67). It is, therefore, a very intriguing fact that the decomposition of some $RCCo_3(CO)_9$ complexes gives acetylenes (68) or acetylenedicobalt hexacarbonyl complexes (1, 69):

$$PhCCo_3(CO)_9 \xrightarrow{\text{diglyme, reflux}} \text{Co metal} + \text{some } PhC\equiv CPh \quad (39)$$

$$ArCCo_3(CO)_9 \xrightarrow{\text{MeOH, reflux}} (ArC_2Ar)Co_2(CO)_6 \quad (40)$$

The preparative sequence shown in Scheme 2, based on the latter reaction,

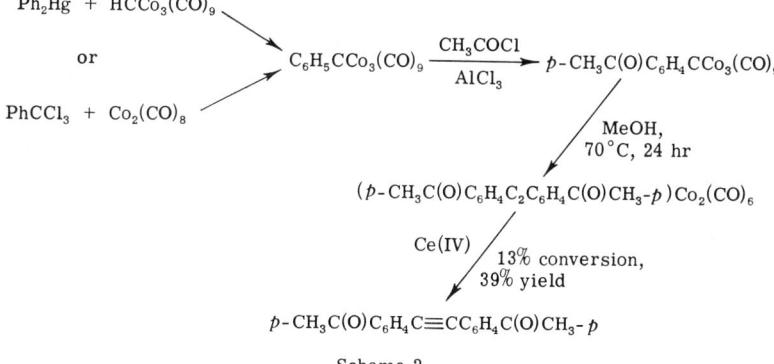

Scheme 2

was carried out in our laboratories (1). However, we know of no successful trapping of the $R\dot{C}$: fragment with an external reagent. Oxidation of two $RCCo_3(CO)_9$ complexes with a limited amount of ceric ammonium nitrate in ethanol gave substantial quantities of acetylenic product (*68*):

$$PhCH_2CCo_3(CO)_9 + \tfrac{1}{2}Ce(IV) \xrightarrow[\text{temp.}]{\text{EtOH, room}} \underset{\text{(3 parts)}}{PhCH_2C\equiv CCH_2Ph} + \underset{\text{(2 parts)}}{PhCH_2CO_2Et} +$$

$$\underset{\text{(1 part)}}{PhCH_2CHO} \quad (41)$$

$$PhCCo_3(CO)_9 + 1Ce(IV) \xrightarrow{\text{EtOH, reflux}} \underset{25\%}{PhC\equiv CPh} + PhCO_2Et \quad (42)$$

However, as we have shown previously (*5, 49a*), when an excess of ceric ammonium nitrate in aqueous acetone is used, oxidation of $RCCo_3(CO)_9$ complexes proceeds cleanly to the carboxylic acid, RCO_2H. Some substituted $ArCCo_3(CO)_9$ were resistant to such complete oxidation by $Ce(IV)$, and in those cases potassium permanganate in aqueous acetone was found to produce the carboxylic acids in 50-60% yields (*5*). In one of the earliest reports on $RCCo_3(CO)_9$ complexes (*3a*), their oxidation to the carboxylic acid with hydrogen peroxide was mentioned. An interesting oxidative degradation of $RCCo_3(CO)_9$ complexes was reported by Japanese workers (*70*):

$$RCCo_3(CO)_9 \xrightarrow{\text{R'OH, 70°C, O}_2\text{ bubbled in}} \underset{40-80\%}{RCH(CO_2R')_2} + Co_2O_3 \quad (43)$$

This reaction has potential for useful application in organic synthesis. Oxidation of alkylidynetricobalt nonacarbonyl complexes with halogens was the key reaction that led Krüerke and Hübel to the correct assignment of their structure (*3a*):

$$PhCH_2CCo_3(CO)_9 + Br_2 \xrightarrow{CCl_4} PhCH_2CBr_3 + 3CoBr_2 + 9CO \quad (44)$$

The action of iodine also causes nearly quantitative release of carbon monoxide from such complexes, but the organic triiodides presumably produced were not isolated (*7*).

A few other reactions of these cluster complexes in which the $R\dot{C}$: fragment is converted to organic products have been reported and are summarized as follows:

$$CH_3CCo_3(CO)_9 \xrightarrow[50\%]{\text{NaOMe/MeOH}} CH_3CH(CO_2Me)_2 \quad (70) \quad (45)$$

[The action of methanolic sodium hydroxide on $RCCo_3(CO)_9$ compounds had been shown previously to result in formation of the $Co(CO)_4^-$ ion $(2, 7)$].

$$CH_3CCo_3(CO)_9 \xrightarrow[C_6H_6]{H_2,\ CO\ (elevated\ P)} C_2H_5CHO \quad (70) \quad 50\% \tag{46}$$

$$CH_3CCo_3(CO)_9 \xrightarrow[100\ atm\ CO]{n\text{-}Bu_3P,\ MeOH} C_2H_5CO_2Me\ + \\ 14\%$$

$$CH_3CH(CO_2Me)_2 + (CH_2CO_2Me)_2 \quad (70) \quad (47) \\ 61\% \qquad\qquad 5\%$$

$$PhCH_2CCo_3(CO)_9 + \text{excess}\ NaBH_4 \xrightarrow{DME} PhCH_2CH_3 \quad (68) \tag{48}$$

$$(OC)_9Co_3CCH_2CH_2CO_2H \xrightarrow{H_2O} HO_2C(CH_2)_3CO_2H \quad (71) \tag{49}$$

$$(OC)_9Co_3CCH{=}CHCO_2H \xrightarrow{H_2O} CH_2{=}CHCH_2CO_2H + HO_2C(CH_2)_3CO_2H\ + \\ HO_2CCH_2CH{=}CHCO_2H \quad (71) \quad (50)$$

The limited number of reactions in which the $\text{R}\dot{\text{C}}$: fragment is converted to organic products, which has been studied thus far, consists of rather simple ones. This area of $RCCo_3(CO)_9$ conversions merits further detailed exploration on a more sophisticated level since its development will provide the basis for synthetic applications that will make these complexes useful to others.

Catalytic applications of $RCCo_3(CO)_9$ complexes have been sought, and two such reactions have been reported. It was found that disubstituted acetylenes are trimerized to hexasubstituted benzenes when heated in the presence of alkylidynetricobalt nonacarbonyls (72):

$$PhCCo_3(CO)_9 + CH_3C{\equiv}CCH_3 \xrightarrow[\text{hexane}]{160°C,\ 35\ hr} (CH_3)_6C_6 \quad (51) \\ 20\%$$

$$CH_3CCo_3(CO)_9 + PhC{\equiv}CPh \xrightarrow{160°C,\ 14\ hr} Ph_6C_6 \quad (52) \\ 72\%$$

The $RCCo_3(CO)_9$ "catalysts" in these reactions, however, only appear to be sources of simpler cobalt carbonyl species through their thermolysis, and it is the latter which cause the acetylene trimerization through well-established routes (73). Indeed, in some cases acetylene-derived cobalt carbonyl complexes are obtained (72):

$$(OC)_9Co_3CCO_2Me + CH_3C{\equiv}CH \xrightarrow{100°C,\ 4\ hr} (CH_3C_2H)Co_2(CO)_6 \tag{53}$$

Various $RCCo_3(CO)_9$ complexes have been shown to be initiators of olefin polymerization $(74, 75)$. With the complexes in which R = H, Cl, or Br, one might speculate that the initiating process involves reaction at the apical carbon atom, in view of the known $HCCo_3(CO)_9$/olefin reactions (*16, 17*) and reported reactions of $BrCCo_3(CO)_9$ with olefins (*17*), e.g.,

$$BrCCo_3(CO)_9 + C_2H_4 \longrightarrow C_2H_5CCo_3(CO)_9 +$$
$$CH_3CH=C(CH_3)CCo_3(CO)_9 \quad (54)$$

$$BrCCo_3(CO)_9 + CH_2=CHCO_2Me \longrightarrow (OC)_9Co_3CCH_2CH_2CCO_2Me \quad (55)$$

However, the fact that complexes in which R = Ph, F, Me_2CH, and C_2F_5 also initiate polymerization of acrylonitrile suggests that chemistry at cobalt may be involved instead. The displacement of carbon monoxide ligands by diolefins to give stable complexes of type $RCCo_3(CO)_7$(norbornadiene) in the case of norbornadiene has been reported (*18, 76*), so that such a process is entirely possible.

VI
CONCLUDING REMARKS

This review cannot be exhaustive in view of space limitations. One would like to discuss other reactions of alkylidynetricobalt nonacarbonyl complexes occurring at the apical carbon atom, such as the remarkable conversion of $ClCCo_3(CO)_9$ to $(OC)_9Co_3C-CCo_3(CO)_9$ by its reaction with triphenylarsine at 100° (*77*), the reaction of $ClCCo_3(CO)_9$ with cobalt tetracarbonyl anion in THF at 70°C, which gives the $[Co_6(CO)_{15}C]^{2-}$ carbide cluster complex (*78*), the Ullmann-type synthesis of benzylidynetricobalt nonacarbonyls from $BrCCo_3(CO)_9$, which unfortunately gives only moderate yields (*5*), and others. Completely omitted from this review have been the discussion of chemistry at cobalt in these complexes and more detailed consideration of the spectroscopic, mass spectrometric, electrochemical, and structural studies of the $RCCo_3(CO)_9$ complexes that have been reported. However, these topics have been covered adequately in reviews by others (*79, 80*).

We have mentioned only in passing other cluster complexes in which a tetrahedral core of 1 carbon and 3 metal atoms is present. Such complexes in which the metal atoms are nickel, ruthenium, and osmium have been prepared: XIII (*81*), XIV (*82, 83*), and XV (*65, 82*). Their chemistry remains largely unexplored, except for the transformations of compound XV in strong acid medium which we mentioned in the previous section.

Related structurally to the $RCCo_3(CO)_9$ complexes are the acetylenedicobalt hexacarbonyls which have the structure with a pseudotetrahedral

(XIII)   (XIV) M = Ru, Os   (XV) M = Ru, Os

C_2Co_2 core shown in complex XVI (*84*). These have been covered quite thoroughly in the review literature (*5, 85*), but the similarities between some of the chemistry of the $RCCo_3(CO)_9$ and $(RC_2R)Co_2(CO)_6$ complexes are worth noting. Thus, $(HC_2H)Co_2(CO)_6$ can be converted to a mixture of $(PhC_2H)Co_2(CO)_6$ and $(PhC_2Ph)Co_2(CO)_6$ by reaction with diphenylmercury (*5*), in a reaction completely analogous to the phenylation of $HCCo_3(CO)_9$ with this reagent. Diphenylacetylenedicobalt hexacarbonyl, like $PhCCo_3(CO)_9$, is readily acetylated in the para position of both of its phenyl rings to give either $[p\text{-}CH_3C(O)C_6H_4C_2C_6H_5]Co_2(CO)_6$ or $[p\text{-}CH_3C(O)C_6H_4C_2C_6H_4C(O)CH_3\text{-}p]Co_2(CO)_6$ in high yield, depending on reaction stoichiometry (*49a, 86*). Finally, evidence has been presented by Nicholas and Pettit (*87*) that carbonium ions derived from dicobalt hexacarbonyl complexes of propargylic alcohols, e.g., compound XVII, are much more readily formed and more stable than analogous carbonium ions derived from the free ligand. Thus, the $Co_2(CO)_6$ group stabilizes α-carbonium ion centers, but not as effectively as the $(OC)_9Co_3C$ group, since compound XVII and other $Co_2(CO)_6$-complexed carbonium ion salts cannot be isolated. The analogy between $RCCo_3(CO)_9$ and $(RC_2R)Co_2$-$(CO)_6$ complexes, however, should not be overstressed.

The acetylenedicobalt hexacarbonyls have been described in terms of

(XVI)   (XVII)

complexes of the first excited state of the acetylene and the $Co_2(CO)_6$ unit (88), and we note that the ^{13}C NMR chemical shifts of the cluster carbon atoms of these complexes are quite different from those of the apical carbon atom in the $RCCo_3(CO)_9$ cluster compounds. As we have seen, δ_C for the apical carbon in neutral alkylidynetricobalt nonacarbonyls is in the range 250–310 ppm. Diphenylacetylenedicobalt hexacarbonyl, on the other hand, shows δ_C for the carbon atoms in its C_2Co_2 core at 89.6 ppm (89), which is within the range of acetylenic carbon atom shieldings (58).

From the discussion of the organic chemistry of the alkylidynetricobalt nonacarbonyl complexes given in this review, it is apparent that this has been a fruitful area of research. We think that it shall continue to be so. Admittedly, the sensitivity of these cluster complexes to diverse bases, nucleophiles, and oxidizing agents will seriously limit the chemistry that can be carried out, but even with these limitations it should be possible to continue a broad development of the organo-functional interconversions of these complexes.

At this point, most of the functional substituents have been introduced into the cluster complexes either at the apical carbon atom, $(OC)_9Co_3CZ$, or at the carbon atom α to it, $(OC)_9Co_3CCZRR'$. Any functional group attached to the apical carbon atom is very sterically hindered and as a result will have rather limited chemical reactivity. Functional groups at the α-carbon atom will be less hindered, but some may be easily lost due to the ease with which α-carbonium ions are formed. Clearly, an investigation of the organo-functional chemistry at carbon atoms more remote from the cluster $(OC)_9Co_3C—Y—CZRR'$ (Y is a difunctional organic unit) should be of interest and, perhaps, more useful in terms of possible applications.

As has been pointed out, very little is known concerning reaction mechanisms in the area of organocobalt cluster chemistry. The main reactions by which these complexes are formed are only poorly understood, as are the reactions by which methylidyne- and halomethylidynetricobalt nonacarbonyls are alkylated and arylated. We can only guess about the mechanism of the remarkable aluminum chloride-induced conversion of $(OC)_9Co_3CCl$ to $(OC)_9Co_3CO^+AlCl_4^-$, and the mechanisms of the decomposition and the oxidation of these clusters are not known. A better knowledge of mechanisms in organocobalt cluster chemistry most certainly would facilitate the development of the chemistry of these complexes.

We were first drawn into studies on alkylidynetricobalt nonacarbonyl complexes because in these one is dealing with a carbon atom in a most unusual environment. We felt that this novel class of complexes would show some rather interesting organic and organometallic chemistry and we have not been disappointed in this expectation. We believe that more interesting chemistry of the $RCCo_3(CO)_9$ and related cluster complexes remains to be uncovered and we are continuing our efforts in this area.

ACKNOWLEDGMENTS

My pre- and postdoctoral co-workers who carried out the research reviewed here, are listed on the title page of this chapter. I am indebted and grateful to them for their dedicated, enthusiastic, and skillful efforts and for their important contributions of original ideas that resulted in rapid development of organo-functional organocobalt cluster chemistry. My co-workers and I are grateful to the National Science Foundation for generous support of this work (NSF Grant GP 31429X).

REFERENCES

1. D. Seyferth, J. E. Hallgren, R. J. Spohn, A. T. Wehman, and G. H. Williams, *Spec. Lect., 23rd Int. Congr. Pure Appl. Chem.*, 1971 **6**, 133–149.
2. R. Markby, I. Wender, R. A. Friedel, F. A. Cotton, and H. W. Sternberg, *J. Amer. Chem. Soc.* **80**, 6529 (1958).
3a. U. Krüerke and W. Hübel, *Chem. Ind. (London)* 1264 (1959).
3b. W. T. Dent, L. A. Duncanson, R. G. Guy, H. W. B. Reed, and B. L. Shaw, *Proc. Chem. Soc.* 169 (1961).
4. P. W. Sutton and L. F. Dahl, *J. Amer. Chem. Soc.* **89**, 261 (1967).
5. D. Seyferth, J. E. Hallgren, R. J. Spohn, G. H. Williams, M. O. Nestle, and P. L. K. Hung, *J. Organometal. Chem.* **65**, 99 (1974).
6. R. S. Dickson and P. J. Fraser, *Advan. Organometal. Chem.* **12**, 323 (1974) (review).
7. R. Ercoli, E. Santambrogio, and G. Tettamanti-Casagrande, *Chim. Ind. (Milano)* **44**, 1344 (1962).
8. G. Bor, L. Markó, and B. Markó, *Acta Chim. Acad. Sci. Hung.* **27**, 395 (1961); *Chem. Ber.* **95**, 333 (1962).
9. D. Seyferth, J. E. Hallgren, and P. L. K. Hung, *J. Organometal. Chem.* **50**, 265 (1973).
10. D. Seyferth, C. S. Eschbach, G. H. Williams, P. L. K. Hung, and Y. M. Cheng, *J. Organometal. Chem.* **78**, C13 (1974).
11. G. Pályi, F. Piacenti, M. Bianci, and E. Benedetti, *Acta Chim. Sci. Hung.* **66**, 127 (1970).
12. R. Dolby and B. H. Robinson, *J. Chem. Soc., Dalton Trans.* 1794 (1973).
13. P. J. Krusic, private communication.
14. T. W. Matheson, B. M. Peake, B. H. Robinson, J. Simpson, and D. J. Watson, *J. Chem. Soc., Chem. Commun.* 894 (1973).
15. R. Dolby and B. H. Robinson, *J. Chem. Soc., Dalton Trans.* 2046 (1972).
16. D. Seyferth and J. E. Hallgren, *J. Organometal. Chem.* **49**, C41 (1973).
17. N. Sakamoto, T. Kitamura, and T. Joh, *Chem. Lett.* 583 (1973).
18. T. Kamijo, T. Kitamura, N. Sakamoto, and T. Joh, *J. Organometal. Chem.* **54**, 265 (1973).
19a. G. Albanese and G. Gavezzotti, *Chim. Ind. (Milano)* **47**, 1322 (1965).

19b. G. Pályi and G. Váradi, *J. Organometal. Chem.* **86**, 119 (1975).
20. B. L. Booth, R. N. Haszeldine, P. R. Mitchell, and J. J. Cox, *J. Chem. Soc. A* 691 (1969).
21. E. Klumpp, G. Bor, and L. Markó, *Chem. Ber.* **100**, 1451 (1967).
22a. F. Klanberg, W. B. Askew, and L. J. Guggenberger, *Inorg. Chem.* **7**, 2265 (1968).
22b. G. Schmid and V. Bätzel, *J. Organometal. Chem.* **46**, 149 (1972).
23. W. M. Ingle, G. Preti, and A. G. MacDiarmid, *J. Chem. Soc., Chem. Commun.* 497 (1973).
24. S. A. Fieldhouse, B. H. Freeland, C. D. M. Mann, and R. J. O'Brien, *J. Chem. Soc., Chem. Commun.* 181 (1970).
25. D. Seyferth, R. J. Spohn, M. R. Churchill, K. Gold, and F. R. Scholer, *J. Organometal. Chem.* **23**, 237 (1970).
26. R. Ercoli, *Chim. Ind. (Milano)* **44**, 565 (1962).
27. D. Seyferth, J. E. Hallgren, and C. S. Eschbach, *J. Amer. Chem. Soc.* **96**, 1730 (1974).
28. (a) L. P. Hammett and A. J. Deyrup, *J. Amer. Chem. Soc.* **55**, 1900 (1933); (b) H. P. Treffers and L. P. Hammett, *ibid.* **59**, 1708 (1937).
29. M. S. Newman, *J. Amer. Chem. Soc.* **63**, 2431 (1941).
30. J. E. Hallgren, Ph.D. Thesis, Mass. Inst. of Technol., 1972.
31. D. Seyferth and G. H. Williams, *J. Organometal. Chem.* **38**, C11 (1972).
32. G. H. Williams, Ph.D. Thesis, Mass. Inst. of Technol., 1973.
33. K. Dey, C. Eaborn and D. R. M. Walton, *Organometal. Chem. Syn.* **1**, 151 (1970/1971).
34. D. Seyferth and G. H. Williams, *J. Organometal. Chem.* **38**, C11 (1972).
35. J. C. Kotz and D. G. Pedrotty, *Organometal. Chem. Rev., Sect. A* **4**, 479 (1969).
36. D. F. Shriver, *Accounts Chem. Res.* **3**, 231 (1970).
37. A. Alich, N. J. Nelson, and D. F. Shriver, *J. Chem. Soc., Chem. Commun.* 254 (1971).
38. R. Dolby, T. W. Matheson, B. K. Nicholson, B. H. Robinson, and J. Simpson, *J. Organometal. Chem.* **43**, C13 (1972).
39. D. Seyferth, G. H. Williams, P. L. K. Hung, and J. E. Hallgren, *J. Organometal. Chem.* **71**, 97 (1974).
40. D. N. Kursanov, Z. N. Parnes, and N. M. Loim, *Synthesis* 633 (1974).
41. H. Gilman and D. Wittenberg, *J. Org. Chem.* **23**, 501 (1958).
42. R. Calas, E. Frainnet and J. Bonastre, *C.R. Acad. Sci.* **251**, 2987 (1960); R. Calas, *Pure Appl. Chem.* **13**, 61 (1966).
43. S. I. Sadykh-Zade and A. D. Petrov, *Zh. Obshch. Khim.* **29**, 3194 (1959).
44. I. Ojima, M. Nihonyanagi, and Y. Nagai, *J. Chem. Soc., Chem. Commun.* 938 (1972).
45. D. Seyferth, G. H. Williams, and J. E. Hallgren, *J. Amer. Chem. Soc.* **95**, 266 (1973).
46. D. Seyferth, C. S. Eschbach, G. H. Williams, P. L. K. Hung, and Y. M. Cheng, *J. Organometal. Chem.* **78**, C13 (1974).
47. C. Eaborn, "Organosilicon Compounds," pp. 140–143. Butterworths, London, 1960.
48. D. A. Young, *Inorg. Chem.* **12**, 482 (1973).
49. (a) D. Seyferth and A. T. Wehman, *J. Amer. Chem. Soc.* **92**, 5520 (1970); (b) D. Seyferth, G. H. Williams, A. T. Wehman, and M. O. Nestle, *ibid.* **97**, 2107 (1975).
50. M. Rosenblum, "Chemistry of the Iron Group Metallocenes," Chapter 4. Wiley (Interscience), New York, 1965.
51. N. Fuson, M.-L. Josien, and E. M. Shelton, *J. Amer. Chem. Soc.* **76**, 2526 (1954).
52. H. W. Sternberg, I. Wender, R. A. Friedel, and M. Orchin, *J. Amer. Chem. Soc.* **75**, 2717 (1953).

53. D. Seyferth, G. H. Williams, and D. D. Traficante, *J. Amer. Chem. Soc.* **96**, 604 (1974).
54. G. A. Olah, E. B. Baker, J. C. Evans, W. S. Tolgyesi, J. S. McIntyre, and I. J. Bastien, *J. Amer. Chem. Soc.* **86**, 1360 (1964).
55. G. A. Olah, *Science* **168**, 1298 (1970).
56. J. J. Dannenberg, M. K. Levenberg, and J. H. Richards, *Tetrahedron* **29**, 1575 (1973), and earlier references cited therein.
57. G. H. Williams, D.D . Traficante, and D. Seyferth, *J. Organometal. Chem.* **60**, C53 (1973).
58. J. B. Stothers, "Carbon-13 NMR Spectroscopy." Academic Press, New York, 1972.
59. D. Seyferth, C. S. Eschbach, and M. O. Nestle, *J. Organometal. Chem.* **97**, C11 (1975).
60. See Stothers (58), pp. 217–238.
61. C. G. Kreiter and V. Formacek, *Angew. Chem.* **84**, 155 (1972).
62. F. H. Köhler, H. J. Kalder, and E. O. Fischer, *J. Organometal. Chem.* **85**, C19 (1975).
63. E. O. Fischer, G. Kreis, C. G. Kreiter, J. Müller, G. Huttner, and H. Lorenz, *Angew. Chem.* **85**, 618 (1973).
64. T. G. Traylor, H. J. Berwin, J. Jerkunica, and M. L. Hall, *Pure Appl. Chem.* **30**, 599 (1972).
65. A. J. Deeming, S. Hasso, M. Underhill, A. J. Canty, B. F. G. Johnson, W. G. Jackson, J. Lewis, and T. W. Matheson, *J. Chem. Soc., Chem. Commun.* 807 (1974).
66. B. E. Mann, *Advan. Organometal. Chem.* **12**, 135 (1974).
67. E. O. Fischer, *Angew. Chem.* **86**, 651 (1974).
68. I. U. Khand, G. R. Knox, P. L. Pauson, and W. E. Watts, *J. Organometal. Chem.* **73**, 383 (1974).
69. A. T. Wehman, unpublished work.
70. K. Tominaga, N. Yamagami, and H. Wakamatsu, *Tetrahedron Lett.* 2217 (1970).
71. G. Albanesi and E. Gavezotti, *Atti Accad. Naz. Lincei, Rend., Cl. Sci. Fis. Mat. Nat.* **41**, 497 (1966).
72. R. S. Dickson and G. R. Tailby, *Aust. J. Chem.* **23**, 229 (1970).
73. C. Hoogzand and W. Hübel, *in* "Organic Syntheses via Metal Carbonyls" (I. Wender and P. Pino, eds.), Vol. 1, pp. 343–371. Wiley (Interscience), New York, 1968.
74. C. H. Bamford, C. G. Eastmond, and W. R. Maltman, *Trans. Faraday Soc.* **60**, 1432 (1964).
75. G. Pályi, F. Baumgartner, and I. Czajlik, *J. Organometal. Chem.* **49**, C85 (1973).
76. P. A. Elder and B. H. Robinson, *J. Organometal. Chem.* **36**, C45 (1972).
77. T. W. Matheson and B. H. Robinson, *J. Chem. Soc. A* 1457 (1971).
78. V. G. Albano, P. Chini, S. Martinengo, M. Sansoni, and D. Strumbolo, *J. Chem. Soc., Chem. Commun.* 299 (1974).
79. G. Pályi, F. Piacenti, and L. Markó, *Inorg. Chim. Acta Rev.* **4**, 109 (1970).
80. B. R. Penfold and B. H. Robinson, *Accounts Chem. Res.* **6**, 73 (1973).
81. T. I. Voyevodskaya, I. M. Pribytkova, and Yu. A. Ustynyuk, *J. Organometal. Chem.* **37**, 187 (1972).
82. A. J. Deeming and M. Underhill, *J. Chem. Soc., Chem. Commun.* 277 (1973); *J. Chem. Soc., Dalton Trans.* 1415 (1974).
83. A. J. Canty, B. F. G. Johnson, J. Lewis, and J. R. Norton, *J. Chem. Soc., Chem. Commun.* 1331 (1972).
84. W. S. Sly, *J. Amer. Chem. Soc.* **81**, 18 (1959).

85. W. Hübel, in "Organic Syntheses via Metal Carbonyls" (I. Wender and P. Pino, eds.), Vol. 1, pp. 273–342. Wiley (Interscience), New York, 1968.
86. D. Seyferth, M. O. Nestle, and A. T. Wehman, *J. Amer. Chem. Soc.* **97,** 7417 (1975).
87. K. M. Nicholas and R. Pettit, *J. Organometal. Chem.* **44,** C21 (1972).
88. (a) Y. Iwashita, F. Tamura, and A. Nakamura, *Inorg. Chem.* **8,** 1179 (1969); (b) Y. Iwashita, *ibid.* **9,** 1178 (1970); (c) Y. Iwashita, A. Ishikawa, and M. Kainosho, *Spectrochim. Acta, Part A* **27,** 271 (1971).
89. L. J. Todd and J. R. Wilkinson, *J. Organometal. Chem.* **80,** C31 (1974).

Ten Years of Metallocarboranes

KENNETH P. CALLAHAN

Metcalf Research Laboratory
Department of Chemistry
Brown University
Providence, Rhode Island

M. FREDERICK HAWTHORNE

*Department of Chemistry**
University of California
Los Angeles, California

I. Introduction	145
II. Metallocarboranes: Synthetic Methods	150
A. Preparation from *nido*-Carborane Anions	150
B. Preparation by Polyhedral Expansion	151
C. Preparation by Polyhedral Contraction	152
D. Preparation by Polyhedral Subrogation	153
E. Preparation by Thermal Metal Transfer	153
III. Twelve-Vertex Metallocarboranes	155
A. Monometallic Complexes with Identical Carborane Ligands	155
B. Monometallic Complexes with Different Carborane Ligands	161
C. Mixed-Ligand Complexes	163
D. Bimetallic Complexes	166
IV. Thirteen-Vertex Metallocarboranes	167
V. Fourteen-Vertex Metallocarboranes	171
VI. Eleven-Vertex Metallocarboranes	171
A. Monometallic Complexes	171
B. Bimetallic Complexes	173
VII. Ten-Vertex Metallocarboranes	175
VIII. Nine-Vertex Metallocarboranes	178
IX. Oxidative Addition to B—H Bonds	180
X. Metallocarboranes in Homogeneous Catalysis	182
References	183

I

INTRODUCTION

In the early 1960s, the chemistry of the boron hydrides had been extended not only to include a remarkable number of new parent boranes having diverse structures, but the polyhedral $B_{10}H_{10}^{2-}$ and $B_{12}H_{12}^{2-}$ ions

---

* Contribution No. 3455.

and two isomeric $C_2B_{10}H_{12}$ carboranes as well.[1] Both $B_{10}H_{10}^{2-}$ and 1,2- (or *ortho-*) $C_2B_{10}H_{12}$ were obtained from decaborane (14), $B_{10}H_{14}$, whereas the 1,7- (or *meta-*) isomer of $C_2B_{10}H_{12}$ was prepared by thermal rearrangement of the 1,2-isomer at 400° to 600C° (Fig. 1) (*31, 39, 57, 65*).

As later work would show (*47, 77*), an entire series of polyhedral $B_nH_n^{2-}$ and isoelectronic $C_2B_{n-2}H_n$ carboranes were synthetically accessible for $n = 6$ through 12. The generally decreased chemical reactivity of these polyhedral species over that of the boron hydrides suggested that they

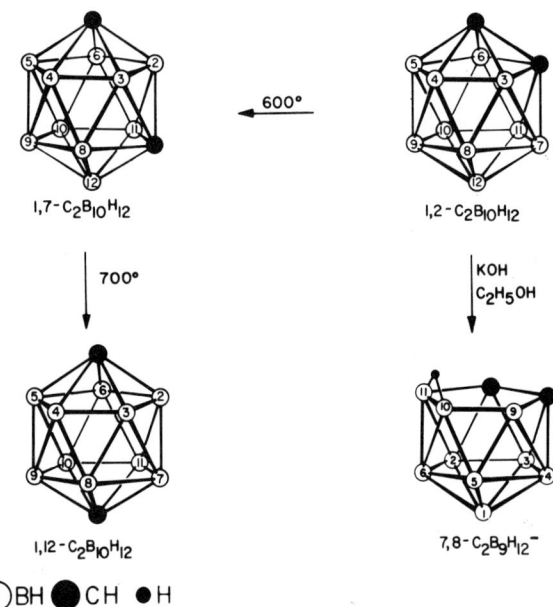

FIG. 1. Structures and numbering of the three isomeric icosahedral carboranes, and the degradation of $1,2\text{-}C_2B_{10}H_{12}$ to $7,8\text{-}C_2B_9H_{12}^-$. The bridging hydrogen is shown in one of the two equivalent bridging positions.

[1] Numbering of polyhedral positions follows the latest IUPAC-approved scheme as published in *Pure Appl. Chem.*, **30**, 683 (1972). The following definitions apply to the various descriptions of polyhedra used in the text: *closo* refers to a borane, carborane, or metallocarborane polyhedron that has a closed, fully triangulated geometry; *nido* refers to a polyhedron the geometry of which can be described as a closo polyhedron from which one vertex (frequently one of high coordination number) has been removed; *commo* refers to metallocarborane complexes in which one vertex, generally a transition metal, is shared between two polyhedra. Coordination numbers of polyhedral vertices are calculated from the number of nearest-neighbor atoms and do not imply the presence of discrete 2-electron chemical bonds between these atoms.

were probably stabilized by three-dimensional electron delocalization and hence were representative "aromatic" members of the boron hydride family. Indeed, Lipscomb (*71, 72*) and co-workers have carried out a series of molecular orbital treatments of selected polyhedral ions and carboranes which adequately explains the bonding present in these species. In all cases, a polyhedron having n number of vertices requires $n + 1$ electron pairs delocalized in an equal number of extended bonding orbitals to achieve polyhedral cage bonding.

Although the icosahedral 1,2- and 1,7-$C_2B_{10}H_{12}$ carboranes were found to be quite stable at high temperatures and toward most common reagents (*31, 41, 57*), strong base in the presence of a protonic solvent caused a specific degradation reaction (*97*) which cleanly removed a BH vertex from the icosahedron to produce the corresponding $C_2B_9H_{12}^-$ ion (Fig. 1):

$$1,2\text{-}C_2B_{10}H_{12} + RO^- + 2ROH \longrightarrow 7,8\text{-}C_2B_9H_{12}^- + B(OR)_3 + H_2 \quad (1)$$

It was correctly assumed and later determined unequivocally (*64*) that in each of these two reactions the BH vertex that was removed was always one of the two equivalent vertices found as the nearest neighbors of the two equivalent CH vertices. The fact that the carbon atoms in polyhedral surfaces of $C_2B_{n-2}H_n$ carboranes are electronic counterparts of boron atoms in the corresponding isoelectronic $B_nH_n^{2-}$ polyhedral ions (*62*) requires these carbon atoms to resemble C^+, a species present in carbonium ions. Consequently, the BH vertices which are nearest neighbors of two such carbon atoms will be activated for nucleophilic attack by base through the advent of a strong inductive effect.

The assumption that the isomeric 7,8- and 7,9-$C_2B_9H_{12}^-$ ions structurally resembled eleven-particle fragments of an icosahedron, coupled with their known empirical formulas, suggested that the twelfth hydrogen atom was present in a three-center B—H—B bridge bond located in the periphery of the open five-membered face (see Fig. 1). Simplified molecular orbital considerations suggested that the removal of this bridge hydrogen atom as a proton would generate a 7,8- or 7,9-$C_2B_9H_{11}^{2-}$ ion having 6 delocalized electrons in the open five-membered face. The orbitals in which these 6 electrons were distributed could, to a first approximation, be considered as sp^3-like and pointed toward the unoccupied vertex of the original carborane icosahedron. This disposition of delocalized electrons (Fig. 2) should closely resemble the ubiquitous cyclopentadienide ion, a constituent in a large number of organometallic compounds. As a result of this observation, the twelfth hydrogen atom in the 7,8-$C_2B_9H_{12}^-$ ion was successfully removed by treatment with strong bases such as sodium hydride (*55*). Reaction of the resulting 7,8-$C_2B_9H_{11}^{2-}$ ion with iron(II) chloride produced

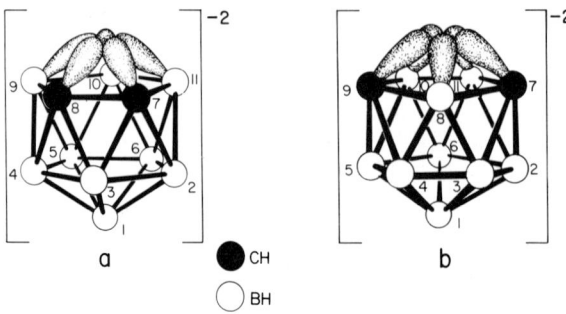

FIG. 2. Schematic representation of the sp^3-like bonding orbitals in (a) 7,8-$C_2B_9H_{11}^{2-}$ and (b) 7,9-$C_2B_9H_{11}^{2-}$.

the first metallocarborane in a manner analogous to the synthesis of ferrocene (56):

$$7,8\text{-}C_2B_9H_{12}^- \xrightarrow[-H_2]{\text{NaH}} 7,8\text{-}C_2B_9H_{11}^{2-} \xrightarrow{\text{Fe(II)}} (1,2\text{-}C_2B_9H_{11})_2\text{Fe(II)}^{2-} \qquad (2)$$

By using the $C_5H_5^-$ ligand in conjunction with the 7,8-$C_2B_9H_{11}^{2-}$ ligand, the mixed-ligand species $C_5H_5Fe(1,2\text{-}C_2B_9H_{11})^-$ was obtained (44). One-electron oxidation of this anion formed the uncharged $C_5H_5Fe(1,2\text{-}C_2B_9H_{11})$ species containing formal iron(III), an analog of the ferricinium ion. Subsequent single-crystal X-ray diffraction studies on this derivative (104) substantiated the belief that the iron atom was indeed occupying the twelfth vertex of an icosahedron. In this manner metallocarborane chemistry was brought into being and the first wedding of transition metals with carboranes into hybrid cluster compounds was accomplished.

Following these events, the chemistry of the metallocarboranes was rapidly expanded through the use of many of the transition metals and certain main group metals such as Be (82, 83), Al, and Ga (102). A boron vertex may be reinserted as well (53). In addition, carborane ligands other than the original 7,8- and 7,9-$C_2B_9H_{11}^{2-}$ ions were attached to metals, often in conjunction with a wide variety of truly organic ligands (47). More recent developments have led to the synthesis of metallocarboranes that contain more than a single transition metal in a polyhedral surface, and these transition metals need not be identical (3).

With very few exceptions, the gross geometries of polyhedral metallocarboranes may be correlated with the total number of vertices present in the polyhedron. Metal, carbon, boron, or other nonhydrogen elements are counted as vertices and their total number equated with the value of n in $B_nH_n^{2-}$ ions. In nearly every case the approximate geometry of the polyhedral metallocarborane will coincide with that of the corresponding $B_nH_n{}^{2-}$ ion, when known (4). Table I lists these geometries as a function of n.

The thermal polyhedral rearrangement of 1,2- to 1,7-$C_2B_{10}H_{12}$ already mentioned (see Fig. 1) is but one example of a reaction commonly observed throughout the polyhedral carborane and metallocarborane families. At the present time, the mechanism of these interesting rearrangements remains obscure, although several schemes have been advanced to explain experimental results (46). Polyhedral rearrangements in metallocarboranes occasionally occur with great facility in comparison to the energy required to effect the 1,2- to 1,7-$C_2B_{10}H_{12}$ isomerization, and are important aspects of the physical and chemical properties of metallocarboranes.

In this review, we treat in depth the synthesis, structures, properties, and reactions of η-bonded metallocarboranes. Our survey is restricted to complexes of 2-carbon carboranes and to species that have between nine and fourteen total polyhedral vertices. Coverage of metal complexes of other heteroboranes is available in Grimes's book (41) and in Todd's review (93). The recent work of Grimes and his group has concentrated on metallocarboranes having fewer than nine vertices (42, 75, 76).

Our approach to the subject has been to divide the metallocarboranes according to the size of the polyhedron. Starting with twelve-vertex compounds, which constitute the majority of the effort, we proceed to the larger polyhedra, so far unknown in the $B_nH_n{}^{2-}$ and $C_2B_{n-2}H_n$ series, and then to the lower polyhedra. Further subdivisions within each polyhedral size include synthesis, structures, and properties of monometallic complexes, reactions of monometallics, bimetallic preparations and reactions, and, in two instances, trimetallic compounds.

TABLE I

CORRELATION OF GROSS POLYHEDRAL GEOMETRY WITH TOTAL NUMBER OF VERTICES

Total vertices, n	Observed geometry
12	Icosahedron
11	Octadecahedron
10	Bicapped square antiprism
9	Tricapped trigonal prism
8	Dodecahedron
7	Pentagonal bipyramid
6	Octahedron

Metallocarboranes have only been known for 10 years, and research into their synthesis and characterization has involved a small number of workers. Consequently, practical applications of these unique compounds have not been rapidly forthcoming. Recent work has shown catalytic activity in certain of these compounds, however, and may signify future commercial value and industrial importance of metallocarboranes.

II

METALLOCARBORANES: SYNTHETIC METHODS

Five major synthetic routes are now available for the preparation of metallocarboranes, although only one was well established in 1969. The recently developed synthetic methods have allowed the preparation of more complex and diverse compounds and have greatly expanded the field of metallocarborane chemistry. These preparative methods are discussed in some detail in this section, for the synthesis of all the known metallocarboranes have been accomplished by one or more of these routes.

A. Preparation from nido-Carborane Anions

As mentioned previously, the first metallocarborane synthesized, $(1,2\text{-}C_2B_9H_{11})_2Fe(II)^{2-}$ (Fig. 3), was prepared in a manner similar to the synthesis of ferrocene, e.g., reaction of anhydrous $FeCl_2$ with the *nido*-carborane dianion $7,8\text{-}C_2B_9H_{11}^{2-}$, which itself was formed from $1,2\text{-}C_2B_{10}H_{12}$

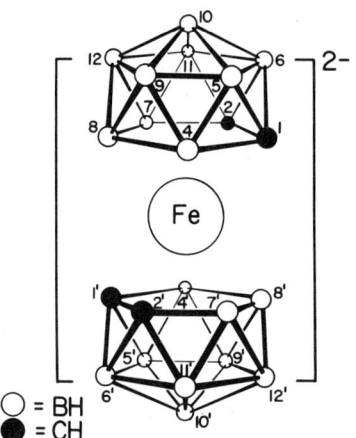

FIG. 3. Structure of the first metallocarborane ever synthesized, $(1,2\text{-}C_2B_9H_{11})_2Fe(II)^{2-}$.

by degradation with strong base [Eq. (2)]. This synthetic method, with some modifications, has been used to prepare a wide variety of monometallic twelve-vertex metallocarboranes.

The reaction conditions for this type of preparation generally involve nonaqueous solvents, such as tetrahydrofuran (THF) or diethyl ether, and rigorous exclusion of air and water. Some metallocarboranes, however, may be prepared in high yield by reaction of a metal salt and the nido monoanion, $C_2B_9H_{12}^-$, in strong aqueous base:

$$7,8\text{-}C_2B_9H_{12}^- + OH^- \longrightarrow H_2O + 7,8\text{-}C_2B_9H_{11}^{2-} \quad (3)$$

$$2(7,8\text{-}C_2B_9H_{11}^{2-}) + M^{n+} \longrightarrow (1,2\text{-}C_2B_9H_{11})_2M^{n-4} \quad (4)$$

In these instances, it is believed that the strong base deprotonates the monoanion to a small extent, permitting complexation to occur.

This synthetic approach proved valuable for the preparation of lower monometallocarboranes as well: the $C_2B_7H_{11}^{2-}$ ion, prepared from 6,8-$C_2B_7H_{13}$, was found to react with metal ions, losing one equivalent of hydrogen gas, to form metallocarboranes of the type $(C_2B_7H_9)_2M^{n-4}$ (38).

B. Preparation by Polyhedral Expansion

Polyhedral expansion, which was first reported in 1970 (Fig. 4) (16), entails the reduction of a *closo*-carborane with a strong reducing agent,

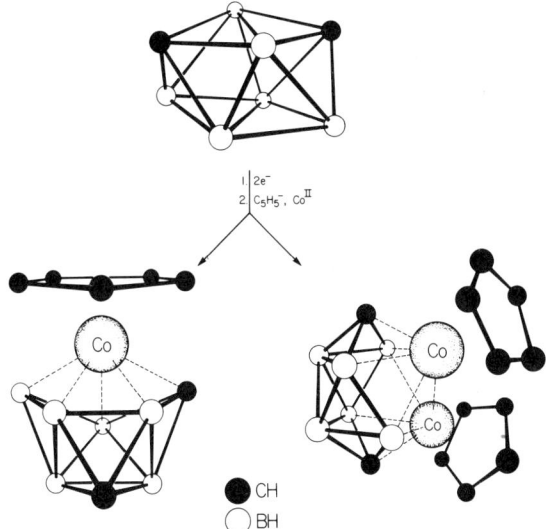

FIG. 4. Polyhedral expansion of 1,7-$C_2B_6H_8$.

such as an alkali metal, followed by reaction with a transition metal reagent. All reactions must be performed in nonaqueous solvents under nitrogen. It is believed (4) that the reduction step results in an opening of the closed carborane polyhedron, producing a nido dianionic species, which then reacts with a transition metal ion in a fashion similar to that just discussed:

$$C_2B_{n-2}H_n + 2Na \longrightarrow C_2B_{n-2}H_n{}^{2-} \tag{5}$$

$$2(C_2B_{n-2}H_n{}^{2-}) + M^{n+} \longrightarrow (C_2B_{n-2}H_n)_2M^{n-4} \tag{6}$$

Although a number of different metallocarboranes have been synthesized in high yield by this reaction (18, 24), the actual chemistry is frequently much more complex than implied by Eqs. (5) and (6). For example, products containing greater and fewer numbers of boron atoms than present in the carborane starting materials have been isolated from polyhedral expansion reactions.

The polyhedral expansion reaction appears to be a general synthetic method for metallocarboranes; all the known *closo*-carboranes have been found to produce metal-containing compounds when subjected to the reduction–complexation operations of this synthetic scheme. Moreover, metallocarboranes containing more than one transition metal may be prepared by the polyhedral expansion of monometallocarboranes. Examples of this synthetic route will be described in following sections.

It should be noted that in the polyhedral expansion process, as idealized in Eqs. (5) and (6), the product metallocarborane has one more vertex than was present in the carborane starting material—hence the origin of the descriptive phrase "polyhedral expansion." By contrast, when metallocarboranes are prepared by reaction with $C_2B_9H_{11}{}^{2-}$ ions, which are prepared from the icosahedral $C_2B_{10}H_{12}$ carboranes, twelve-vertex metallocarboranes result.

C. Preparation by Polyhedral Contraction

The polyhedral contraction route to metallocarboranes consists of the degradative removal of a polyhedral boron atom of a metallocarborane followed by oxidative closure of the resulting *nido*-metallocarborane complex to a closo species having one fewer vertex than present in the starting material (68):

$$C_5H_5Co(1,2\text{-}C_2B_9H_{11}) \xrightarrow[\text{2. }H_2O_2]{\text{1. OR}^-} C_5H_5Co(2,4\text{-}C_2B_8H_{10}) \tag{7}$$

The polyhedral contraction process is thus complementary to polyhedral

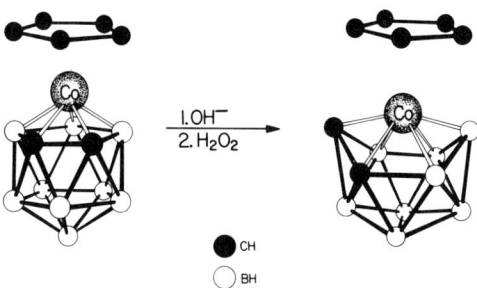

FIG. 5. Polyhedral contraction of $C_5H_5Co(1,2-C_2B_9H_{11})$ to $C_5H_5Co(2,4-C_2B_8H_{10})$.

expansion, in that the former decreases polyhedral size (Fig. 5) whereas the latter increases the number of polyhedral vertices. Polyhedral contraction is not as general a synthetic method as is polyhedral expansion, since some metallocarboranes undergo complete decomposition upon attempted partial degradation, and side reactions are frequently a difficulty in polyhedral contraction. Nevertheless, this is a valuable route to new complexes if the proper reaction conditions can be effected.

D. Preparation by Polyhedral Subrogation

Synthetic polyhedral subrogation for the preparation of polymetallocarboranes from monometallocarboranes is an offshoot of polyhedral contraction in that, after degradative removal of a BH vertex, a transition metal ion is reacted with the *nido*-metallocarborane produced rather than with an oxidizing agent. In this way, a new transition metal vertex is incorporated into the polyhedral framework without a change in the number of vertices between reactant and product (Fig. 6):

$$C_5H_5Co(C_2B_{10}H_{12}) \xrightarrow[\text{2. Co(II), } C_5H_5^-]{\text{1. OH}^-} (C_5H_5)_2Co_2(C_2B_9H_{11}) \quad (8)$$

This method is useful for the synthesis of metallocarboranes containing two similar or different transition metal vertices (*19, 23*) but has not yet been explored to determine its full potential.

E. Preparation by Thermal Metal Transfer

A newly discovered synthetic route (*27*) to bimetallocarboranes, thermal metal transfer may prove to be a highly valuable synthetic method but, as

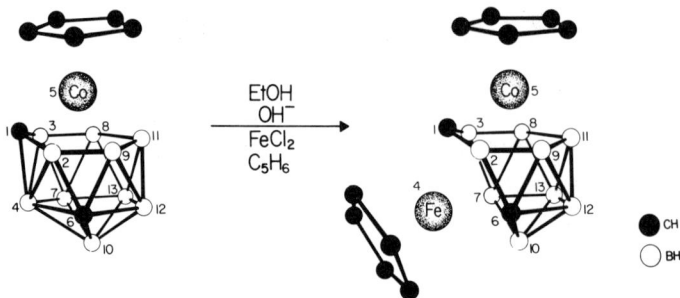

Fig. 6. Polyhedral subrogation of $C_5H_5Co(1,6\text{-}C_2B_{10}H_{12})$ with Fe(II).

yet, has been solidly established for only one system (Fig. 7). The method, which involves the pyrolysis of a metallocarborane through a hot tube *in vacuo*, was found to produce bimetallocarboranes having one vertex greater than was present in the starting material:

$$C_5H_5Co(C_2B_8H_{10}) \xrightarrow{\text{heat}} (C_5H_5)_2Co_2(C_2B_8H_{10}) \qquad (9)$$

Even more intriguing was the observation that the same products were formed upon pyrolysis of ionic cobalticinium salts of *commo*-metallocarborane ions:

$$(C_5H_5)_2Co^+(C_2B_8H_{10})_2Co^- \xrightarrow{\text{heat}} (C_5H_5)_2Co_2(C_2B_8H_{10}) \qquad (10)$$

This production of neutral bimetallocarboranes from ionic monometallic

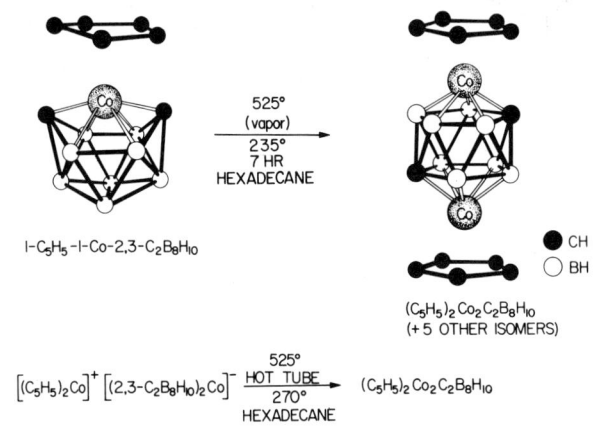

Fig. 7. Thermal metal transfer as a synthetic route to bimetallocarboranes.

precursors was unprecedented, and further examples of this type of reaction are being sought.

Spencer, Green, and Stone (88) have recently described a new synthetic approach to metallocarboranes in which an organometallic transition metal complex is thermally reacted with a neutral *closo*-carborane, resulting in the incorporation of the metal into the polyhedron:

$$1,8\text{-}(CH_3)_2C_2B_9H_9 + (1,5\text{-}C_8H_{12})_2Ni \longrightarrow (1,5\text{-}C_8H_{12})Ni(1,7\text{-}(CH_3)_2C_2B_9H_9) \quad (11)$$

Grimes and co-workers (76) subsequently reported the synthesis of several other metallocarboranes in a similar fashion. This method may proceed in a manner similar to thermal metal transfer and is, thus, included here. A new synthetic classification may be warranted if this preparative route proves to have widespread utility.

III

TWELVE-VERTEX METALLOCARBORANES

Although several thirteen- and fourteen-vertex metallocarboranes have been prepared and investigated, we have chosen to begin our closer examination of metallocarboranes with twelve-vertex species for several reasons. The first metallocarborane prepared was a twelve-vertex complex (56), and much of our understanding of metallocarboranes in general has arisen from studies on twelve-vertex compounds. In the early days of metallocarborane research, almost all the work was performed on twelve-vertex systems (54), and there have been more "icosahedral" metallocarboranes reported than of any other geometry. Thus the foundations of metallocarborane chemistry are twelve-vertex systems, and the properties of metallocarboranes of nonicosahedral geometry may be predicted on the basis of the behavior of twelve-vertex analogs.

A. Monometallic Complexes with Identical Carborane Ligands

1. *Synthesis, Structure, and Physical Properties*

The d^3 chromium(III) complex $(1,2\text{-}C_2B_9H_{11})_2Cr^-$ was prepared (85) from $7,8\text{-}C_2B_9H_{11}{}^{2-}$ and $CrCl_3$ in THF solution. An X-ray crystal structure determination on the Cs^+ salt, performed by St. Clair, Zalkin, and Templeton (90) showed it to have a symmetrical sandwich structure in which the metal atom is nearly equidistant from 3 boron atoms and 2 carbon atoms of each icosahedron but is distinguished by relatively long metal–ligand bond distances. No other significant difference in structure between this overall 15 valence-electron complex and 18 valence-electron complexes such as $C_5H_5Fe(1,2-C_2B_9H_{11})$ (104) were noticeable. This Cr(III) compound was the first electron-deficient metallocarborane prepared.

The first metallocarboranes ever synthesized were the iron(II) and iron(III) species, $(1,2-C_2B_9H_{11})_2Fe^{2-,1-}$, which were prepared in the manner described previously (56). The pink diamagnetic Fe(II) compound was observed to be air-sensitive; it is rapidly oxidized to the maroon, paramagnetic Fe(III) monoanionic complex $(1,2-C_2B_9H_{11})_2Fe^-$. By contrast, the ferricinium ion, $(C_5H_5)_2Fe^+$, exhibits a limited lifetime (84). This stabilization of a high metallic oxidation state by carborane ligands is observed in other metal systems (see below), and may be due to the ability of the "aromatic" carborane ligands to donate electron density to the metal. The electrostatic effects due to the presence of dianionic carborane moieties rather than monoanionic cyclopentadienyl ligands may also be important in this stabilization of high metallic oxidation states. The yellow, diamagnetic, 18-electron complex anion $(1,2-C_2B_9H_{11})_2Co^-$ may be prepared from $7,8-C_2B_9H_{11}^{2-}$ and $CoCl_2$ by either aqueous or nonaqueous routes; the yields are comparable. The 1,7-isomer may be prepared from $7,9-C_2B_9H_{12}^-$ in a similar fashion (54). An X-ray diffraction study of the Cs^+ salt of this latter complex confirmed symmetrical sandwich bonding, although the polyhedral carbon atoms could not be precisely located due to some disordering (103).

A significant change is observed in the analogous complexes of nickel, palladium, copper, and gold. Reaction of $7,8-C_2B_9H_{12}^-$ with Ni(II) salts, using either aqueous or nonaqueous techniques, results in the isolation of a d^7 Ni(III) complex when the workup is performed in air (54). This paramagnetic complex exhibits a geometry similar to that observed in the Co(III) and Fe(II) complexes (43). It undergoes reversible 1-electron oxidation and reduction reactions, forming complexes of the formulas $(1,2-C_2B_9H_{11})_2Ni$, which contains formal d^6 Ni(IV), and $(1,2-C_2B_9H_{11})_2Ni^{2-}$, which contains formal d^8 Ni(II). The latter compound is highly susceptible to air oxidation, forming the Ni(III) species. Analogous compounds are formed with palladium (95). Nonaqueous reaction of $7,8-C_2B_9H_{11}^{2-}$ with cupric salts produces complexes of Cu(II) and Cu(III), e.g., $(1,2-C_2B_9H_{11})_2Cu^{2-}$, which contains formal d^9 Cu(II), and $(1,2-C_2B_9H_{11})_2Cu^-$, which contains formal d^8 Cu(III). Similar compounds are formed using gold salts (94). Complexes of Ni(II), Ni(III), Pd(II), Pd(III), Cu(II), Cu(III), Au(II), and Au(III) all have more than 18 valence electrons.

Crystallographic studies performed by Wing (99, 100) have shown that these electron-rich metallocarboranes exhibit severe distortions from the symmetrical sandwich structures of the d^3 Cr(III) (90), d^5 Fe(III) (103), d^6 Co(III) (104), and d^7 Ni(III) (43) compounds. The d^8 complexes of Ni(II), Cu(III), Pd(II), and Au(III) are isomorphous (99) and have a "slipped sandwich" structure in which the metal atom is in a position closer to the 3 boron atoms in the bonding face of each carborane ligand than to the 2 polyhedral carbon atoms also nearest neighbors to the metal vertex. The d^9 compounds exhibit similar distortions: the Cu(II) complex, for example, shows an average Cu—B distance of 2.20 Å and an average Cu—C distance of 2.57 Å whereas for the Cu(III) species, Cu—B averages 2.11 Å and Cu—C is 2.52 Å (99, 100).

Two explanations for the adoption of "slipped sandwich" structures by these d^8 and d^9 metal complexes have been offered. Wing suggested (99, 100) that the complexes be considered as carborane analogs of organometallic π-allyl complexes, in which 3 ligand atoms are coordinated to the metal, thus reducing the effective atomic number at the metal vertex. Warren and Hawthorne (95) have suggested that the slipped sandwich mode of coordination is required by orbital symmetry considerations, similar to those invoked (15, 78) in explaining the slipped configuration of the Ag(I)–benzene complex.[2]

Complexes of d^6 Ni(IV) and Pd(IV), although isoelectronic with the Co(III) and Fe(II) species, exhibit structures different from these latter anionic compounds (91). Although there is no slip distortion, as present in the electron-rich d^8 and d^9 complexes, the Ni(IV) and Pd(IV) species are distorted from the icosahedral geometries of the other d^6 metallocarboranes. The polyhedral carbon atoms are arranged in a cisoid fashion in the Ni(IV) and Pd(IV) compounds and are tucked in toward the center of the icosahedron rather than occupying the normal positions in a twelve-vertex closed polyhedron. This geometrical configuration produces a sizable dipole moment: the value measured for $(1,2\text{-}C_2B_9H_{11})_2$Ni(IV) is 6.16 D in cyclohexane solution (95). The polyhedral C—H protons in the Ni(IV) and Pd(IV) compounds show a greater degree of acidity than normally observed in metallocarboranes (95); electronic attraction by the highly positive transition metal atom is doubtless a major influence in producing this increase in acidity. The structural changes occasioned by change in oxidation state are shown in Fig. 8 for the nickel complexes.

The most intriguing property of these tetravalent metal complexes was

---

[2] Molecular orbital theory has recently been employed by Wegner to account for these structural changes [P. A. Wegner, *Inorg. Chem.*, **14**, 212 (1975)].

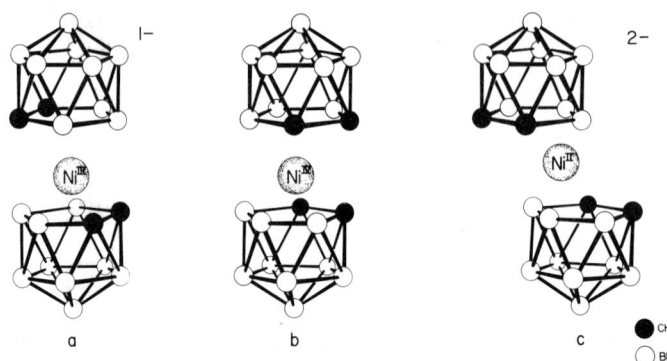

FIG. 8. (a) Symmetrical sandwich structure of $(1,2\text{-}C_2B_9H_{11})_2Ni(III)^-$ ion; (b) cisoid sandwich structure of $(1,2\text{-}C_2B_9H_{11})_2Ni(IV)$; and (c) slipped sandwich structure of $(1,2\text{-}C_2B_9H_{11})_2Ni(II)^{2-}$.

observed in the compounds in which all four of the polyhedral carbon atoms were substituted with methyl groups (95). The steric hindrance caused by the proximity of these bulky substituents in cisoid positions produced the first example of a low-energy polyhedral rearrangement in a metallocarborane.

Reaction of nickel(II) acetylacetonate with $7,8\text{-}(CH_3)_2\text{-}7,8\text{-}C_2B_9H_9^{2-}$ under rigorous exclusion of air and water produced an Ni(II) complex, denoted as $[(Me_2)_2Ni(II)]^{2-}$, which could be reversibly oxidized to a nickel(III) complex, $[(Me_2)_2Ni(III)]^-$. This latter compound underwent irreversible oxidation to the neutral $(Me_2)_2Ni(IV)$, which upon reduction afforded an Ni(III) species different from that originally prepared. Further study showed the existence of another isomeric $(Me_2)_2Ni(IV)$ complex, two additional Ni(II) isomers, and a third Ni(III) species. A combination of spectral and crystallographic information determined that these isomers arose by migration of polyhedral C—CH$_3$ vertices to positions distant from each other, thereby decreasing steric crowding. These polyhedral migrations occur in addition to the structural changes occasioned in the Ni(II)–Ni(III)–Ni(IV) oxidation series, in which the complexes have slipped sandwich, symmetrical transoid, and distorted cisoid configurations, respectively. The structures established (10, 95) for the three positionally isomeric $\{(Me_2)_2Ni\}$ series are depicted in Fig. 9, where the migration of polyhedral C—CH$_3$ vertices on each icosahedron explain the existence of the three isomeric families. That steric hindrance is responsible for the low activation energy required to effect the observed rearrangements was shown by attempts to effect similar migration of polyhedral C—H vertices in the unsubstituted complex $(1,2\text{-}C_2B_9H_{11})_2Ni$; a mixture

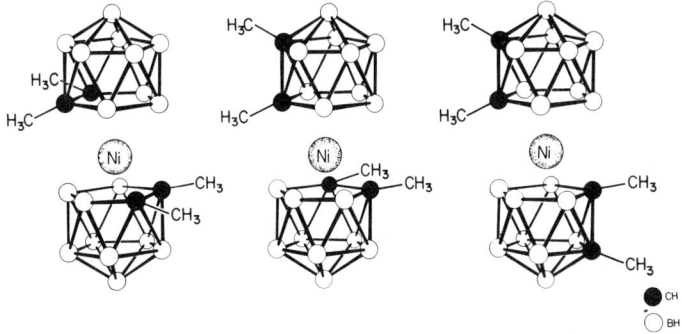

FIG. 9. The three positional isomers of $[(CH_3)_2C_2B_9H_9]_2Ni(III)^-$.

of isomeric compounds could be detected only after pyrolysis in the vapor phase at 360° to 400°C, whereas rearrangements of the methylated compounds occurred between room temperature and 100°C.

2. Reactions

Limited investigations of the chemical reactions of metallocarboranes have been reported. In general, these complexes are much less reactive than the analogous metallocenes, yet several unique new species have been prepared from the unsubstituted metallocarboranes.

It has been known for a considerable time that ferrocene is protonated by strong acids to produce the cationic species $(C_5H_5)_2FeH^+$; there is some evidence that the proton is directly coordinated to the iron atom, although some interaction of this proton with the cyclopentadienyl rings must be invoked to explain the rapid acid-catalyzed deuteration of ferrocene (13). The metallocarborane analog of protonated ferrocene may be prepared by reaction of $(1,2-C_2B_9H_{11})_2Fe(II)^{2-}$ with strong acids such as HCl or $HClO_4$, and the resulting species has been isolated as an air- and water-sensitive crystalline solid (52). A band in the infrared spectrum of this compound at 1885 cm^{-1} has been tentatively assigned as ν_{FeH}, although no signal attributable to this proton was observed by ^{1}H NMR.

This protonated metallocarborane is unique in that it undergoes ready substitution at polyhedral boron atoms, whereas the unprotonated species is unreactive in the absence of acid. Reaction of $(1,2-C_2B_9H_{11})_2FeH^-$ with Lewis bases such as dialkyl sulfides results in the loss of H_2 and the formation of boron-substituted complexes in good yields:

$$HFe(1,2-C_2B_9H_{11})_2^- + R_2S \longrightarrow (1,2-C_2B_9H_{11})(1,2-C_2B_9H_{10}SR_2)Fe^- + H_2 \quad (12)$$

It is not necessary to start with the protonated Fe(II) complex to isolate these substituted products, however. Reaction of a slurry of the unsubstituted Fe(III) complex with anhydrous HCl in diethyl sulfide produced a mixture of substitution products, including the species shown in Eq. (12). Although no evidence has been found for a protonated cobalt metallocarborane analog, boron-substituted complexes of cobalt(III) may also be obtained by reaction of the $(1,2\text{-}C_2B_9H_{11})_2Co^-$ ion with R_2S and HCl (52).

The cobalt(III) complex, $(1,2\text{-}C_2B_9H_{11})_2Co^-$, undergoes bromination in glacial acetic acid solution to afford a hexabromo derivative, $(1,2\text{-}C_2B_9H_8Br_3)_2Co^-$ (54). An X-ray crystallographic study of this product (14) showed that bromination occurred on the boron atoms farthest from the polyhedral carbon atoms, those boron atoms expected to have the greatest electron density, as predicted for an electrophilic attack mechanism.

Treatment of the potassium salt $K[(1,2\text{-}C_2B_9H_{11})_2Co]$ with CS_2 and HCl in the presence of $AlCl_3$ afforded a novel neutral compound, $Co(1,2\text{-}C_2B_9H_{10})_2S_2\overset{+}{C}H$ in which boron atoms on the two icosahedra are substituted by sulfur and the polyhedra are linked by the $S_2\overset{+}{C}H$ moiety (11). The structure of this compound is depicted in Fig. 10. It contains an electron-deficient carbon atom presumably stabilized by the nonbonding electron pairs of its adjacent sulfur atoms. A similar complex may be formed using

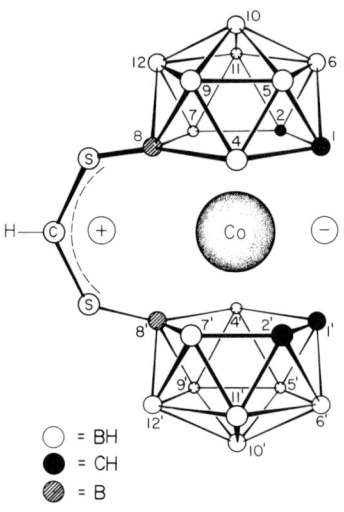

FIG. 10. Zwitterionic structure of $(1,2\text{-}C_2B_9H_{10})_2Co(S_2CH)$.

acetic acid–acetic anhydride in place of CS_2; in this case, the intercage bridging group is the $H_3C\overset{+}{-}CO_2$ moiety (*32*).

The diazonium salts $(C_6H_4RN_2)^+ (1,2-C_2B_9H_{11})_2Co^-$ undergo an interesting free-radical reaction upon pyrolysis. Nitrogen gas is liberated and the product is found to consist of a symmetrical sandwich *commo*-metallocarborane with an ortho-disubstituted phenyl ring bonded to polyhedral boron atoms and bridging between the two icosahedra (*35*).

The neutral complex $(1,2-C_2B_9H_{11})_2Ni(IV)$ behaves as a rather strong Lewis acid and forms stable crystalline 1:1 addition complexes with electron-donating molecules and ions. Complexes with halide and pseudohalide ions, as well as with certain large aromatic hydrocarbons, have been reported (*95*). Preliminary crystallographic results (*101*) indicate that the donor molecule approaches the metallocarborane on the side holding the cisoid polyhedral carbon atoms, as might be expected for charge-transfer interactions.

B. Monometallic Carboranes with Different Carborane Ligands

Degradation of the icosahedral 1,2- and 1,7-$C_2B_{10}H_{12}$ isomers with strong base to produce the nido eleven-vertex anions 7,8- and 7,9-$C_2B_9H_{12}^-$ has been previously discussed and is an important route to the preparation of twelve-vertex monometallocarboranes. The discovery that similar reactions could be performed on metallocarboranes led to the isolation of novel chains of metal atoms bridged by carborane groups and to the development of the polyhedral contraction and polyhedral subrogation reactions.

The first indication that metallocarboranes could undergo base degradation was found upon investigation of the colored products remaining after isolation of the $(1,2-C_2B_9H_{11})_2Co^-$ ion from an aqueous preparation (*32, 34*). The high molecular weight complexes found in the reaction residue contained species formed by partial degradation of the icosahedral complex, which had undergone further reaction to produce polymetallic compounds. The structures (*12, 89*) of the two reaction products isolated are shown in Fig. 11.

The following reaction sequence leading to these complexes is believed to be similar to the stepwise reactions in which metallocarborane complexes are formed from neutral *closo*-carboranes:

$(1,2-C_2B_9H_{11})_2Co^- + 2OH^- + H_2O \longrightarrow$

$(1,2-C_2B_9H_{11})(C_2B_8H_{10})Co^{3-} + H_2 + B(OH)_3$ (13)

$(1,2-C_2B_9H_{11})(C_2B_8H_{10})Co^{3-} + Co(II) + 7,8-C_2B_9H_{11}^{2-} \xrightarrow{\text{[O]}}$

$(1,2-C_2B_9H_{11})Co(C_2B_8H_{10})Co(1,2-C_2B_9H_{11})^{2-}$ (14)

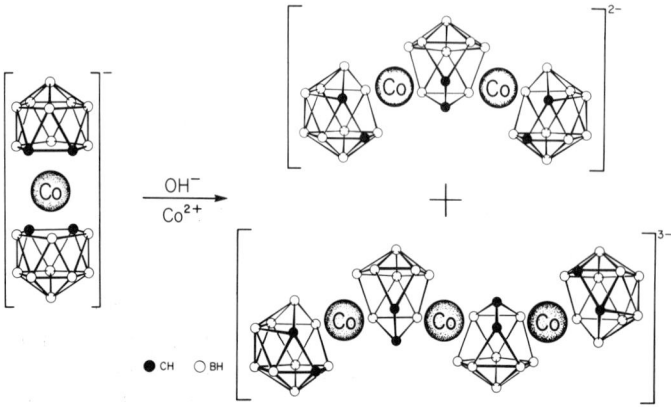

FIG. 11. Degradation of $(1,2\text{-}C_2B_9H_{11})_2Co^-$ to bi- and trimetallic compounds.

$(1,2\text{-}C_2B_9H_{11})Co(C_2B_8H_{10})Co(1,2\text{-}C_2B_9H_{11})^{2-} + OH^- + H_2O \longrightarrow$

$\qquad (1,2\text{-}C_2B_9H_{11})Co(C_2B_8H_{10})Co(C_2B_8H_{10})^{4-}$ (15)

$(1,2\text{-}C_2B_9H_{11})Co(C_2B_8H_{10})Co(C_2B_8H_{10})^{4-} + Co(II) + 7,8\text{-}C_2B_9H_{11}^{2-} \xrightarrow{[O]}$

$\qquad (1,2\text{-}C_2B_9H_{11})Co(C_2B_8H_{10})Co(C_2B_8H_{10})Co(1,2\text{-}C_2B_9H_{11})^{3-}$ (16)

This reaction sequence may be considered the first example of polyhedral subrogation in that a boron vertex of a metallocarborane has been degradatively removed from the polyhedron and replaced by a transition metal vertex. No overall change in size of the polyhedron occurs in the reaction, however. Evidence substantiating this proposed mechanism has been gathered by the success of polyhedral contraction and subrogation reactions on this same substrate (66–68). Further examples of polyhedral subrogation reactions in the synthesis of polymetallocarboranes will be discussed in Section III, D.

Mixed *commo*-metallocarboranes of the general formula $(C_2B_9H_{11})Co(C_2B_8H_{10})^-$ have been prepared by the polyhedral contraction of isomeric $(C_2B_9H_{11})_2Co^-$ ions (68). These species react with pyridine and other Lewis bases to afford *nido*-metallocarboranes that have the Lewis base attached to a boron atom on the 8-boron cage. A crystallographic structure determination has been performed on one of these ligand adducts (9). Spectral evidence indicates that the carbon atom locations in the $C_2B_8H_{10}Co$ polyhedron of the $(1,2\text{-}C_2B_9H_{11})Co(C_2B_8H_{10})^-$ ion are identical with those observed in the product from polyhedral contraction of $C_5H_5Co(1,2\text{-}C_2B_9H_{11})$ (Section VI, A) (66). Reaction of this B_8—B_9 *closo*-metallocarborane anion with $FeCl_3$ in ethanol afforded $(1,2\text{-}C_2B_9H_{11})Co(1,6\text{-}$

$C_2B_7H_9)^-$, which could be rearranged to $(1,2\text{-}C_2B_9H_{11})Co(1,10\text{-}C_2B_7H_9)^-$ (*66*). No intermediate nido complexes were noted in this latter transformation, in contrast to the cyclopentadienyl analog (see Section VII) although these species have been prepared (*67*) and appear to have both B—H—B and B—H—Co bridge bonds.

C. Mixed-Ligand Complexes

The class of compounds known as monometallocarboranes is not limited to species that contain two carborane ligands bonded to a transition metal, but also encompasses compounds that contain both carborane fragments and other, predominantely organic, ligands.

The first cyclopentadienyl metallocarborane to be synthesized emphasized the analogy between metallocene chemistry and metallocarborane chemistry. This preparation simply involved the reaction of stoichiometric amounts of the two ligands with a transition metal halide:

$$7,8\text{-}C_2B_9H_{11}{}^{2-} + C_5H_5{}^- + Fe(II) \xrightarrow{[O]}$$
$$C_5H_5Fe(1,2\text{-}C_2B_9H_{11}) + (C_5H_5)_2Fe + (1,2\text{-}C_2B_9H_{11})_2Fe^- \quad (17)$$

As would be expected from this synthetic route, considerable amounts of symmetrically substituted products were obtained along with the desired mixed-sandwich complex (*49*). This synthetic route was successful for the preparation of other mixed-sandwich cyclopentadienyl metallocarboranes with the transition metals cobalt(III) (*54*) and nickel(III) (*98*), although yields were relatively low. Modified preparative schemes, developed over the past several years, now permit the synthesis of $C_5H_5Co(1,2\text{-}C_2B_9H_{11})$ in high yield, starting with either the potassium salt of $7,8\text{-}C_2B_9H_{12}{}^-$ or with $1,2\text{-}C_2B_{10}H_{12}$ itself (*69, 81*). These new routes have been of great value in recent developments of metallocarborane chemistry, since neutral diamagnetic $C_5H_5Co(1,2\text{-}C_2B_9H_{11})$ is an ideal starting material for the preparation of other metallocarboranes.

As observed with the icosahedral carborane $1,2\text{-}C_2B_{10}H_{12}$, the mixed-sandwich metallocarborane $C_5H_5Co(1,2\text{-}C_2B_9H_{11})$ undergoes thermal polyhedral rearrangement to produce eight other isomeric species. All of these possible positional isomers of this complex have been prepared and isolated by pyrolysis and chromatographic separation (*70*). The thermal rearrangements are depicted in Fig. 12. Isomeric identification was effected primarily by ^{11}B NMR spectroscopy and electrochemical and chromatographic behavior. As was found in the rearrangement of $1,2\text{-}C_2B_{10}H_{12}$, the polyhedral carbon atoms in $C_5H_5Co(1,2\text{-}C_2B_9H_{11})$ migrate away from each

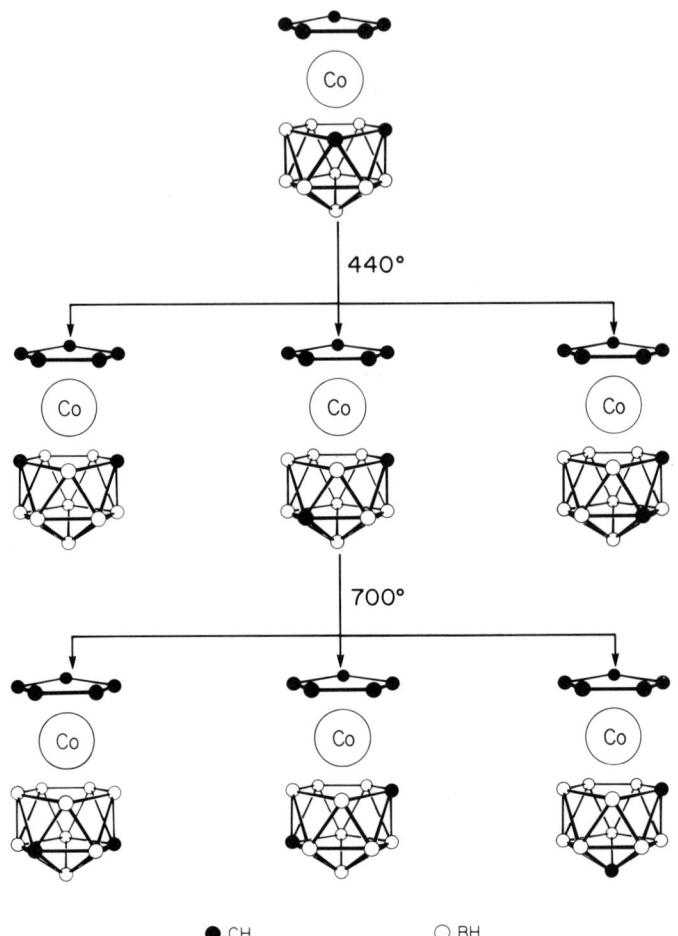

FIG. 12. Isomers of $C_5H_5Co(C_2B_9H_{11})$ prepared by thermal rearrangement.

other during thermal rearrangement. In order to isolate examples of all the possible isomers, it was, therefore, necessary to link the carbon atoms with a trimethylene bridge, forcing them to stay adjacent while still permitting migration over the polyhedral surface. Another route to some of these isomeric complexes is through the polyhedral expansion reaction: two $C_5H_5CoC_2B_9H_{11}$ isomers were prepared by polyhedral expansion of 2,3-$C_2B_9H_{11}$ (24).

Although mixed carborane–cyclopentadienyl complexes, predominantly those of cobalt, are the metallocarboranes that have received the most study (due primarily to their air and water stability, neutrality, and diamagnetism), other mixed-ligand metallocarboranes may be prepared as well. Metal carbonyl compounds react with carborane dianions to afford metallocarborane carbonyl complexes (45, 54), for example,

$$7,8\text{-}C_2B_9H_{11}^{2-} + BrMn(CO)_5 \longrightarrow Br^- + \{C_2B_9H_{11}Mn(CO)_5^-\} \longrightarrow$$
$$(1,2\text{-}C_2B_9H_{11})Mn(CO)_3^- + 2CO \quad (18)$$

$$7,8\text{-}C_2B_9H_{11}^{2-} + W(CO)_6 \xrightarrow{h\nu} 3CO + (1,2\text{-}C_2B_9H_{11})W(CO)_3^{2-} \quad (19)$$

The existence of the $Mn(CO)_5$–carborane complex was inferred (54) by the observation of rapid precipitation of NaBr from the reaction mixture, whereas CO evolution did not occur until the resulting solution was heated to reflux.

The photochemical synthesis of the tungsten carbonyl metallocarborane [Eq. (18)] is also effective in the preparation of the molybdenum carbonyl analog. The resulting air-sensitive complexes show chemical behavior similar to the cyclopentadienyl analogs, $C_5H_5M(CO)_3^-$, in that they undergo protonation with anhydrous HCl and methylation with CH_3I. They also react further with metal hexacarbonyls to afford bimetallic complexes (54):

$$(1,2\text{-}C_2B_9H_{11})Mo(CO)_3^{2-} + W(CO)_6 \longrightarrow (1,2\text{-}C_2B_9H_{11})Mo(CO)_3W(CO)_5^{2-} + CO$$
$$(20)$$

Several other twelve-vertex metallocarborane carbonyl complexes have been prepared (51). In general, the chemistry and structures of these species, when investigated, have been found to parallel the analogous cyclopentadienyl metal carbonyls.

Until recently, little work had been reported on the preparation of metallocarboranes containing more diverse counterligands. The tetraphenylcyclobutadiene complex $[(C_6H_5)_4C_4]Pd(1,2\text{-}C_2B_9H_{11})$ was reported in 1966 (96), and a cyclooctadiene complex of platinum(II) was subsequently prepared (95). Only in the past 2 years have the existence of phosphine-substituted *closo*-metallocarboranes been established, and a rich new area of metallocarborane chemistry thereby developed.

Reaction of the *nido*-carborane monoanion $7,8\text{-}C_2B_9H_{12}^-$ with phosphine-substituted transition metal complexes afforded novel hydridophosphinometallocarboranes (80) according to

$$[(C_6H_5)_3P]_3RhCl + 7,8\text{-}C_2B_9H_{12}^- \longrightarrow HRh[(C_6H_5)_3P]_2(1,2\text{-}C_2B_9H_{11}) \quad (21)$$

This complex, along with its iridium analog, has proven to be an effective catalyst for the isomerization and hydrogenation of olefins (80), for H—D exchange at BH vertices (60), and for other organic reactions. Catalysis by metallocarboranes is discussed in detail in Section X.

Spencer, Green, and Stone (88) have prepared nickelacarboranes with phosphine and other counterligands by reacting the *closo*-carborane 2,3-$C_2B_9H_{11}$ with organometallic nickel compounds. Grimes (76) found that the reaction of 1,2-$C_2B_{10}H_{12}$ with $C_5H_5Co(CO)_2$ in a sealed tube afforded a mixture of the isomeric $C_5H_5Co(C_2B_9H_{11})$ compounds.

D. Bimetallic Complexes

Examples of bimetallic twelve-vertex metallocarboranes have been provided by various synthetic efforts. The preparation of bi- and trimetallic chain complexes by polyhedral subrogation of $(1,2-C_2B_9H_{11})_2Co^-$ has been mentioned earlier (33, 34), and the $\{Co(C_2B_8H_{10})Co\}$ fragment present therein may be considered as an example of this type of bimetallic complex.

Cyclopentadienyl analogs of the bi- and trimetallic chain complexes were prepared recently (69). These compounds, as well as the carborane-capped compounds, exhibit reversible 1-electron redox couples, indicating that the electrons are added to and removed from delocalized molecular orbitals rather than localized metal centers.

A second route to bimetallic twelve-vertex complexes is via polyhedral expansion. The starting material may be either an eleven-vertex monometallocarborane,

$$C_5H_5Co(C_2B_8H_{10}) \xrightarrow[\text{2. Co(II), C}_5\text{H}_5^-, \text{[O]}]{\text{1. 3e}^-} (C_5H_5)_2Co_2C_2B_8H_{10} + \text{other products} \quad (22)$$

or a ten-vertex neutral carborane,

$$1,6\text{-}C_2B_8H_{10} \xrightarrow[\text{2. Co(II), C}_5\text{H}_5^-, \text{[O]}]{\text{1. 2e}^-} (C_5H_5)_2Co_2C_2B_8H_{10} + \text{other products} \quad (23)$$

but yields are substantially higher in the former case. Both routes have been used (24) to prepare the icosahedral complex $2,3\text{-}(C_5H_5)_2\text{-}2,3\text{-}Co_2\text{-}1,7\text{-}C_2B_8H_{10}$, which has been shown by X-ray crystallography to have the metals in adjacent positions only 2.39 Å apart (7). At 250°C, this complex thermally rearranges to an isomer in which the metals are no longer adjacent (30). The identical product is obtained upon pyrolysis of the bimetallic product prepared by double polyhedral subrogation (69) (Fig. 13).

The discovery of the thermal metal transfer reaction (27) afforded a third preparative route to bimetallic twelve-vertex metallocarboranes. This method involves the pyrolysis of eleven-vertex *closo*-metallocarboranes or cobalticinium salts of eleven-vertex *commo*-metallocarboranes and results in the production of several isomeric, closed, neutral, twelve-vertex bimetallocarboranes [Eqs. (9) and (10); Fig. 7]. Yields are reasonable in

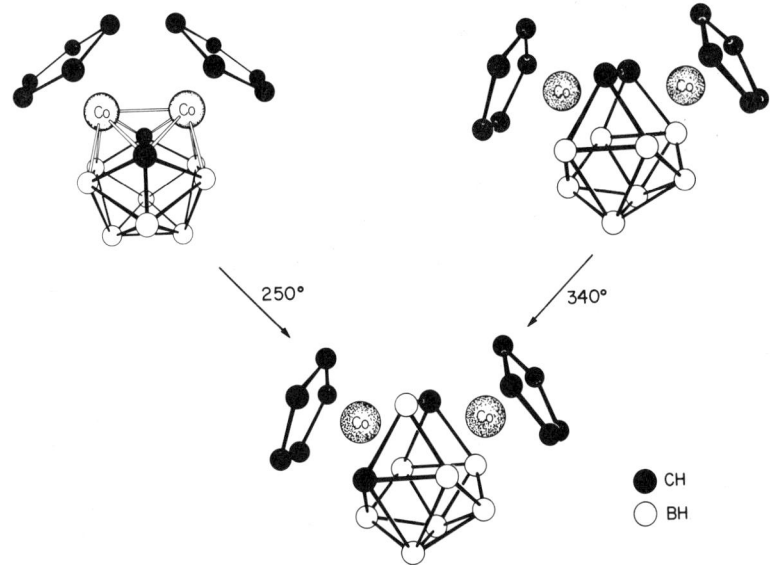

FIG. 13. Structures and interconversions of three isomeric $(C_5H_5)_2Co_2C_2B_8H_{10}$ compounds.

this reaction, and thermal metal transfer could prove to be a valuable new synthetic tool.

Only one trimetallic twelve-vertex metallocarborane has been reported. This species, $(C_5H_5)_3Co_3C_2B_7H_9$, arose as a side product during the polyhedral expansion of $2\text{-}C_5H_5\text{-}2\text{-}Co\text{-}1,6\text{-}C_2B_7H_9$ with Co(II) and $C_5H_5^-$ (25, 28). The isolation of this trimetallic complex suggests that the polyhedral expansion reaction may be extended to bimetallic substrates to produce novel metal-rich polyhedra.

IV

THIRTEEN-VERTEX METALLOCARBORANES

The first supraicosahedral monometallocarborane was prepared by polyhedral expansion of $1,2\text{-}C_2B_{10}H_{12}$ with $CoCl_2$ and $C_5H_5^-$ (17, 18):

$$1,2\text{-}C_2B_{10}H_{12} \xrightarrow[\text{2. Co(II), C}_5\text{H}_5^-,\text{[O]}]{1.\ 2e^-} C_5H_5CoC_2B_{10}H_{12} \qquad (24)$$

Three isomeric complexes of this formula have been prepared, which are

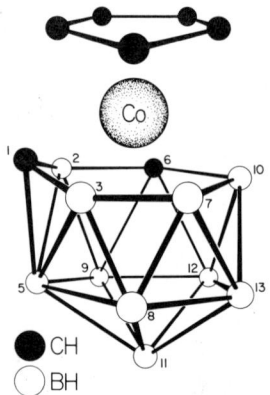

Fig. 14. Structure of the red isomer of $C_5H_5CoC_2B_{10}H_{12}$.

colored red, orange, and amber. All three exhibit the same melting point, 250°–251°C, but upon heating, the red complex is seen to change color to orange, then amber before melting, and the orange isomer turns amber before it melts. The red isomer is quantitatively converted to the orange isomer by boiling in hexane (68°C), and the orange complex rearranges to the amber isomer on heating in benzene solution (80°C). These rearrangements occur at surprisingly low temperatures: recall that rearrangement of 1,2- to 1,7-$C_2B_{10}H_{12}$ required a temperature in excess of 400°C. The thermochemical parameters of those rearrangements have been determined, and reflect an intramolecular mechanism, as expected (20).

The structure of the red isomer of $C_5H_5CoC_2B_{10}H_{12}$ (Fig. 14), determined by Churchill (8), indicated that the complex was asymmetric, although ^{11}B NMR spectroscopy indicated mirror symmetry for this species in solution at room temperature. Low-temperature ^{11}B NMR spectral investigations showed that the molecule was fluxional in solution at room temperature. It has been postulated (18) that the complex undergoes rapid diamond–square–diamond rearrangement (Fig. 15) in which enantiomeric complexes interconvert.

Thirteen-vertex complexes of metals other than cobalt have been prepared, as have commo thirteen-vertex anions and metal carbonyl derivatives (18). In general, these complexes exhibit stabilities similar to or less than their twelve-vertex counterparts.

The chemistry of the thirteen-vertex monometallocarboranes, although less extensively studied, is similar to that of the twelve-vertex analogs. Thus, $C_5H_5CoC_2B_{10}H_{12}$ undergoes polyhedral subrogation to produce thirteen-vertex bimetallocarboranes [the metals may be identical (23) or

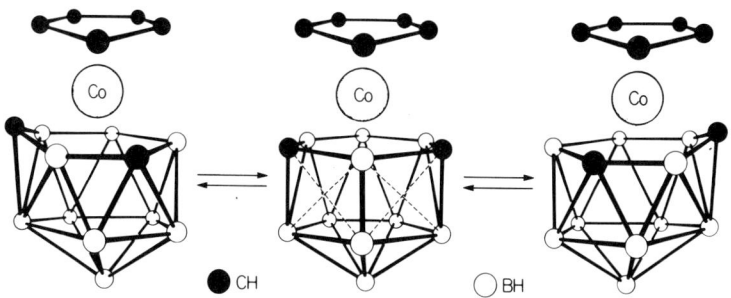

FIG. 15. Proposed mechanism of racemization of the red isomer of $C_5H_5CoC_2B_{10}H_{12}$ through a diamond–square–diamond intermediate.

they may differ (19)]:

$$C_5H_5CoC_2B_{10}H_{12} + OH^- + Fe(II) + C_5H_5^- \xrightarrow{[O]} (C_5H_5)_2CoFeC_2B_9H_{11} \quad (25)$$

Polyhedral expansion of the thirteen-vertex monometallocarborane with Co(II) and $C_5H_5^-$ produced the novel fourteen-vertex bimetallocarborane $(C_5H_5)_2Co_2C_2B_{10}H_{12}$ (29). These species are discussed in Section V.

Attempted polyhedral contraction of $C_5H_5CoC_2B_{10}H_{12}$, expected to produce one or more of the isomeric twelve-vertex $C_5H_5CoC_2B_9H_{11}$ complexes, instead resulted in extensive disruption of the polyhedron. Three boron vertices and one carbon vertex were extruded from the starting material, and a novel monocarbon metallocarborane ion, $C_5H_5CoCB_7H_8^-$, was isolated (21, 22). As expected for a nine-vertex complex, this species has the geometry of a tricapped trigonal prism, with the carbon atom in a low-coordinate capping site (Fig. 16) (6).

When subjected to the action of alcoholic base and excess Co(II) in the absence of additional cyclopentadienide ion, the thirteen-vertex monometallocarborane reacts to produce a trimetallic complex that contains two bridging thirteen-vertex frameworks (23). The proposed structure of this species is shown in Fig. 17. This reaction is similar to that used to prepare the twelve-vertex bi- and trimetallic chain complexes discussed in Section III, B.

Polyhedral expansion of $1,2\text{-}C_2B_{10}H_{12}$ in the presence of early transition metal halides has been found to produce the first examples of metallocarborane complexes of these elements. Expansion using $TiCl_4$ afforded a titanacarborane in which the metal has a formal oxidation state of 2+ and only 14 valence electrons (86).

$$1,2\text{-}C_2B_{10}H_{12} \xrightarrow[\text{2. TiCl}_4]{\text{1. } 2e^-} (C_2B_{10}H_{12})_2Ti^{2-} \quad (26)$$

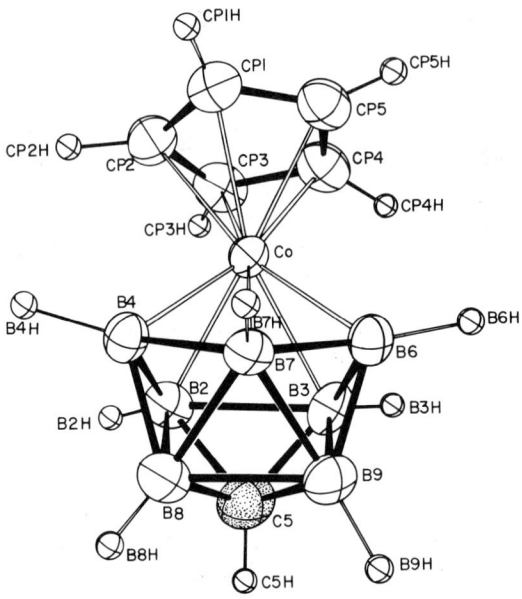

FIG. 16. Structure of $C_5H_5CoCB_7H_8^-$.

This electron deficiency is not structurally significant, however; crystallographic investigation of the complex formed between 1,2-$(CH_3)_2$-1,2-$C_2B_{10}H_{10}$ and $TiCl_4$ showed the thirteen-vertex polyhedra to be symmetrically disposed about the metal (74) and to exhibit less distortion than observed in the $C_5H_5CoC_2B_{10}H_{12}$ polyhedron (8). Complexes of V(II) and

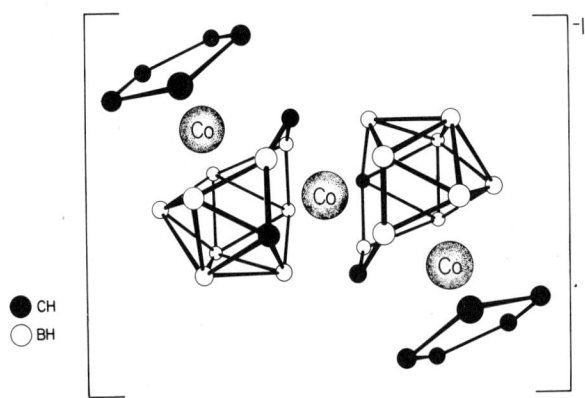

FIG. 17. Proposed structure of the trimetallocarborane prepared by polyhedral subrogation.

Zr(II) have been similarly prepared and appear spectrally comparable to the Ti(II) compound (*86*).

V
FOURTEEN-VERTEX METALLOCARBORANES

The polyhedral expansion of the orange and amber thirteen-vertex closed metallocarborane isomers with Co(II) and $C_5H_5^-$ was found to afford complex mixtures of products, the primary species in each case being the desired isomeric $(C_5H_5)_2Co_2C_2B_{10}H_{12}$ complexes (*29*). Spectral data are consistent with a closed fourteen-vertex geometry for these compounds, but definitive structure determination by X-ray crystallography will be required to determine the overall geometry of this large polyhedron. A possible structure of one of the isolated isomers which is consistent with the spectral properties (*29*) is presented in Fig. 18.

VI
ELEVEN-VERTEX METALLOCARBORANES

A. Monometallic Complexes

The first reported preparation of an eleven-vertex closed monometallocarborane involved the polyhedral expansion of the ten-vertex *closo*-car-

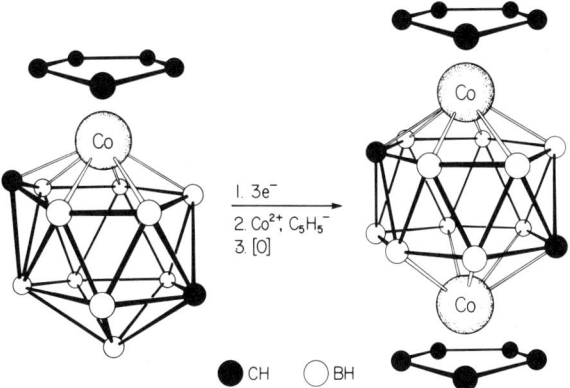

FIG. 18. Preparative method and proposed structure of one isomer of a fourteen-vertex metallocarborane.

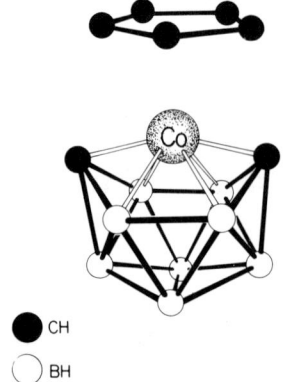

FIG. 19. Structure of $C_5H_5CoC_2B_8H_{10}$ prepared by polyhedral expansion.

borane 1,6-$C_2B_8H_{10}$ with Co(II) and $C_5H_5^-$ (24, 26):

$$1,6\text{-}C_2B_8H_{10} \xrightarrow[\text{2. Co(II), } C_5H_5^-, \text{[O]}]{\text{1. } 2e^-} C_5H_5CoC_2B_8H_{10} \qquad (27)$$

The structure of this compound, inferred from ^{11}B NMR spectroscopy (26) is shown in Fig. 19. The metal resides at the high-coordinate vertex, while the 2 polyhedral carbon atoms occupy the two low-coordinate positions in the closed eleven-vertex geometry. The analogous Fe(III) compound was formed using Fe(II) in the polyhedral expansion reaction (24).

Another eleven-vertex complex, isomeric with the cobalt compound just mentioned, may be prepared (66) by the polyhedral contraction of $C_5H_5Co(1,2\text{-}C_2B_9H_{11})$. Reaction of this substrate with strong aqueous hydroxide ion in the absence of additional Co(II), followed by hydrogen peroxide oxidation, resulted in the isolation of a new $C_5H_5CoC_2B_8H_{10}$ isomer in which both polyhedral carbon atoms occupy adjacent sites, one in a low-coordinate position and one in a six-coordinate site (68). This isomer is thermally unstable and rearranges (26) to the isomer produced in the polyhedral expansion of 1,6-$C_2B_8H_{10}$ with Co(II) and $C_5H_5^-$, when heated to 150°C. The corresponding commo complexes have also been isolated as side products from the polyhedral expansion of 1,6-$C_2B_8H_{10}$ (24).

The nido eleven-vertex complex, $C_5H_5CoC_2B_8H_{12}$, has been prepared, and spectral data indicate the presence of B—H—B and B—H—Co bridges (67). The commo eleven-vertex metallocarborane, $(C_2B_8H_{10})_2Co^-$, prepared by polyhedral expansion of 1,6-$C_2B_8H_{10}$ with Co(II) in the absence of cyclopentadienide ion, has carbon atoms in positions identical to the sites in the $C_5H_5CoC_2B_8H_{10}$ isomer prepared by polyhedral expansion in the presence of $C_5H_5^-$, i.e., at low-coordinate vertices (26).

Reactions of monometallic eleven-vertex metallocarboranes have been discussed in previous sections and may be summarized briefly as (*a*) polyhedral expansion to bimetallic twelve-vertex complexes and (*b*) thermal metal transfer to bimetallic twelve-vertex compounds. Polyhedral contraction to ten-vertex monometallocarboranes is discussed in Section VII.

B. Bimetallic Complexes

The eleven-vertex bimetallocarboranes have primarily been synthesized by application of polyhedral expansion techniques. One obvious synthetic route, polyhedral expansion of a ten-vertex *closo*-monometallocarborane, has been investigated in detail (*24*). The products arising from this reaction are shown in Fig. 20. Three isomeric eleven-vertex cobaltacarboranes have

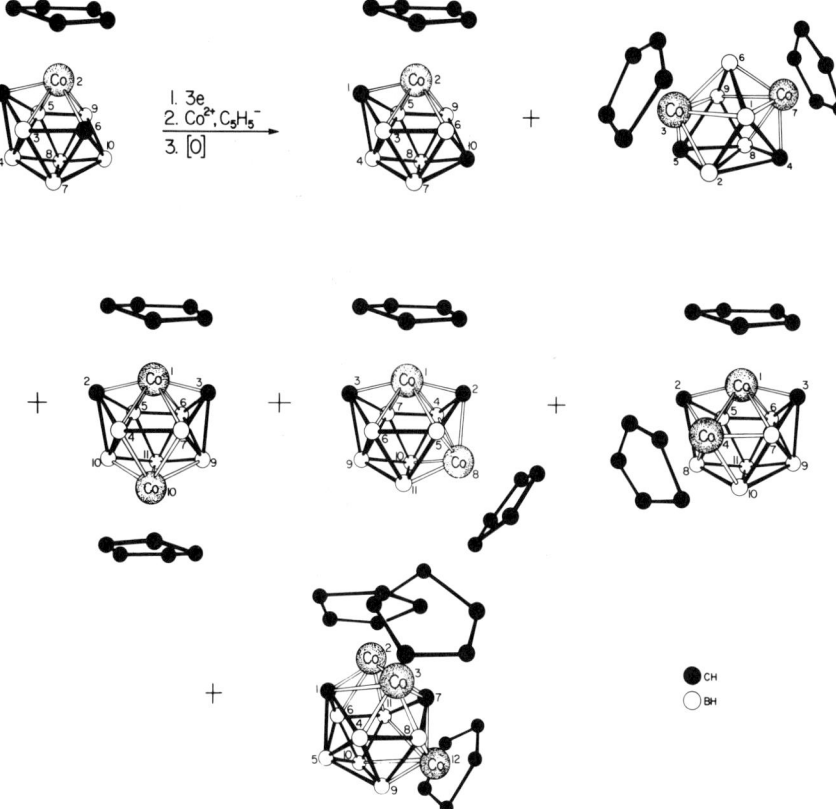

FIG. 20. Polyhedral expansion of $C_5H_5Co(1,6\text{-}C_2B_7H_9)$.

been isolated in addition to the trimetallic icosahedral complex mentioned previously, as well as a positional isomer of the starting ten-vertex monometallic species and a bimetallic nine-vertex compound resulting from polyhedral degradation.

Polyhedral expansion of the same ten-vertex starting material, employing Fe(II) rather than Co(II) as the entering transition metal, led to the preparation (19) of the heterobimetallocarborane $(C_5H_5)_2CoFeC_2B_7H_9$:

$$C_5H_5CoC_2B_7H_9 \xrightarrow[\text{2. Fe(II), C}_5\text{H}_5^-, \text{[O]}]{\text{1. }3e^-} (C_5H_5)_2CoFeC_2B_7H_9 \qquad (28)$$

The structure of this complex has recently been determined by X-ray crystallography (73) and shown to be that depicted in Fig. 21. Interestingly, the iron atom occupies the high-coordinate vertex, a result consistent

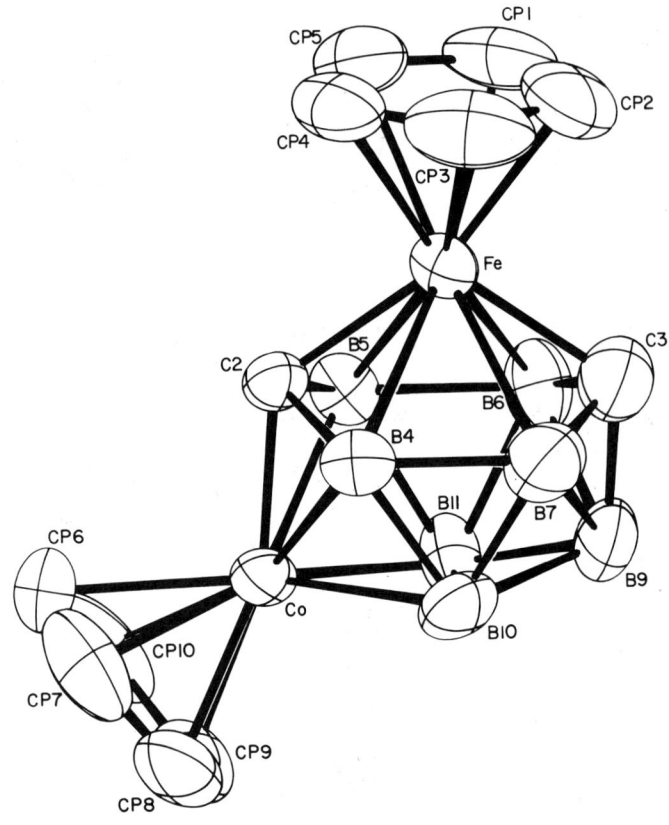

FIG. 21. Structure of $(C_5H_5)_2CoFeC_2B_7H_9$.

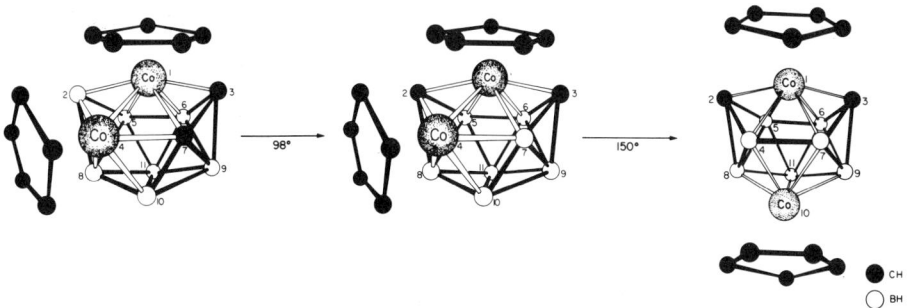

FIG. 22. Polyhedral rearrangements in the $(C_5H_5)_2Co_2C_2B_7H_9$ system.

with the fact that it formally possesses a d^5 electronic configuration and is thus more electron-deficient than the d^6 Co(III) vertex.

Other isomeric eleven-vertex bimetallocarboranes have been prepared, and the thermal rearrangements of these compounds have been studied in depth (30). There appears to be a major driving force, in all cases, for the polyhedral carbon atoms to occupy the low-coordinate vertices of the octadecahedron and for a metal atom to occupy the high-coordinate site. Once this arrangement has been achieved, further thermal isomerization results in migration of the second metal vertex over the polyhedral surface to positions at greater distance from the high-coordinate metal (Fig. 22). These rearrangement tendencies have been rationalized in terms of decreasing electrostatic repulsions and more favorable electronic environments (46).

VII
TEN-VERTEX METALLOCARBORANES

The original preparation of ten-vertex metallocarboranes involved the deprotonation of the *nido*-carborane 6,8-$C_2B_7H_{13}$, followed by reaction with a transition metal halide. Additional hydrogen gas was liberated during the complexation, and *closo*-metallocarborane compounds were formed (38):

$$6,8\text{-}C_2B_7H_{13} \xrightarrow[]{2\text{NaH}} C_2B_7H_{11}^{2-} \xrightarrow[\text{[O]}]{\text{Co(II)}} (C_2B_7H_9)_2Co^- + H_2 \qquad (29)$$

Mixed-sandwich complexes could be synthesized by adding cyclopentadienide ion to the reaction mixture (38, 48).

These ten-vertex monometallic complexes exhibit thermal polyhedral rearrangements similar to those mentioned previously. In fact, they were

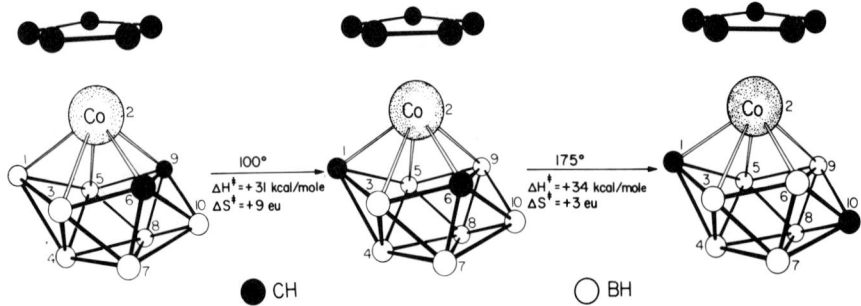

FIG. 23. Polyhedral rearrangements in the $C_5H_5CoC_2B_7H_9$ system.

the first metallocarboranes found to isomerize thermally (37). Both the commo salts and the cyclopentadienyl compounds underwent carbon atom migration when heated to 315°C. Rearrangements in this geometry (20) consist of movements of the polyhedral carbon atoms to the low-coordinate capping positions of the bicapped square antiprism. The metal has always been found to occupy a high-coordinate equatorial position in these complexes (Fig. 23) (46).

Another route to ten-vertex monometallocarboranes is through polyhedral expansion of $4,5\text{-}C_2B_7H_9$. A new isomer of $C_5H_5CoC_2B_7H_9$ with carbon atoms in positions 3 and 10 is produced in this reaction. Side products included the nine-vertex species $C_5H_5CoC_2B_6H_8$ (Section VIII) and the eleven-vertex $C_5H_5CoC_2B_8H_{10}$ (24).

Polyhedral contraction of $C_5H_5Co(2,4\text{-}C_2B_8H_{10})$ proceeds through an unusual intermediate on the way to the closo ten-vertex complex $C_5H_5Co\text{-}C_2B_7H_9$ (66). This intermediate has been shown (5) to be a *nido*-metallocarborane, $C_5H_5CoC_2B_7H_{11}$, which has both B—H—B and B—H—Co bridge bonds (Fig. 24). On heating this compound, hydrogen gas is evolved and the polyhedral contraction product $C_5H_5CoC_2B_7H_9$ is isolated.

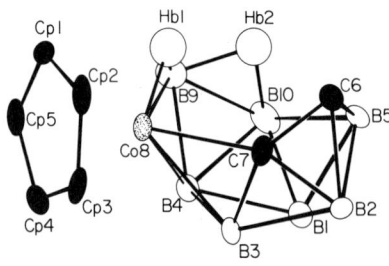

FIG. 24. Structure of the *nido*-metallocarborane $C_5H_5CoC_2B_7H_{11}$.

A novel ten-vertex bimetallocarborane, $(C_5H_5)_2Co_2C_2B_6H_8$, is produced upon the polyhedral expansion of $1,7-C_2B_6H_8$ (16). The expected monometallic nine-vertex species, $C_5H_5CoC_2B_6H_8$, is also produced in this reaction (see Fig. 4). An X-ray crystal structure of the bimetallic product (61) showed the 2 cobalt atoms to occupy adjacent positions on the two equatorial belts of the bicapped square antiprism. The carbon atoms occupy the low-coordinate caps. Thermal rearrangement (30) produced an isomeric complex in which the metals are separated from each other, but the carbon atoms remain in the low-coordinate sites.

Polyhedral expansion of $4,5-C_2B_7H_9$ with Fe(II) and $C_5H_5^-$ led to the isolation of the ten-vertex bimetallic iron complex $(C_5H_5)_2Fe_2C_2B_6H_8$ (2). This compound differs from the cobalt complex discussed above in that it has 2 fewer electrons [d^5 Fe(III) as opposed to d^6 Co(III)]. A profound structural change results from this electron deficiency: the complex adopts a geometry different from the bicapped square antiprism, although it is still classified as a *closo*-metallocarborane since it has a fully triangulated closed polyhedral structure. The iron atoms are bonded to each other at a distance of 2.571 Å (Fig. 25). A paramagnetic isomer of this compound has been reported, and it was suggested that it may have a nido geometry in which the iron atoms are no longer interacting, but definitive structural data are still lacking.

Ten-vertex metallocarboranes containing both cobalt and nickel in the same polyhedron have been synthesized from the nine-vertex anionic species $C_5H_5CoCB_7H_8^-$ (Section VIII) by polyhedral expansion (87). The resulting neutral complexes, of formula $(C_5H_5)_2CoNiCB_7H_8$ and con-

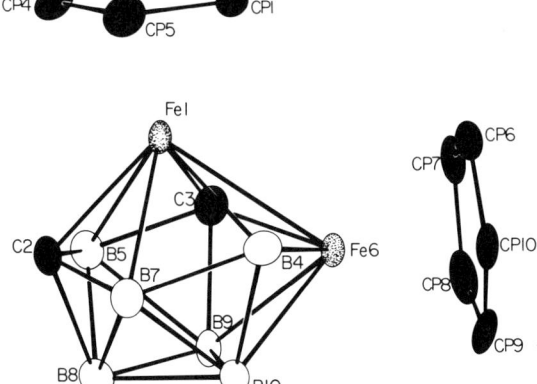

FIG. 25. Structure of $(C_5H_5)_2Fe_2C_2B_6H_8$.

taining formal Ni(IV), exhibit thermal rearrangements to isomeric species. The structure of one isomer has been crystallographically determined: the cobalt and nickel atoms occupy positions on the same equatorial belt, whereas the polyhedral carbon atom is at a capping site *(73)*. These complexes are isoelectronic with the $(C_5H_5)_2Co_2C_2B_6H_8$ isomers previously mentioned.

By heating $2,4$-$C_2B_5H_7$ at 260° in the presence of $C_5H_5Co(CO)_2$, two isomeric trimetallic complexes formulated as $(C_5H_5)_3Co_3C_2B_5H_7$ were isolated *(76)*. One isomer has one Co-Co interaction with a high-coordinate carbon atom, whereas the proposed structure of the other isomer has the 3 cobalt atoms positioned on the same equatorial belt of the bicapped square antiprism, bound to each other, and the carbon atoms occupying the apical vertices.

Polyhedral expansion of $2,4$-$C_2B_5H_7$ afforded, among other products, a new $(C_5H_5)_2Co_2C_2B_6H_8$ isomer in which the cobalt atoms occupy nonadjacent vertices on the same belt of the ten-vertex polyhedron. One of the carbon atoms occupied an equatorial position between the cobalt vertices *(42)*. The other carbon atom is located at the apical position most distant from the metal-containing belt.

The chemical reactions of monometallic ten-vertex metallocarboranes have been examined *(40)*. Friedel-Crafts acylation of $C_5H_5Co(1,6$-$C_2B_7H_9)$ produced a monosubstitution product. Attack occurred on the boron atom farthest removed from the polyhedral carbon vertices. No substitution on the cyclopentadienyl ring was observed.

VIII

NINE-VERTEX METALLOCARBORANES

The first metallocarborane of this geometry to be synthesized was an unexpected product. In an attempt to prepare a ten-vertex manganese carbonyl complex, the $C_2B_7H_{11}{}^{2-}$ ion, discussed in Section VII, was reacted with $BrMn(CO)_5$. Surprisingly, the only metal-containing compound isolated from the reaction mixture had just 6 boron atoms *(36, 50)*. The course of the reaction may be outlined as follows:

$$6,8\text{-}C_2B_7H_{13} \xrightarrow{2NaH} C_2B_7H_{11}{}^{2-} \xrightarrow{BrMn(CO)_5} C_2B_6H_8Mn(CO)_3{}^- \quad (30)$$

Similar results were observed when $ClMn(CO)_5$ or $Mn_2(CO)_{10}$ were used as metal sources and when C-phenyl- or C,C'-dimethyl-substituted carboranes were used as starting materials. The structure proposed for this complex and later confirmed by a crystallographic study *(63)* was based on tricapped trigonal prismatic geometry with the $Mn(CO)_3$ group occupying a high-coordinate position and the 2 carbon atoms in positions adjacent to the metal in low-coordinate capping vertices.

The first reported example of polyhedral expansion also resulted in the isolation of a nine-vertex monometallocarborane (16). Polyhedral expansion of $1,7\text{-}C_2B_6H_8$ in the presence of $CoCl_2$ and $C_5H_5^-$ resulted in the isolation of $C_5H_5CoC_2B_6H_8$ as well as the ten-vertex bimetallic $(C_5H_5)_2Co_2\text{-}C_2B_6H_8$ mentioned in Section VII. The monometallic complex thus prepared has both carbon atoms in low-coordinate capping positions but only 1 carbon atom is adjacent to the metal vertex. The isomeric species with both carbon atoms adjacent to the cobalt atom was isolated from the mixture of products obtained from the polyhedral expansion of $4,5\text{-}C_2B_7H_9$ with $CoCl_2$ and $C_5H_5^-$ (24) and was also prepared by thermal rearrangement of the first isomer (20). Another isomer is formed by expanding $2,4\text{-}C_2B_5H_7$ (75).

A more unusual structure has been established for the complex $C_5H_5Fe\text{-}C_2B_6H_8$, which was synthesized by polyhedral expansion of $1,7\text{-}C_2B_6H_8$ in the presence of $FeCl_2$ (24). The ^{11}B NMR spectra of the Fe(III) and Fe(II) compounds [the latter generated by Na/Hg reduction of the Fe(III) complex *in situ*] and the ^1H NMR spectrum of the Fe(II) species indicate that the carbon atoms occupy high-coordinate positions and the iron vertex is at a low-coordinate site. The structure proposed on this basis is shown in Fig. 26. An isomeric product, believed to have a structure similar to the $C_2B_6H_8Mn(CO)_3^-$ ion discussed in the foregoing, was also isolated from this reaction mixture.

Commo complexes containing two fused nine-vertex polyhedra may be prepared by the polyhedral expansion of $1,7\text{-}C_2B_6H_8$ with $CoCl_2$ in the

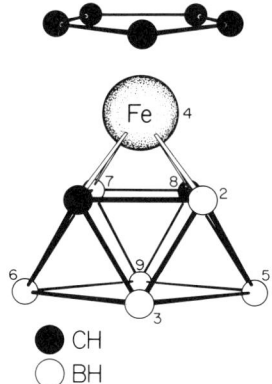

FIG. 26. Proposed structure of the ferracarborane $C_5H_5FeC_2B_6H_8$.

absence of cyclopentadienide ion (24). The carbon atom positions in the two polyhedra are identical to those in the product generated in the presence of $C_5H_5^-$.

Although it is not within our limitation of two-carbon metallocarboranes, mention has been made of the $C_5H_5CoCB_7H_8^-$ ion, which results from attempted polyhedral contraction of orange or amber $C_5H_5CoC_2B_{10}H_{12}$ (21, 22). The structure of this anion (6), previously presented in Fig. 17, is again based on tricapped trigonal prismatic geometry. The lone carbon atom is not adjacent to the metal but occupies a low-coordinate capping vertex.

Two isomeric bimetallic $(C_5H_5)_2Co_2C_2B_5H_7$ compounds were formed by heating $2,4-C_2B_5H_7$ at 260°C in the presence of $C_5H_5Co(CO)_2$. One of these complexes exhibits a metal-metal interaction (76). The other isomer, originally prepared by polyhedral expansion of $2,4-C_2B_5H_7$ with $CoCl_2$ (75), is also formed in the polyhedral expansion of $C_5H_5Co(1,6-C_2B_7H_9)$ (25).

Reaction of the $C_2B_4H_7^-$ monoanion with $NiBr_2$ and $C_5H_5^-$ afforded, among other products, the dinickelacarborane $(C_5H_5)_2Ni_2C_2B_5H_7$ (42). This electron-rich complex is diamagnetic, indicating intermetallic interaction, but a lengthening of the Ni—Ni bond was suggested as a possible geometric distortion, as expected in this compound because of its excess electron density.

IX

OXIDATIVE ADDITION TO B—H BONDS

Early in the development of polyhedral borane and carborane chemistry the need arose to degrade these materials under protolytic conditions for analytical purposes. Such reactions, for example,

$$B_nH_n^{2-} + 2H^+ + 3n\text{-RCOOH} \xrightarrow{Pt} n\text{-B(OCOR)}_3 + (2n+1)\text{-H}_2 \quad (31)$$

were useful in establishing complete empirical formulas based on total hydrogen evolution. Oddly enough, reactions of this sort would proceed at a reasonable rate only in the presence of metallic platinum or palladium catalysts. It appears as though the protolytic degradation reactions may proceed through a series of oxidative addition reactions in which the metal catalyst is inserted into B—H bonds and, possibly, into the borane cage itself. The recent results described in the following support this view.

The now well-known ortho-metalation reaction observed in certain low-valent transition metal complexes of arylphosphines (1, 79) provided a model for the observation of oxidative addition to terminal B—H bonds.

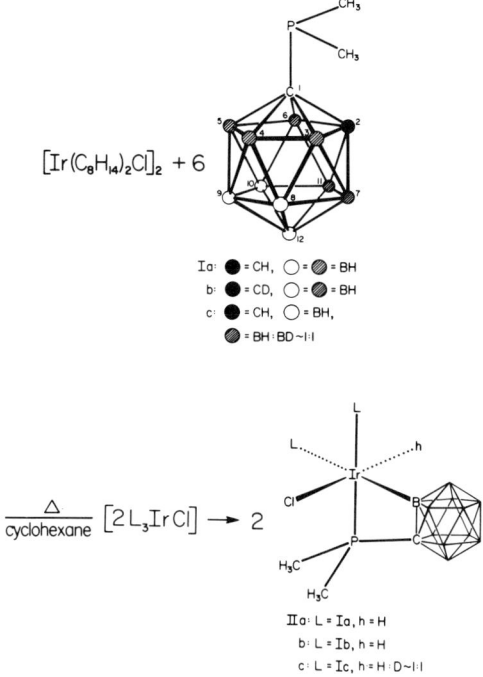

FIG. 27. Oxidative addition to a polyhedral BH vertex.

The phosphine 1-$(1,2$-$C_2B_{10}H_{11})P(CH_3)_2$ was prepared and reacted with an iridium(I) complex, as shown in Fig. 27. Spectroscopic evidence strongly supported the view that the complexed Ir(I) had inserted into a B—H bond with the formation of an iridium–boron bond and an iridium–hydride link (58). Specific deuterium labels attached to the carborane moiety of the phosphine clearly proved this point. Thus, the B-H vertices of the icosahedral carborane group that are nearest the carbon atom bearing phosphorus were shown to be involved, although a distinction could not be made between the 3,6 or 4,5 sets of BH groups.

In another series of experiments (59), toluene solutions of 1-$(1,2$-$C_2B_{10}H_{11})P(CH_3)_2$ were reacted with deuterium gas in the presence of $(Ph_3P)_3RuHCl$. Mass spectral evidence proved that up to eight deuterium atoms had been introduced into the phosphine. Since only four BH exchange sites could possibly be involved in deuterium exchange which proceeded only through ortho-metalation intermediates, an intermolecular, reversible oxidative addition of catalyst to terminal B—H bonds was indicated. Indeed, further work proved that under the same conditions 1,2-, 1,7-, and 1,12-$C_2B_{10}H_{12}$ carboranes could be totally deuterated at boron

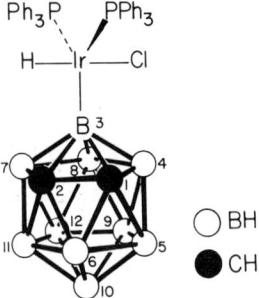

Fig. 28. Structure of the product of oxidative addition to $1,2\text{-}C_2B_{10}H_{12}$.

vertices, as could the metallocarborane $C_5H_5Co(1,2\text{-}C_2B_9H_{11})$. In fact, every B—H-containing compound examined to date was found to be susceptible to catalyzed deuterium exchange [i.e., $B_{10}H_{14}$, $B_{10}H_{10}^{2-}$, and $(CH_3)_3NBH_3$], and no exchange was observed at carbon. These results lead one to the somewhat surprising conclusion that terminal BH groups are much more reactive with low-valent transition metal complexes than are ordinary CH groups.

With the foregoing results in hand the reactions of the 1,2-, 1,7-, and 1,12-$C_2B_{10}H_{12}$ carboranes were examined with Ir(I) compounds. In each case iridium was inserted into terminal B—H bonds to form stable σ-bonds between boron and iridium (60):

$\frac{1}{2}[Ir(C_8H_{14})_2Cl]_2 + 2Ph_3P + 1,2\text{-}C_2B_{10}H_{12} \longrightarrow 3\text{-}[(Ph_3P)_2IrHCl]\text{-}1,2\text{-}C_2B_{10}H_{11}$ (32)

The relative reactivities of the various BH groups present in each carborane were such that those BH groups nearest the carbon vertices were favored for attack. Figure 28 illustrates this point. Reaction of the B–σ-bonded metallocarboranes with CO led to the regeneration of the carborane and the formation of Vaska's compound, $trans\text{-}IrCl(CO)(PPh_3)_2$,

$3\text{-}[(Ph_3P)_2IrHCl]\text{-}1,2\text{-}C_2B_{10}H_{11} + CO \longrightarrow 1,2\text{-}C_2B_{10}H_{12} + trans\text{-}IrCl(CO)(PPh_3)_2$

(33)

X

METALLOCARBORANES IN HOMOGENEOUS CATALYSIS

Although the chemistry described in the foregoing deals with the oxidative addition of low-valent transition metal complexes to terminal B—H bonds, we now describe the formal addition of the same reagents to the open face of the isomeric $C_2B_9H_{12}^-$ ions with the formation of an Ir—H bond *(80)*. In a sense this represents oxidative addition to a B—H—B bridge linkage. The most thoroughly studied system of this sort was generated from Rh(I):

$$(Ph_3P)_3RhCl + 7,8\text{-}C_2B_9H_{12}^- \longrightarrow Cl^- + 3,3\text{-}(Ph_3P)_2\text{-}3\text{-}H\text{-}3\text{-}Rh\text{-}1,2\text{-}C_2B_9H_{11} \quad (34)$$

The structure of the product has been confirmed by an X-ray diffraction study *(92)*. The iridium analog was also prepared in a similar fashion.

Although the hydridorhodacarborane is formally a rhodium(III) derivative, it functions as a facile catalyst in alkene isomerization, hydrogenation, hydroformylation, and hydrosilylation reactions *(80)*. This catalyst system is extremely stable and may be recovered quantitatively from alkene isomerization and hydrogenation reactions. In addition to these reactions, the hydridorhodacarborane is very effective in the catalysis of deuterium exchange at terminal BH positions *(59)*. These discoveries may soon lead to industrially useful metallocarborane catalysts.

REFERENCES

1. Bennett, M. A., and Milner, D. L., *J. Amer. Chem. Soc.* **91**, 6983 (1969).
2. Callahan, K. P., Evans, W. J., Lo, F. Y., Strouse, C. E., and Hawthorne, M. F., *J. Amer. Chem. Soc.* **97**, 296 (1975).
3. Callahan, K. P., and Hawthorne, M. F., *J. Organometal. Chem.* **100**, 97 (1975), and references therein.
4. Callahan, K. P., and Hawthorne, M. F., *Pure Appl. Chem.* **39**, 475 (1974) and references therein.
5. Callahan, K. P., Lo, F. Y., Sims, A. L., Strouse, C. E., and Hawthorne, M. F., *Inorg. Chem.* **13**, 2842 (1974).
6. Callahan, K. P., Sims, A. L., Strouse, C. E., and Hawthorne, M. F., *Inorg. Chem.* **13**, 1393 (1974).
7. Callahan, K. P., Sims, A. L., Strouse, C. E., and Hawthorne, M. F., *Inorg. Chem.* **13**, 1397 (1974).
8. Churchill, M. R., and DeBoer, B. G., *Inorg. Chem.* **13**, 1411 (1974).
9. Churchill, M. R., and Gold, K., *Inorg. Chem.* **12**, 1157 (1973).
10. Churchill, M. R., and Gold, K., *J. Amer. Chem. Soc.* **92**, 1180 (1970).
11. Churchill, M. R., Gold, K., Francis, J. N., and Hawthorne, M. F., *J. Amer. Chem. Soc.* **91**, 1222 (1969).
12. Churchill, M. R., Reis, A. H., Jr., Francis, J. N., and Hawthorne, M. F., *J. Amer. Chem. Soc.* **92**, 4993 (1970).
13. Curphey, T. J., Santer, J. O., Rosenblum, M., and Richards, J. H., *J. Amer. Chem. Soc.* **82**, 5249 (1960).
14. DeBoer, B. G., Zalkin, A., Hopkins, T. E., and Templeton, D. H., *Inorg. Chem.* **7**, 2288 (1968).
15. Dewar, M. J. S., *Bull. Soc. Chim. France* **18**, 79 (1951).
16. Dunks, G. B., and Hawthorne, M. F., *J. Amer. Chem. Soc.* **92**, 7213 (1970).

17. Dunks, G. B., McKown, M. M., and Hawthorne, M. F., *J. Amer. Chem. Soc.* **93**, 2541 (1971).
18. Dustin, D. F., Dunks, G. B., and Hawthorne, M. F., *J. Amer. Chem. Soc.* **95**, 1109 (1973).
19. Dustin, D. F., Evans, W. J., and Hawthorne, M. F., *J. Chem. Soc., Chem. Commun.*, 805 (1973).
20. Dustin, D. F., Evans, W. J., Jones, C. J., Wiersema, R. J., Gong, H., Chan, S., and Hawthorne, M. F., *J. Amer. Chem. Soc.* **96**, 3085 (1974).
21. Dustin, D. F., and Hawthorne, M. F., *Inorg. Chem.* **12**, 1380 (1973).
22. Dustin, D. F., and Hawthorne, M. F., *J. Chem. Soc., Chem. Commun.* 1329 (1972).
23. Dustin, D. F., and Hawthorne, M. F., *J. Amer. Chem. Soc.* **96**, 3462 (1974).
24. Evans, W. J., Dunks, G. B., and Hawthorne, M. F., *J. Amer. Chem. Soc.* **95**, 4565 (1973).
25. Evans, W. J., and Hawthorne, M. F., *Inorg. Chem.* **13**, 869 (1974).
26. Evans, W. J., and Hawthorne, M. F., *J. Amer. Chem. Soc.* **93**, 3063 (1971).
27. Evans, W. J., and Hawthorne, M. F., *J. Amer. Chem. Soc.* **96**, 301 (1974).
28. Evans, W. J., and Hawthorne, M. F., *J. Chem. Soc., Chem. Commun.* 706 (1973).
29. Evans, W. J., and Hawthorne, M. F., *J. Chem. Soc., Chem. Commun.* 38 (1974).
30. Evans, W. J., Jones, C. J., Štibr, B., and Hawthorne, M. F., *J. Organometal. Chem.* **60**, C27 (1973).
31. Fein, M. M., Bobinski, J., Mayes, N., Schwartz, N., and Cohen, M. S., *Inorg. Chem.* **2**, 1111 (1963).
32. Francis, J. N., and Hawthorne, M. F., *Inorg. Chem.* **10**, 594 (1971).
33. Francis, J. N., and Hawthorne, M. F., *Inorg. Chem.* **10**, 863 (1971).
34. Francis, J. N., and Hawthorne, M. F., *J. Amer. Chem. Soc.* **90**, 1663 (1968).
35. Francis, J. N., Jones, C. J., and Hawthorne, M. F., *J. Amer. Chem. Soc.* **94**, 4878 (1972).
36. George, A. D., and Hawthorne, M. F., *Inorg. Chem.* **8**, 1801 (1969).
37. George, T. A., and Hawthorne, M. F., *J. Amer. Chem. Soc.* **90**, 1661 (1968).
38. George, T. A., and Hawthorne, M. F., *J. Amer. Chem. Soc.* **91**, 5475 (1969).
39. Grafstein, D., and Dvorak, J., *Inorg. Chem.* **2**, 1128 (1963).
40. Graybill, B. M., and Hawthorne, M. F., *Inorg. Chem.* **8**, 1799 (1969).
41. Grimes, R. N., "Carboranes." Academic Press, New York, 1970, and references therein.
42. Grimes, R. N., Beer, D. C., Sneddon, L. G., Miller, V. R., and Weiss, R., *Inorg. Chem.* **13**, 1138 (1974).
43. Hanson, F. V., Hazell, R. G., Hyatt, C., and Stucky, G. D., *Acta Chem. Scand.* **27**, 1210 (1973).
44. Hawthorne, M. F., and Andrews, T. D., *Chem. Commun.* 443 (1965).
45. Hawthorne, M. F., and Andrews, T. D., *J. Amer. Chem. Soc.* **87**, 2196 (1965).
46. Hawthorne, M. F., Callahan, K. P., and Wiersema, R. J., *Tetrahedron* **30**, 1795 (1974), and references therein.
47. Hawthorne, M. F., and Dunks, G. B., *Science* **178**, 462 (1972), and references therein.
48. Hawthorne, M. F., and George, T. A., *J. Amer. Chem. Soc.* **89**, 7114 (1967).
49. Hawthorne, M. F., and Pilling, R. L., *J. Amer. Chem. Soc.* **87**, 3987 (1965).
50. Hawthorne, M. F., and Pitts, A. D., *J. Amer. Chem. Soc.* **89**, 7115 (1967).
51. Hawthorne, M. F., and Ruhle, H. W., *Inorg. Chem.* **8**, 176 (1969).
52. Hawthorne, M. F., Warren, L. F., Jr., Callahan, K. P., and Travers, N. F., *J. Amer. Chem. Soc.* **93**, 2407 (1971).

53. Hawthorne, M. F., and Wegner, P. A., *J. Amer. Chem. Soc.* **90,** 896 (1968).
54. Hawthorne, M. F., Young, D. C., Andrews, T. D., Howe, D. V., Pilling, R. L., Pitts, A. D., Reintjes, M., Warren, L. F., Jr., and Wegner, P. A., *J. Amer. Chem. Soc.* **90,** 879 (1968).
55. Hawthorne, M. F., Young, D. C., Garrett, P. M., Owen, D. A., Schwerin, S. G., Tebbe, F. N., and Wegner, P. A., *J. Amer. Chem. Soc.* **90,** 862 (1968).
56. Hawthorne, M. F., Young, D. C., and Wegner, P. A., *J. Amer. Chem. Soc.* **87,** 1818 (1965).
57. Heying, T. L., Ager, J. W., Jr., Clark, S. L., Mangold, D. J., Goldstein, H. L., Hillman, M., Polak, R. J., and Szymanski, J. W., *Inorg. Chem.* **2,** 1089 (1963).
58. Hoel, E. L., and Hawthorne, M. F., *J. Amer. Chem. Soc.* **95,** 2712 (1973).
59. Hoel, E. L., and Hawthorne, M. F., *J. Amer. Chem. Soc.* **96,** 4676 (1974).
60. Hoel, E. L., and Hawthorne, M. F., *J. Amer. Chem. Soc.* **96,** 6770 (1974).
61. Hoel, E. L., Strouse, C. E., and Hawthorne, M. F., *Inorg. Chem.* **13,** 1388 (1974).
62. Hoffmann, R. and Lipscomb, W. N., *J. Chem. Phys.* **36,** 3489 (1962).
63. Hollander, F. J., Templeton, D. H., and Zalkin, A., *Inorg. Chem.* **12,** 2262 (1973).
64. Howe, D. V., Jones, C. J., Wiersema, R. J., and Hawthorne, M. F., *Inorg. Chem.* **10,** 2516 (1971).
65. Hughes, R. L., Smith, I. C., and Lawless, E. W., "Production of the Boranes and Related Research" (R. T. Holzmann, ed). Academic Press, New York, 1967.
66. Jones, C. J., Francis, J. N., and Hawthorne, M. F., *J. Amer. Chem. Soc.* **94,** 8391 (1972).
67. Jones, C. J., Francis, J. N., and Hawthorne, M. F., *J. Amer. Chem. Soc.* **95,** 7633 (1973).
68. Jones, C. J., Francis, J. N., and Hawthorne, M. F., *J. Chem. Soc., Chem. Commun.* 900 (1972).
69. Jones, C. J., and Hawthorne, M. F., *Inorg. Chem.* **12,** 608 (1973).
70. Kaloustian, M. K., Wiersema, R. J., and Hawthorne, M. F., *J. Amer. Chem. Soc.* **94,** 6679 (1972).
71. Lipscomb, W. N., *Accounts Chem. Res.* **6,** 257 (1973), and references therein.
72. Lipscomb, W. N., "Boron Hydrides," Benjamin, New York, 1963.
73. Lo, F. Y., and Strouse, C. E., personal communication.
74. Lo, F. Y., Strouse, C. E., Callahan, K. P., Knobler, C. B., and Hawthorne, M. F., *J. Amer. Chem. Soc.* **97,** 428 (1975).
75. Miller, V. R., and Grimes, R. N., *J. Amer. Chem. Soc.* **95,** 2830 (1973).
76. Miller, V. R., Sneddon, L. G., Beer, D. C., and Grimes, R. N., *J. Amer. Chem. Soc.* **96,** 3090 (1974).
77. Muetterties, E. L., and Knoth, W. H., "Polyhedral Boranes." Dekker, New York, 1968.
78. Mulliken, R. S., *J. Amer. Chem. Soc.* **74,** 811 (1952).
79. Parshall, G. W., *Accounts Chem. Res.* **3,** 139 (1970).
80. Paxson, T. E., and Hawthorne, M. F., *J. Amer. Chem. Soc.* **96,** 4674 (1974).
81. Plešek, J., Štibr, B., and Heřmánek, S., *Syn. Inorg. Metal-Org. Chem.* **3,** 291 (1973).
82. Popp, G., and Hawthorne, M. F., *Inorg. Chem.* **10,** 391 (1971).
83. Popp, G., and Hawthorne, M. F., *J. Amer. Chem. Soc.* **90,** 6553 (1968).
84. Rosenblum, M., "Chemistry of the Iron Group Metallocenes." Wiley, New York, 1965.
85. Ruhle, H. W., and Hawthorne, M. F., *Inorg. Chem.* **7,** 2279 (1968).
86. Salentine, C. G., and Hawthorne, M. F., *J. Amer. Chem. Soc.* **97,** 426 (1975).
87. Salentine, C. G., and Hawthorne, M. F., *J. Chem. Soc., Chem. Commun.* 560 (1973).

88. Spencer, J. L., Green, M., and Stone, F. G. A., *J. Chem. Soc., Chem. Commun.* 1178 (1972).
89. St. Clair, D., Zalkin, A., and Templeton, D. H., *Inorg. Chem.* **8**, 2080 (1969).
90. St. Clair, D., Zalkin, A., and Templeton, D. H., *Inorg. Chem.* **10**, 2587 (1971).
91. St. Clair, D., Zalkin, A., and Templeton, D. H., *J. Amer. Chem. Soc.* **92**, 1173 (1970).
92. Strouse, C. E., and Hardy, G., personal communication.
93. Todd, L. J., *Advan. Organometal. Chem.* **8**, 87 (1970).
94. Warren, L. F., Jr., and Hawthorne, M. F., *J. Amer. Chem. Soc.* **90**, 4823 (1968).
95. Warren, L. F., Jr., and Hawthorne, M. F., *J. Amer. Chem. Soc.* **92**, 1157 (1970).
96. Wegner, P. A., and Hawthorne, M. F., *Chem. Commun.* 861 (1966).
97. Weisboeck, R. A., and Hawthorne, M. F., *J. Amer. Chem. Soc.* **86**, 1642 (1964).
98. Wilson, R. J., Warren, L. F., Jr., and Hawthorne, M. F., *J. Amer. Chem. Soc.* **91**, 758 (1969).
99. Wing, R. M., *J. Amer. Chem. Soc.* **89**, 5599 (1967).
100. Wing, R. M., *J. Amer. Chem. Soc.* **90**, 4828 (1968).
101. Wing, R. M., personal communication.
102. Young, D. A. T., Wiersema, R. J., and Hawthorne, M. F., *J. Amer. Chem. Soc.* **93**, 5687 (1971).
103. Zalkin, A., Hopkins, T. E., and Templeton, D. H., *Inorg. Chem.* **6**, 1911 (1967).
104. Zalkin, A., Templeton, D. H., and Hopkins, T. E., *J. Amer. Chem. Soc.* **87**, 3988 (1965).

Recent Advances in Organoantimony Chemistry

ROKURO OKAWARA and YOSHIO MATSUMURA*

Department of Applied Chemistry
Osaka University
Yamadakami, Suita, Osaka, Japan

I. Introduction 187
II. Hexacoordinate Mono- and Diorganoantimony Compounds . . 188
III. Triorganostibine Sulfide 192
   A. Nature of the Stibine–Sulfur Bond 193
   B. Reactions 195
IV. Tertiary Stibines 197
   A. Cleavage Reactions of the Phenyl–Antimony Bond . . . 197
   B. Properties of Asymmetrical Tertiary Stibines and Resolution of Quaternary Stibonium Iodide 198
   C. Coordination Behavior with Transition Metal Carbonyls . . 200
   References 202

I
INTRODUCTION

Organoantimony compounds were synthesized for the first time by Löwig in 1850, and between 1910 and 1930 many new compounds were prepared, largely in the hope of finding pharmacologically active substances similar to those of arsenic. However, further progress was slow, and organoantimony chemistry has been one of the backward areas compared with the remarkable progress since 1960 in other fields of organometallic chemistry. In 1970, there appeared an extensive monograph on organometallic compounds of arsenic, antimony, and bismuth by Doak and Freedman (1), covering the pertinent literature on organoantimony compounds through the end of 1967. In the past 5 years, there has been increased activity in this area of chemistry. Here we describe some recent advances in the field, emphasizing mainly structural aspects and interesting new reactions.

* Present address: Japan Synthetic Rubber Co., Ltd., Research Laboratory, 7569 Ikuta, Tama, Kawasaki, Japan.

II
HEXACOORDINATE MONO- AND DIORGANOANTIMONY COMPOUNDS

Pentavalent compounds, R_5Sb, R_4SbX, and R_3SbX_2, are the best-known organoantimony compounds, and the structure of these stable compounds has been the subject of several investigations (1–4). On the other hand, mono- and dialkylantimony halides, $RSbX_4$ and R_2SbX_3, are thermally unstable and decompose even at room temperature with evolution of RX forming SbX_3 and $RSbX_2$, respectively, so that very few studies had been done on these compounds (5, 6). However, we have succeeded in stabilizing these halides by substituting an X atom with a chelating acetylacetonate group or by adding an oxygen donor molecule, such as hexamethylphosphoric triamide, and have studied the structures of resulting hexacoordinate alkylantimony compounds (7–9) together with those of corresponding arylantimony derivatives (8–11). At almost the same time a research group in Utrecht independently studied hexacoordinate organoantimony compounds including acetylacetonate (12–14), oxinate (15), carboxylate (16), and alkoxide (16) derivatives.

One of the most interesting features in these hexacoordinate compounds is the existence of two isomers for dichlorodiaryl(acetylacetonato)antimony in solution, deduced from the IR and PMR spectra of $(C_6H_5)_2SbCl_2(acac)$[1] in solution. The research group in Utrecht (12) initially suggested that two forms existed in equilibrium in $CHCl_3$: a hexacoordinate *trans*-dichloro structure with a chelating acetylacetonate ligand and a pentacoordinate structure with the antimony atom bonded to the γ-carbon atom of the acetylacetonate group. However, we proposed (10) that the two isomers in various solvents might be *cis*- and *trans*-diphenyl forms. This proposal was confirmed by detailed studies (11) of dihalodiaryl(acetylacetonato)-antimony in solution. Recently, in agreement with our conclusion, the Utrecht group reported further studies (17) on the trans–cis isomerization process and the influence of these two forms on β-diketonate ligand exchange reactions in dichlorodiphenyl(β-diketonato)antimony.

Some new mono- and diorganoantimony chloride derivatives were obtained according to the following reactions.[2]

---

[1] Acac = acetylacetonate group.

[2] Compounds $(p\text{-}YC_6H_4)_2SbBr_2(acac)$ (Y = NO_2, Cl, H, CH_3, CH_3O) were prepared from $p\text{-}YC_6H_4SbO(OH)_2$ (11, 18b) and $p\text{-}YC_6H_4SbCl_3(acac)$ (Y = NO_2, H, CH_3) were obtained from corresponding stibonic acid (18a, b). HMPA = hexamethylphosphoric triamide; DMSO = dimethyl sulfoxide; PyO = pyridine N-oxide; 4-CH_3PyO = γ-picoline N-oxide; TPPO = triphenylphosphine oxide.

```
                          Na(acac)
                         ┌─────────→ R₂SbCl₂(acac)   (7)
           SO₂Cl₂        │
R₂SbCl ──────────→ R₂SbCl₃─┤
                         │  L
                         └─────────→ R₂SbCl₃L        (7, 8)

         (acac)H
         SO₂Cl₂
        ┌─────────→ RSbCl₃(acac)    (7)
        │  −70°
RSbCl₂─┤
        │  SO₂Cl₂
        │  L
        └─────────→ RSbCl₄L         (9)
           −70°
```

(R = CH₃, C₆H₅, p-CH₃C₆H₄; L = HMPA, DMSO, PyO, 4-CH₃PyO, TPPO)

These compounds are stable at room temperature in the solid state, but in solution some monomethylantimony tetrachloride adducts, CH_3SbCl_4L (L = HMPA, at room temperature; L = PyO or 4-CH₃PyO above 70°), decompose into CH_3Cl and $SbCl_3L$.

All these compounds mentioned contain a hexacoordinate antimony atom. From the PMR spectrum of $C_6H_5SbCl_3(acac)$ in CHCl₃, we have suggested a structure I in which the phenyl and the acetylacetonate groups are in the same plane (*18a*). Recently, a similar structure has been established for the methyl analog, $CH_3SbCl_3(acac)$ (*19*), by X-ray crystallography.

```
              Cl
              │       CH₃
         Cl---┼---O═C
              │ \   \
              Sb     CH
         /    │ \   /
      H₅C₆---┼---O═C
              │       CH₃
              Cl
```

(I)

Configurational isomers in equilibrium in solution were also found for the two series of hexacoordinate antimony compounds, $R_2SbX_2(acac)$ (R = aryl) and CH_3SbCl_4L (L = PyO or 4-CH₃PyO). The PMR spectrum of $(C_6H_5)_2SbCl_2(acac)$ at 22° in CDCl₃ shows two sets of acetylacetonate methyl (A,B) and γ-protons (C,D), as shown in Fig. 1. However, the spectrum at −30° of the solution freshly prepared below −30° in dichloromethane shows only one methyl resonance (A in Fig. 2). With increasing temperature, a new peak (B in Fig. 2) begins to appear and the intensity ratio B/A increases rapidly to the equilibrium value. From these observations, together with the results of benzene-induced solvent shifts of methyl protons of both tolyl- and acetylacetonate groups in (p-CH₃C₆H₄)₂-

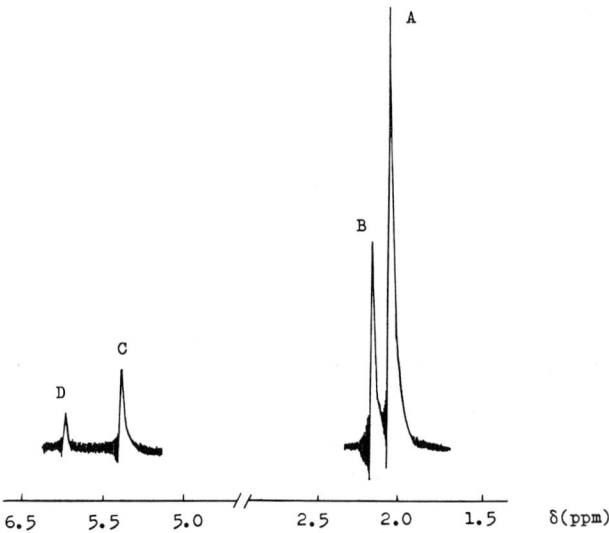

FIG. 1. Proton magnetic resonance spectrum of $(C_6H_5)_2SbCl_2(acac)$ in $CDCl_3$ at 22°.

$SbCl_2(acac)$, we proposed that isomerization of the *trans*-phenyl (II) (peak A) to *cis*-phenyl (III) (peak B) isomer takes place and that the two isomers are in equilibrium in solution.[3] In the solid state, this compound gave two crystal forms. Both forms show the same PMR spectral properties on solution as described in the foregoing. The X-ray structure determination revealed, however, that both have *trans*-phenyl structures. The only difference is in the dihedral angle of the two phenyl ring planes as shown in Fig. 3 (*20, 21*). A *trans*-methyl structure was also indicated by an X-ray structural study for the methyl analog, $(CH_3)_2\text{-}SbBr_2(acac)$ (*22*).

---

[3] The effect of substituents X and Y on the equilibrium was also studied for $(p\text{-}YC_6H_4)_2SbX_2(acac)$ (X = F, Cl, Br; Y = NO_2, Cl, CH_3, CH_3O) (*11*).

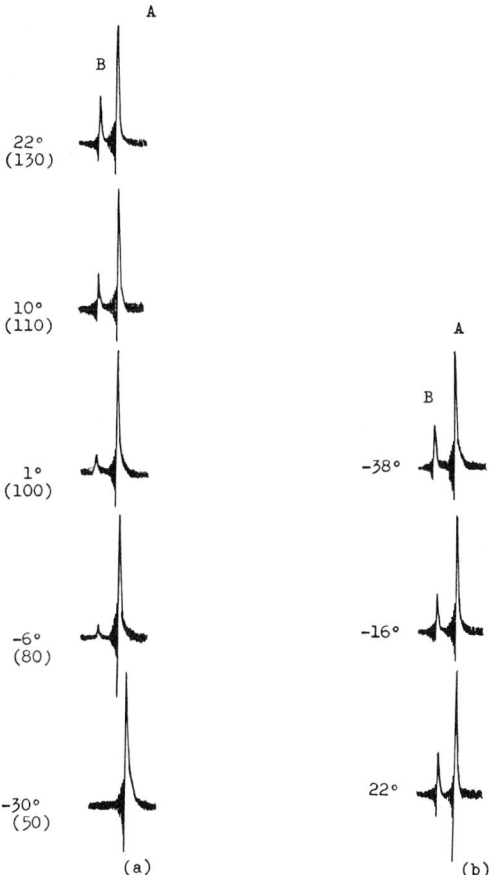

FIG. 2. Methyl proton resonances of $(C_6H_5)_2SbCl_2(acac)$ in dichloromethane at various temperatures. (a) The solution was freshly prepared below $-30°$. The figures in parentheses represent the time (in minutes) elapsed from the preparation of the solution to the measurements. (b) The solution was kept at room temperature for 24 hours, and ample time was allowed for each measurement.

For monomethylantimony adducts, CH_3SbCl_4L (L = PyO or 4-CH_3PyO) in solution, two isomeric forms (VI) and (VII) were also suggested to be in equilibrium from the solvent-dependent PMR spectra (9).

From the PMR and IR spectra of $(CH_3)_2SbCl_3L$ (L = DMSO, HMPA, TPPO, or PyO) in solution, an octahedral geometry with *trans*-methyl configuration (VIII) was suggested (8).

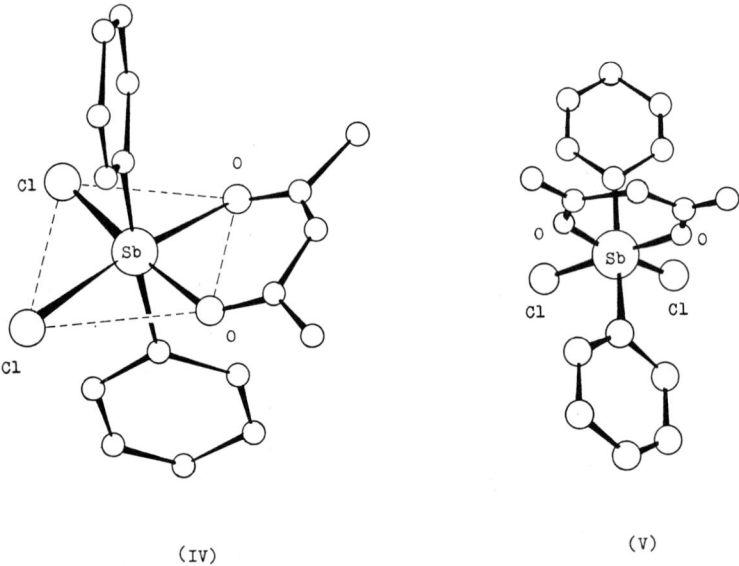

FIG. 3. The molecular structure of $(C_6H_5)_2SbCl_2(acac)$. The dihedral angle of two phenyl ring planes: (IV), 84.6° (20); (V), 38° (21).

It is notable that the existence of configurational isomers in solution has been established for these hexacoordinate organoantimony compounds, since among hexacoordinate organotin compounds, the structures of which have been extensively studied, there are few reports of such isomers.[4]

III
TRIORGANOSTIBINE SULFIDE

In contrast to the well-studied triorganophosphine oxide or sulfide, very little work had been done on the analogous organoantimony compounds.

[4] Recently two configurational isomers have been isolated from dimethyltin N,N'-bis(salicylaldehyde)ethylenediiminate (23).

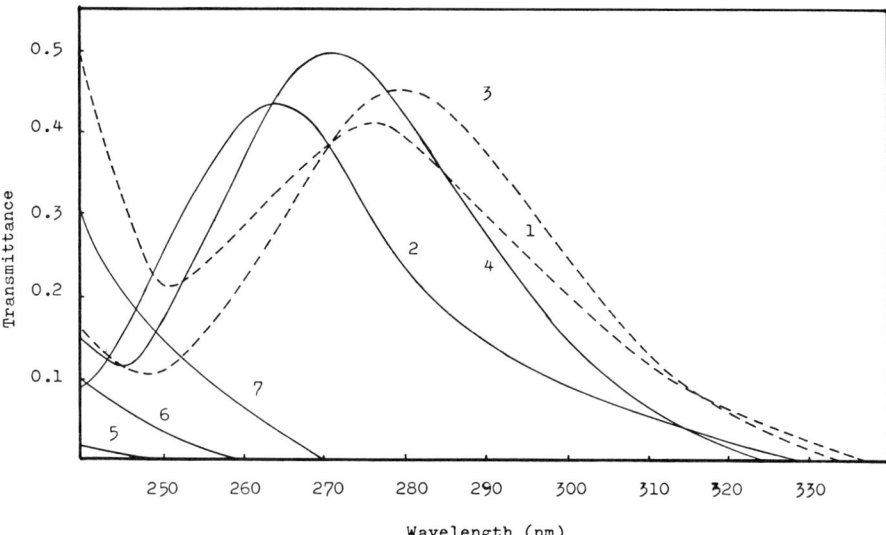

FIG. 4. Ultraviolet spectra of R_3SbS and R_3SbX_2. Dashed curve, in n-hexane; solid curve, in acetonitrile. (1) $(CH_3)_3SbS$ (λ_{max}, 279 nm, because of its poor solubility, the supernatant solution obtained after shaking for 30 minutes was used); (2) $(CH_3)_3SbS$ (1.10×10^{-4} mole/liter, λ_{max} 267 nm, log ϵ 3.6); (3) $(C_6H_{11})_3SbS$ (0.893×10^{-4} mole/liter, λ_{max} 282 nm, log ϵ 3.7); (4) $(C_6H_{11})_3SbS$ (0.839×10^{-4} mole/liter, λ_{max} 274 nm, log ϵ 3.8); (5) $(CH_3)_3SbCl_2$ (1.27×10^{-4} mole/liter); (6) $(CH_3)_3SbBr_2$ (1.06×10^{-4} mole/liter); (7) $(C_6H_{11})_3SbBr_2$ (1.01×10^{-4} mole/liter).

Zingaro and Cheremos (24) reported that triorganostibine oxide in solution is an equilibrium mixture of monomer and polymer. However, triorganostibine sulfide is monomeric and is expected to have a semipolar bond like that of R_3PO or R_3PS. We investigated the nature of this bond by spectroscopic methods (25, 26). Also, in the hope of finding a particular reactivity of this bond, the reactions with alkyl halides (25), acid halides (27), or some organometallic compounds (28–30) were carried out.

A. Nature of the Stibine–Sulfur Bond

The UV spectra of trimethyl- and tricyclohexylstibine sulfides have absorption maxima at ca. 280 nm in n-hexane which give a blue shift in acetonitrile, as shown in Fig. 4. These bands may be assigned to intra-

TABLE I

The ^{13}C—H Coupling Constants of Methylantimony and Methyl Phosphorus Compounds

Compound	Solvent	J(Hz)^a	fpc^b
$(CH_3)_3Sb$	Neat	130.5	0
$(CH_3)_3SbS$	$CHCl_3$	136.5	0.67
$(CH_3)_4Sb^+$ ^{c}	H_2O	139.5	1
$(CH_3)_3P$	Neat	127^d	0
$(CH_3)_3PO$	D_2O	129^d	0.29^e
$(CH_3)_4P^+$	D_2O	134^d	1

^{a} Values are considered reliable to ±0.5 Hz.
^{b} fpc = formal positive charge.
^{c} $(CH_3)_4SbI$ was employed as a source of $(CH_3)_4Sb^+$.
^{d} Haake et al. (31).
^{e} This value is comparable with that obtained from a dipole moment measurement (0.36) [see Phillips et al. (32)].

molecular charge transfer from the semipolar structure (IX) to the double bond structure (X).

$(CH_3)_3\overset{+}{Sb}-\overset{-}{S}$ $(CH_3)_3Sb=S$

(IX)   (X)

This suggests that the semipolar structure makes a major contribution to the resonance hybrid of the sulfide in its ground state (25).[5]

In Table I the ^{13}C—H coupling constants of three methylantimony compounds are shown, together with those of methyl phosphorus analogs for comparison. From the formal positive charge, calculated on the assumption that the ^{13}C—H coupling constants are dependent linearly on the amount of positive charge on the central atom, the Sb—S bond in $(CH_3)_3SbS$ may be regarded as a semipolar bond (IX), although some contribution of the $d\pi$-$p\pi$ bonding may also be involved (26).[6] This is consistent with the results obtained by the UV spectral investigation.

[5] The Sb—S stretching frequency of $(CH_3)_3SbS$ is observed at 433 cm^{-1}. The estimated frequencies for Sb—S pure single and double bonds are 337 and 483 cm^{-1}, respectively (28).

[6] The P—O bond in $(CH_3)_3PO$ has been reported to be well described as a double bond, since the ^{13}C—H coupling constant of this compound is closer to that of $(CH_3)_3P$ than to that of $(CH_3)_4P^+$ (31).

B. Reactions

1. With Alkyl and Acid Halides

Reactions of trimethylstibine sulfide with alkyl halides proceed smoothly under mild conditions to give dialkyl disulfides and organoantimony compounds in quantitative yields (25):

$$2\,(CH_3)_3SbS + 2\,RX \longrightarrow 2\begin{bmatrix}(CH_3)_3Sb-S\\ |\quad\quad\;|\\ X-R\end{bmatrix} \longrightarrow 2\begin{bmatrix}(CH_3)_3Sb\diagdown_{X}^{SR}\end{bmatrix}$$

$$\quad\quad\quad\quad\quad\quad\quad\quad\quad\quad\quad\quad (XI) \quad\quad\quad\quad\quad\quad (XII)$$

$$(CH_3)_3Sb + (CH_3)_3SbX_2 + R_2S_2$$

$(RX = CH_3I,\ C_2H_5I,\ C_6H_5CH_2I,\ C_6H_5CH_2Br)$

The semipolar Sb—S bond may easily allow formation of trimethylantimony halide thioalkoxide (XII) via a four-centered intermediate (XI).

In the reaction of trimethylstibine sulfide with acid halide, trimethylantimony halide thiocarboxylates (XIII), corresponding to compound XII in Eq. (1) were isolated (27):

$$(CH_3)_3SbS + RCOX \longrightarrow (CH_3)_3SbX(SCOR) \quad\quad (2)$$
$$\quad\quad\quad\quad\quad\quad\quad\quad\quad\quad\quad (XIII)$$

$(R = CH_3,\ or\ C_6H_5;\ X = Cl,\ or\ Br)$

The difference in stability between compounds XII and XIII may be attributed to a less reactive Sb—S bond in XIII. This is consistent with the fact that trimethylantimony bis(thiocarboxylates) is thermally more stable than the bis(thioalkoxides) (33, 34).

2. With Some Organometallic Compounds

The reactions of R_3SbS ($R = CH_3,\ C_6H_{11}$) with organotin compounds are summarized in Table II. The results can be classified into three types: recovery of the reactants, complex formation, and sulfur–halogen exchange (28, 35). Complexes $R_2SnX_2\cdot 2(CH_3)_3SbS$ ($R = CH_3,\ C_2H_5;\ X = Cl,\ or\ Br$) were obtained only in a few cases. These complexes exist as monomeric species in toluene, but in $CHCl_3$ the IR and PMR spectra indicate that these complexes change into three species, $(CH_3)_3SbS$, $(CH_3)_3SbX_2$, and $[(CH_3)_2SnS]_3$. Since the complexes can be prepared from the mixture of

TABLE II

REACTIONS OF R_3SbS AND TIN COMPOUNDS IN METHANOL, ACETONE, OR CHLOROFORM

Tin chloride[a]	$(C_6H_{11})_3SbS$[b]	$(CH_3)_3SbS$[b]
R_3SnCl		
$R = n\text{-}C_4H_9$	—[b]	—
$n\text{-}C_3H_7$	—	—
C_2H_5	—	—
CH_3	—	—
C_6H_5	Ex[c]	Ex
$n\text{-}C_4H_9SnCl_3$	Ex	Ex
R_2SnCl_2		
$R = n\text{-}C_4H_9$	—	—
$n\text{-}C_3H_7$	Ex	—
C_2H_5	Ex	Complex
CH_3	Ex	Complex
C_6H_5	Ex	Ex
$SnCl_4$	Ex	Ex

[a] The results obtained with bromide are quite similar to those of chloride.
[b] Dash (—) = starting materials are recovered.
[c] Ex = exchange reaction.

these three in $CHCl_3$, the presence of the following equilibrium has been confirmed:

$$R_2SnX_2 \cdot 2(CH_3)_3SbS \rightleftharpoons (CH_3)_3SbS + (CH_3)_3SbX_2 + \tfrac{1}{3}(R_2SnS)_3 \quad (3)$$

Reactions of trimethylstibine sulfide with alkylindium halides are quite different from those with organotin compounds described above (30). As shown in the following equation, reaction of R_2InX with $(CH_3)_3SbS$ gives stibonium salts:

$$(CH_3)_3SbS + R_2InX \longrightarrow R(CH_3)_3SbX + RInS \quad (4)$$

$$(R = CH_3, C_2H_5; \quad X = Cl, Br, I)$$

Under mild reaction conditions, complexes $R_2InX \cdot (CH_3)_3SbS$ (X = Cl, R = CH_3, C_2H_5; X = Br, R = C_2H_5) were isolated. Alkyl migration from indium to antimony probably occurs via this type of intermediate complexes.

Triorganostibine sulfide reacts with some organotin compounds containing a stable tin–tin bond (29):

$$R_3SbS + R_3'SnSnR_3' \longrightarrow R_3Sb + (R_3'Sn)_2S \qquad (5)$$

$(R = CH_3; \quad R' = C_6H_5, C_6H_5CH_2 \quad \text{or} \quad R = C_6H_5; \quad R' = C_6H_5)$

This result is particularly interesting since no reaction occurred when such ditins were heated with elemental sulfur under similar conditions. The enhanced reactivity of the triorganostibine sulfide in these reactions may be attributed to the semipolar Sb—S bond.

IV
TERTIARY STIBINES

Various tertiary stibines, R_3Sb, are easily obtained by the reaction of $SbCl_3$ with the Grignard reagent. However, it is difficult to prepare $RR_2'Sb$ or $RR'R''Sb$ by this method, for the materials $R_2'SbCl$ or $R'R''SbCl$ are not easily obtained. We have succeeded in preparing these types of compounds in fairly good yields via cleavage of the phenyl–antimony bond of $(C_6H_5)_3Sb$ or $(C_6H_5)_2RSb$ (R = alkyl) with sodium in liquid ammonia.[7] By this method, some asymmetrical tertiary stibines $(CH_3)(C_6H_5)RSb$ were prepared (37, 38), and optically active stibonium compounds have been obtained for the first time from these asymmetrical stibines (39). A selective cleavage reaction of the phenyl–antimony bonds of $(C_6H_5)_2RSb$ with dry hydrogen chloride to give $RSbCl_2$ was also found to be applicable to the preparation of compounds $RR_2'Sb$ (40, 41).

One of the notable features of tertiary stibines is that they have coordinating ability toward transition metals. However, the coordination behavior of R_3Sb had been investigated only in connection with analogous phosphine or arsine ligands. We have prepared some interesting tertiary stibines, such as $(R_2Sb)_2CH_2$ (41–43) and $RR'(CH_2=CHCH_2)Sb$ (44), and studied their coordination behavior (44–47).

A. Cleavage Reactions of the Phenyl–Antimony Bond

1. *By Sodium in Liquid Ammonia*

As shown in the following reaction scheme, tertiary stibines $R(C_6H_5)_2Sb$ and $[(C_6H_5)_2Sb]_2(CH_2)_n$ were obtained in good yields by a cleavage re-

---

[7] Recently, Meinema et al. (36) reported that an alkyl–antimony bond of R_3Sb (R = alkyl) is also cleaved by sodium in liquid ammonia.

action of one phenyl–antimony bond of $(C_6H_5)_3Sb$ by sodium in liquid ammonia and subsequent reaction with RCl and $(CH_2)_nCl_2$,[8] respectively:

$$(C_6H_5)_3Sb \xrightarrow[\text{in liq. NH}_3]{\text{Na}} [(C_6H_5)_2SbNa]$$

$$\begin{array}{c} \xrightarrow{\text{RCl}} R(C_6H_5)_2Sb \\ (R = CH_3, C_2H_5, i\text{-}C_3H_7, C_6H_5CH_2) \quad (6) \\ \xrightarrow{(CH_2)_nCl_2} [(C_6H_5)_2Sb]_2(CH_2)_n \\ (n = 1, 3) \end{array}$$

Selective cleavage of one phenyl–antimony bond of $CH_3(C_6H_5)_2Sb$ was also found to occur,[9] and, from the sodium-containing intermediate, hitherto unknown[10] asymmetrical tertiary stibines, $(CH_3)(C_6H_5)R'Sb$ $(R' = C_2H_5, i\text{-}C_3H_7, C_6H_5CH_2)$, and $[(CH_3)(C_6H_5)Sb]_2(CH_2)_n$ $(n = 1, 3, 4)$ can be prepared easily in about 60% yield (37).

2. *By Dry Hydrogen Chloride*

It was also found that the phenyl–antimony bonds of $(CH_3)(C_6H_5)_2Sb$ were cleaved selectively by dry hydrogen chloride in chloroform to give CH_3SbCl_2 (40). This cleavage reaction was successfully applied to the preparation of new ligands,[11] bis(diorganostibino)methanes (41), as follows:

$$[(C_6H_5)_2Sb]_2CH_2 \xrightarrow[\text{in CHCl}_3]{\text{HCl}} (Cl_2Sb)_2CH_2 \xrightarrow{\text{RMgX}} (R_2Sb)_2CH_2 \quad (7)$$

$$(R = CH_3, C_2H_5, p\text{-}CH_3C_6H_4)$$

B. Properties of Asymmetrical Tertiary Stibines and Resolution of Quaternary Stibonium Iodide

As shown in Table III, the PMR spectra of phenylmethylbenzylstibine and phenylmethylisopropylstibine in $CDCl_3$ at 23° show an AB quartet

---

[8] Reaction of $(C_6H_5)_2SbNa$ with $(CH_2)_2Cl_2$ gave the distibine $[(C_6H_5)_2Sb]_2$ (48).

[9] In the reaction of $R(C_6H_5)_2Sb$ ($R = C_2H_5$, $i\text{-}C_3H_7$, $C_6H_5CH_2$), cleavage of the alkyl–antimony bond took place (38).

[10] Asymmetrical triarylstibines had been prepared (49, 50) by Campbell from the reaction of phenylmagnesium halides with diarylchlorostibines $RR'SbCl$, which were obtained by the diazonium reaction. However, the alkyl-substituted analogs were not identified.

[11] Compound $[(CH_3)_2Sb]_2CH_2$ could be obtained from the reaction of $(CH_3)_2SbNa$ with CH_2Cl_2 in liquid ammonia, but the purification is difficult (43).

TABLE III

PROTON MAGNETIC RESONANCE DATA OF ASYMMETRICAL TERTIARY STIBINES[a]

Tertiary stibines	δ (ppm)	J (Hz)	Assignment
$(C_6H_5)(CH_3)RSb$			
$R = C_2H_5$	0.90 (s)	—	$SbCH_3$
	1.30 (m)	—	SbC_2H_5
$(CH_3)_2CH$	0.95 (s)	—	$SbCH_3$
	1.21 (d)	7.5[b]	$C(CH_3)_2$
	1.26 (d)	7.5[b]	
	1.90 (m)	—	SbCH
$C_6H_5CH_2$	0.95 (s)	—	$SbCH_3$
	2.96 (d)	12.0[c]	$SbCH_2$
	3.06 (d)	12.0[c]	
$[(C_6H_5)(CH_3)Sb]_2(CH_2)_n$			
$n = 1$[d]	0.91 (s)	—	$SbCH_3$
	1.45 (d)	11.3[c]	$SbCH_2Sb$
	1.50 (d)	11.3[c]	
	1.50 (s)	—	
3	0.83 (s)	—	$SbCH_3$
	1.64 (m)	—	$Sb(CH_2)_3Sb$
4	0.84 (s)	—	$SbCH_3$
	1.63 (m)	—	$Sb(CH_2)_4Sb$

[a] In $CDCl_3$ at 23°; δ(ppm) downfield from internal TMS. Aromatic protons were observed at about 7.0–7.6 ppm. Multiplicity: (s) singlet; (d) doublet; (m) multiplet.
[b] $J(CH_3—CH)$.
[c] $J(gem)$.
[d] At 100 MHz.

for the benzyl methylene group and two doublets for the isopropyl methyl groups, which are expected for the asymmetrical compounds. In bis(phenylmethylstibino)methane, a singlet and an AB quartet for the methylene group are observed. These are reasonably assigned to the racemic (XIV) and the meso (XV) forms, respectively. These observations suggest that the rates of the pyramidal inversion of our asymmetrical tertiary stibines

```
        C₆H₅                    CH₃
     Sb                       Sb
    /   CH₃                  /   C₆H₅
  H                        H
   \C                       \C
  H/  \                    H/  \
       Sb   C₆H₅                Sb   C₆H₅
          \CH₃                    \CH₃

     meso                    racemic
     (XIV)                    (XV)
```

are very slow on the PMR time scale.[12] This is consistent with a lower limit to the barrier of pyramidal inversion for diisopropyl-p-tolylstibine, which was determined to be 26 kcal/mole from the PMR spectrum (53).

The resolution of phosphonium (54) and arsonium salts (1) containing four different organic groups has been the subject of considerable investigation. We have succeeded in resolving a quaternary stibonium iodide for the first time (39). A racemic mixture of the quaternary stibonium iodide, $(CH_3)(C_2H_5)(i\text{-}C_3H_7)(C_6H_5)SbI$ was obtained by the reaction of $(CH_3)(i\text{-}C_3H_7)(C_6H_5)Sb$ with $(C_2H_5)_3OBF_4$ in methylene chloride, followed by treatment with KI in methanol. One enantiomer of this stibonium iodide was obtained by way of Ag–(−)-dibenzoylhydrogentartrate (DBHT):

$(CH_3)(C_2H_5)(i\text{-}C_3H_7)(C_6H_5)SbI \xrightarrow{\text{Ag—(−)-DBHT}}$

$[\alpha]_D^{18} - 66.5°$ (c, 1.02 in methanol) $\xrightarrow{\text{KI}} [\alpha]_D^{26} + 4.10°$ (c, 0.63 in methanol)   (8)

This enantiomer is optically stable both in the solid and in solution.

C. Coordination Behavior with Transition Metal Carbonyls

1. Reaction of Bis(diorganostibino)methane with Metal Carbonyls

In the substitution reactions of the metal hexacarbonyls, $M(CO)_6$ (M = Cr, Mo, W), with bis(diorganostibino)methane, we found that bis(diarylstibino)methane acts as a monodentate ligand and bis(dimethylstibino)methane as a bridging ligand to form pentacarbonyl complexes XVI and XVII (45). Furthermore, doubly bridged tetracarbonyl complexes (XVIII) were obtained from the reaction of $M(CO)_4$(diene) with bis-(diarylstibino)methane (46).[13] This coordination behavior of bis(diorganostibino)methane seems to be different from that of the phosphine analog

---

[12] The calculated barrier to the pyramidal inversion for trimethylstibine was reported to be 26.7 (51) and 25 kcal/mole (52).

[13] Diene = 1,5-cyclooctadiene (C_8H_{12}) or 2,5-norbornadiene (C_7H_8).

(XVI) (R = C_6H_5, p-$CH_3C_6H_4$; M = Cr, Mo, W)

(XVII) (R = CH_3, M = Mo, W)

(XVIII) (R = C_6H_5, p-$CH_3C_6H_4$; M = Cr, Mo)

in the sense that bis(diphenylphosphino)methane acts as a chelating ligand in Mo(CO)$_4$[(C_6H_5)$_2$P]$_2$CH$_2$ (55). However, a recent report (56) suggests the possibility that one bis(diphenylstibino)methane acts as a chelating ligand in new monomeric compounds, Mo(CO)$_4$[(C_6H_5)$_2$Sb]$_2$CH$_2$ and Mo(CO)$_3$ {[(C_6H_5)$_2$Sb]$_2$CH$_2$}$_2$.

Bis(diorganostibino)methane reacted with dicobalt octacarbonyl, under mild conditions, to give a dinuclear complex (XIX) containing both bridging carbonyl groups and bridging stibinomethane. The substitution reaction of compound XIX with diarylacetylene gave complex XX (47).

(XIX) (R = CH_3, C_2H_5, C_6H_5, p-$CH_3C_6H_4$; R' = C_6H_5, p-$CH_3C_6H_4$)

(XX)

Recently, Hartwell (57) reported a rhodium complex, {[(C_6H_5)$_2$Sb]$_2$-CH$_2$RhCOCl}$_2$, in which stibinomethane also acts as a bridging ligand.

2. *Reaction of Tertiary Allylstibines with $Cp(CO)_2FeCl$*

In the reaction of tertiary allylstibines with $Cp(CO)_2FeCl$,[14] a cleavage of one stibine–allyl bond took place and yellow unstable compounds (XXI) were obtained (*44*):

$Cp(CO)_2FeCl + (CH_2=CHCH_2)RR'Sb \longrightarrow$

$\{[Cp(CO)_2Fe]_2SbRR'\}Cl + CH_2=CHCH_2Cl + \textit{trans}\text{-}CH_3CH=CHCl$ (9)

(XXI)

Compounds XXI were characterized as stable tetraphenylborate salts, which are thought to be a type of stibonium that contains Fe—Sb σ-bonds (XXII). Similar compounds containing halogen groups bonded to the antimony atom were prepared recently by the reaction of $[Cp(CO)_2Fe]_2$ or $Cp(CO)_2FeNa$ with SbX_3 (*58, 59*).

$$\begin{bmatrix} Cp(CO)_2Fe & & R \\ & Sb & \\ Cp(CO)_2Fe & & R' \end{bmatrix}^+ \quad B(C_6H_5)_4^-$$

(R = R' = CH_3, C_6H_5, $CH_2=CHCH_2$;
R = $CH_2=CHCH_2$, R' = CH_3, C_6H_5)

(XXII)

Reaction of $(CH_2=CHCH_2)(C_6H_5)_2E$ (E = P, As) with $Cp(CO)_2FeCl$ gave the product $[Cp(CO)_2FeE(CH_2=CHCH_2)(C_6H_5)_2]Cl$, and, even in the presence of excess $Cp(CO)_2FeCl$, cleavage of the E–allyl bond was not observed. The difference in behavior of these allyl-substituted Group V elements toward $Cp(CO)_2FeCl$ may be related to the strength of the bond from these elements to carbon.

References

1. Doak, G. O., and Freedman, L. D., "Organometallic Compounds of Arsenic, Antimony and Bismuth." Wiley, New York, 1970.
2. Shindo, M., and Okawara, R., *J. Organometal. Chem.* **5**, 537 (1966).
3. Shindo, M., and Okawara, R., *Inorg. Nucl. Chem. Lett.* **5**, 77 (1969).
4. Matsumura, Y., Shindo, M., and Okawara, R., *J. Organometal. Chem.* **27**, 357 (1971).
5. Morgan, G. T., and Davies, G. R., *Proc. Roy. Soc., Ser. A* **110**, 523 (1926).
6. Doak, G. O., and Long, G. G., *Trans. N.Y. Acad. Sci.* **28**, 402 (1966).
7. Nishii, N., Shindo, M., Matsumura, Y., and Okawara, R., *Inorg. Nucl. Chem. Lett.* **5**, 529 (1969).

---

[14] $Cp = \pi\text{-}C_5H_5$.

8. Nishii, N., Matsumura, Y., and Okawara, R., *J. Organometal. Chem.* **30,** 59 (1971).
9. Nishii, N., Hashimoto, K., and Okawara, R., *J. Organometal. Chem.* **55,** 133 (1973).
10. Nishii, N., Matsumura, Y., and Okawara, R., *Inorg. Nucl. Chem. Lett.* **5,** 703 (1969).
11. Nishii, N., and Okawara, R., *J. Organometal. Chem.* **38,** 335 (1972).
12. Meinema, H. A., and Noltes, J. G., *J. Organometal. Chem.* **16,** 257 (1969).
13. Meinema, H. A., and Noltes, J. G., *J. Organometal. Chem.* **37,** C31 (1972).
14. Meinema, H. A., and Noltes, J. G., *J. Organometal. Chem.* **37,** 285 (1972).
15. Meinema, H. A., Rivarola, E., and Noltes, J. G., *J. Organometal. Chem.* **17,** 71 (1969).
16. Meinema, H. A., and Noltes, J. G., *J. Organometal. Chem.* **36,** 313 (1972).
17. Meinema, H. A., and Noltes, J. G., *J. Organometal. Chem.* **55,** C77 (1973).
18a. Kawasaki, Y., and Okawara, R., *Bull. Chem. Soc. Jap.* **40,** 428 (1967).
18b. Kawasaki, Y., Ito, T., and Okawara, R., *Int. Symp. Decomposition Organometal. Comp. Refractory Ceramics, Metal and Metal Alloys, 1967,* p. 47.
19. Kanehisa, N., Kai, Y., and Kasai, N., *Inorg. Nucl. Chem. Lett.* **8,** 375 (1972).
20. Onuma, K., Kai, Y., and Kasai, N., *Inorg. Nucl. Chem. Lett.* **8,** 143 (1972).
21. Kroon, J., Hulscher, J. B., and Peerdeman, A. F., *J. Organometal. Chem.* **37,** 297 (1972).
22. Uda, S., Kai, Y., Yasuoka, N., and Kasai, N., *Cryst. Struct. Commun.* **3,** 257 (1974).
23. Kawakami, K., Miyauchi, M., and Tanaka, T., *J. Organometal. Chem.* **70,** 67 (1974).
24. Cheremos, G. N., and Zingaro, R. A., *J. Organometal. Chem.* **22,** 637 (1970).
25. Otera, J., and Okawara, R., *J. Organometal. Chem.* **16,** 335 (1969).
26. Otera, J., and Okawara, R., *Inorg. Nucl. Chem. Lett.* **6,** 855 (1970).
27. Otera, J., and Okawara, R., *J. Organometal. Chem.* **17,** 353 (1969).
28. Shindo, M., Matsumura, Y., and Okawara, R., *J. Organometal. Chem.* **11,** 299 (1968).
29. Otera, J., Kadowaki, T., and Okawara, R., *J. Organometal. Chem.* **19,** 213 (1969).
30. Maeda, T., Yoshida, G., and Okawara, R., *J. Organometal. Chem.* **44,** 237 (1972).
31. Haake, P., Miller, W. B., and Tyssee, D. A., *J. Amer. Chem. Soc.* **86,** 3577 (1964).
32. Phillips, G. M., Hunter, J. S., and Sutton, L. E., *J. Chem. Soc.*, 146 (1945).
33. Matsumura, Y., Shindo, M., and Okawara, R., *Inorg. Nucl. Chem. Lett.* **3,** 219 (1967).
34. Schmidtbauer, H., and Mitschke, K. H., *Chem. Ber.* **104,** 1842 (1971).
35. Shindo, M., Matsumura, Y., and Okawara, R., *Bull. Chem. Soc. Jap.* **42,** 265 (1969).
36. Meinema, H. A., Martens, H. F., and Noltes, J. G., *J. Organometal. Chem.* **51,** 223 (1973).
37. Sato, S., Matsumura, Y., and Okawara, R., *J. Organometal. Chem.* **43,** 333 (1972).
38. Sato, S., Matsumura, Y., and Okawara, R., *Inorg. Nucl. Chem. Lett.* **8,** 837 (1972).
39. Sato, S., Matsumura, Y., and Okawara, R., *J. Organometal. Chem.* **60,** C9 (1973).
40. Matsumura, Y., and Okawara, R., unpublished work.
41. Matsumura, Y., and Okawara, R., *Inorg. Nucl. Chem. Lett.* **7,** 113 (1971).
42. Matsumura, Y., and Okawara, R., *Inorg. Nucl. Chem. Lett.* **5,** 449 (1969).
43. Matsumura, Y., and Okawara, R., *J. Organometal. Chem.* **25,** 439 (1970).
44. Matsumura, Y., Harakawa, M., and Okawara, R., *J. Organometal. Chem.* **71,** 403 (1974).
45. Fukumoto, T., Matsumura, Y., and Okawara, R., *J. Organometal. Chem.* **37,** 113 (1972).
46. Fukumoto, T., Matsumura, Y., and Okawara, R., *Inorg. Nucl. Chem. Lett.* **9,** 711 (1973).
47. Fukumoto, T., Matsumura, Y., and Okawara, R., *J. Organometal. Chem.* **69,** 437 (1974).
48. Hewertson, W., and Watson, H. R., *J. Chem. Soc.*, 1490 (1962).

49. Campbell, I. G. M., *J. Chem. Soc.*, 3116 (1955).
50. Campbell, I. G. M., and White, A. W., *J. Chem. Soc.*, 1184 (1958).
51. Weston, R. E., Jr., *J. Amer. Chem. Soc.* **76,** 2645 (1954).
52. Keopple, G. W., Sagatys, D. S., Krishnamurthy, G. S., and Miller, S. I., *J. Amer. Chem. Soc.* **89,** 3396 (1967).
53. Jacobus, J., *J. Chem. Soc. D*, 1058 (1971).
54. Gallagher, M. J., and Jenkins, I. D., *in* "Topics in Stereo-Chemistry" (E. L. Eliel and N. L. Allinger, eds.), Vol. 3, p. 1. Wiley (Interscience), New York, 1968.
55. Cheung, K. K., Lai, T. F., and Mok, K. S., *J. Chem. Soc. A*, 1644 (1971).
56. Beall, T. W., and Houk, L. W., *J. Organometal. Chem.* **56,** 261 (1973).
57. Garrou, P. E., and Hartwell, G. E., Private communication.
58. Trinh-toan and Dahl, L. F., *J. Amer. Chem. Soc.* **93,** 2654 (1971).
59. Cullen, W. R., Patmore, D. J., and Sams, J. R., *Inorg. Chem.* **12,** 867 (1973).

Pentaalkyls and Alkylidene Trialkyls of the Group V Elements

HUBERT SCHMIDBAUR

Anorganisch-chemisches Laboratorium
Technische Universität München
Munich, West Germany

I. Introduction	205
II. Simple Nitrogen Ylides	207
III. Phosphorus Ylides and Pentaalkylphosphoranes	209
A. Alkylidene Trialkylphosphoranes	209
B. Pentaalkylphosphoranes	214
IV. Arsenic Ylides and Pentaalkylarsoranes	224
A. Alkylidene Trialkylarsoranes	224
B. Silylated Ylides of Phosphorus and Arsenic	228
C. Pentamethylarsorane	229
V. Antimony Ylides and Pentaalkylstiboranes	231
A. Antimony Ylides	231
B. Pentaalkylstiboranes	232
VI. Bismuth Compounds	236
VII. Related Compounds of Vanadium, Niobium, and Tantalum	236
A. Vanadium Ylides?	236
B. Niobium Compounds	237
C. Tantalum Pentaalkyls and Tantalum Ylides	238
D. Metal Ylides or Metal Carbene Complexes?	239
References	240

I

INTRODUCTION

With the exception of nitrogen, all of the Group Vb elements are expected to form pentacoordinate compounds in their 5+ oxidation state, and this is, indeed, the case with some of the halides, alkoxides, etc. It was not until the pioneering work of Georg Wittig and his collaborators, however, that the first examples of pentaorganyls of these borderline elements between metals and metalloids were reported (95–99, 102, 104). In this early investigation, a complete set of the pentaphenyls could be obtained and characterized (95–99, 102), but apart from the pentamethylantimony case, all attempts for the preparation of pentaalkyl derivatives failed (104).

Wittig's work was later confirmed and extended by Russian, Dutch, and Japanese groups, who were able to prepare other pentaalkylantimony compounds using his method, which consisted of the reaction of trialkyl (or tetraalkyl) antimony halides with alkyllithium reagents. Table IV lists some of the important examples.

When this method is applied to compounds of phosphorus and arsenic, the components undergo an entirely different reaction, and alkylidene trialkyl derivatives of these elements (ylides) are obtained (*102, 104*). Thus, instead of the introduction of an additional alkyl substituent at the central element, as observed with antimony, a deprotonation at one of the α-carbon atoms adjacent to phosphorus and arsenic occurs:

$$[(CH_3)_4P]Cl \xrightarrow[-LiCl]{LiCH_3} (CH_3)_3P=CH_2 + CH_4 \quad (1)$$

$$[(CH_3)_4As]Cl \xrightarrow[-LiCl]{LiCH_3} \begin{cases} (CH_3)_3As=CH_2 + CH_4 & (2a) \\ (CH_3)_5As & (2b) \end{cases}$$

$$[(CH_3)_4Sb]Cl \xrightarrow[-LiCl]{LiCH_3} (CH_3)_5Sb \quad (3)$$

Although the reaction path of Eq. (2a) is typical for phosphorus, and pentaalkylation is characteristic for antimony [Eq. (3)], arsenic has recently been found to be the only element to supply both the alkylidenetrialkyl (*73*) and the pentaalkyl (*37*) derivative, depending very critically on the reaction conditions [Eq. (2b)].

The revived interest in this area led to new efforts for improved methods of synthesis for pure alkylidene trialkyls (ylides), for the detection and possible characterization of a pentaalkylphosphorane species, and for a better understanding of the state of structure and bonding in ylides and pentaalkyls. This recent work is summarized here with the main focus directed to some of the very simple compounds and their silicon derivatives, which have turned out to be of special significance.

These derivatives have not only provided new synthetic pathways but have shown improved thermal stability (as in the case of arsenic ylides) and a modified pattern of chemical reactivity. The donor properties of ylides (*55, 24*), and most of their synthetic applications (*103*), have been covered in other reviews and articles (*3, 26*) and are not duplicated here. The general organometallic chemistry of arsenic, antimony, and bismuth is the subject of the invaluable monograph by Doak and Freedman (*11*). The broad scope of phosphorus ylide and pentaorganophosphorane chemistry was covered in the leading multivolume series on organophosphorus chemistry edited by Kosolapoff and Maier (*3, 21*). Finally, the recent success of efforts to synthesize pentaalkyls [Nb(CH$_3$)$_5$ (83)] and even ylides ([(CH$_3$)$_3$CCH$_2$]$_3$TaCHC(CH$_3$)$_3$ (82a)) of the subgroup metals vanadium, niobium, and tantalum add a new dimension to the scope of this article.

II
SIMPLE NITROGEN YLIDES

In an attempt to demonstrate the existence of pentavalent nitrogen, Schlenk and Holtz studied the reaction of triphenylmethyl sodium with tetramethylammonium chloride (52). The highly colored material was strongly conducting in polar solvents and could be identified as a salt, the stability of which is due to the resonance stabilization of the triphenylmethide anion. In the absence of such stabilizing substituent effects (53), as with n-butyl or another alkyllithium reagent, a metalation of the tetramethylammonium cation occurs, which leads to type I products (18):

$$(CH_3)_4N^\oplus Cl^\ominus + (C_6H_5)_3C^\ominus Na^\oplus \xrightarrow{-NaCl} (CH_3)_4N^\oplus C(C_6H_5)_3^\ominus \quad (4)$$

$$(CH_3)_4N^\oplus Br^\ominus + RLi \xrightarrow{-RH} (CH_3)_3\overset{\oplus}{N}-CH_2LiBr^- \quad (5)$$
$$(I)$$

These species can be formulated as alkali halide complexes of the nitrogen ylide $(CH_3)_3\overset{\oplus}{N}-\overset{\ominus}{C}H_2$, and many of their chemical reactions have been interpreted according to this structure (9, 15, 40, 91–93, 100–102, 105).

The addition compounds (I) are insoluble in diethyl ether, and the slurries obtained are quite stable. In more strongly solvating media, such as tetrahydrofuran or dimethoxyethane, the compounds are soluble but show rapid decomposition, with trimethylamine and polymethylene as the main products. These experiments indicate (9, 40) that when the lithium salt is trapped by donor solvent molecules, the free ylide quickly undergoes decomposition (40). No free trialkylammonium ylide has yet been prepared, even under very mild conditions (35). On the other hand, it has been shown, that the tetramethylammonium cation can even be metalated twice by organolithium reagents (102) to afford dimethylammonium bismethylides:

$$(CH_3)_4N^\oplus Br^\ominus \xrightarrow[-2C_6H_6]{2C_6H_5Li} \underset{(II)}{\overset{CH_3 \quad CH_2-Li}{\underset{CH_3 \quad CH_2-Li}{N^\oplus}}} Br^\ominus \quad (6)$$

Chemical reactions of these materials have also been carefully investigated (101, 102), and their properties closely resemble those of other typical carbon nucleophiles, as those of the monometalated analogs do. From this work it is apparent that these ammonium ylides, which are only stable in the presence of lithium salts, are similar in their characteristics to other organolithium reagents and have very little in common with other ylides (26). The nitrogen plays no role beyond stabilizing the neighboring carbanion, largely through its inductive effect.

As will be mentioned again later, the introduction of trialkylsilyl groups as substituents on the carbanion of an ylide generally leads to significant stabilization. Inherently unstable ylides, e.g., of arsenic, could thus be converted into strikingly stable derivatives through silylation (see Section IV,B). However, similar efforts have not been successful with nitrogen ylides. The landmark experiments by Miller (35) have shown that no ylide (III) can be isolated as the product of a reaction according to Eq. (7). Instead decomposition products (III'), Eq. (8), are observed (Scheme 1). As a consequence the desilylation process (74), normally an important method for generating free ylides, is not applicable with nitrogen ylides. The situation is somewhat different in the case of pyridinium ylides $C_5H_5N^{\oplus}$—$CH_2^{\ominus}$, but even compounds of this type are not very stable thermodynamically in the absence of substituents offering some resonance participation to the carbanion or to the onium group (26).

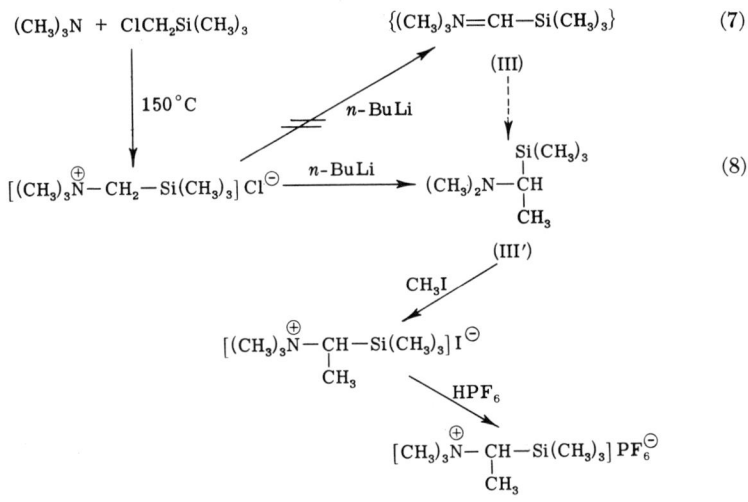

Scheme 1

As already mentioned (105), no pentavalent nitrogen compound NR_5 has ever been reported. In view of the present knowledge of the valence properties of second-row elements generally, and of CX_5^- and related isosters in particular, this is not surprising.

III
PHOSPHORUS YLIDES AND PENTAALKYLPHOSPHORANES

Alkylidene triorganophosphoranes (phosphorus ylides) are known with a large variety of substituents both at the phosphonium center and at the carbanionic function:

$$\begin{array}{c}R^1\\ \diagdown\overset{\oplus}{P}-\overset{\ominus}{C}\diagup\\ R^2\diagup\diagdown R^5\\ R^3\end{array} \longleftrightarrow \begin{array}{c}R^1\\ \diagdown P=C\diagup\\ R^2\diagup\diagdown R^5\\ R^3\end{array} \qquad (9)$$

Although it is not the purpose of this chapter to review the chemistry of this important class of compounds, it is necessary to recall the main methods of synthesis and the characteristic properties of these species in order to point out some relationships between these and the more elusive pentaalkylphosphoranes.

A. Alkylidene Trialkylphosphoranes

1. *Preparation*

Virtually all of the important methods (3) of synthesis for phosphorus ylides are based on the reaction of a phosphonium salt with a strong base:

$$(R^1R^2R^3)\overset{\oplus}{P}-\overset{R^4}{\underset{R^5}{CH}}\quad X^\ominus\xrightarrow[-HX]{\text{base}}(R^1R^2R^3)P=\overset{R^4}{\underset{R^5}{C}} \qquad (10)$$

In the presence of electron-withdrawing substituents (R^1—R^5), even relatively weak deprotonating agents may be used. If however R^1—R^3, and in particular R^4—R^5, are alkyl groups, only the strongest bases are able to produce the ylides (102). Typically, alkali alkyls or amides are used. With these reagents an alkali halide is, thus, formed, and alkane or ammonia is evolved.

The presence of the alkali salt in the reaction mixture was a major problem in these syntheses because most of the ylides were found to form very stable coordination compounds with the smaller alkali cations. Two procedures are now available to avoid this difficulty. The first one was derived from the observation (*34, 72*) that silyl substituents preclude the formation of addition compounds with the salt through steric and electronic effects. Therefore salt-free silylated ylides can be readily obtained, purified, and finally desilylated by alcohols (*74*) [Eqs. (11a,b)]. The second method is based on the fact that sodium chloride does not form ylide complexes in tetrahydrofuran (*28*), in which sodium amide reacts sufficiently smoothly with phosphonium salts [Eq. (12)]:

$$[(CH_3)_3\overset{\oplus}{P}-CH_2Si(CH_3)_3]Cl^{\ominus} \xrightarrow[-LiCl, -RH]{+LiR} (CH_3)_3P=CH-Si(CH_3)_3 \quad (11a)$$

$$\downarrow CH_3OH \quad (11b)$$

$$[(CH_3)_4P]Cl \xrightarrow[-NaCl, -NH_3]{+NaNH_2, THF} (CH_3)_3P=CH_2 \quad (12)$$

With bulky substituents attached to the phosphorus, none of these special measures are necessary because complex formation is precluded through steric effects (*74*). Therefore ethylidene triethylphosphorane may be obtained in high yield without complications from the tetraethylphosphonium salt and an alkyllithium reagent (*74*):

$$[(C_2H_5)_4P]Cl \xrightarrow{LiCH_3} \begin{array}{c} \xrightarrow{-CH_4, -LiCl} (C_2H_5)_3P=CHCH_3 \quad (13a) \\ \xrightarrow[-LiCl]{\quad\quad /\!/\!\!\longrightarrow} (C_2H_5)_4PCH_3 \quad (13b) \end{array}$$

Although the reaction conditions have been greatly modified, no evidence whatsoever has been found for the generation of pentaalkylphosphorane according to Eq. (13b). This is also true for other reactions of noncyclic alkylphosphonium salts with lithium alkyls.

2. *Properties, Structure, and Bonding*

Alkylidene trialkylphosphoranes $R_3P=CR_2'$ (R = alkyl, R′ = hydrogen, alkyl) are colorless mobile liquids that can be crystallized at low temperatures in most cases and distilled under reduced pressure (*74, 28*) (Table I lists melting and boiling points.). The compounds are extremely reactive, spontaneously igniting when finely divided on paper, and vigorously fuming when exposed to air (*74*). Miscibility with and solubility in most aprotic organic solvents is excellent and the materials dissolve as monomers. The reaction with alcohols (*57*), which leads to tetraalkylalkoxyphosphoranes and related secondary products (*70, 71*), is highly exothermic; even in cold water hydrolysis occurs explosively.

TABLE I
Physical Constants of Some Simple Alkylidene Trialkylphosphoranes

Compound	bp (°C/mm Hg)	mp (°C)	Ref.
$(CH_3)_3P=CH_2$	122/760	13–14	28, 74
$C_2H_5(CH_3)_2P=CH_2$	143–145/760	(−16)−(−14)	74
$CH_3(C_2H_5)_2P=CH_2$	60–62/12	(−45)−(−43)	74
$(C_2H_5)_3P=CH_2$	26/0.2	(−17)−(−15)	28, 74
	80–83/12	—	
$(CH_3)_2C_3H_7P=CH_2$	25/0.2	—	28
$(i\text{-}C_3H_7)_3P=CH_2$	38/0.001	—	28
$(n\text{-}C_4H_9)_3P=CH_2$	58/0.001	—	28, 60
$(c\text{-}C_6H_{11})_3P=CH_2$	—	128	28
$(C_2H_5)_3P=CHCH_3$	25/0.01	(−38)−(−36)	28, 74
	86–87/12	—	
$(C_2H_5)_2(n\text{-}C_3H_7)P=CH\text{—}CH_3$	30/0.001	—	28
$(i\text{-}C_3H_7)_3P=CH\text{—}CH_3$	45/0.001	—	28
$(C_2H_5)_3P=CHC_2H_5$	30/0.001	—	28
$(i\text{-}C_3H_7)_3P=CHC_2H_5$	50/0.001	—	28
$(n\text{-}C_4H_9)_3P=CHC_2H_5$	80/0.001	—	28
$(C_2H_5)_3P=CHC_3H_7$	33/0.001	—	28
$C_2H_5(n\text{-}C_4H_9)_2P=CHC_3H_7$	65/0.001	—	28
$(CH_3)_3P=CH\text{—}CH=CH_2$	—	—	62

Crystal structure determinations have not been carried out for a simple peralkylated phosphorus ylide, but a whole series of arylated species and more complicated systems have been carefully investigated (2, 6–8, 32, 84, 85, 94). The results of these studies can be summarized as follows:

1. Phosphorus atoms in ylides largely retain the tetrahedral ligand geometry of the phosphonium cation precursor, but the bond to the ylene–ylide carbon atom is shortened, indicating an increase in bond order.

2. The ylide–ylene carbon atom is found to be the center of a trigonal-planar arrangement of ligands. This planar configuration may be part of a π system, thus giving rise to restriction in bond rotation and to the existence of geometrical isomers.

Infrared and Raman spectroscopy has been applied primarily to provide additional information on the nature of the ylidic function. An evaluation of the data obtained through a normal coordinate analysis gave a bond order of 1.65 in $(CH_3)_3PCH_2$ (51) and of 1.3 in $(C_6H_5)_3PCH_2$ (30). This result complements the bond shortenings observed by X-ray analyses.

Dipole moments of ylides, although measured only with some of the more complicated species (*31*), indicate a high degree of charge separation in the ylidic function, ranging between 5 and 9 D. These results strongly favor the zwitterionic (ylide) over the double-bond (ylene) formula.

Nuclear magnetic resonance data, which are now essentially complete, also clearly support this description of the bonding in alkylidene phosphoranes. Early ^{1}H NMR results had already shown a very considerable shielding of hydrogens attached to the ylidic function, and ^{31}P NMR values pointed to a quasi-onium-type character of the phosphorus atoms (*22, 28, 58, 75, 80*). Whereas all the chemical shifts and coupling constants of the alkylidene groups differed completely from those of the alkyl ligands attached to phosphorus, the shielding of the phosphorus and its coupling to the alkyl hydrogens were similar to that observed for related phosphonium salts. The more recently reported ^{13}C NMR spectra (*22, 58, 80*) have finally shown that the ylidic carbon itself is very strongly shielded and that its couplings with the phosphorus and hydrogen atoms indicate an sp^2 state of hybridization. On the other hand, the values for the alkyl carbon atoms and for the phosphorus exhibited the expected "normal" parameters for a tetrahedral sp^3 configuration. Some of these results are summarized in Table II.

3. *Theory and Photoelectron Spectra*

The set of physical data compiled so far for the alkylidene trialkylphosphoranes seems to allow us to draw a consistent picture of the structure and bonding in these molecules. A valence bond description thus still has to contain the classic Wittig formulas, but it is clear that the ylide form is the most significant for the ground-state phenomena of these molecules. Semiempirical, extended Hückel molecular orbital calculations on the hypothetical $H_3P=CH_2$ molecule with and without the inclusion of d orbitals have shown, according to this model (*23*), that both the charge distribution and the P=C bond order are affected significantly by d-orbital participation. The extended Hückel method employed in this LCAO study still gave a high negative charge at the ylidic carbon even when d orbitals were included! The extended Hückel calculations also indicate a preference for a planar geometry at the ylidic carbon, but it is not obvious whether this result is sufficiently reliable because other geometries have been wrongly predicted by this method.

As $(CH_3)_3P=CH_2$ is still too formidable a problem for theoretical chemists, the related but hypothetical $H_3P=CH_2$ molecule was the subject of the *ab initio* LCAO-MO-SCF studies published in 1972 (*1*). Using two uncontracted Gaussian basis sets, one with and the other without phos-

TABLE II

^{13}C NUCLEAR MAGNETIC RESONANCE SPECTRA OF THREE SIMPLE ALKYLIDENE PHOSPHORANES[a]

Compound	\C=			CH$_3$—C			CH$_{3(2)}$P		
	δ (ppm)[b]	^1J(PC) (Hz)	^1J(HC) (Hz)	δ (ppm)[b]	^2J(PCC) (Hz)	^1J(HC) (Hz)	δ (ppm)[b]	^1J(PC) (Hz)	^1J(HC) (Hz)
(CH$_3$)$_3$P=CH$_2$	−2.3	90.5	149.0	—	—	—	18.9	56.0	127
(C$_2$H$_5$)$_3$P=CH$_2$	−14.2	86.8	147.5[c]	6.6	2.9	129.5	20.3	55.9	125
(C$_2$H$_5$)$_3$P=CHCH$_3$	−7.9	113.2	—[d]	6.0 (C$_2$H$_5$) 9.5 (C$_2$H$_4$)	—	—[d]	27.6	54.4	—[d]

[a] From Schmidbaur et al. (80).
[b] Value of δ is measured relative to C$_6$D$_6$, and converted relative to TMS.
[c] Broad; ±3 Hz.
[d] Not observed. Off-resonance: dd.

phorus d orbitals, two conformations of this molecule have been investigated. An essentially zero barrier to C—P bond rotation was calculated in either basis set. Allowing d orbitals to participate, the energy change was such that the picture is consistent with π feedback in the ylidic function. Again, however, a high negative atomic charge at the ylidic carbon was obtained. This quantity "charge on the atom" is sensitive to the choice of basis set and is to be taken as a formal rather than as a physical property. Inspection of the various orbital sets shows that the ylide has π-type orbitals even when d character is disallowed, but when d character is included the bonding in these molecular orbitals taken as a group is particularly enhanced.

Photoelectron studies of $(CH_3)_3P=CH_2$ have provided data on the energy of the highest occupied molecular orbital and some of the lower-lying states (49). The frontier orbital energy of 6.81 eV is very low and seems to illustrate the carbanionic nature of the carbon, where this orbital is largely localized. This is borne out by semiquantitative CNDO/2 calculations on this molecule. The orbital sequence obtained is satisfactory according to *ab initio* results.

B. *Pentaalkylphosphoranes*

It has already been pointed out in the Introduction that no simple PR_5 compound, in which all R's were independent alkyl groups, has been reported to date. However, mixed species, e.g., Hellwinkels $CH_3P(C_6H_4)_4$, have been synthesized (19) as has the perarylated analog $P(C_6H_5)_5$ (102). These partly or fully arylated compounds are very stable thermally, and their properties and reactions have been studied (21). From these investigations it seems that further attempts to prepare a pentaalkylphosphorane might be promising. The successful synthesis of pentamethylarsenic (37) and of some derivatives of pentamethylphosphorane, such as $(CH_3)_4POCH_3$ (71), provided the necessary stimulus, but the ultimate goal has not yet been reached.*

In the meantime, new concepts for the synthesis of cyclic pentaalkylphosphoranes have been developed (88), which have proved the existence of a phosphorus atom bearing 5 aliphatic carbon atoms, 2 of which are

---

* *Note added in proof*: In May 1975 The and Cavell have reported ($86b$) the successful synthesis of dimethyltris(trifluoromethyl) and trimethylbis(trifluoromethyl) phosphorane from the reaction of tetramethyllead with suitable fluorophosphorane precursors. The compounds are stable white solids of trigonal bipyramidal structure.

part of a ring system. In the pioneering work by Katz and Turnblom (27, 87), the concept of ring strain was employed and ingeniously applied to this problem. These authors obtained the first stable pentaalkylphosphorane.

The investigations of the group of the present author led to cyclic pentaalkylphosphoranes through insertion of ylides into the strained silacyclobutane system (78), but the initial products of this reaction could not be isolated. Their transitory existence has been demonstrated, however, by a careful study of the reaction mechanism and the products of the insertion and rearrangement processes.

1. *Polycyclic Pentaalkylphosphorane*

It was assumed (80) that there were two possible reasons for the formation of a pentaalkylphosphorane: either there must be a peculiar instability preventing formation of the ylide, or there must be a peculiar stability enhancing formation of the pentavalent phosphorus. Inspection of certain bridged model phosphonium salts has shown that a bridged ring system might, indeed, enhance formation of the pentavalent phosphorane by straining the internal angle at phosphorus in the phosphonium salt. This effect is analogous to that suggested to explain the accelerated rate of hydrolysis of small-ring phosphate esters (90). The following reaction scheme summarizes a series of reactions, in which various phenyl, phenyl/methyl, and fully methyl-substituted homocubylphosphoranes are generated (27, 87, 88):

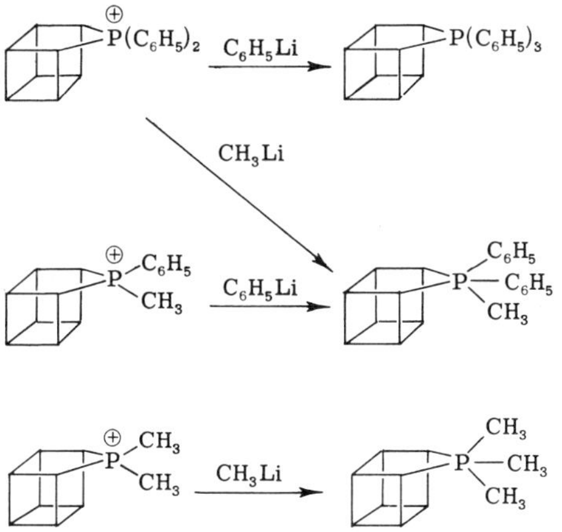

(IV)

In all of these examples the phosphorus atom is a member of two doubly fused five-membered rings in the phosphonium salt. This configuration gives rise to considerable strain because the five-membered rings are themselves part of the homocubyl system. This strain would not be changed if one of the *exo*-methyl groups was converted into an ylidic function as this basically does not alter the geometry at the phosphorus center (*88*). However, the strain is relieved if the coordination number at phosphorus is increased to five, because the trigonal-bipyramidal and square-pyramidal polyhedra can accommodate a whole range of valence angles between 120° and 90°.

Previous work by Hellwinkel had already made use of this and related concepts in his synthesis of the methyl bisbiphenylenephosphorane (*19*):

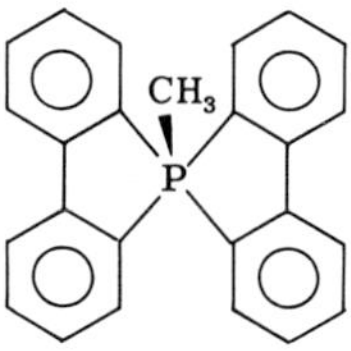

The structure of pentaalkylphosphorane has been proved by a complete set of spectroscopic and analytical data, including ^{1}H, ^{13}C, and ^{31}P NMR and IR and mass spectroscopy (*88*). Among these the chemical shift of the phosphorus atom is particularly diagnostic as it is very close to the values reported for authentically pentacoordinate species. The ^{1}H–^{13}C, ^{1}H–^{31}P, and ^{13}C–^{31}P coupling constants are also characteristic of methyl groups attached to pentacoordinate phosphorus. Therefore there is no doubt that compound IV "is a reality." Its existence has encouraged efforts to synthesize related compounds including pentamethylphosphorane.

Furthermore, the general properties of compound IV indicate that pentaalkylphosphoranes, once prepared, can be quite stable and easy to handle. The compound is a colorless liquid (bp. 20°C at 10^{-6} mmHg) that fragments thermally to trimethylphosphine and cyclooctatetraene at 75°C:

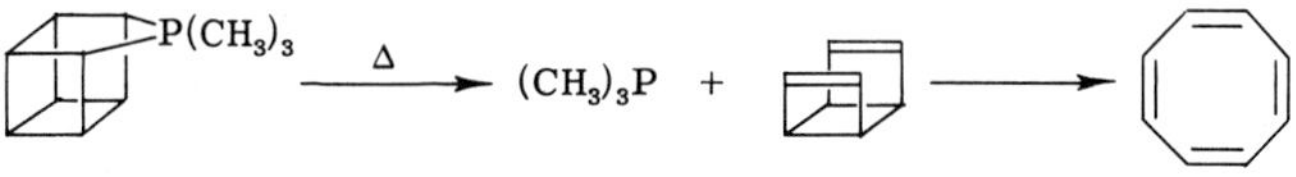

(17)

After 108 hours at this temperature, only 50% of the material was decomposed. The thermal decomposition of the arylated species is much faster and, hence, the pentaalkylphosphorane seems to exhibit the most satisfactory properties within this series [Eqs. (14)–(16)]. This result is in agreement with the striking properties of pentamethylarsenic (*37*).

2. *Evidence for Existence of Pentaalkylphosphorane Intermediates*

The failure of attempts to generate a PR_5 species via routes according to

$$[R_4P]X + RLi \not\rightarrow LiX + PR_5 \qquad (18)$$

i.e., through alkylation of a phosphonium salt,[1] led to reconsideration of various other methods of synthesis.

A promising alternative was found in the insertion of alkylidene phosphoranes into small carbocyclic systems, such as cyclopropane or -butane:

$$R_3PCH_2 \longrightarrow \begin{cases} \triangle \longrightarrow R_3P\!\!-\!\!\bigcirc & (19a) \\ \square \longrightarrow R_3P\!\!-\!\!\bigcirc & (19b) \end{cases}$$

Both of these reactions were expected to profit from relief of strain in the starting material and from a strain-free heterocycle in the product. Owing to the polar nature of the ylide it was desirable, however, to introduce at least 1 heteroatom into the cyclobutane ring in order to facilitate heterolytic cleavage of the system. The silacyclobutanes seemed to be an excellent choice, and, consequently, a project on ylide cleavage reaction of mono- and disilacyclobutanes was initiated. Although no stable cyclic pentaalkylphosphorane was obtained, it was possible to confirm the appearance of

---

[1] Ylides are obtained instead [see Eq. (1)].

these species as primary products in these reactions:

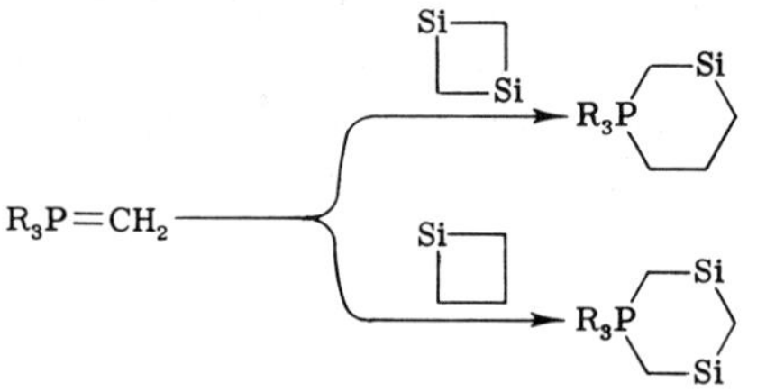

$$R_3P=CH_2 \longrightarrow \begin{cases} \text{[Si-Si 4-ring]} \longrightarrow R_3P\text{[Si-Si 6-ring]} & (20a) \\ \text{[Si 4-ring]} \longrightarrow R_3P\text{[Si-Si 6-ring]} & (20b) \end{cases}$$

(No silacyclopropane is available for similar studies at the present time.[2])

a. Ylide Cleavage of 1,3-Disilacyclobutanes (78). 1,1,3,3-Tetramethyl-1,3-disilacyclobutane is readily cleaved by trimethyl and triethyl methylenephosphorane to yield silylated ylides of type V (a–c) (Scheme 2). The

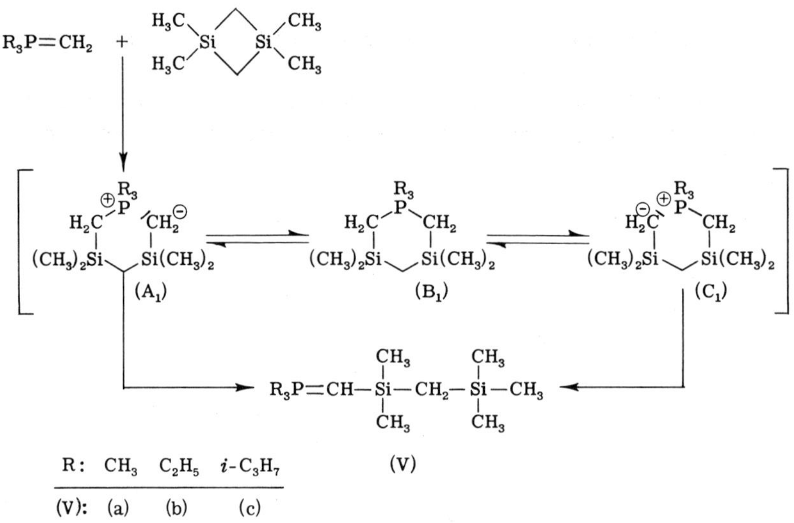

[Scheme 2 showing:

$R_3P=CH_2$ + [1,1,3,3-tetramethyl-1,3-disilacyclobutane ring with H₃C, CH₃ groups on Si atoms]

↓

$\left[\begin{array}{ccc} \overset{\oplus}{P}R_3 & \overset{\oplus}{P}R_3 & \overset{\oplus}{P}R_3 \\ H_2C\diagup\ \diagdown CH_2^{\ominus} & H_2C\diagup\ \diagdown CH_2 & H_2^{\ominus}C\diagup\ \diagdown CH_2 \\ | \qquad | & | \qquad | & | \qquad | \\ (CH_3)_2Si\diagdown\ \diagup Si(CH_3)_2 & (CH_3)_2Si\diagdown\ \diagup Si(CH_3)_2 & (CH_3)_2Si\diagdown\ \diagup Si(CH_3)_2 \\ (A_1) & (B_1) & (C_1) \end{array} \right]$

↓

$R_3P=CH-\underset{\underset{CH_3}{|}}{\overset{\overset{CH_3}{|}}{Si}}-CH_2-\underset{\underset{CH_3}{|}}{\overset{\overset{CH_3}{|}}{Si}}-CH_3$

(V)

R:	CH₃	C₂H₅	i-C₃H₇
(V):	(a)	(b)	(c)

Scheme 2]

reaction with $(i\text{-}C_3H_7)_3PCH_2$ is slower due to a steric effect, but an analogous product is obtained. It is not known at this stage if this reaction actually involves an insertion product of the pentaalkylphosphorane form (B_1) because compound V may be formed either via (A_1) alone or via both (A_1) and (C_1). If, however, (B_1) is formed as an intermediate, it obviously undergoes facile ring opening, whereby silicon-stabilized car-

---

[2] Seyferth et al. (29) have been able to synthesize a first member of this series, but its set of substituents was unfortunately not suitable for our purposes.

banions (A_1) and (C_1) can be generated with equal probability. These are then isomerized through proton migration, resulting in the silylated ylides (60). Similarly, no decision as to the existence of a pentaalkylphosphorane intermediate could be made in the corresponding reaction with 1,3-dimethyl-1,3-disilacyclobutane (Scheme 3). The final product,

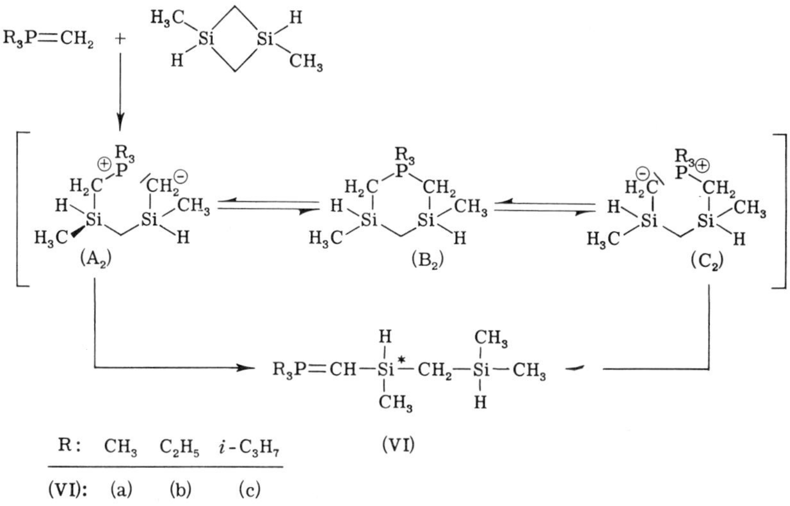

$$R_3P=CH_2 + \underset{\text{1,3-dimethyl-1,3-disilacyclobutane}}{\text{diagram}} \longrightarrow [(A_2) \rightleftarrows (B_2) \rightleftarrows (C_2)] \longrightarrow R_3P=CH-\underset{CH_3}{\overset{H}{Si}}-CH_2-\underset{H}{\overset{CH_3}{Si}}-CH_3$$

(VI)

R: CH_3 C_2H_5 i-C_3H_7

(VI): (a)   (b)   (c)

Scheme 3

isolated in quantitative yield, again consisted of a silylated ylide only, and there was no evidence for a PR_5 species.

b. Ylide Cleavage of a Monosilacyclobutanes (78). The situation was much more favorable in the monosilacyclobutanes, where possible intermediates no longer have any symmetry with respect to the distribution of the heteroatom. Thus 1,1-dimethylsilacyclobutane, although also yielding only an ylide (VII) as the sole product of the reaction with $(CH_3)_3$-PCH_2, makes it possible to demonstrate clearly the transitory existence of a cyclic pentaalkylphosphorane B_3 (Scheme 4). The structure of the material obtained can only be explained on the basis of the equilibria $(A_3) \rightleftarrows (B_3) \rightleftarrows (C_3)$ because there is obviously no mechanism for a direct conversion of A_3 into compound VII. The B_3 differs significantly from the intermediates B_2 and B_1 as in this case it is only carbon atom 5 that may form a silicon-stabilized carbanion, whereas carbon atom 4 lacks this possibility. The pentaalkylphosphorane ring B_3 will, therefore, open selectively to C_3, avoiding the formation of the "hot" carbanion A_3.

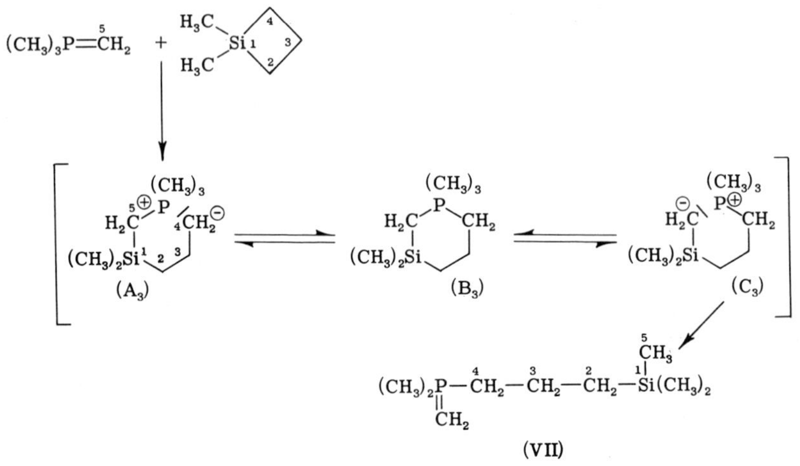

$$(CH_3)_3P=\overset{5}{C}H_2 \ + \ \underset{H_3C}{\overset{H_3C}{\diagdown}}\overset{4}{\underset{2}{\overset{3}{Si\,1}}}$$

$$\left[\begin{array}{ccc} \underset{(A_3)}{\underset{(CH_3)_2\overset{1}{Si}\diagdown\!\!\underset{2}{\ \ }\!\!\overset{3}{\ \ }}{H_2\overset{5\oplus}{C}\diagup\overset{(CH_3)_3}{P}\diagdown\overset{4}{C}H_2^\ominus}} & \underset{(B_3)}{\underset{(CH_3)_2Si\diagdown\!\!\!\diagup}{H_2C\diagup\overset{(CH_3)_3}{P}\diagdown CH_2}} & \underset{(C_3)}{\underset{(CH_3)_2Si\diagdown\!\!\!\diagup}{H_2\overset{\ominus}{C}\diagup\overset{(CH_3)_3}{P^\oplus}\diagdown CH_2}} \end{array} \right]$$

$$(CH_3)_2\underset{\underset{CH_2}{\|}}{\overset{4}{P}}-\overset{3}{C}H_2-\overset{2}{C}H_2-\overset{1|}{\underset{}{C}}H_2-\overset{\overset{5}{C}H_3}{\underset{}{Si(CH_3)_2}}$$

(VII)

Scheme 4

The observation that the proton migration of reaction $C_3 \rightarrow$ (VII) affects one of the methyl hydrogens is in agreement with previous findings (74). An inspection of formula VII shows that one of the methyl carbons attached to silicon in the product is supposed to originate from the alkylene group of the ylide.[3] In order to provide further support for this conclusion, which was based on mechanistic grounds only, it was desirable to introduce different substituents both at the ylide and at silicon, as well as to locate their position in the final product. The combinations of the simple dihydrosilacyclobutane and various methyl/ethyl ylides were considered to be an attractive selection of compounds for this purpose. With these components the reactions were complicated by an unexpected cyclisation process, which tended to obscure the course of the reactions. However, unambiguous conclusions were reached, offering convincing proof for the above mechanism.

c. *Cleavage with Successive Recyclization.* Monosilacyclobutane reacts with $(CH_3)_3PCH_2$ even at temperatures as low as $-60°C$. An NMR study of the reaction mixture shows the appearance of a product analogous to compound VI, but this compound undergoes a further reaction with

---

[3] Similarly, it was assumed that the ylidic carbon and the carbon of one of the methyl groups of the terminal silicon atoms should stem from equal parts of carbon atom 5 of the R_3PCH_2 starting material and from one of the ring carbons (⁴C) in the product of the reaction according to Eq. (21).

evolution of hydrogen above −50°C. The final product isolated in high yields was shown to be the cyclic ylide (VIII), which contains the moieties of the starting materials in a very peculiar arrangement, as indicated by the numbering of the carbon and silicon atoms (Scheme 5). The structure of this ylide was proved by detailed analytical and spectroscopic studies.

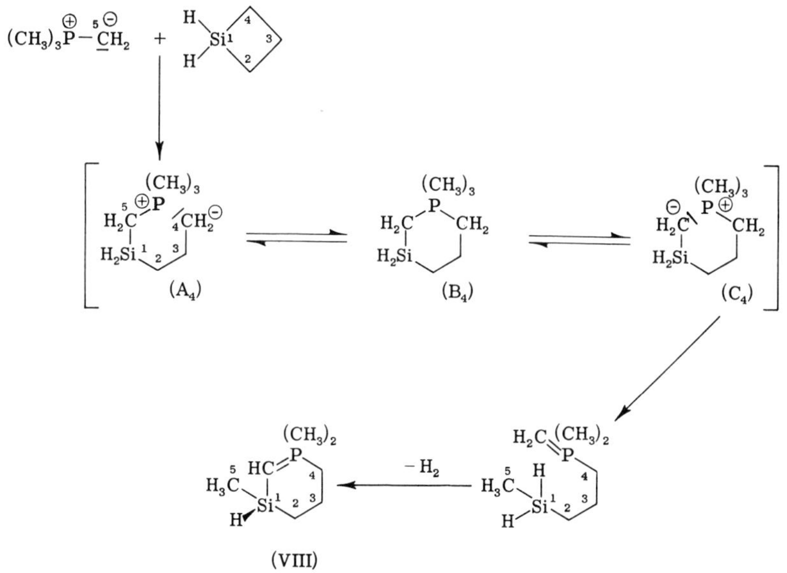

Scheme 5

Again the silicon atom of the ylide obtained bears a methyl group, which must originate from the ylide carbon of $(CH_3)_3PCH_2$ if the reaction scheme (Scheme 5) is correct.

In agreement with this assumption it was demonstrated that $(C_2H_5)_3PCH_2$ and $H_2Si\diamondsuit$ form a product with the three ethyl groups in the expected positions on the six-membered ring:

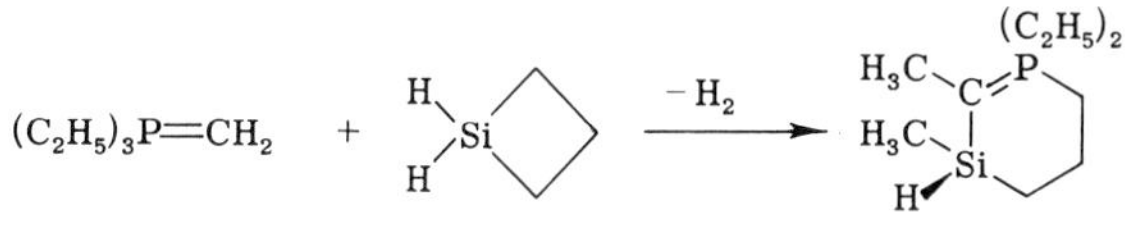

(25)

From ethylidene triethylphosphorane and $H_2Si\diamondsuit$, a cyclic ylide was obtained, in which an ethyl group was, indeed, attached to the silicon atom.

$$(C_2H_5)_3P=CHCH_3 \; + \; \text{[silacyclobutane with SiH}_2\text{]} \longrightarrow$$

$$\left[\text{(B}_5\text{)} \right] \longrightarrow \text{intermediate} \xrightarrow{-H_2} (X) \tag{26}$$

The formation of this species cannot be explained on the basis of any other obvious mechanism. Neither does the initially considered alternative, a C—C bond cleavage of the silacyclobutane instead of a Si—C bond cleavage (77), account for the ethylation of silicon that occurs in this crucial test experiment [Eq. (26)].

The structure of compound X is of prime importance in understanding a mechanism involving a pentaalkylphosphorane, so that some of its reactions were studied in addition to its physical and spectral properties. Among these the reaction with methanol was very valuable because this process first yields a well-defined methoxy derivative XI and then leads to a complete degradation with formation of ethyltrimethoxysilane:

$$(X) \xrightarrow{CH_3OH} (XI) \xrightarrow[CH_3OH]{\text{excess}} H_5C_2-Si(OCH_3)_3 \; + \; \text{by-products} \tag{27}$$

All of these substitution and degradation products were identified by their ^1H, ^{13}C, ^{31}P NMR and mass spectra, with special decoupling techniques being employed in the former, in order to confirm the assignments. The cleavage of Si—C—P bonds in ylides by methanol is in agreement with previous findings (74), and the same is true for the degradation of the SiCCCP linkage (79).

1-Monomethyl-1-silacyclobutane undergoes analogous reactions with ylides, which may be summarized as follows:

$$(CH_3)_3P=CH_2 \; + \; \text{[1-methyl-1-silacyclobutane]} \xrightarrow{-H_2} (XII) \tag{28}$$

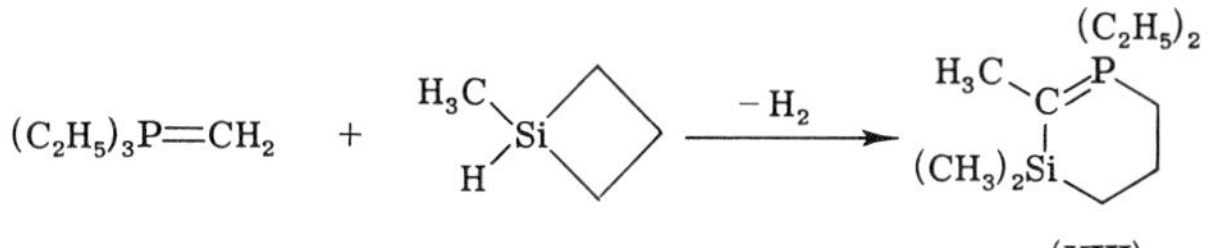

$$(C_2H_5)_3P=CH_2 \;+\; \underset{H}{\overset{H_3C}{\diagup}}Si\diagdown \triangle \xrightarrow{-H_2} \underset{(CH_3)_2Si}{\overset{H_3C\diagdown C\diagup P(C_2H_5)_2}{}} \qquad (29)$$

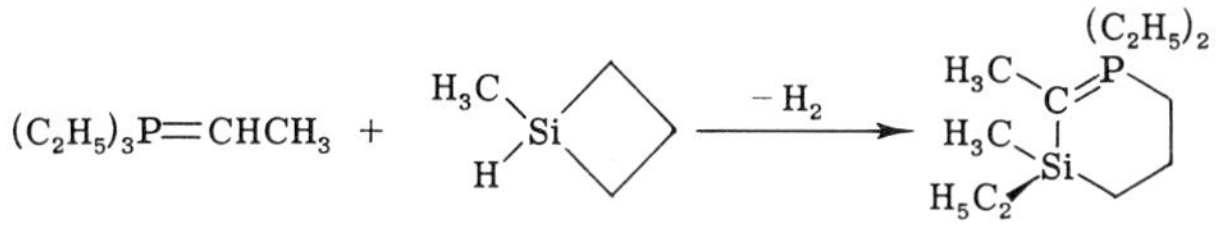

(XIII)

$$(C_2H_5)_3P=CHCH_3 \;+\; \underset{H}{\overset{H_3C}{\diagup}}Si\diagdown \triangle \xrightarrow{-H_2} \begin{array}{c} H_3C\diagdown \\ H_3C\diagdown \\ H_5C_2\diagup \end{array}\overset{(C_2H_5)_2}{\underset{Si}{C=P}} \qquad (30)$$

(XIV)

There can be no doubt that the products are formed via the same mechanism as proposed for the earlier examples. Detection of the $CH_3(C_2H_5)Si$ moiety in compound XIV again demonstrated the partition of moieties of the starting materials in the final product.

Still another group of reactions, the mechanism of which requires the intermediacy of a pentaalkylphosphorane, was found with difluorosilacyclobutane. The cleavage of the ring by an ylide and the closing and reopening of the cyclic phosphorane are followed by a dehydrofluorination caused by a second ylide equivalent. The HF abstraction leads to a recyclization analogous to the H_2 elimination in the previous examples (Scheme 6). The HF is trapped by $(C_2H_5)_3PCHCH_3$ to form the covalent tetraethyl

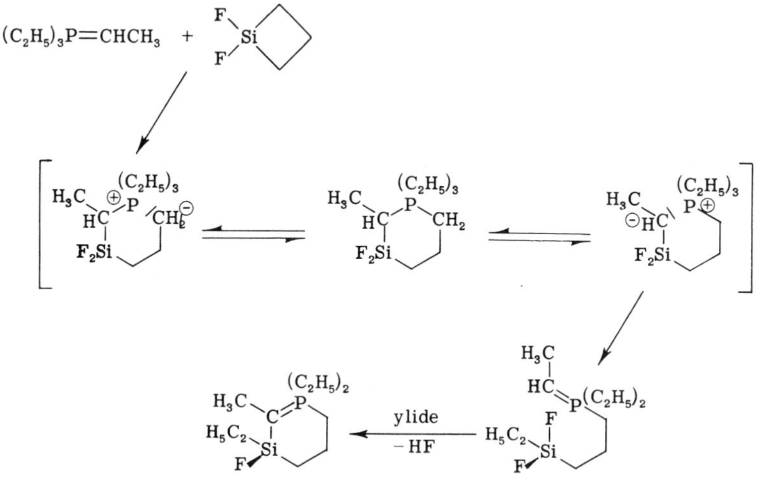

$$(C_2H_5)_3P=CHCH_3 \;+\; \underset{F}{\overset{F}{\diagup}}Si\diagdown \triangle$$

$$\left[\begin{array}{c} H_3C \\ HC \\ F_2Si \end{array} \overset{(C_2H_5)_3}{\underset{\oplus}{P}} \diagup CH^{\ominus} \right] \rightleftarrows \left[\begin{array}{c} H_3C \\ HC \\ F_2Si \end{array} \overset{(C_2H_5)_3}{\underset{P}{\diagdown}} CH_2 \right] \rightleftarrows \left[\begin{array}{c} H_3C \\ \ominus HC \\ F_2Si \end{array} \overset{(C_2H_5)_3}{\underset{P \oplus}{\diagdown}} \right] \qquad (31)$$

$$\underset{F}{\overset{H_3C}{\underset{H_5C_2}{\diagdown}}}\overset{(C_2H_5)_2}{\underset{Si}{C=P}} \xleftarrow[-HF]{ylide} \underset{F}{\overset{H_3C}{\underset{H_5C_2}{\diagdown}}}\overset{HC}{\underset{Si}{\diagdown}}p(C_2H_5)_2$$

(XV)

Scheme 6

fluorophosphorane, $(C_2H_5)_4PF$ (64). In the reaction with $(CH_3)_3PCH_2$ the ionic tetramethylphosphonium fluoride is isolated (67):

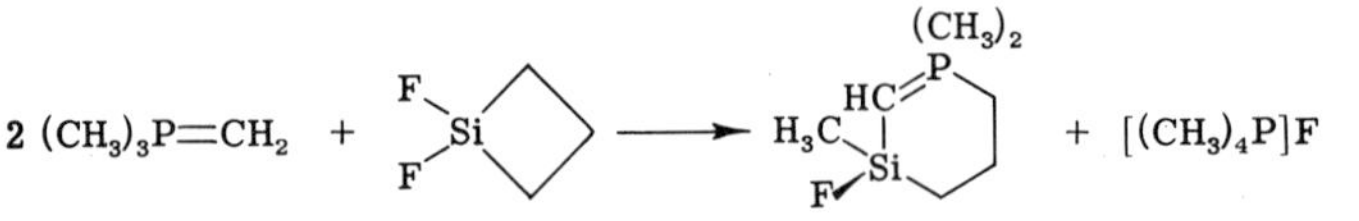

$$2\ (CH_3)_3P=CH_2\ +\ \underset{F}{\overset{F}{Si}}\!\!\diamondsuit\ \longrightarrow\ \underset{F}{\overset{(CH_3)_2}{\underset{Si}{\overset{P}{\underset{|}{H_3C}}}}}\!\!\diamondsuit\ +\ [(CH_3)_4P]F \tag{32}$$

(XVI)

With the corresponding chlorosilacyclobutanes the substitution of chlorine predominates and no ring cleavage is observed (78).

In conclusion, the reactions of silacyclobutanes with ylides have provided unambiguous evidence for the existence of cyclic pentaalkylphosphorane intermediates. The presence of silicon atoms in an α-position to the phosphorus atom of these ring systems does, however, facilitate the reopening of the heterocycle to such an extent that there is little chance for an isolation of these intermediates. It is characteristic of the ring reopening that it occurs exclusively through heterolytic breaking of the P—C—Si linkage, which leads to the silicon-stabilized carbanion. The synthesis of a simple phosphacarbocyclic system without silicon may be possible, perhaps, by using reaction conditions under which the nonpolar cyclobutane itself can be attached to an ylide without inducing secondary reactions [Eq. (19b)]. The reopening of the pentaalkylphosphorane ring must be of much higher activation energy than in the sila-substituted case and the desired species should be more easily trapped, although there is little doubt about the unfavorable thermodynamics of pentaalkylphosphoranes with respect to the ylidic analogs.

IV

ARSENIC YLIDES AND PENTAALKYLARSORANES

A. Alkylidene Trialkylarsoranes

1. *Preparation*

Only one alkylidene trialkylarsorane has been prepared in a pure state and fully characterized (73). This compound, $(CH_3)_3AsCH_2$, can be obtained through desilylation of a trimethylsilyl precursor (34) (first synthesized by Miller in 1965):

$$(CH_3)_3As + ClCH_2Si(CH_3)_3 \longrightarrow \begin{array}{c} (CH_3)_3As=CH_2 \\ \uparrow CH_3OH \end{array}$$

$$\downarrow$$

$$[(CH_3)_3AsCH_2Si(CH_3)_3]Cl \xrightarrow{n-C_4H_9Li} (CH_3)_3As=CH-Si(CH_3)_3$$

(33)

Methylene trimethylarsorane (the term "trimethylarsonium methylide" is equally correct) had already been formulated in 1953 by Wittig and Torssell (104) who studied the reaction of tetramethylarsonium salts with organolithium compounds. Although this method cannot be used for the preparation of the salt-free material due to the strong complexation by the lithium cations, it is clear from reactions of the product mixture that the ylide is present in solution. The formation of arsonium salts upon addition of alkyl halides is a typical example:

$$[(CH_3)_4As]Br \xrightarrow[-RH]{LiR} (CH_3)_3AsCH_2 \cdot Li^\oplus Br^\ominus$$

$$\downarrow CH_3I$$

$$[(CH_3)_3AsCH_2CH_3]I + LiBr$$

(34)

2. Properties, Structure, and Bonding

At room temperature, methylene trimethylarsorane (73) is a colorless crystalline compound, mp 33°C, which sublimes in a vacuum at 20°C/0.1 mm Hg. It is rapidly decomposed above 33°C; trimethylarsine and polymethylene are the main products formed. Blue and brown colors develop and some gas is evolved, predominantly ethylene. With trimethylphosphine, a methylene transfer reaction takes place yielding trimethylarsine and methylene trimethylphosphorane:

$$(CH_3)_3As=CH_2 \longrightarrow (CH_3)_3As + (CH_2)_n \quad (35)$$

$$\downarrow (CH_3)_3P$$

$$(CH_3)_3As + (CH_3)_3P=CH_2 \quad (36)$$

The compound is extremely sensitive to oxygen and moisture and must be handled with great care.

Thick fumes are observed when the material is exposed to air and spontaneous ignition may occur. A vigorous reaction with water leads to a strongly alkaline solution of some tetramethylarsonium hydroxide, but trimethylarsenic oxide is also formed (104). Tetramethyl methoxyarsorane can be obtained in a quantitative yield (68) with methanol under carefully controlled conditions:

$$(CH_3)_3As=CH_2 \xrightarrow{H_2O} [CH_3)_4As]OH, (CH_3)_3AsO \quad (37)$$

$$\downarrow CH_3OH$$

$$H_3C-\underset{\underset{H_3C}{\overset{O}{|}}}{\overset{\overset{CH_3}{|}}{As}}\overset{\cdots CH_3}{\underset{CH_3}{\diagdown}} \quad (38)$$

(XVII)

Compound XVII is a colorless distillable liquid of surprisingly high thermal stability. Low temperature ^{1}H and ^{13}C NMR spectroscopy revealed a trigonal-bipyramidal geometry of this molecule, which is nonrigid at room temperature on the NMR time scale (68). As expected from current theory, the methoxy group occupies an axial position at the low-temperature limit.

With traces of protic species (H_2O, CH_3OH), a rapid proton exchange is induced in $(CH_3)_3AsCH_2$ samples (73), rendering all hydrogen atoms equivalent on the NMR time scale:

$$CH_3-\underset{\underset{CH_3}{|}}{\overset{\overset{CH_3}{|}}{As}}=CH_2 \rightleftarrows CH_3-\underset{\underset{CH_2}{||}}{\overset{\overset{CH_3}{|}}{As}}-CH_3 \rightleftarrows CH_2=\underset{\underset{CH_3}{|}}{\overset{\overset{CH_3}{|}}{As}}-CH_3 \rightleftarrows CH_3-\underset{\underset{CH_3}{|}}{\overset{\overset{CH_2}{||}}{As}}-CH_3 \quad (39)$$

The same phenomenon is well documented for phosphorus ylides (75).

Methylene trimethylarsorane is a monomer in benzene solution. Study of its ^{1}H and ^{13}C NMR spectra led to some unexpected conclusions, distinctly different from the findings with the phosphorus analog (80). As shown in Table III, the δ and J characteristics of the CH_2 nuclei are not nearly as different from those of the CH_3 groups as are those of CH_2 and CH_3 in the phosphorus ylide (58). Whereas $J(H_2C) = 149$ of the latter indicates an sp^2 carbon (the value for ethylene is 150 Hz), $J(H_2C) = 131$ in the former rather points to an sp^3 carbon for the carbanion in $(CH_3)_3$-$AsCH_2$. (Methane, 125 Hz; CH_3 in the ylide, 133 Hz!)

In the absence of X-ray data for an arsenic ylide, it was, therefore, predicted that the carbanion in arsenic ylides should possess pyramidal geometry as opposed to the planar geometry in the phosphorus ylides.

$$\underset{\text{ylide}}{R_3\overset{\oplus}{As}-\overset{\ominus}{CR_2'}} \longleftrightarrow \underset{\text{ylene}}{R_3As=CR_2'} \quad (40)$$

This implies that the ylene formula is even less important for arsenic

TABLE III

^{13}C Nuclear Magnetic Resonance Spectra of $(CH_3)_3P=CH_2$, $(CH_3)_3As=CH_2$, and Their Silyl Derivatives[a]

Compound	\C=			CH$_3$P(As)			CH$_3$Si		
	δ (ppm)[b]	^1J(PC) (Hz)	^1J(HC) (Hz)	δ (ppm)[b]	^1J(PC) (Hz)	^1J(HC) (Hz)	δ (ppm)[b]	^3J(PC) (Hz)	^1J(HC) (Hz)
(CH$_3$)$_3$P=CH$_2$	−2.3	90.5	149	18.9	56.0	127	—	—	—
(CH$_3$)$_3$P=CH—Si(CH$_3$)$_3$	+0.7	88.2	134.6	20.0	57.3	128	+4.8	4.4	117.7
(CH$_3$)$_3$P=C[Si(CH$_3$)$_3$]$_2$	+0.3	63.3	—	21.4	57.3	127	+6.7	4.4	118
(CH$_3$)$_3$As=CH$_2$	7.6	—	130.9	15.6	—	133.8	—	—	—
(CH$_3$)$_3$As=CH—Si(CH$_3$)$_3$	8.0	—	141.2	17.1	—	133.0	+5.0	—	116.2
(CH$_3$)$_3$As=C[Si(CH$_3$)$_3$]$_2$	9.3	—	—	19.1	—	134	+6.9	—	118

[a] From Schmidbaur et al. (69).
[b] Value of δ is measured relative to C$_6$D$_6$, and converted relative to TMS.

then it was for phosphorus. The very limited stability of arsenic ylides and their high chemical reactivity are in agreement with this proposal.

The photoelectron spectrum of $(CH_3)_3AsCH_2$ shows an energy of only 6.72 eV for the highest occupied molecular orbital, which is predominantly a lone pair localized at the carbanion (69). This energy is lower than in the phosphorus compound (6.81 eV). This result is at least qualitatively in agreement with the picture drawn above.

Very little is known about chemical reactions of arsenic ylides, and the examples reported in the literature have been conducted with arylated species often bearing stabilizing substituents (11, 26). Even from these examples it appears that often the course of the reaction is entirely different from analogous processes with phosphorus ylides (50).

B. Silylated Ylides of Phosphorus and Arsenic

It has been pointed out previously that silylation of ylides leads to stabilized products and that this is only one example of the very general phenomenon of carbanion stabilization through silicon (34, 61, 72). This effect was also found for arsenic ylides (34, 73), and is the basis for the preparation of other compounds of this series. The influence of silicon is by no means solely an electronic effect. In many cases, where alkylsilyl substituents are introduced, a steric effect may well dominate, which may reduce lattice energies for salts in transylidation reactions, preventing intermolecular contacts in decomposition processes, and rendering the formation of salt adducts unfavorable. This steric effect is reduced to a minimum, but not eliminated, if simple SiH_3 groups are employed (61). Even then, however, a pronounced silicon effect is found, which must be based on electronic influences (49, 60, 61).

Of the two explanations advanced, one is the usual π-bonding concept involving d orbitals of the third-row elements (34), whereas the other refers to a more electrostatic picture (60). The lone pair of electrons at the carbanion is supposed (a) to interact with the suitable empty orbitals of silicon and (b) to suffer much less repulsive interactions from bond pairs in the presence of the larger elements. Both arguments call for the planar or quasi-planar geometry of the carbanion, which is observed in phosphorus ylides.

In arsenic ylides, As—C multiple bonding seems to be strongly reduced and the nonbonding interactions are also likely to be less important. The proposed nonplanar geometry may well account for this. Because in silylated arsenic ylides the silicon atoms would meet with unfavorable bonding conditions, it was, therefore, intriguing to find that the ^{1}H and ^{13}C NMR data and the photoelectron spectra again indicated a different situation (*69, 80*). Whereas with $(CH_3)_3PCH_2$ the silylation is known to cause a decrease in $^1J(^1H-^{13}C)$ in the carbanion by 15 Hz, this coupling constant was found to increase by more than 10 Hz (Table III) in $(CH_3)_3AsCH_2$. A first plausible explanation for these opposite effects has been offered by postulating a flattening of the carbanion pyramid in the arsenic ylide upon silylation, with favorable consequences for the electronic interactions of the carbon with silicon:

$$(CH_3)_3\overset{\oplus}{P}-\overset{\ominus}{C}\overset{H}{\diagdown H} \qquad (CH_3)_3\overset{\oplus}{P}-\overset{\ominus}{C}\overset{-Si(CH_3)_3}{\diagdown H}$$

$$(CH_3)_3\overset{\oplus}{As}-\overset{\ominus}{C}\overset{}{\diagdown H}\diagdown_H \qquad (CH_3)_3\overset{\oplus}{As}-\overset{}{C}\overset{-Si(CH_3)_3}{\underset{\ominus}{\diagdown H}}$$

Photoelectron spectra have shown that the energy of the HOMO in $(CH_3)_3AsCH_2$ is, indeed, lowered much more by silylation than that in $(CH_3)_3PCH_2$ (*69*).

C. Pentamethylarsorane

1. *Preparation*

In his study of the reaction of tetramethylarsonium salts with dimethylzinc, Cahours (*5*) originally claimed to have obtained pentamethylarsorane. However, subsequent attempts to repeat this preparation have met with failure (*16, 104*). Very recent experiments have led to successful synthesis of this compound through reaction of $(CH_3)_3AsCl_2$ with methyllithium under very mild conditions (*37*). If these precautions are taken no ylide is formed in this reaction, which was first tried by Wittig and Torssell (*104*).

Deprotonation seems to be a very slow process at low temperature in ether as a solvent so that the attack of the methyl carbanion at the arsonium center can compete successfully. Once formed, pentamethylarsenic is stable and easily recovered in the work-up of the reaction mixture:

$$(CH_3)_3AsCl_2 \xrightarrow[-LiCl]{LiCH_3} [(CH_3)_4As]Cl \xrightarrow[-LiCl]{LiCH_3} (CH_3)_5As \quad (41a)$$

$$\downarrow {\scriptstyle -CH_4, \; LiCH_3 \atop -LiCl}$$

$$(CH_3)_3As{=}CH_2 \quad (41b)$$

2. Properties, Spectra, and Structure

At room temperature, pentamethylarsorane is a colorless liquid of a characteristic odor which resembles the analogous antimony compound. It crystallizes below −6°C and can be sublimed under reduced pressure at −10°C. The $(CH_3)_5As$ is a monomer in benzene solution and shows a molecular ion in the mass spectrum with very low intensity. The vibrational spectra, infrared and Raman, could be assigned to a trigonal-bipyramidal skeleton. There are striking similarities to the spectra of $Sb(CH_3)_5$ (13).

The proton NMR spectrum appears as a singlet over a temperature range of +35° to −95°C and indicates nonrigid behavior. This pseudo-rotation phenomenon is characteristic for pentacoordinate molecules. The ^1H signal of the methyl resonance of pentakis-p-tolylarsorane also remains unsplit even at −90°C (20).

Compound $(CH_3)_5As$ is slowly hydrolyzed by water, and tetramethylarsonium hydroxide and some trimethylarsenic oxide are formed (37). With methanol, tetramethylmethoxyarsorane is generated with evolution of methane (68). It is interesting to note that $(CH_3)_5As$ and $(CH_3)_3AsCH_2$ both lead to identical products with these protic reagents:

$$(CH_3)_5As \begin{cases} \xrightarrow[-CH_4]{H_2O} [(CH_3)_4As]OH \xleftarrow{H_2O} \\ \\ \xrightarrow[-CH_4]{CH_3OH} (CH_3)_4AsOCH_3 \xleftarrow{CH_3OH} \end{cases} (CH_3)_3As{=}CH_2 \quad \begin{matrix}(42a)\\ \\ (42b)\end{matrix}$$

More generally speaking, these are only two examples for a larger series of reactions which are currently under investigation (89).

$$(CH_3)_5As + HX \xrightarrow{-CH_4} [(CH_3)_4As]X \quad (43)$$
$$(CH_3)_3AsCH_2 + HX \xrightarrow{}$$

The thermal decomposition of $(CH_3)_5As$ at 100°C leads to quantitative yields of trimethylarsine, methane, and ethylene, as followed by gas chromatography. Only traces of ethane were detectable. It is, therefore, assumed that the compound is decomposed via the ylide, which is known to be unstable under these conditions:

$$2(CH_3)_5As \xrightarrow[\text{slow}]{\text{heat}} 2(CH_3)_3As{=}CH_2 + 2CH_4 \quad (44)$$

$$\xrightarrow[\quad]{\text{fast}} 2(CH_3)_3As + C_2H_4 \quad (45)$$

with a heat / very slow pathway to $2C_2H_6$.

Thus, $(CH_3)_5As$ is at least kinetically much more stable than the corresponding ylide, and this relationship should give pentamethylphosphorane a good possibility for isolation if it can be prepared. There is no information available, however, on whether the decomposition occurs via a polar or a radical mechanism.

V
ANTIMONY YLIDES AND PENTAALKYLSTIBORANES

A. *Antimony Ylides*

It is obvious from the literature summarized in recent reviews (*11, 26*) that very little information is available on ylidic compounds of antimony. Moreover, the few compounds synthesized were all taken from the aryl series, whereas those of the alkyl series appear to be completely unknown. Even the aryl antimony ylides were of very limited thermal stability and could only be isolated in special cases (*11*).

It is, therefore, to be expected that a compound such as trimethylmethylene stiborane, $(CH_3)_3SbCH_2$, should be relatively unstable and attempts to synthesize this species would have to be carried out at low temperature. Another difficulty arises from the experimental fact (*104*) that all conventional methods of synthesis for stibonium ylides lead to pentaalkylstiboranes instead of ylides (see Introduction). Thus methylation of $(CH_3)_4SbX$ or $(CH_3)_3SbX_2$ halides by organometallic reagents of lithium, magnesium, aluminum, or zinc invariably yield $(CH_3)_5Sb$ as the sole product.

In a search for other possible routes to trimethylmethylene stiborane the transylidation method in the combination with phosphorus ylides was investigated very recently (59). Although $(CH_3)_3PCH_2$ was found to convert tetramethylstibonium salts into $(CH_3)_5Sb$, the corresponding ethylidene phosphorane reacted in a different manner—$(CH_3)_3Sb$ was the only antimony-containing product. As expected for a transylidation reaction, tetraethylphosphonium salt was found in a quantitative yield. This result can be explained in terms of

$$[(CH_3)_4Sb]Cl + (C_2H_5)_3P=CHCH_3 \longrightarrow (CH_3)_3Sb=CH_2 + [(C_2H_5)_4P]Cl$$
$$\downarrow \text{heat}$$
$$(CH_3)_3Sb + (CH_2)_n \tag{46}$$

in which the transitory existence of antimony ylide is proposed. Unfortunately it has not yet been possible to obtain more direct evidence for this intermediate, even when the reaction was carried out at $-25°C$. In principle, however, processes of this type seem to be very promising because they use ylides as very strongly basic but nonalkylating reactants, which are to be preferred over the strongly alkylating organometallic reagents. Further studies are required, at even lower temperatures, with other solvents and a variety of ylides.

This section is not complete without mentioning the hydrogen–deuterium exchange experiment by Doehring and Hoffmann (12), which indicated an enhanced acidity of the hydrogens in a tetramethylstibonium salt. In the corresponding equilibrium, stibonium ylides play the role of strong bases:

$$(CH_3)_4Sb^\oplus \rightleftharpoons \left\{ \begin{matrix} (CH_3)_3Sb^\oplus-CH_2^\ominus \\ \updownarrow \\ (CH_3)_3Sb=CH_2 \end{matrix} \right\} + H^\oplus \tag{47}$$

As would be expected, the proton exchange is much slower than in tetramethylphosphonium and -arsonium salts (12).

B. Pentaalkylstiboranes

1. *Preparation*

Pentamethylantimony was the first R_5Sb species to be obtained by Wittig and Torssell in 1955 (104). Various other derivatives of this type have since been synthesized, and a range of preparative methods has been described. Among these are the alkylation of R_4SbX, R_3SbX_2, and SbX_5 species with organometallic compounds of lithium, magnesium, aluminum, and zinc, as follows:

$$\left.\begin{array}{l}R_4SbX \\ R_3SbX_2 \\ SbX_5\end{array}\right\} \xrightarrow[R_2'Zn]{R'Li,\ R_2'Mg,\ R_3'Al} R_nR'_{5-n}Sb \qquad (48)$$

$$X = \text{halogen}$$
$$R,R' = \text{alkyl}$$

If $R = R'$, isoleptic pentaalkylstiboranes are obtained, but if $R \neq R'$, a whole range of mixed alkylated compounds is present in the product. This is due to a facile alkyl exchange process between R_5Sb compounds, as demonstrated by Meinema and Noltes (33). This exchange is probably catalyzed by organometallic reagents.

Pentamethylantimony is also generated in the reaction of pentaarylstiboranes with methyllithium; mixed species are the intermediates (54):*

$$(C_6H_5)_5Sb \xrightarrow[-C_6H_5Li]{+CH_3Li} (CH_3)_5Sb \qquad (49)$$

The chemistry of pentaalkenylstiboranes has been carefully studied by Nesmeyanov et al. (41–48). Thus, pentavinylantimony was prepared from trivinylantimony dichloride and vinylmagnesium bromide (42). The synthesis of these compounds obviously presents no problem, even when more complicated alkenyl groups are to be introduced (44).

2. Properties

Pentaalkylstiboranes are pale yellow liquids that can be distilled under reduced pressure (Table IV). Although distillation at atmospheric pressure is also possible, it is to be avoided. A violent explosion was reported in one case when a larger quantity of $(CH_3)_5Sb$ was rectified (still temperature 160°C) (33). When small amounts of R_5Sb compounds are heated, slow thermal decomposition is observed, which leads to trialkylantimony compounds, alkane and alkene (86a):

$$(C_2H_5)_5Sb \longrightarrow (C_2H_5)_3Sb + C_2H_6 + C_2H_4 \qquad (50)$$

---

* *Noted added in proof*: More recent experiments have shown (59) that up to five trimethylsilylmethyl groups can also be introduced at the antimony-V center, and, e.g., compounds of the type $(CH_3)_nSb[CH_2Si(CH_3)_3]_{5-n}$ can be easily obtained. Among the decomposition products of pentakis-trimethylsilylmethyl-stiborane ($n = 0$) a high yield of tetramethylsilane has been detected. This observation suggests that an antimony ylide may be among the primary products of this decomposition, but such a species could not be isolated:

$$[(CH_3)_3SiCH_2]_5Sb \rightarrow (CH_3)_4Si + [(CH_3)_3SiCH_2]_3Sb=CHSi(CH_3)_3$$
$$\downarrow$$
$$\text{decomposition}$$

The mechanism of this reaction is not yet fully understood, but it is likely that ylidic species are involved.

Vibrational spectra of $(CH_3)_5Sb$ have been measured and assigned on the basis of a D_{3h} structure for this molecule (13). Valence force constants have later been calculated and large differences are found for axial (1.40 mdyn/Å) and equatorial (1.90 mdyn/Å) Sb—C bonds of the trigonal-bipyramidal structure (17a, 106). The IR spectrum of $(C_2H_5)_5Sb$ has also been reported (86a).

By using proton magnetic resonance, however, one is not able to distinguish between axial and equatorial methyl groups, and only one singlet has been recorded even at $-100°C$. The usual "pseudorotational" process is invoked to account for this result (39). Pentaethylantimony and mixed methyl–ethyl species show similar PMR characteristics. The intermolecular alkyl exchange is slow on the NMR time scale, however, and separate signals are recorded in the mixture of $(CH_3)_n(C_2H_5)_{5-n}Sb$ compounds (33).

The electronic absorption spectrum of cyclohexane solutions of $(CH_3)_5Sb$ reveals at least two absorption bands (13).

3. Reactions

The chemical reactions of pentaalkylstiboranes have only recently been studied by a number of groups. The bulk of the reactions, some of which

TABLE IV
Boiling Points Reported for Some Pentaalkylstiboranes

Compound	bp (°C/mm Hg)	Ref.
$(CH_3)_5Sb$	126–127/760	104
	130–131/760	33
	[mp −16 to −18]	104
$(CH_3)_4SbC_2H_5$	53–54/16	33
$(CH_3)_3Sb(C_2H_5)_2$	71–74/16	33
$(CH_3)_2Sb(C_2H_5)_3$	42/0.03	33
$CH_3Sb(C_2H_5)_4$	55/0.4	33
$(C_2H_5)_5Sb$	64/0.4	33
	55.8/0.19	86a
$(CH_3)_4SbCH_2Si(CH_3)_3$	68/5.5	59
$(CH_3)_3Sb[CH_2Si(CH_3)_3]_2$	65/0.1	59
$(CH_3)_2Sb[CH_2Si(CH_3)_3]_3$	30/0.001 (subl)	59
$CH_3Sb[CH_2Si(CH_3)_3]_4$	60/0.001	59
$Sb[CH_2Si(CH_3)_3]_5$	50/0.001 (subl)	59
	[mp 93]	

had already been considered by Wittig and his collaborators (104), may be classified into three or four groups.

a. With Brønsted Acids, pentaalkylstiboranes undergo cleavage of one or two Sb—C bonds, with evolution of alkane and formation of R_4SbX or R_3SbX_2 derivatives, respectively:

$$R_5Sb \xrightarrow[-RH]{HX} R_4SbX \xrightarrow[-RH]{HX} R_3SbX_2 \quad (51)$$

Any further reaction of the R_3SbX_2 products is very slow even with strong acids. In the first step in Eq. (51), very weak acids can be employed. Thus, alcohols, water, mercaptans, and carboxylic acids are now known to react smoothly with pentamethylantimony under mild conditions. The following tabulation cites compounds prepared from $(CH_3)_5Sb$ and free acids:

$$(CH_3)_5Sb \xrightarrow[-CH_4]{HX} (CH_3)_4SbX \quad (52)$$

X:	HO	RO	RS	RCOO	C_6H_5O	F	N_3, SCN, etc.
Ref.	(76)	(76)	(63)	(66)	(36)	(76)	(65)
X:	O_2PR_2	O_2PF_2	O_2PH_2	$OSPR_2$	$NOSR_2$	$OSiR_3$	
Ref.:	(89)	(89)	(89)	(89)	(36)	(56)	

Depending on the nature of X, the $(CH_3)_4SbX$ products were found to be either covalent molecular species (OH, OR, SR, F) or saltlike materials (azide, thiocyanide, chloride, dimethylphosphinate). The state of bonding may be different in solution from the solid state, and it may vary with the nature of the solvent (carboxylates). Only the stronger acids, if employed in excess, will afford the R_3SbX_2 species. This is true with the halogen hydrides, halosulfuric or -phosphonic acids, etc.

b. With oxidizing reagents, a similar sequence of reactions is observed, which may be represented by

$$R_5Sb \xrightarrow[-RX]{X_2} R_4SbX \xrightarrow[-RX]{X_2} R_3SbX_2 \quad (53)$$

A typical example (95) is the reaction of $(CH_3)_5Sb$ with bromine, which leads to $(CH_3)_4SbBr$ or $(CH_3)_3SbBr_2$, depending on the molar ratio of the reagents. This process has not yet been widely used for preparative purposes.

c. With Lewis acids, an alkyl carbanion may be abstracted from the R_5Sb species with formation of stibonium salts. Such a reaction was demonstrated with aluminum alkyls R_3Al (and alkyl halides R_nAlX_{3-n})

and with antimony pentachloride (36, 86):

$$(C_2H_5)_5Sb + Al(C_2H_5)_3 \longrightarrow [(C_2H_5)_4Sb]^+[Al(C_2H_5)_4]^- \qquad (54)$$

$$(CH_3)_5Sb + SbCl_5 \longrightarrow [(CH_3)_4Sb]^+[CH_3SbCl_5]^- \qquad (55)$$

With thallium tribromide, a tetrabromothallate complex was obtained with formation of an alkyl bromide and thallium(I) bromide (47):

$$\begin{array}{c} R_5Sb + TlBr_3 \longrightarrow [R_4Sb]^+[RTlBr_3]^- \\ \qquad\qquad\qquad\quad\;\; \Big\downarrow TlBr_3 \\ [R_4Sb]^+[TlBr_4]^- \longleftarrow [R_4Sb]^+Br^- + RBr + TlBr \end{array} \qquad (56)$$

Only toward the strongest nucleophiles do the pentaalkylstiboranes act as Lewis acids. This reaction principle is involved in the interaction of pentamethylantimony with methyllithium, which is believed to yield a hexamethylantimonate complex:

$$(CH_3)_5Sb + LiCH_3 \longrightarrow Li^+[Sb(CH_3)_6]^- \qquad (57)$$

The structure of this product is not known, however, and may well be much more complex (104).

VI
BISMUTH COMPOUNDS

No pentaalkylbismuth compounds R_5Bi and no trialkylbismuthonium ylides $R_3Bi{=}CR_2'$ have been recorded in the literature (11, 26). If there have been attempts for the preparation of these compounds, they must have met with complete failure. Although the strong oxidizing properties of bismuth(V), and the extremely unfavorable situation for π interactions with carbon, make it highly unlikely that such compounds can ever be isolated, very recent findings with niobium and tantalum call for caution in making such a prediction.

VII
RELATED COMPOUNDS OF VANADIUM, NIOBIUM, AND TANTALUM

A. Vanadium Ylides?

Recent investigations directed toward the synthesis of homoleptic vanadium alkyls (10) have not been successful for VR_5 species, and only

VR_4, VR_3, and VR_3O compounds have been obtained (38), even when the promising novel ligand systems R = $(CH_3)_3SiCH_2$, $(CH_3)_3CCH_2$, or $[(CH_3)_3Si]_2CH$ were employed.

It might have been expected that the reactions could lead to vanadium ylides instead of the pentaalkyls, but this was obviously not the case. From the literature it is not clear whether or not true vanadium(V) compounds have been used as starting materials in at least some of the reactions to warrant the appropriate oxidation state. The problems associated with the vanadium pentahalides (except for the fluoride) and other VX_5 compounds may well be the reason for the difficulties arising in the syntheses. If these difficulties can be overcome, there should be a chance for a successful preparative procedure according to one of the following pathways:

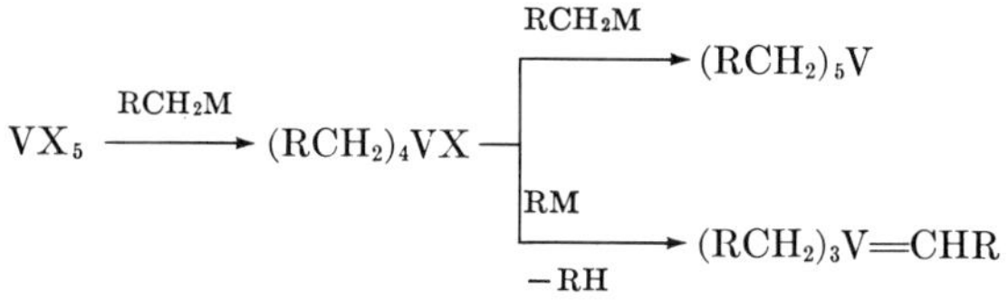

$$VX_5 \xrightarrow{RCH_2M} (RCH_2)_4VX \begin{array}{l} \xrightarrow{RCH_2M} (RCH_2)_5V \quad (58a) \\ \xrightarrow[-RH]{RM} (RCH_2)_3V=CHR \quad (58b) \end{array}$$

M = metal
X = halogen, etc.
R = alkyl, silyl, hydrogen

Considering the presently available knowledge on vanadium(V) compounds, reaction (58b) seems to be more attractive, as compounds of types $(RO)_3VO$, $(RO)_3V=NR$, etc., are well documented for both R = alkyl and R = silyl (4, 81), whereas only very few VX_5 species are known. The pertinent chemistry of the congeners niobium and tantalum offers examples for pentacoordination as well as for ylide formation.

B. Niobium Compounds

It was only in a 1974 report (83) that some evidence was provided for the possible existence of pentamethylniobium through the isolation of its addition compounds with a ditertiary phosphine and with methyllithium. These compounds were synthesized according to the following scheme:

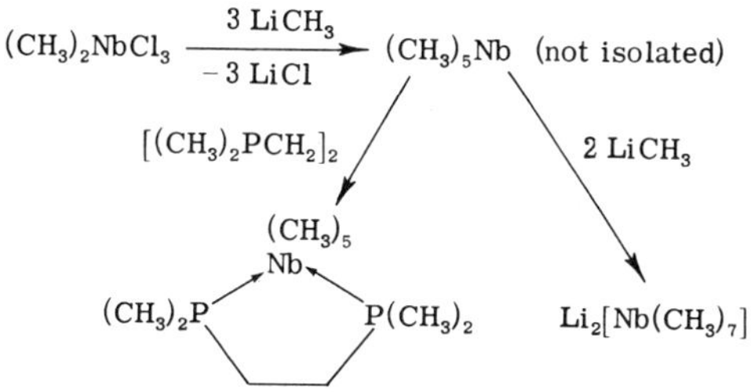

$$(CH_3)_2NbCl_3 \xrightarrow[-3\,LiCl]{3\,LiCH_3} (CH_3)_5Nb \text{ (not isolated)} \qquad (59)$$

with $[(CH_3)_2PCH_2]_2$ giving the chelate $(CH_3)_5Nb[(CH_3)_2P\frown P(CH_3)_2]$ and with $2\,LiCH_3$ giving $Li_2[Nb(CH_3)_7]$.

From these results it appears that the stepwise alkylation of $(CH_3)_2NbCl_3$, which should involve the tetramethylniobium chloride stage, leads finally to the polymethylated species $Nb(CH_3)_n^{(n-5)\ominus}$ where $n = 5, 6, 7$, and does not yield ylidic species.

$$(CH_3)_2NbCl_3 \xrightarrow{LiCH_3} (CH_3)_3NbCl_2 \xrightarrow{LiCH_3} (CH_3)_4NbCl \begin{array}{c} \nearrow (CH_3)_5Nb \\ \searrow\!\!\!\!\!\!/ \ (CH_3)_3Nb=CH_2 + CH_4 \end{array} \quad (60)$$

In this respect Nb resembles Sb, which is also exclusively converted into $Sb(CH_3)_n^{(n-5)\ominus}$ structures.

With the more bulky $(CH_3)_3SiCH_2$ ligands, completely different products are obtained (25). From the reaction of $NbCl_5$ with $(CH_3)_3SiCH_2MgCl$, a binuclear complex of formula $[(CH_3)_3SiCH_2]_4Nb_2[CSi(CH_3)_3]_2$ was isolated. The crystal structure of this complex has been determined and the niobium centers were shown to be bridged by two trimethylsilylmethylidyne moieties. The diamagnetism of the material requires spin pairing through direct or indirect metal–metal interactions to account for the formal d^1 configuration of the niobium atoms:

$$\begin{array}{c}
 Si(CH_3)_3 \\
 | \\
(CH_3)_3Si-CH_2 C CH_2-Si(CH_3)_3 \\
 \diagup \ \diagdown \\
 Nb \cdots Nb \\
 \diagdown \ \diagup \\
(CH_3)_3Si-CH_2 C CH_2-Si(CH_3)_3 \\
 | \\
 Si(CH_3)_3
\end{array}$$

From this result it may be concluded that successive alkylation of Nb(V) with $(CH_3)_3SiCH_2$ groups does not proceed to the NbR_5 stage, as observed in the methylation, but that deprotonation of the CH_2 groups does occur, and that ylidic species may therefore be intermediates. Ultimately, oligomerization with loss of alkyl and silyl groups leads to dimers, however, and no mononuclear compounds have been characterized to date.

C. Tantalum Pentaalkyls and Tantalum Ylides

Tantalum is the first element of subgroup Va for which both pentaalkyl and ylidic compounds could be completely characterized and fully investigated. Not only was it possible to prove the existence of $Ta(CH_3)_5$ and some of its complexes (83), but the "tantalum ylide," trisneopentyl neopentylidene tantalum, could also be isolated (82a). This ylide is the product of a reaction between trisneopentyltantalum dichloride and neopentyllithium in pentane at room temperature. Two equivalents of LiCl and 1 equivalent of neopentane are observed as the sole by-products.

$$[(CH_3)_3CCH_2]_3TaCl_2 \xrightarrow[-2(LiCl, -C(CH_3)_4]{2LiCH_2C(CH_3)_3} [(CH_3)_3CCH_2]_3Ta\!\!=\!\!CHC(CH_3)_3 \quad (61)$$

The ylide was identified by its ^{1}H and ^{13}C NMR and mass spectra, by chemical analysis, and by a cryoscopic determination of molecular mass in solution. It could also be obtained, in lower yields, from $TaCl_5$ and 5 equivalents of the neopentyl Grignard reagent. It has a melting point of 71°C and can be distilled (!) at 75°C under vacuum. The compound is very sensitive to oxygen and moisture, but may be stored indefinitely at room temperature in an inert atmosphere.

The mechanism of the formation of the ylide has been studied by deuterium-labeling experiments, and it is assumed that the pentakisneopentyl tantalum, which is formed first, is "decomposed" with elimination of neopentane. This idea parallels the findings for pentamethylarsenic (37), which has also been found to decompose via an ylide intermediate.

In the light of these results it is surprising that the reaction of the silaneopentyl analogs with $TaCl_5$ does not afford a similar trimethylsilylmethylide species. Again, a binuclear complex, as already described for niobium, is formed (25):

$$TaCl_5 + 5(CH_3)_3SiCH_2MgCl \longrightarrow [(CH_3)_3SiCH_2]_4Ta_2[CSi(CH_3)_3]_2 \quad (62)$$

It is not clear how this difference in reaction behavior arises as both ligands have at least similar steric requirements.*

D. Metal Ylides or Metal Carbene Complexes?

In this article the author has been using the traditional ylide–ylene nomenclature as well as the modern phosphorane–arsorane–stiborane formalism. No attempt has been made to discriminate between one or the other of these modalities because it is felt that both offer a sufficiently clear description of at least those species that are derived from the main Group Vb elements. (Fortunately no NR_5 or BiR_5 had to be given names.). However, when this procedure was followed with the vanadium, niobium, and tantalum compounds, an interesting problem arose because the authors

---

* *Note added in proof*: Very recent work by Schrock has extended this tantalum ylide chemistry very considerably and some cyclopentadienyl tantalum methylenes and benzylidenes could be isolated and fully characterized (82b). Among the new compounds the complex $(C_5H_4CH_3)_2Ta(CH_3)(CH_2)$ is particularly noteworthy. Its crystal structure has been determined and the temperature dependence of its nmr spectra has been carefully investigated (17b). The ylidic carbon is in a trigonal planar configuration!

of the original communications had baptized some of their compounds "carbene complexes" (*82*). Indeed, an $R_3Ta\!\!=\!\!CR_2'$ compound may well be thought of as a carbene adduct of a (hypothetical) R_3Ta component. The niobium and tantalum compounds of this type would thus merely add to the ever-increasing plethora of transition metal carbene complexes (*14*).

It is perhaps inopportune to elaborate on the nomenclature, but some of the data reported for the "tantalum ylide" indicate that there may be a fundamental difference between this transition metal compound and the formally related ylides of the Group Vb elements. The most significant discrepancy is found with the ^{13}C NMR shift of the carbene/ylide carbon atoms, which typically is downfield for the Va element, but upfield for the Vb element derivatives. Ylidic carbon atoms may, therefore, possibly bear a much higher negative charge.

Apart from this feature there are many similarities between ylides and carbene complexes, primarily among the structural criteria. The carbene carbon may be, but not necessarily, in a planar configuration, and the M—C bonding indicates some multiple bonding character just as in most of the ylides. On the other hand, carbene transfer reactions have been observed with ylides [e.g., Eq. (36)], indicating that the carbene complex formalism can, indeed, be successfully applied with ylides. There is hope, therefore, for a fruitful symbiosis of ylide and carbene complex chemistry, which may soon become complementary as more data become available from this new area of transition metal chemistry.

REFERENCES

1. Absar, I., and Van Wazer, J. R., *J. Amer. Chem. Soc.* **94**, 2382 (1972).
2. Bart, J. C. J., *J. Chem. Soc. B* 350 (1969).
3. Bestmann, H. J., and Zimmermann, R., in "Organic Phosphorus Compounds" (G. M. Kosolapoff and L. Maier, eds.), Vol. 3, p. 1. Wiley (Interscience), New York, 1972.
4. Bürger, H., Sawodny, W., and Wannagat, U., *Monatsh. Chem.* **95**, 292 (1964).
5. Cahours, A., *Justus Liebigs Ann. Chem.* **122**, 329 (1862).
6. Chioccola, G., and Daly, J. J., *J. Chem. Soc. A* 568 (1968).
7. Daly, J. J., *J. Chem. Soc. A* 1913 (1967).
8. Daly, J. J., and Wheatley, P. J., *J. Chem. Soc. A* 1703 (1966).
9. Daniel, H., and Paetsch, J., *Chem. Ber.* **98**, 1915 (1965).
10. Davison, P. J., Lappert, M. F., and Pearce, R., *Accounts Chem. Res.* **7**, 209 (1974).
11. Doak, G. O., and Freedman, L. D., "Organometallic Compounds of Arsenic, Antimony and Bismuth." Wiley, New York, 1970.
12. Doering, W. v. E., and Hoffmann, A. K., *J. Amer. Chem. Soc.* **77**, 521 (1955).
13. Downs, A. J., Schmutzler, R., and Steer, J. A., *Chem. Commun.* 221 (1966).
14. Fischer, E. O., *Angew. Chem.* **86**, 651 (1974); *Angew. Chem., Int. Ed. Engl.* (1974).
15. Franzen, V., and Wittig, G., *Angew. Chem.* **72**, 417 (1960).

16. Friedrich, M. E. P., and Marvel, C. S., *J. Amer. Chem. Soc.* **52**, 376 (1930).
17a. Goel, R. G., Maslowsky, E., and Senoff, C. V., *Inorg. Chem.* **10**, 2592 (1971).
17b. Guggenberger, L. J., and Schrock, R. R., *J. Amer. Chem. Soc.* **97**, 6578 (1975).
18. Hager, F. D., and Marvel, C. S., *J. Amer. Chem. Soc.* **48**, 2689 (1926).
19. Hellwinkel, D., *Chem. Ber.* **98**, 576 (1965).
20. Hellwinkel, D., *Angew. Chem.* **78**, 749 (1966).
21. Hellwinkel, D., *in* "Organic Phosphorus Compounds" (G. M. Kosolapoff and L. Maier, eds.), Vol. 3, p. 185. Wiley (Interscience), New York, 1972.
22. Hildenbrand, K., and Dreeskamp, H., *Z. Naturforsch. B* **28**, 226 (1973).
23. Hoffmann, R., Boyd, D. B., and Goldberg, S. Z., *J. Amer. Chem. Soc.* **92**, 3929 (1970).
24. Houssain, S., and Schmidbaur, H., *J. Organometal. Chem.* in prep.
25. Huq, F., Mowat, W., Skapki, A. C., and Wilkinson, G., *Chem. Commun.* 1477 (1971).
26. Johnson, A. W., "Ylid Chemistry." Academic Press, New York, 1966.
27. Katz, T. J., and Turnblom, E. W., *J. Amer. Chem. Soc.* **92**, 6701 (1970).
28. Köster, R., Grassberger, M. A., and Simić, D., *Justus Liebigs Ann. Chem.* **739**, 211 (1970).
29. Lamberth, R. L., and Seyferth, D., *J. Amer. Chem. Soc.* **94**, 9246 (1972).
30. Lüttke, W., and Wilhelm, K., *Angew. Chem.* **77**, 867 (1965).
31. Lumbroso, H., Bertin, D. M., and Froyen, P., *Bull. Soc. Chim. Fr.* 819 (1974), and earlier papers quoted therein.
32. Mak, T. C. W., and Trotter, J., *Acta Crystallogr.* **18**, 81 (1965).
33. Meinema, H. A., and Noltes, J. G., *J. Organometal. Chem.* **22**, 653 (1971).
34. Miller, N. E., *J. Amer. Chem. Soc.* **87**, 390 (1965).
35. Miller, N. E., *Inorg. Chem.* **4**, 1458 (1965).
36. Mitschke, K. H., Dissertation, University of Würzburg, 1972.
37. Mitschke, K. H., and Schmidbaur, H., *Chem. Ber.* **106**, 3645 (1973).
38. Mowat, W., Shortland, A. J., Yagupsky, G., Hill, N. J., Yagupsky, M., and Wilkinson, G., *J. Chem. Soc., Dalton Trans.* 533 (1972).
39. Muetterties, E. L., Mahler, W., Packer, K. J., and Schmutzler, R., *Inorg. Chem.* **3**, 1298 (1964).
40. Musker, K. W., and Stevens, R. R., *Inorg. Chem.* **8**, 255 (1969), and earlier papers quoted therein.
41. Nesmeyanov, A. N., Borisov, A. E., and Novikova, N. V., *Izvest. Akad. Nauk SSSR, Otdel. Khim. Nauk* 147 (1960).
42. Nesmeyanov, A. N., Borisov, A. E., and Novikova, N. V., *Izvest. Akad. Nauk SSSR, Otdel. Khim. Nauk* 952 (1960).
43. Nesmeyanov, A. N., Borisov, A. E., and Novikova, N. V., *Izvest. Akad. Nauk SSSR, Otdel. Khim. Nauk* 612 (1961).
44. Nesmeyanov, A. N., Borisov, A. E., and Novikova, N. V., *Izvest. Akad. Nauk SSSR, Otdel. Khim. Nauk* 730 (1961).
45. Nesmeyanov, A. N., Borisov, A. E., and Novikova, N. V., *Izvest. Akad. Nauk SSSR, Otdel. Khim. Nauk* 1578 (1961).
46. Nesmeyanov, A. N., Borisov, A. E., and Novikova, N. V., *Izvest. Akad. Nauk SSSR, Otdel. Khim. Nauk* 1197 (1964).
47. Nesmeyanov, A. N., Borisov, A. E., and Novikova, N. V., *Izvest. Akad. Nauk SSSR, Otdel. Khim. Nauk* 1202 (1964).
48. Nesmeyanov, A. N., Borisov, A. E., and Novikova, N. V., *Tetrahedron Letters* 1960, 23.

49. Ostoja-Starzewski, K. A., Tom Dieck, H., and Bock, H., *J. Organometal. Chem.* **65**, 311 (1974); see also Ostoja-Starzewski, K. A., and Bock, H., *J. Amer. Chem. Soc.* submitted.
50. Piskala, A., Zimmermann, M., Fouquet, G., and Schlosser, M., *Collection Czech. Chem. Soc.* **36**, 1482 (1971).
51. Sawodny, W., *Z. Anorg. Allg. Chem.* **368**, 284 (1969).
52. Schlenk, W., and Holtz, J., *Chem. Ber.* **49**, 603 (1916).
53. Schlenk, W., and Holtz, J., *Chem. Ber.* **50**, 274 (1917).
54. Schlosser, M., Kadibelban, and Steinhoff, G., *Justus Liebigs Ann. Chem.* **743**, 25 (1971).
55. Schmidbaur, H., *Accounts Chem. Res.* **8**, 62 (1975).
56. Schmidbaur, H., Arnold, H. S., and Beinhofer, E., *Chem. Ber.* **97**, 449 (1964).
57. Schmidbaur, H., Buchner, W., and Köhler, F. H., *J. Amer. Chem. Soc.* **96**, 6208 (1974).
58. Schmidbaur, H., Buchner, W., and Scheutzow, D., *Chem. Ber.* **106**, 1251 (1973).
59. Schmidbaur, H., and Hasslberger, G., unpublished results, 1974.
60. Schmidbaur, H., and Malisch, W., *Chem. Ber.* **103**, 97 (1970).
61. Schmidbaur, H., and Malisch, W., *Chem. Ber.* **103**, 3007 (1970).
62. Schmidbaur, H., Malisch, W., and Rankin, D., *Chem. Ber.* **104**, 145 (1971).
63. Schmidbaur, H., and Mitschke, K. H., *Chem. Ber.* **104**, 1837 (1971).
64. Schmidbaur, H., Mitschke, K. H., Buchner, W., Stühler, H., and Weidlein, J., *Chem. Ber.* **106**, 1226 (1973).
65. Schmidbaur, H., Mitschke, K. H., Weidlein, J., and Cradock, S., *Z. Anorg. Allg. Chem.* **386**, 139 (1971).
66. Schmidbaur, H., Mitschke, K. H., and Weidlein, J., *Z. Anorg. Allg. Chem.* **386**, 147 (1971).
67. Schmidbaur, H., Mitschke, K. H., and Weidlein, J., *Angew. Chem.* **84**, 166 (1972); *Angew. Chem., Int. Ed. Engl.* **11**, 145 (1972).
68. Schmidbaur, H., and Richter, W., *Angew. Chem.* **87**, 204 (1975); *Angew. Chem., Int. Ed. Engl.* **14**, 183 (1975).
69. Schmidbaur, H., Richter, W., and Ostoja-Starzewski, K. A., *Chem. Ber.* **109**, 473 (1976).
70. Schmidbaur, H., and Stühler, H., *Chem. Ber.* **107**, 1420 (1974).
71. Schmidbaur, H., Stühler, H., and Buchner, W., *Chem. Ber.* **106**, 1238 (1973).
72. Schmidbaur, H., and Tronich, W., *Chem. Ber.* **100**, 1032 (1967).
73. Schmidbaur, H., and Tronich, W., *Inorg. Chem.* **7**, 168 (1968).
74. Schmidbaur, H., and Tronich, W., *Chem. Ber.* **101**, 595 (1968).
75. Schmidbaur, H., and Tronich, W., *Chem. Ber.* **101**, 604 (1968).
76. Schmidbaur, H., Weidlein, J., and Mitschke, K. H., *Chem. Ber.* **102**, 4136 (1969).
77. Schmidbaur, H., and Wolf, W., *Angew. Chem.* **85**, 345 (1974); *Angew. Chem., Int. Ed. Engl.* **12**, 321 (1973).
78. Schmidbaur, H., and Wolf, W., *Chem. Ber.* **108**, 2834, 2842, 2851 (1975).
79. Schmidbaur, H., and Wolf, W., unpublished results, 1974.
80. Schmidbaur, H., Wolf, W., Richter, W., and Köhler, F. H., *Chem. Ber.* **108**, 2649 (1975).
81. Schmidt, M., and Schmidbaur, H., *Angew. Chem.* **71**, 220 (1959).
82a. Schrock, R. R., *J. Amer. Chem. Soc.* **96**, 6796 (1974).
82b. Schrock, R. R., *J. Amer. Chem. Soc.* **97**, 6577 (1975).
83. Schrock, R. R., and Meakin, P., *J. Amer. Chem. Soc.* **96**, 5288 (1974).
84. Speziale, A. J., and Ratts, K. W., *J. Amer. Chem. Soc.* **87**, 5603 (1965).

85. Stephens, F. S., *J. Chem. Soc.* 5640 and 5658 (1965).
86a. Takashi, Y., *J. Organometal. Chem.* **8**, 225 (1967).
86b. The, K. I., and Cavell, R. G., *J. Chem. Soc., Chem. Commun.* 716 (1975).
87. Turnblom, E. W., and Katz, T. J., *J. Amer. Chem. Soc.* **93**, 4065 (1971).
88. Turnblom, E. W., and Katz, T. J., *J. Amer. Chem. Soc.* **95**, 4292 (1973).
89. Weidlein, J., Ott, R., Mitschke, K.-H., and Eberwein, B., *Chimia* **29**, 262 (1975); in press.
90. Westheimer, F. H., *Accounts Chem. Res.* **1**, 70 (1968).
91. Weygand, F., and Daniel, H., *Chem. Ber.* **94**, 3147 (1961).
92. Weygand, F., Daniel, H., and Simon, H., *Chem. Ber.* **91**, 1691 (1958); *Justus Liebigs Ann. Chem.* **654**, 111 (1962).
93. Weygand, F., Daniel, H., and Schroll, A., *Chem. Ber.* **97**, 1217 (1964).
94. Wheatley, P. J., *J. Chem. Soc. A* 5785 (1965).
95. Wittig, G., and Clauss, K., *Justus Liebigs Ann. Chem.* **577**, 26 (1952).
96. Wittig, G., and Clauss, K., *Justus Liebigs Ann. Chem.* **578**, 136 (1952).
97. Wittig, G., and Geissler, G., *Justus Liebigs Ann. Chem.* **580**, 44 (1953).
98. Wittig, G., and Hellwinkel, D., *Angew. Chem.* **74**, 76 (1962).
99. Wittig, G., and Kochendörfer, E., *Angew. Chem.* **70**, 506 (1958).
100. Wittig, G., and Krauss, D., *Justus Liebigs Ann. Chem.* **679**, 34 (1964).
101. Wittig, G., and Polster, R., *Justus Liebigs Ann. Chem.* **599**, 1 (1956); **599**, 13 (1956).
102. Wittig, G., and Rieber, M., *Naturwissenschaften* **35**, 345 (1948); *Justus Liebigs Ann. Chem.* **562**, 187 (1949).
103. Wittig, G., and Schöllkopf, U., *Chem. Ber.* **87**, 1318 (1954).
104. Wittig, G., and Torssell, K., *Acta Chem. Scand.* **7**, 1293 (1955).
105. Wittig, G., and Wetterling, M., *Justus Liebigs Ann. Chem.* **557**, 193 (1947).
106. Woods, J., and Long, G. G., *J. Mol. Spectrosc.* **38**, 387 (1971).

Acetylene and Allene Complexes: Their Implication in Homogeneous Catalysis

SEI OTSUKA AND AKIRA NAKAMURA

Department of Chemistry
Faculty of Engineering Science
Osaka University
Toyonaka, Osaka, Japan

I. Introduction	245
II. Acetylene Complexes	246
A. Structure and Bonding	246
B. Insertion Reactions	251
C. Catalytic Reactions	261
III. Allene Complexes	265
A. Structure and Bonding	265
B. Cyclooligomerization	270
References	279

I

INTRODUCTION

The field of acetylene complex chemistry continues to develop rapidly and to yield novel discoveries. A number of recent reviews (1–10) covers various facets including preparation, structure, nature of bonding, stoichiometric and catalytic reactions, and specific aspects with particular metals. The first part of this account is confined to those facets associated with the nature of the interactions between acetylenes and transition metals and to the insertion reactions of complexes closely related to catalysis. Although only scattered data are available, attempts will be made to give a consistent interpretation of the reactivities of coordinated acetylene in terms of a qualitative molecular orbital picture.

The second part deals with the complex chemistry and catalytic oligomerizations of allene. We emphasize the importance of the role played by auxiliary ligands of transition metals in determining the paths of catalytic oligomerizations. Recent reviews (11–12) covering most of the literature published up to 1972 lack a perspective view on catalysis. The most recent review (13) describes mainly oligomerization and cooligomerization.

II

ACETYLENE COMPLEXES

A. Structure and Bonding

The side-on coordination of acetylene may be conveniently described by the four orbital interactions shown in Fig. 1. These interactions have already been discussed by Maitlis (7) and Jonassen et al. (14). The extent of overlap and the energy level difference between these interacting orbitals are important in determining the nature of bonding. The overlap decreases in the order: (a) > (b) > (c) > (d). The energy levels are functions of the effective oxidation state of the metal, the nature of auxiliary ligands, and the substituents on the acetylenic carbons. Interaction (a) is usually bonding. Interaction (b) is in most cases bonding, since most transition metal ions or atoms have d-electrons which may occupy $\pi_{||}$-orbitals. In early transition metal complexes such as $Cp_2W=O(RC\equiv CR)$ (15) or $CpNb(CO)(PhC\equiv CPh)_2$ (16) having vacant $d\pi_\perp$-orbitals interaction (c) is bonding, but the contribution to the bond strength is less than that of (a) or (b). In later transition metal complexes, especially d^{10} complexes the interaction should be repulsive and antibonding (c'). Interaction (d) can be neglected because of the poor overlap. We consider that interactions (a), (b), and (c) or (c') are important in determining the overall bond strength. The qualitative molecular orbital scheme in Fig. 1, although

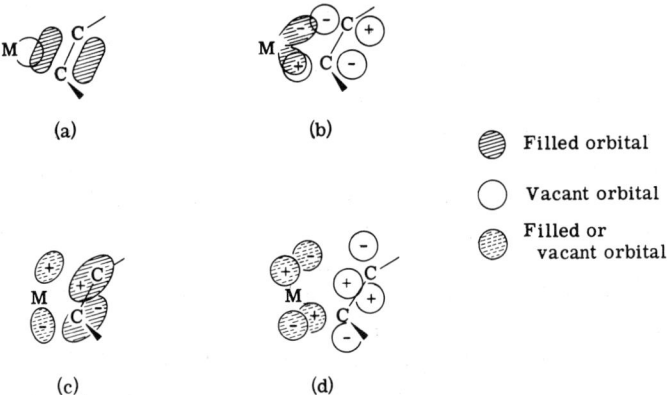

FIG. 1. Molecular orbital interactions for coordinated acetylene. (a) Metal σ-orbital (s, $p\sigma$, or $d\sigma$) and $\pi_{||b}$; (b) metal $\pi_{||}$-orbital ($p\pi_{||}$ or $d\pi_{||}$) and $\pi_{||}$*; (c) metal $\pi_\perp$-orbital ($p\pi_\perp$ or $d\pi_\perp$) and $\pi_{\perp b}$; and (d) metal δ-orbital ($d\delta$) and $\pi_\perp$*.

naive, is useful for our present purpose in discussing the reactivity of coordinated acetylenes.

A number of acetylene complexes have been prepared in recent years and structural and spectroscopic data accumulated. Here the nature of bonding will be discussed on the basis of *inter alia* structural parameters and infrared frequencies. The C≡C bond lengths in various types of acetylene complexes are summarized in Table I. The C≡C bond elongates upon π-coordination. The extent of the lengthening and of the bending of C≡C—C angle reflects a major contribution by π^* back donation [interaction (b) in Fig. 1] and a minor contribution of σ and π donation [interactions (a) and (c)]. The coordinated C≡C length of 1.24 Å in a typical Pt(II) complex with an electron-donating acetylene is definitely shorter than the values in Pt(0) complexes with π-acidic acetylenes, implying importance of back donation. The data in Table I also indicate that the majority lies in a relatively small range, i.e., 1.28 ± 0.02 Å. In an electron-deficient Nb complex (16), however, a long C≡C bond is observed (1.35 Å). This may be accounted for by an effective π-donor interaction [(c)

TABLE I

THE C≡C BOND LENGTHS AND DEFORMATION ANGLES IN ACETYLENE COMPLEXES

Acetylene complex	C≡C (Å)[a]	α C≡C—R	Ref.
Pt(cyclo-C$_7$H$_{10}$)(PPh$_3$)$_2$	1.294(17)	137°, 143°	17
Pt(cyclo-C$_6$H$_8$)(PPh$_3$)$_2$	1.289(17)	127°, 128°	17
Pt(CF$_3$C≡CCF$_3$)(PPh$_3$)$_2$	1.255(9)	140.1°(5)	18
Pt(PhC≡CPh)(PPh$_3$)$_2$	1.32(9)	140°	19
Pt(NCC≡CCN)(PPh$_3$)$_2$	1.40	140°	20
Pt(CH$_3$)Cl(CF$_3$C≡CCF$_3$)(AsMe$_3$)$_2$	1.32(4)	—	21
Pt(CH$_3$)(tripyrazolylborate)(CF$_3$C≡CCF$_3$)	1.292(12)	145.6°	22
PtCl$_2$(p-toluidine)(t-BuC≡C—t-Bu)	1.24(2)	162°-165°	23
Pd(CH$_3$O$_2$CC≡CCO$_2$CH$_3$)(PPh$_3$)$_2$	1.28(1)	145°	24
[Pd(PPh$_2$C≡CCF$_3$)(PPh$_3$)]$_2$	1.28(2)		25
	1.29(2)	138° (1)	
Ni(PhC≡CPh)(t-BuNC)$_2$	1.28(2)	149°	26
Nb(C$_5$H$_5$)(CO)(PhC≡CPh)$_2$	1.35(2)	138°	16
Nb(C$_5$H$_5$)(CO)(PhC≡CPh)(Ph$_4$C$_4$)	1.26(4)	142°	27
WO(C$_5$H$_5$)$_2$(PhC≡CPh)	1.29(3)	143.5°(4)	15
Ir(*trans*-C(CN)=CHCN)(CO)(PPh$_3$)$_2$ (NC-C≡C—CN)	1.29(2)	139°	28
W(PhC≡CPh)$_3$(CO)	1.30	140°	29

[a] Standard deviation in parentheses.

in Fig. 1]. A further structural study of related electron-deficient acetylene complexes is desirable to confirm this inference. The deformation angles are also in a small range, 143° ± 5°. Larger values (162°–165°) in [PtCl$_2$-(toluidine)(t-BuC≡Ct-Bu)] suggest an extensive π-donor interaction [(a) in Fig. 1]. Quantitative interpretation of these structural parameters seems difficult at the present stage (18).

The infrared frequencies (30) associated with metal–acetylene bonding (MC$_2$) can be factored into three fundamentals ($2a_1 + b_1$) if the system is regarded as a vibrationally isolated, triatomic, isosceles (C_{2v} local symmetry):

$$\nu_1(a_1) \qquad \nu_2(a_1) \qquad \nu_3(b_1)$$

Coupling is expected between these vibrations and other vibrations of the rest of the molecule. Fundamentals of the essentially metal–carbon vibrations (ν_2 and ν_3) have never been assigned due to the extensive coupling. Vibration $\nu_1(a_1)$ is remarkably high (1600–2050 cm^{-1}) and relatively isolated (Tables II to IV). Therefore, ν_1 values can be used in discussing metal–acetylene bonding. Thus, Pt(II)Cl$_2$(RC≡CR)L (38) or [Pt(II)(CH$_3$)(RC≡CR)L$_2$]PF$_6$ (38) exhibits a band at about 2000 cm^{-1} (ca. 250 cm^{-1} lower than for the free acetylene; see Table IV), whereas Pt(0)(RC≡CR)L$_2$ shows a strong band in the range, 1680–1850 cm^{-1} (see Tables II and III). The MC$_2$ frequency (a_1) in isostructural iridium complexes decreases with an increase in the electron-donating ability of the auxiliary ligands, as clearly shown in [Ir(CO)$_2$L$_2$(RC≡CR)]$^+$ (see Table III); however, the trend in Pt(CF$_3$C≡CCF$_3$)L$_2$ is obscure (Table III). It appears that the extent of π back donation is so great that the MC$_2$ frequencies become insensitive to a change in the phosphine ligands.

The data in Table II suggest that the apparent thermal stability of the isostructural acetylene complexes of zero-valent metals (Pt > Ni > Pd) is mainly governed by the backbonding effect as reflected in the decrease ($\Delta\nu$) in ν(MC$_2$) (ν_1 vibration) frequencies ($\Delta\nu$: Pt > Ni > Pd). This correlation does not occur among complexes of different structures (see Tables II to IV). Because of the varying extent of mixing, a meaningful comparison of metal–acetylene bond strengths requires, first of all, assignment of the a_1 vibrations, particularly ν_2, through isotopic (e.g., ^{13}C) infrared data.

The IR N≡C frequencies of a series of complexes, M(Un)(t-BuNC)$_2$ (M = Ni, Pd), are useful for discussing the nature of metal–acetylene

TABLE II

COMPARISON OF THE $MC_2(a_1)$ FREQUENCIES IN ISOSTRUCTURAL ACETYLENE COMPLEXES

$M(R{-}C{\equiv}C{-}R)(PPh_3)_2$[a]	Ph (cm^{-1})	CO$_2$Me (cm^{-1})	CN (cm^{-1})	CF$_3$ (cm^{-1})
M = Pt	1740, 1768	1782	1682[b]	1775
Pd	[c]	1830, 1845	[d]	1811, 1838
Ni	1800	[d]	[d]	1790
Free acetylene ligand	2223	2256	2218	2300

$M(R{-}C{\equiv}C{-}R)(t\text{-}BuNC)_2$[e]	Ph (cm^{-1})	CO$_2$Me (cm^{-1})		
M = Ni	1810	1830		
Pd	1825	1804		

$M(\eta\text{-}C_5H_5)_2(R{-}C{\equiv}C{-}R)$[f]	H (cm^{-1})	CH$_3$ (cm^{-1})	Ph (cm^{-1})	CF$_3$ (cm^{-1})	CO$_2$CH$_3$ (cm^{-1})
M = Mo	1613	1830	1774	1782	—
V	—	—	—	1800	1821

$Pt(RC{\equiv}CR')(PPh_3)_2$[g]:

RC≡CR'	PtC$_2(a_1)$ (cm^{-1})	RC≡CR	PtC$_2(a_1)$ (cm^{-1})
MeC≡CH	1712	EtC≡CEt	1805
EtC≡CH	1705	p-tolyl—C≡C—p-tolyl	1755
PhC≡CD	1642		

[a] Data from Davidson (*30*) and Wilke and Herrmann (*31*).
[b] Data from McClure and Baddley (*32*) and Baddley (*33*).
[c] The acetylene complex is too unstable to be readily isolated.
[d] Thermal oligomerisation of the acetylene prevents characterization.
[e] Data from Otsuka et al. (*34*).
[f] Data from Nakamura and Otsuka (*35*), from Brintzinger and Thomas (*36*), and from Tsumura and Hagihara (*37*).
[g] Data from Mann et al. (*38*).

bonding. The correlation between the N≡C stretching values and the electron density on the metal provides a measure of metal–acetylene donation and/or back donation. In general, high N≡C stretching values are associated with a high effective oxidation state of the metal. Table V appears to suggest that dimethyl acetylenedicarboxylate and dimethyl

TABLE III

EFFECT OF THE PHOSPHINE SUBSTITUENT ON THE LIGAND INFRARED FREQUENCIES IN ISOSTRUCTURAL ACETYLENE COMPLEXES

(a) $[Ir(CO)_2L_2(MeO_2C-C{\equiv}C-CO_2Me)]^+$ [a]:

L	ν_{CO} (cm^{-1})	ν_{IrC_2} (cm^{-1})	$\nu_{C=O}$ of CO_2Me (cm^{-1})
PPh$_3$	2091, 2049	1814	1708
PMePh$_2$	2086, 2038	1808	1704
PEt$_3$	2088, 2044	1791	1697
PCy$_3$	2068, 2018	1794	1703

(b) $PtL_2(CF_3C{\equiv}CCF_3)$ [b]:

L	ν_{PtC_2} (cm^{-1})	L	ν_{PtC_2} (cm^{-1})
PPh$_3$	1765, 1775	PMe$_2$Ph	1767
PEt$_3$	1771	P(n-Bu)$_3$	1758

[a] Data from Church et al. (39).
[b] Data from Davidson (30) and Cherwirski et al. (40).

maleate in nickel complexes are comparable in $d\pi_{\parallel}$-accepting property. This is rather peculiar since an acetylenic carbon is more electronegative than the corresponding olefinic carbon. In analogous Pd complexes (Table V), we observe the expected trend for the N≡C values. The relatively lower $\nu_{N{\equiv}C}$ values in the nickel complexes are explicable in terms of an enhanced interaction (c') that donates electrons to the isocyanide bonds.

TABLE IV

SELECTED VIBRATIONAL FREQUENCIES OF $trans$-$[PtCH_3(RC{\equiv}CR')L_2]^+PF_6^-$ [a]

RC≡CR'	L	ν_{Pt-CH_3}	$\nu_{PtC_2(a_1)}$
CH$_3$C≡CCH$_3$	PMe$_2$Ph	547	2114
CH$_3$C≡CC$_2$H$_5$	PMe$_2$Ph	551	2116
C$_2$H$_5$C≡CC$_2$H$_5$	PMe$_2$Ph	547	2101
PhC≡CPh	PMe$_2$Ph	556	2087
PhC≡CCH(OH)Ph	PMe$_2$Ph	547	2051
PhC≡CPh	AsMe$_3$	549	2024

[a] Data from Chisholm and Clark (41).

TABLE V

Comparison of N≡C Stretching Frequencies in
$M(Un)(t\text{-BuNC})_2 (M = Ni, Pd)^a$

M	Un	$\nu_{N\equiv C}$ (cm^{-1})	$\Delta\nu^b$	
Ni	(cis-CH$_3$O$_2$C—CH=CH—CO$_2$CH$_3$)	2154, 2120	—	—
	(CH$_3$O$_2$C—C≡C—CO$_2$CH$_3$)	2160, 2123	+6	+3
	(PhC≡CPh)	2138, 2100	−16	−20
Pd	(cis-CH$_3$O$_2$C—CH=CH—CO$_2$CH$_3$)	2160, 2140	—	—
	(CH$_3$O$_2$C—C≡C—CO$_2$CH$_3$)	2177, 2158	+17	+18
	(PhC≡CPh)	2150, 2125	−10	−15

[a] Data from Otsuka et al. (34).
[b] Difference in the frequencies from the values of the corresponding dimethyl maleate complexes.

B. Insertion Reactions

Insertion of acetylenes into transition metal–hydrogen, –σ-carbon, –η^2-acetylene, or –halogen bonds is an important elementary step in catalytic hydrogenation, oligomerization (linear or cyclic), or polymerization. In recent years, considerable information has become available in this field, particularly with regard to stereochemistry and mechanism. Kinetic data, however, still remain scarce. The bonding between metals and acetylenes is considerably complicated by the existence of the two mutually orthogonal π-orbitals on the ligand, and insertion reactions exhibit complex features reflected in the varied regio- and/or stereoselectivity. A short account of the stereochemistry and mechanism of the insertion into an H—M bond will be given to illustrate the situation.

A delicate dependence of the cis/trans stereochemistry on the identity of the metal, oxidation state, auxiliary ligands, and the nature of the substituents on the acetylene has already been observed (42, 43–65). The complexity of the problem, however, has prevented general interpretation of the mechanism (42). The stereochemistry determined after the decomposition by hydrogen (or sometimes by water) of a labile σ-alkenyl complex is sometimes ambiguous (43) because of possible geometrical isomerization during decomposition. Fluoroacetylenes, CF$_3$C≡CH or CF$_3$C≡CCF$_3$, may be used with advantage because of their ability to form stable σ-alkenyl complexes which can be easily examined by ^{1}H and ^{19}F NMR spectroscopy. The observed coupling constants [e.g., $J(\text{CF}_3\text{—CF}_3)$] unambiguously reveal the stereochemistry about the double bond in the σ-alkenyl complexes. The stereochemistry of insertion determined with the fluoroacetylenes may be depicted as follows.

$CF_3C\equiv CCF_3$

    Trans insertion to HML_n: ML_n: $Mn(CO)_5$ (44), $Re(CO)_5$ (45), $Mo(H)Cp_2$ (46), $W(H)Cp_2$ (47), $ReCp_2$ (47)

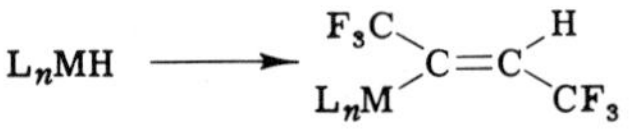

$$L_nMH \longrightarrow \underset{L_nM}{\overset{F_3C}{>}}C=C\underset{CF_3}{\overset{H}{<}}$$

    Cis insertion to HML_n: ML_n: $PtCl(PEt_3)_2$ (73), $IrCl_2(PMe_2Ph)_3$ (49), $RuCp(PPh_3)_2$ (50), $Rh(CO)(PPh_3)_3$ (51)

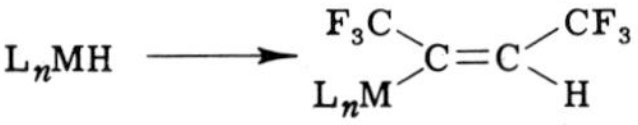

$$L_nMH \longrightarrow \underset{L_nM}{\overset{F_3C}{>}}C=C\underset{H}{\overset{CF_3}{<}}$$

$CF_3C\equiv CH$

    Trans insertion to HML_n: ML_n: $Re(CO)_5$ (52, 53), $FeCp(CO)_2$ (53)

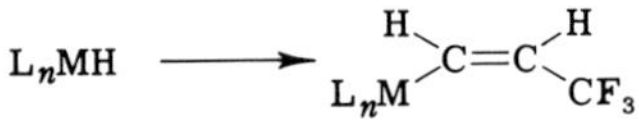

$$L_nMH \longrightarrow \underset{L_nM}{\overset{H}{>}}C=C\underset{CF_3}{\overset{H}{<}}$$

    Cis insertion to HML_n: ML_n: $Mo(H)Cp_2$ (47)

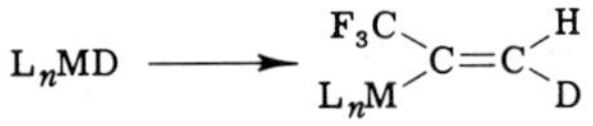

$$L_nMD \longrightarrow \underset{L_nM}{\overset{F_3C}{>}}C=C\underset{D}{\overset{H}{<}}$$

In addition to these, α-trifluoromethyl products were obtained regioselectively from $CF_3C\equiv CH$ and $PtHCl(PEt_3)_2$ (53) or $MnH(CO)_5$ (53) without stereochemical information.

For the insertion of diphenylacetylene or dimethyl acetylenedicarboxylate the steric course is sometimes uncertain because of the meager structural information available for the alkenyl complex. Hopefully ^{13}C NMR data will provide the necessary information.

$PhC\equiv CPh$

    Cis insertion to HML_n: ML_n = $Mo(H)Cp_2$ (46), $IrCl_2(dmso)_2$ (54), $PtCl(PEt_3)_2$ (55), $Co(dmg)_2$ (56)

    Trans insertion to HML_n: ML_n = $Rh(CO)(PPh_3)_3$ (52), $CoCp(PPh_3)$ (57)

$MeO_2CC\equiv CCO_2Me$

    Cis insertion to HML_n: ML_n = $Mo(H)Cp_2$ (46), $W(H)Cp_2$ (46), $Mn(CO)_4PPh_3$ (58), $ReCp_2$ (59), $RuCp(PPh_3)_2$ (50), $Rh(CO)(PPh_3)_3$ (43), $Pd(C\equiv CPh)(PEt_3)_2$ (60), $PtCl(PEt_3)_2$ (60)

Trans insertion to HML_n: ML_n = $Mn(CO)_5$ (58, 61), $Rh(CO)(PPh_3)_3$ (51)

Various mechanisms for the insertion reaction are conceivable: (a) ionic stepwise, (b) radical (chain or nonchain), and (c) concerted. Generally, ionic or radical mechanisms give a mixture of products with cis and trans stereochemistry. In some special cases of the ionic reaction, however, exclusive formation of a trans product has been observed (62, 66). Therefore, stereoselectivity does not necessarily imply a concerted mechanism; other evidence, e.g., regioselectivity, kinetic data, solvent effects, and substituent effects, must be sought out.

The addition of fluoroacetylenes to hydrogen compounds of some typical elements, e.g., R_2EH (E = N, P, As) (66) or RXH (X = O, S) (67), proceeds via an ionic stepwise or radical mechanism; for example,

$$CF_3-C\equiv CH + R_2EH \longrightarrow \underset{R_2E}{\overset{H}{>}}C=C\underset{H}{\overset{CF_3}{<}} + \underset{R_2E}{\overset{H}{>}}C=C\underset{CF_3}{\overset{H}{<}} + \underset{R_2E}{\overset{F_3C}{>}}C=C\underset{H}{\overset{H}{<}}$$

E = N, P, As

The distribution of the products depends on the electronic properties of R_2E and the reaction conditions (66). By contrast, the reaction of $CF_3C\equiv CH$ with Cp_2MoH_2 proceeds with high regio- and stereoselectivity to give cis insertion over a range of reaction temperatures (47):

$$CF_3C\equiv CH + Cp_2MoD_2 \longrightarrow \underset{Cp_2Mo}{\overset{F_3C}{>}}C=C\underset{D}{\overset{H}{<}}$$

Similar cis insertion was observed for diphenylacetylene or dimethyl acetylene dicarboxylate (47).

A mechanism analogous to the related olefin cis insertion (68) has been proposed for the reaction with Cp_2MH_2 (M = Mo, W) (Scheme 1), on the grounds of the same stereochemistry and comparable reaction temperature (47). In view of the failure of $CF_3C\equiv CH$ to react with the moderately σ-basic Cp_2WH_2 (pK_b = 8.6) at ambient temperature, the facile reaction of the weakly σ-basic Cp_2MoH_2 (pK_b = 12.5) with $CF_3C\equiv CH$ cannot be ascribed to the σ-basicity at the metal. Intervention of a thermally excited, parallel metallocene molecule with high π-basicity is thus assumed. The excited molecule will then receive a π-acidic acetylenic bond forming an acetylene π-complex that smoothly gives the observed cis-insertion product. The stereoselective cis insertion is best accounted for by a four-centered

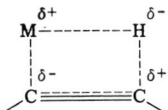

transition state with some polarity:

The polarity and contribution of d-orbital character seems to assist this formally forbidden $[\sigma_{2s} + \pi_{2s}]$ reaction, and the process may not be rigorously synchronous (69).

Recently, trans insertion of hexafluorobutyne into one of the M—H bonds in some metallocene hydrides, Cp_2MH_n, was studied in some detail (47). Experiments carried out in the presence of various radical-sensitive reagents such as N-phenyl-α-naphthylamine suggested that a free radical mechanism was unlikely. A stepwise ionic mechanism, involving a zwitterionic intermediate, $Cp_2(H_2)M^+$—$C(CF_3)=\bar{C}CF_3$, is improbable, since (i) the stereochemistry and the apparent rate are not influenced by the polarity of the solvents, (ii) no deuterium is incorporated in the reaction in EtOD, and (iii) the trend in reactivity (Mo > W) does not reflect the trend in σ-basicity or M—C bond stability (W > Mo). An essentially concerted trans-insertion mechanism is inferred, which is supported *inter alia* by the low kinetic deuterium isotope effect ($k_H/k_D = 1$).

The concerted trans insertion formally belongs to a thermally allowed [$\sigma_{2s} + \pi_{2a}$] reaction utilizing the acetylene $\pi_\perp$ orbital. A nonpolar four-centered heteroatomic transition state with a skewed disposition of participating σ- and π-bonds may be postulated:

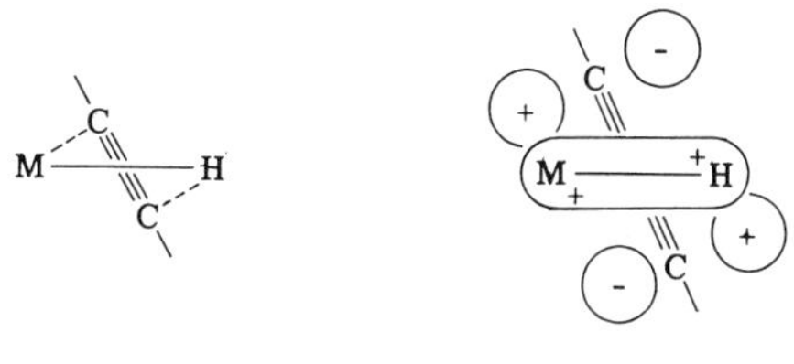

The geometry in the transition state readily explains the preferential formation of the conformational isomer a (*46*) (Scheme 2).

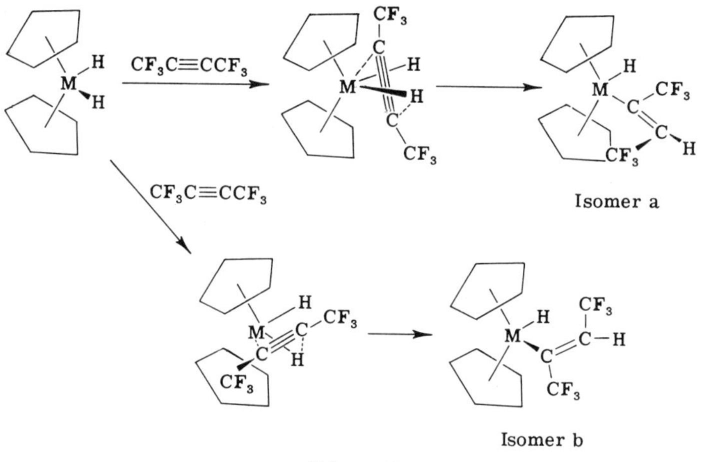

Scheme 2

It is interesting to observe different mechanisms for the apparently similar insertion reactions between Cp$_2$MoH$_2$ and fluoroacetylenes. Thus, Cp$_2$MoH$_2$ acts as a π-base for CF$_3$C≡CH with a polar triple bond, but in the essentially concerted trans insertion it may behave as a σ_{2s} component against a nonpolar bulky fluoroacetylene, CF$_3$C≡CCF$_3$. The observed discriminating behavior toward these fluoroacetylenes contrasts sharply with the nondiscrimination of typical N-bases (R$_2$NH or R$_2$PH) (*66*) in reactions of the fluoroacetylenes. Specificity of transition metal complexes in reactions with apparently similar organic substrates is thus of interest and deserves further study.

1. Metalocyclization

Metal η^2-acetylene complexes react with further molecules of acetylenes in two different ways, namely ligand exchange or substitution (Scheme 3).

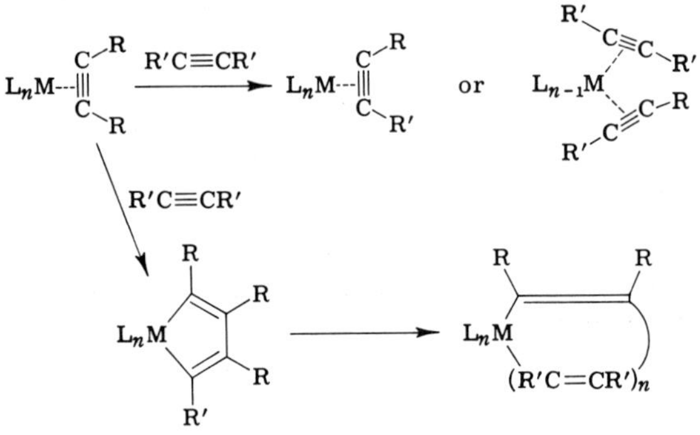

Scheme 3

Insertion initially gives metalocyclopentadienes which may further give rise to larger metalocyclic complexes. Factors determining the reaction paths are not clear at present. In general many metal acetylene complexes of d^8–d^{10} metals (70–89), e.g., Fe (3, 70), Co (10, 71), Ni (72), Ru (73, 74), Rh (75–78), Pd (83–84), and Ir (86), react with excess acetylene to give metalocyclopentadiene complexes or acetylene oligomers (87). The following are some examples:

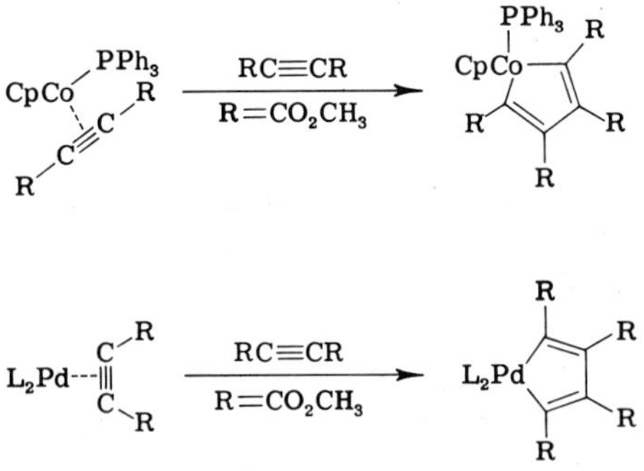

In these complexes, repulsive interaction (c′) is operating in addition to interactions (a) and (b) (see Fig. 1). Thermal excitation then either causes these complexes to lose the acetylene or gives rise to a thermally excited molecule in which a change in coordination state (to a monohapto state) may occur (73, 89). A similar thermal activation has been proposed (90) for some insertion reactions of dioxygen complexes, L_2MO_2 (M = Ni, Pd, Pt), where the orbital interaction scheme is similar except for the occupancy of the dioxygen $\pi_\perp^*$ orbital:

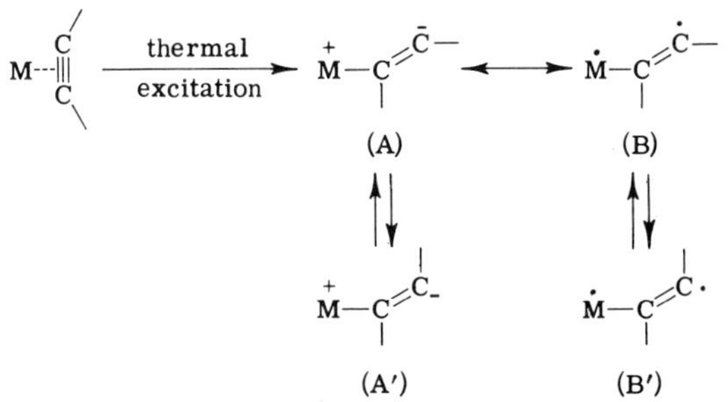

The radical or ionic character depends on the identity of the metal, the effective metal oxidation state, and the auxiliary ligands. These combined effects determine the reactivity, stereo- and regioselectivity toward acetylene insertion. For example, the reaction of $CpCo(PPh_3)(RC\equiv CR')$ with asymmetric acetylenes, $RC\equiv CR'$, gives a mixture of isomeric products:

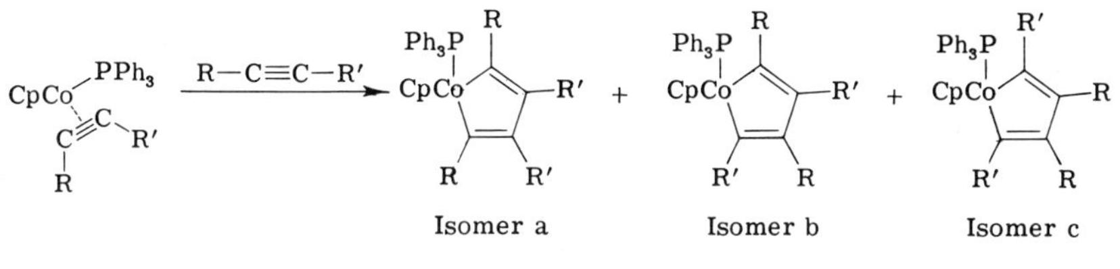

(i) R = CO_2CH_3, R' = CH_3 : a. 9%; b. 50%
(ii) R = Ph, R' = CO_2CH_3 : a. 13%; b. 20%

A predominance of isomer b and the absence of isomer c indicates the direction of polarization in the metal η^1-acetylene moiety (91). A zwitterionic intermediate, $Cp(PPh_3)Co^+$—$C(CH_3)=\bar{C}$—CO_2CH_3, is implied for case (i); much less polarization with some radical character would account for the isomer distribution in case (ii). Recently, a similar polar monohapto-acetylene intermediate was invoked to explain the novel addition of

$CF_3C\equiv CCF_3$ to a C—H bond of an Ru–alkenyl complex (92)[1]:

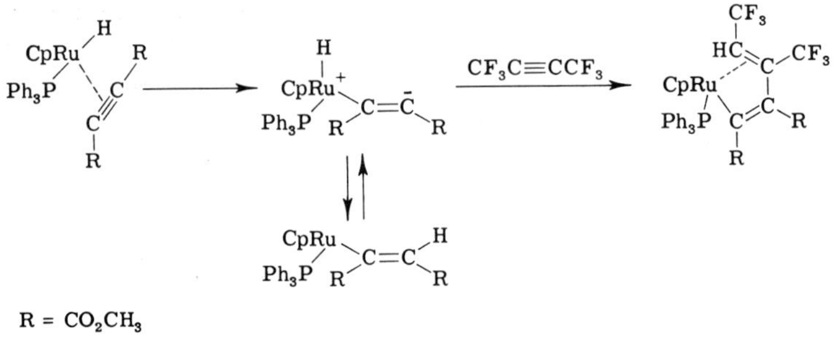

R = CO_2CH_3

For reactions of cationic Pt–acetylene complexes (93), another polar monohaptoacetylene complex (C) may be postulated:

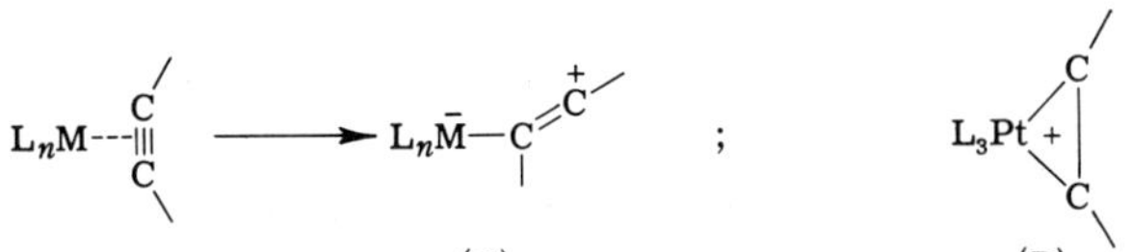

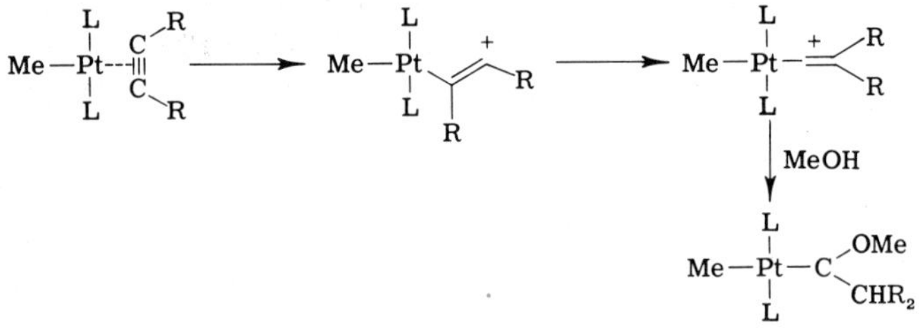

Thus, most of the electrophilic reactions of $[PtCH_3(L)_2(RC\equiv CR)]^+$ can be explained with the intermediate mechanistically indistinguishable from the platinized carbonium ion model (D) proposed by Chisholm and Clark. Since the positive β-carbon has a vacant $p\pi$-orbital, the rearrangement to alkoxycarbene complexes can be regarded as a carbonium ion rearrangement.

In contrast to the later transition metal complexes, electron-deficient complexes of the earlier transition metals, e.g., $CpV(CO)_2(RC\equiv CR)$ (94), are mostly inert to acetylene cyclization. Thus, bis- or trisacetylene com-

---

[1] The molecular structure of the product $[\overline{Ru \cdot C(CO_2Me):C(CO_2Me)C(CF_3):C(CF_3)}H$-$(PPh_3)(\eta^5-C_5H_5)]$ has been fully confirmed by an X-ray diffraction study (L. E. Smart, J. Chem. Soc., Dalton Trans., in press).

plexes, e.g., $M(RC{\equiv}CR)_3(CO)$ (95–97), are prepared by thermal and photochemical reactions. The absence of stable bis or tris complexes of d^8-d^{10} metals may be attributed to the facile metalocyclization. Thus, interaction (c) (see Fig. 1) or (c′) appears to impart this distinction. In molybdenum acetylene complexes, $Cp_2Mo(RC{\equiv}CR)$ (46, 47), intermediate between the two cases, the relevant $d\pi_\perp$ orbital is partially occupied by interaction with the E_{1g} (Cp) orbital,

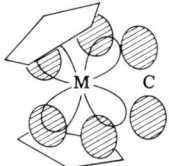

resulting in a weak attractive interaction (c). Indeed, these acetylene complexes are inert to acetylene oligomerization.

A further reaction of metalocyclopentadiene complexes with acetylenes leads to metalocycloheptatriene complexes by metalocyclic enlargement (3, 10, 98) or to benzene derivatives by reductive elimination (57, 70, 73, 77, 82, 98):

For example, reaction of excess $CF_3C{\equiv}CCF_3$ with $Pt(PEt_3)_4$ gave $(Et_3P)_2$-$Pt[\eta^2$-$C_6(CF_3)_6]$, which probably formed from a η^2-acetylene complex through a platinacyclopentadiene complex (98).[2]

[2] Recently, Stone et al. [*J. Chem. Soc., Chem. Commun.*, 723 (1975)] have established a reverse pathway for reactions of this kind. Certain complexes of Pt(0) react with $C_6(CF_3)_6$ with cleavage of a C—C bond of the benzene derivative to give a platinacyclohepta-*cis, cis, cis*-triene.

A related reaction with Ni[P(OMe)$_3$]$_4$ gave unexpectedly a *cis,trans,cis*-nickelacycloheptatriene complex that must be constructed with a trans-insertion process at some stage of its formation *(98)*:

Ni[P(OMe)$_3$]$_4$ + CF$_3$C≡CCF$_3$ ⟶ [(MeO)$_3$P]$_2$Ni(C$_6$R$_6$)

(R = CF$_3$)

2. *Metalococyclization of Acetylenes with CO or RNC*

Metalocycles are also formed by the reaction of acetylenes with metal carbonyls or with isonitrile complexes *(3, 73, 99–102)*. Their formation may involve monohaptoacetylene intermediates.

Some of these metalocycles have been confirmed by spectroscopic data and by an X-ray analysis *(103)* and are important intermediates in catalytic cocyclization with carbon monoxide or with isonitriles (see following).

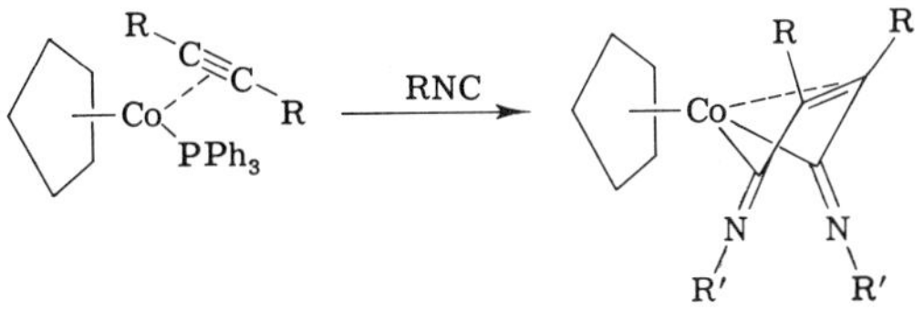

C. Catalytic Reactions

1. Activation of Acetylene by Complexation

Studies on elementary reactions of acetylenes with metal complexes are now beginning to shed some light on the nature of "activation" caused by complexation. This activation is not a simple process. Many low-valent d^8-d^{10} metal complexes and also some early transition metal compounds with higher oxidation state (d^1-d^2 complexes) are capable of activating acetylenes. As already described, in the former complexes, interaction (c') would lead to activation of an η^2-acetylene ligand to an η^1-acetylene having some radical as well as some anionic character:

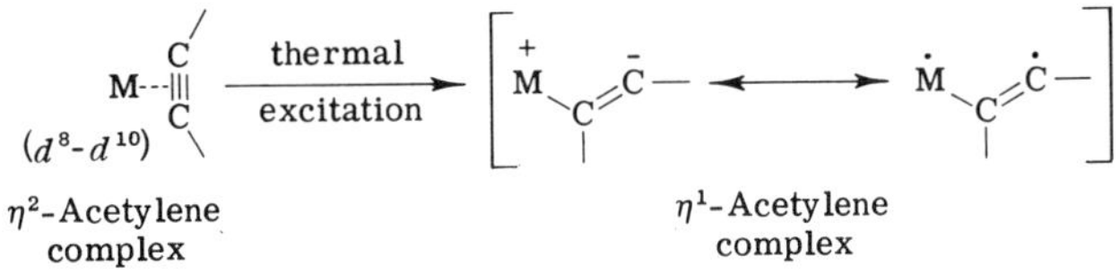

In the latter complexes, strong σ-donor interaction (a) and weak π-back donation (b) (see Fig. 1) would lead to the formation of apparently similar η^1-acetylene complexes by thermal activation. Here the species, however, have some cationic character as manifested by their preferential reactions with electron-donating acetylenes (63, 104):

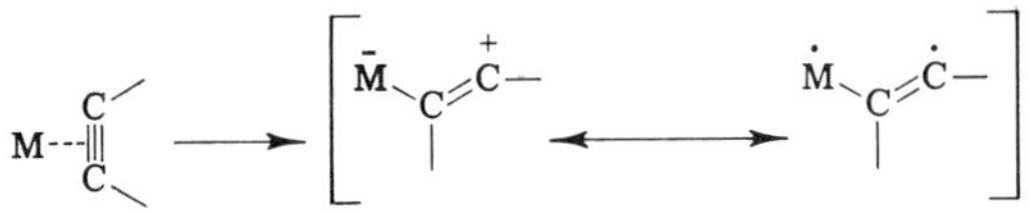

In sharp contrast to these activations, an η^2-acetylene complex is stabilized when all the interactions [(a), (b), and (c)] are bonding, as in some electron-deficient d^6 complexes, e.g., $W(RC\equiv CR)_3(CO)$ (95–97).

2. Catalytic Cyclooligomerization

Acetylenes are catalytically cyclized to benzenes and cyclooctatetraenes (70, 105–107). Small amounts of styrenes, vinylcyclooctatetraenes, naphthalenes, and azulenes are also formed in some instances (108–110). Some elementary steps in these reactions have already been discussed. A plausible reaction path for the cyclization is in Scheme 4 (111).

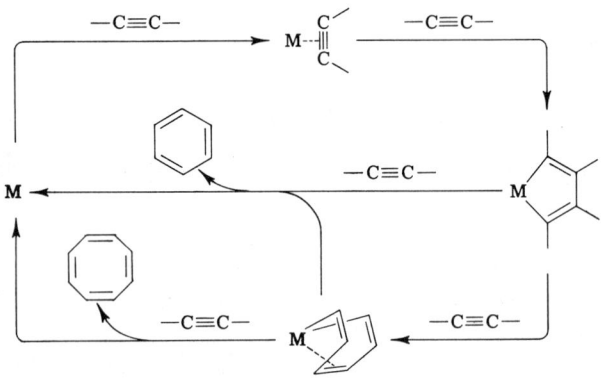

Scheme 4

3. Linear Oligomerization

Acetylenes are also oligomerized to mono- or divinylacetylenes, or dienylacetylenes by Ni(0) (112), Rh(I) (113), or Pd(II) (114) complexes (Scheme 5). Meriwether et al. (112) proposed hydrido-σ-alkynylnickel complexes as active intermediates in the catalytic linear oligomerization. Subsequent insertion of acetylene into an M-σ-alkynyl bond has been assumed.

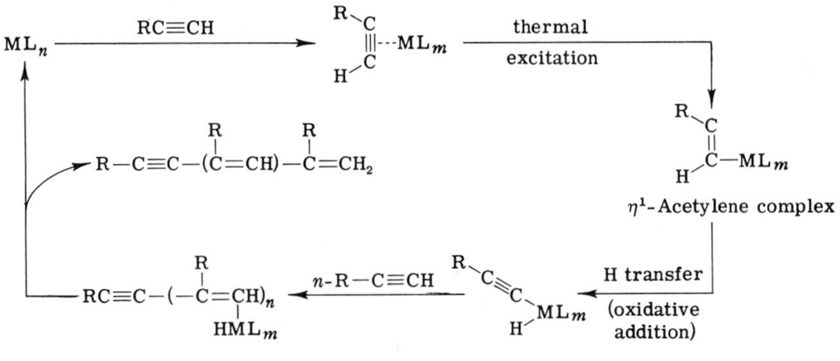

Scheme 5

Conversion of an η^2-acetylene complex to the hydridoalkynyl complex will lead to linear oligomerization or polymerization. The tendency of some Rh or Pd complexes to form hydridooalkynyl complexes explains their catalytic activity toward linear oligomerization. Recently, Hagihara et al. (115) examined the reaction of preformed hydrido-σ-alkynyl complexes, $MH(\sigma\text{-}C\equiv CPh)L_2$, with $MeO_2CC\equiv CCO_2Me$ and found cis insertion into the M—H bond.

$$trans\text{-}MH(C\equiv CPh)L_2 + RC\equiv CR \longrightarrow trans\text{-}M(C\equiv CPh)(-\overset{R}{\underset{|}{C}}=\overset{R}{\underset{|}{C}}H)L_2$$

$$M = Pd, Pt; L = PEt_3; R = CO_2Me$$

They also found a novel stereoselective linear trimerization of $PhC\equiv CH$ with a $Pd(Ph)_2(PBu_3)_2$ catalyst.

4. Catalytic Cocyclization with Isocyanides

Cocyclization of acetylene with isocyanides gives interesting new cyclic compounds (103, 116). The reaction patterns are generally similar to the cocyclization with carbon monoxide which is already known (103, 117). Low-valent nickel, palladium, or cobalt complexes are active in the following reactions (102, 103) for which intervention of acetylene complexes has been suggested:

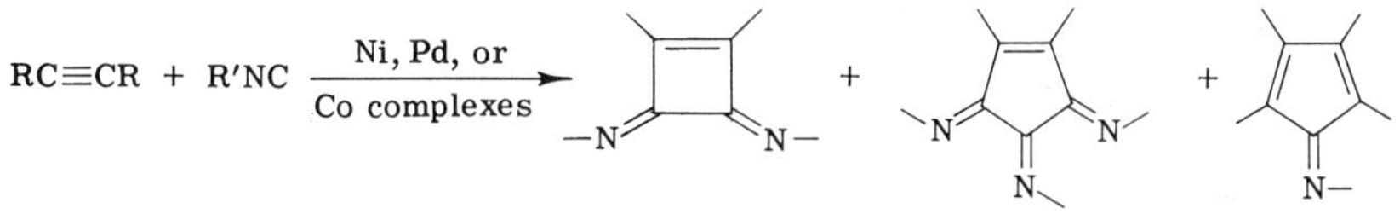

$$RC\equiv CR + R'NC \xrightarrow{\text{Ni, Pd, or Co complexes}} \text{(products)}$$

Recently, Yamazaki et al. (103) carried out stoichiometric reactions of cobalt–acetylene complexes with isocyanides and isolated the expected intermediate metalocyclic complexes (Scheme 6).

Another interesting reaction is the formation of aminopyrrole derivatives from t-BuNC and various acetylenes (118). The catalysts include various Ni(II) and Ni(0) phosphine complexes:

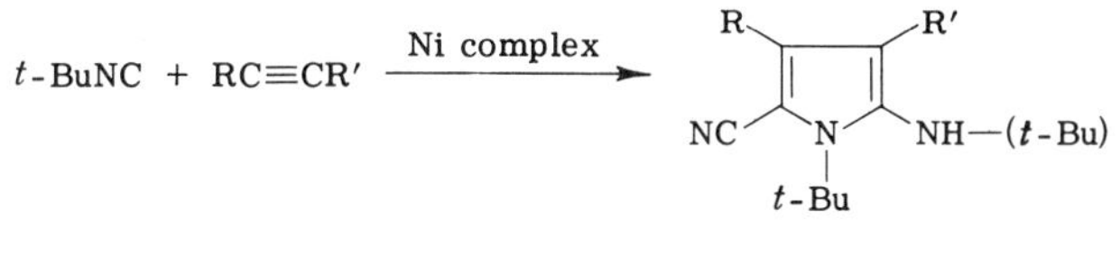

$$t\text{-BuNC} + RC\equiv CR' \xrightarrow{\text{Ni complex}} \text{(aminopyrrole product)}$$

Based on the formation of a $CpCo(CNR')_2(RC\equiv CR)$ complex from acetylene and isocyanides (103), the paths shown in Scheme 7 are proposed.

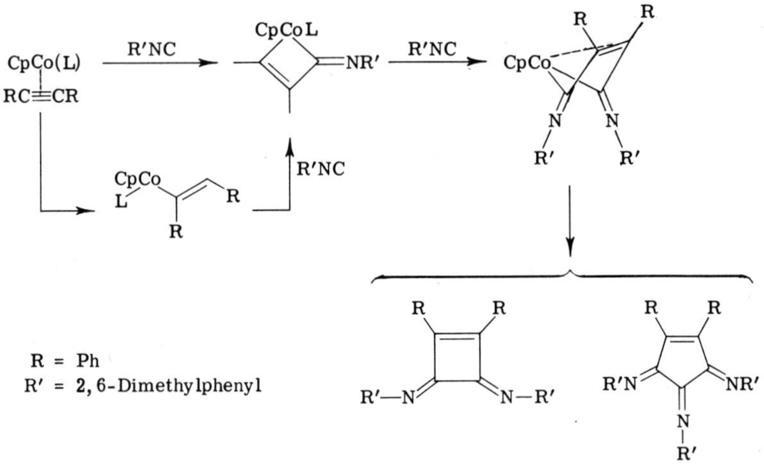

R = Ph
R' = 2,6-Dimethylphenyl

Scheme 6

5. Catalytic Cocyclization with Heterounsaturation

Yamazaki et al. (91, 119) and Bönnemann et al. (120) have recently reported catalytic syntheses of substituted pyridines from acetylenes and nitriles. Various cobalt complexes serve as active catalysts, in particular, $CpCo(PPh_3)_2$ (91) or $Co(C_8H_{12})(C_8H_{13})$ (120). Similar reactions of acetylenes with CS_2 or RNCS also give new heterocycles (91):

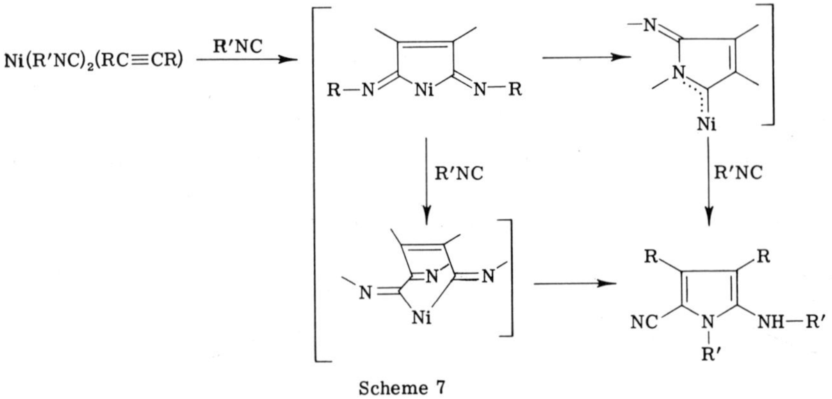

Scheme 7

RC≡CR + R'CN ⟶ [pyridine product]

RC≡CR + CS$_2$ ⟶ [thiopyran product]

RC≡CR + R'NCS ⟶ [thiazine product]

The following intermediate complex has also been isolated (*121*):

CpRh(RC≡CR)(PPh$_3$) $\xrightarrow{CS_2}$ CpRh[metallacycle with S–C(=S)–R and R groups]   R = CO$_2$CH$_3$

This structure gives support for the proposal of an ionic monohaptoacetylene complex, Cp(PPh$_3$)$\overset{+}{Co}$—C(R)=$\bar{C}$R, as an activated precursor for the reaction with CS$_2$.

III
ALLENE COMPLEXES

A. *Structure and Bonding*

Since the compilation of complexes in 1972 (*11–13*), a few have been reported: M(PPh$_3$)$_2$(allene) (*122, 123*) (M = Ni, Pd, Pt; allene is CH$_2$=C=CH$_2$, CH$_2$=C=CMe$_2$, PhCH=C=CHPh); Cp$_2$Fe(CO)-[CH$_2$=C=C(CH$_3$)SnCl] (*124, 125*); and PtCH$_3$(HBpz$_3$)(R$_2$C=C=CR'$_2$) (HBpz$_3$ = tripyrazolylborato; R = R' = CH$_3$ or R = H, R' = CH$_3$) (*126*). Single-crystal X-ray diffraction data are shown in Table VI. As in olefin complexes, there are two types of coordination—one containing an allene double bond perpendicular to the equatorial molecular plane and the other containing the ligand in the plane. In the former type, we find labile complexes, e.g., RhI(PPh$_3$)$_2$(CH$_2$=C=CH$_2$) (*136*) and fluxional

TABLE VI
Structural Parameters of Transition Metal–Allene Complexes[a]

Compound	Structure	C(1)–C(2) (Å)	C(2)–C(3) (Å)	C(1)–C(2)–C(3) (°)	M–C(1)	M–C(2)	Ref.
Rh(acac)(CH$_2$=C=CMe$_2$)$_2$	Perpendicular[b]	1.40(1)	1.30(1)	153.3(6)	2.13(1)	2.07(1)	127
		1.41(1)	1.29(1)	152.6(6)	2.13(1)	2.06(1)	
Rh(acac)(Me$_2$C=C=CMe$_2$)$_2$	Perpendicular[b]	1.373(8)	1.325(8)	147.2(6)	2.177(6)	2.027(5)	128
		1.377(8)	1.321(9)	148.9(6)	2.176(6)	2.033(5)	
Rh$_2$(acac)$_2$(CO)$_2$(CH$_2$=C=CH$_2$)[c]	Perpendicular[b]	1.37(1)		144.5(6)	2.12(1)	2.05(1)	127
		1.41(1)			2.14(1)	2.06(1)	
RhI(PPh$_3$)$_2$(CH$_2$=C=CH$_2$)[d]	Perpendicular[b]	1.35(6)	1.34(7)	158(4)	2.17(4)	2.04(4)	129
[PtCl$_2$(Me$_2$C=C=CMe$_2$)]$_2$	Perpendicular[b]	1.37(3)	1.36(3)	151(2)	2.25(2)	2.07(2)	128
Pt(PPh$_3$)$_2$(CH$_2$=C=CH$_2$)	In-plane[b]	1.48(5)	1.31(5)	142(3)	2.13(3)	2.03(3)	130
Pt(PPh$_3$)$_2$(CH$_2$=C=CHMe)	In-plane[b]	1.44(4)	1.32(4)	146(3)	2.12(3)	2.05(3)	131
Pt(PPh$_3$)$_2$(CH$_2$=C=CMe$_2$)	In-plane[b]	1.430(11)	1.316(11)	140.8(8)	2.107(8)	2.049(7)	132
Pd(PPh$_3$)$_2$(CH$_2$=C=CH$_2$)	In-plane[b]	1.44(2)	1.32(2)	148(1)	2.12(1)	2.07(1)	133
		S(1)–C(2) (Å)	C(2)–S(3) (Å)	S(1)–C(2)–S(3) (°)	M–S(1)	M–C(2)	
Pt(PPh$_3$)$_2$(CS$_2$)	—	1.72(5)	1.54(5)	136(1.5)	2.33(1)	2.06(4)	134
Pd(PPh$_3$)$_2$(CS$_2$)	—	1.65(3)	1.63(3)	140(2)	2.305(11)	2.00(3)	135

[a] Carbon disulfide complexes are included for comparison. The C=C bond length of free molecules ranges from 1.309 to 1.312 Å.
[b] *Perpendicular* refers to the molecular structure containing a coordinated double bond perpendicular to the molecular plane, and *in-plane* to that containing the double bond lying in the molecular plane.
[c] Each of the two double bonds acts as a monodentate olefin and thus the bent allene bridges the two metal atoms.
[d] The bromo analog, RhBr(PPh$_3$)$_2$(CH$_2$=C=CH$_2$), has quite similar structural parameters (Kasai *et al.*, unpublished).

ones, e.g., $Pt_2Cl_4(Me_2C=C=CMe_2)_2$ (*137, 138*). In the latter are found both labile, e.g., $Pd(PPh_3)_2(CH_2=C=CH_2)$ (*139*) and inert compounds, e.g., $Pt(PPh_3)_2(CH_2=C=CH_2)$ (*122*) (see following).

In general the degree of elongation of the double bond upon coordination parallels the degree of bending of the allene molecule. The Dewar-Chatt-Duncanson molecular orbital model of the metal–olefin bond would account for these features. The central carbon–metal distance is shorter than the other carbon–metal distance and may be explained by an additional interaction between a filled metal d-orbital and the olefin π^*-orbital with the uncoordinated double bond. However, this view, although attractive, appears not to be supported by the fact that the M–C(2) distance remains nearly constant, within standard error, regardless of the number of methyl substituents at the uncoordinated double-bond carbon in $Pt(PPh_3)_2$-(allene). The X-ray bond distances may not be sensitive to the electronic variation or they may simply be a reflection of the atomic radii susceptible to the change in hybridization. In the case of $Pt(PPh_3)_2(CH_2=C=CMe_2)$, a nonbonding repulsion exists between the phosphine ligand and one of the methyl substituents, as the difference Fourier map indicates the particular methyl group to be a hindered rotator (*132*). The repulsion is also reflected in the two P-metal-C angles. This steric factor may be responsible for the apparent irregularity in the bending [C(1)–C(2)–C(3) angle] and, hence, for the absence of a linear correlation between the angle and the distance of the coordinated double-bond [C(1)–C(2)] in $Pt(PPh_3)_2$(allene).

A series of nickel triad complexes ML_2(allene) (Table VII) were prepared (*122, 123, 139*) and studied in solution by means of ^1H NMR spectroscopy. Consistent with a planar molecular structure,

the ^1H NMR spectrum shows three signals with the proton signal at site a highest field, and the signal at site c lowest field. The following features are conspicuous: (i) large Pt—H coupling constants, (ii) fairly strong P—H coupling, in particular the long-range coupling $J_{P^a-H^c}$ (23–37 Hz); (iii) $J_{P^a-H^a} \doteq J_{P^b-H^a}$ (~10 Hz), $J_{P^b-H^b}$ (<3 Hz) < $J_{P^a-H^b}$ (10–20 Hz), and $J_{P^b-H^c}$ (<3 Hz) ≪ $J_{P^a-H^c}$; and (iv) $J_{H^a-H^b}$ (complexes) < $J_{H^a-H^b}$ (free allene). The ^1H NMR spectra revealed that monosubstituted allenes form only one isomer in which the substituent occupies site c. 1,3-Disubstituted

TABLE VII
ALLENE COMPLEXES OF THE NICKEL TRIAD[a]

M	PR_3	Allene	Mp(°C)	Color
Ni	PPh_3	$PhCH=C=CHPh$	150–152 (dec)	Yellow
	PPh_3	$(CH_3)_2C=C=CH_2$		Yellow
	$P(O\text{-}o\text{-}tolyl)_3$	$PhCH=C=CHPh$	138–141 (dec)	Yellow
	$P(O\text{-}o\text{-}tolyl)_3$	$(CH_3)_2C=C=CH_2$	117–120 (dec)	Yellow
Pd	PPh_3	$CH_2=C=CH_2$	83–85 (dec)	Colorless
	PPh_3	$PhCH=C=CHPh$	160–162 (dec)	Pale yellow
	$P(OPh)_3$	$CH_2=C=CH_2$	110 (dec)	Colorless
	$P(OPh)_3$	$PhCH=C=CHPh$	134–137 (dec)	Colorless
	$P(OMe)_3$	$PhCH=C=CHPh$	Liquid	Pale yellow
Pt	PPh_3	$CH_2=C=CH_2$	152–154 (dec)	Colorless
	PPh_3	$CH_2=C=CHCH_3$	142–146 (dec)	Colorless
	PPh_3	$CH_2=C=C(CH_3)_2$	130–135 (dec)	Colorless
	PPh_3	$CH_2=C=CHPh$	170 (dec)	Yellow
	PPh_3	$PhCH=C=CHPh$	170–182 (dec)	Yellow
	PPh_3	$(CH_3)_2C=C=C(CH_3)_2$	118 (dec)	Colorless
	$P(OPh)_3$	$CH_2=C=CH_2$	89–91	Pale yellow
	$P(OPh)_3$	$PhCH=C=CHPh$	167–168	Yellow

[a] Data from Otsuka et al. (122, 123, 139).

allenes also form only one isomer with the two substituents occupying site c and one of the a positions.

The Pt(0) complexes are in general rather stable, except for Pt(PPh$_3$)$_2$-(TMA) (TMA = tetramethylallene). In solution, the latter shows a signal due to free TMA in addition to those of the complexed ligand, indicating dissociation:

$$Pt(PPh_3)_2(TMA) \rightleftharpoons Pt(PPh_3)_2 + TMA$$

A dissociation constant of about 1.5×10^{-3} (25°C) was assessed from the intensity ratio. The complex does not show intramolecular fluxional behavior up to +50°C. Of the three methyl resonances the chemical shift of that at position b occurs at highest field owing to the influence of the phenyl ring current. Consistent with a corollary of the X-ray study (132) on Pt(PPh$_3$)$_2$(CH$_2$=C=CMe$_2$), we may infer the existence of a steric compression between the methyl and the phenyl groups which may be primarily responsible for the ready dissociation of TMA.

The Pd(0) complexes are rather labile. Complex Pd(PPh$_3$)$_2$(CH$_2$=C=CH$_2$) shows some broadening of the three proton signals even at −74°C, and above 22°C complete equilibration of the signals to a sharp singlet is observed. Complex Pd(PPh$_3$)$_2$(PhCH=C=CHPh) shows a limiting spectrum of the olefinic protons below −40°C [H^a, δ = 6.04 (m); H^b, δ = 4.75 (d)]. On raising the temperature, a coalesced signal appears (δ = 5.45) which becomes a very sharp singlet (δ = 5.50) above 60°C. The disappearance of the coupling with the phosphorous atoms excludes an intramolecular process for the equilibration. An exchange process between the complex and the free allene is indicated by broadening of the signal of the free PhCH=C=CHPh upon addition of the allene. Since the half-height widths of the two signals before the coalescence are not affected by addition of free PhCH=C=CHPh, a dissociative mechanism is concluded:

$$Pd(PPh_3)_2(PhCH=C=CHPh) \rightleftharpoons Pd(PPh_3)_2 + PhCH=C=CHPh$$

Addition of free PPh$_3$ to the complex leads to the formation of Pd(PPh$_3$)$_4$. Dissociation of the allene ligand is also observed for Pd[P(OPh)$_3$]$_2$-(PhCH=C=CHPh) and Pd[P(OMe)$_3$]$_2$(PhCH=C=CHPh), but dissociation of phosphine or phosphite ligands is not observed for any of these Pd(0) compounds.

The Ni(0) compounds are in general reactive. It seems impossible to isolate Ni(PPh$_3$)$_2$(CH$_2$=C=CH$_2$). Complex Ni(PPh$_3$)$_2$(PhCH=C=CHPh) is isolable and shows a limiting spectrum for the olefinic protons below −15°; i.e., δ = 6.15 (H^b complex doublet, J = 8.9 Hz) and δ = 4.17 (H^a multiplet). At higher temperature (+76°C), coupling with the phosphorus atoms is lost; two doublets at δ = 6.07 and δ = 4.13 (J = 4 Hz) indicate only coupling between the allenic protons, implying an intermolecular phosphorus ligand exchange. The widths of the two signals due to H^a and H^b become narrower upon addition of free PPh_3. In this case the exchange of phosphine ligands occurs through an associative mechanism:

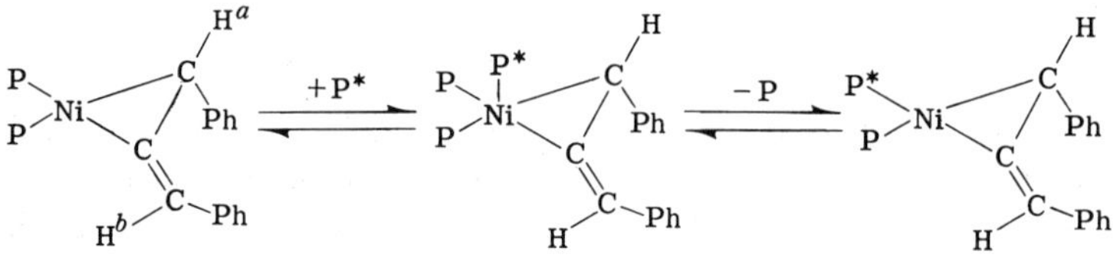

The exchange of H^a and H^b does not occur up to 76°C. The two methyl resonances of the phosphite ligands in $Ni[P(O—o\text{-tolyl})_3]_2$(PhCH=C=CHPh) coalesce at low temperature (22.5°C) retaining the P—H coupling with allenic protons. This indicates an intramolecular exchange of the phosphite ligands prior to the intermolecular exchange. At higher temperatures (>50°C) the latter process is clearly shown by a sharpening of the H^a signal.

In summary, (i) PtL_2(allene) has a rigid planar structure in the temperature range −50° to +50°C and, except for PtL_2(TMA), no dissociation of ligands is observable, (ii) PdL_2(allene) assumes a similar structure at low temperature although rapid allene exchange, via a dissociative mechanism, may be observed at high temperature, and (iii) NiL_2(allene) assumes a planar structure at low temperature but undergoes a rapid configurational change at high temperature leading to equilibration of the two phosphorus ligands. Dissociation of the allene ligand does not occur before the decomposition of the complex. Addition of excess PPh_3 causes rapid ligand exchange via an associative mechanism, whereas addition of allene leads to oligomerization.

B. Cyclooligomerization

In general the thermal reaction of allene gives a complex mixture of dimers, trimers, and higher oligomers including small amounts of spiro compounds (140). A highly selective dimerization to 1,2-dimethylenecyclobutane is achieved by thermal reaction of dilute solutions (141). Theoretically the process [2s + 2a] of allenes and related cumulenes may be facilitated by participation of the orthogonal $p\pi$-orbital in the addition (142). However, the concertedness of the cyclodimerization is still in dispute. Selective cyclotrimerization and pentamerization of allene have not been achieved thermally, although the concerted processes are "thermally allowed." In metal-catalyzed reactions, allene is cyclized selectively to methylene-substituted compounds ranging from four- to twelve-membered rings. Studies of the mechanism are of interest both for its own sake and for its pertinence to general questions regarding the role of ligands in transition metal–olefin catalysis.

1. *Nickel Oligomerization*

Nickel(0)–allene complexes are characterized by configurational instability and a propensity to assume a high coordination number. It may not be surprising to find that the Ni(0) species is the most catalytically active of the triad. The cyclic dimers, 1,3- and 1,2-dimethylenecyclobutane, are formed only in the vapor phase reaction of allene with $Ni(CO)_2(Ph_2PC_6H_4PPh_2)$ (*143*). The liquid phase reaction with Ni(0) complexes selectively produces the trimer (I), tetramer (II), and pentamer (III) (Table VIII) (*123*). Several intermediate Ni(0) complexes (IV–VI) were isolated.

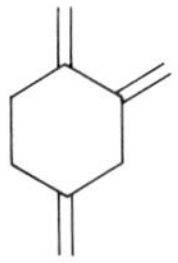

(I) 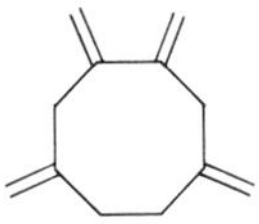 (II) 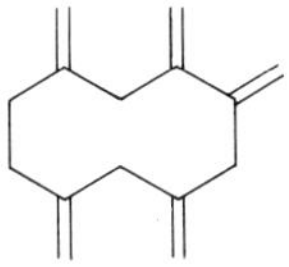 (III)

Their structures and relative reactivities provide most important information as to mechanism. The allene–trimer complex, $NiL(C_9H_{12})$ (IV) (*123, 144, 145*) is readily obtained as rather stable crystals by treating a mixture of $Ni(cod)_2$ and a phosphorus ligand (e.g., PPh_3, $P(OPh)_3$, or $P(OC_6H_4-o\text{-Ph})_3$) with allene in solution. The structure was deduced from IR and NMR data and confirmed by a single-crystal X-ray diffraction study (*145*). The intermediacy of compound IV (L = PPh_3) in the formation of compound II is confirmed by monitoring (^1H NMR) the reaction of IV with allene at 50°C:

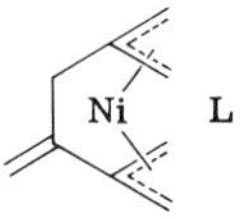

(IV)

$$IV + 4C_3H_4 \xrightarrow{50°C} II + IV \tag{1}$$

TABLE VIII

CYCLOOLIGOMERIZATION OF ALLENE WITH NICKEL(0) CATALYSTS IN BENZENE[a,b]

Catalyst	Added ligand[c]	Concentration[d] (mole %)	Reaction conditions (t/°C; time/hr)	Conversion (%)	Selectivity (%)			
					Trimer[e] (I)	Tetramer (II)	Pentamer (III)	Higher oligomers and polymer
[Ni(cod)$_2$]	—	0.20	40, 40	100	Trace	Trace	52	48
[Ni(cod)$_2$]	4PPh$_3$	2.5	70, 20	93	28	66	6	Trace
[Ni(cod)$_2$]	PBu$_3^n$	2.5	50, 72	100	13	60	18	10
[Ni(cod)$_2$]	2P(OPh)$_3$	0.25	70, 20	15	48	21	Trace	31
[Ni(cod)$_2$]	P(O—o-tolyl)$_3$	2.5	70, 24	100	38	9	20	33
[Ni(cod)$_2$]	P(O—o-biphenylyl)$_3$	2.5	70, 24	96	39	11	5	44
[Ni(C$_3$H$_4$)(PPh$_3$)$_2$]	—	2.5	70, 26	100	20	53	10	17
[Ni(cod)(PBu$_3^n$)$_2$]	—	2.5	60, 48	100	17	83	Trace	Trace
[Ni(P(OPh)$_3$)$_4$]	—	2.5	70, 20	81	72	25	3	Trace

[a] Data from Otsuka et al. (123).
[b] Allene (2.1 gm, 52.5 mmoles); benzene (10 cm^3).
[c] Added in situ to the reaction mixture at −78°C.
[d] Mole % of Ni(0) complexes based on allene.
[e] About 5% of 1,3,5-trimethylenecyclohexane was present.
[f] Cod = cycloocta-1,5-diene.

Allene–tetramer complexes, $NiL(C_{12}H_{16})$ (V) (*123*) and $Ni_2L_3(C_{12}H_{16})$ (VI) (*123*), were isolated from the low-temperature treatment of $Ni(cod)_2$ with allene followed by addition of PPh_3. These tetramer complexes, being more reactive than complex IV, readily react with allene even at room temperature in benzene to give IV and tetramer II:

$$V + 3C_3H_4 \longrightarrow IV + II \tag{2}$$

It is noteworthy that the reaction does not produce pentamer III. Their structures are deduced from the IR and NMR data and support is obtained from the following reactions:

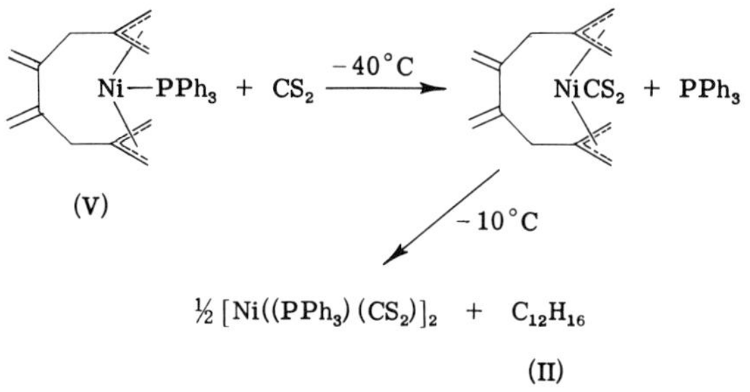

$$\text{(V)} \xrightarrow{-40°C} NiCS_2 + PPh_3 \tag{3}$$

$$\downarrow -10°C$$

$$\tfrac{1}{2}[Ni((PPh_3)(CS_2))]_2 + C_{12}H_{16}$$
$$\text{(II)}$$

$$\text{(VI)} \longrightarrow Ni-PPh_3 + 2CS_2$$

$$\downarrow -40°C \tag{4}$$

$$[Ni(C_{12}H_{16})(CS_2)] + \tfrac{1}{2}[Ni(CS_2)(PPh_3)]_2 + 2PPh_3$$

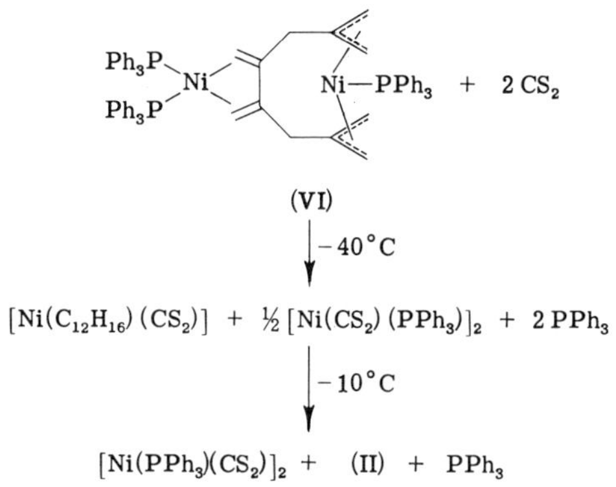

$$\downarrow -10°C$$

$$[Ni(PPh_3)(CS_2)]_2 + \text{(II)} + PPh_3$$

Kinetic studies (*123*) on the three types of oligomerization indicate that the rate of reaction increased in the order $Ni(0)-P(OPh)_3 < Ni(0)-PPh_3 < Ni(0)$. Interestingly, the rates for both pentamerization and tetramerization are first order with respect to the allene concentration, whereas the trimerization rate is nearly zero order. These results are accommodated by the following reaction sequences.

Trimerization:

$$\text{Ni-P(OAr)}_3 + 3\text{C}_3\text{H}_4 \xrightarrow{\text{fast}} \text{Ni(C}_9\text{H}_{12})\text{P(OAr)}_3$$

$$\text{Ni(C}_9\text{H}_{12})\text{P(OAr)}_3 \xrightarrow{\text{slow}} \text{Ni-P(OAr)}_3 + \text{C}_9\text{H}_{12}$$

Tetramerization:

$$\text{Ni-PR}_3 + 3\text{C}_3\text{H}_4 \xrightarrow{\text{fast}} \text{Ni(C}_9\text{H}_{12})\text{PR}_3$$

$$\text{Ni(C}_9\text{H}_{12})\text{PR}_3 + \text{C}_3\text{H}_4 \xrightarrow{\text{slow}} \text{Ni(C}_{12}\text{H}_{16})\text{PR}_3$$

$$\text{Ni(C}_{12}\text{H}_{16})\text{PR}_3 \xrightarrow{\text{fast}} \text{Ni-PR}_3 + \text{C}_{12}\text{H}_{16}$$
$$\text{(II)}$$

Pentamerization[3]:

$$\text{Ni*} + 4\text{C}_3\text{H}_4 \xrightarrow{\text{fast}} \text{Ni(C}_{12}\text{H}_{16})$$

$$\text{Ni(C}_{12}\text{H}_{16}) + \text{C}_3\text{H}_4 \xrightarrow{\text{slow}} \text{Ni(C}_{15}\text{H}_{20})$$

$$\text{Ni(C}_{15}\text{H}_{20}) \xrightarrow{\text{fast}} \text{Ni} + \text{C}_{15}\text{H}_{20}$$
$$\text{(III)}$$

The reaction of Ni(cod)_2 with allene below $-30°\text{C}$ produces a mixture of $\text{Ni(C}_9\text{H}_{12})$ and $\text{Ni(C}_{12}\text{H}_{16})$ in roughly equal amounts, indicating that the potential barriers are quite low up to the trimer and to the tetramer stage in the absence of a phosphorus ligand. At higher temperature the reaction mixture yields pentamer III. The qualitative reaction potential profile may be deduced as the solid line in Fig. 2.

In the presence of 1 mole of PPh_3, the low-temperature reaction of Ni(cod)_2 with allene produces exclusively the trimer complex (IV). At 50°C, reaction (1) takes place producing tetramer II. Apparently the role of PPh_3 is to facilitate ring closure of the linear tetramer ligand in complex V or VI. These results suggest the potential energy profile depicted (the broken line in Fig. 2). The Ni(0)-phosphite system yields the most stable trimer complex (IV) and, as the kinetics indicate, the unimolecular thermal decomposition of complex IV constitutes the rate-determining step.

All attempts to identify the initial stage of the reaction have failed due to the fast rate even at very low temperature. Appropriate substituents on allene are effective in retarding the first few steps, and the monomer complexes $\text{NiL}_2(\text{allene})$ are isolated using both 1,3-diphenylallene and 1,1-dimethylallene. A linear dimerization involving hydrogen migration was observed for the reaction of 1,1-dimethylallene with NiL_2 (L =

[3] Ni*: labile ligands such as cycloocta-1,5-diene are abbreviated.

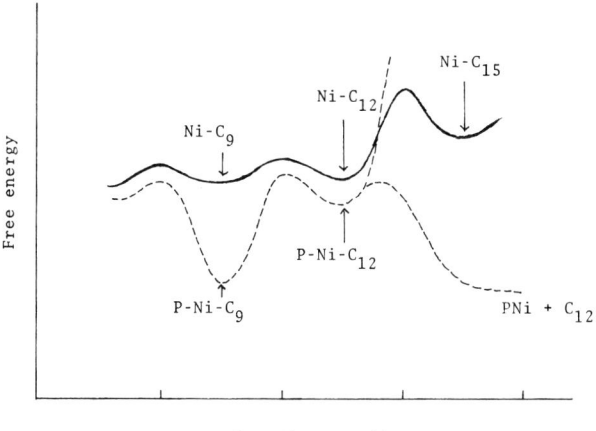

FIG. 2. Partial reaction potential profiles for cyclooligomerization reactions.

$P(OC_6H_4\text{—}o\text{-Ph})_3)$ (143):

Not only this Ni(0)-catalyzed reaction but also all reported allene dimer complexes, e.g., hexacarbonyl-μ-[1-3:1'-3'-η-(2,2'-biallyl)]diiron (Fe-Fe) (146), hexacarbonyl-μ-[1-3:1'-3'-η-(1,1'-diphenyl-2,2'-biallyl)]diiron (Fe-Fe) (147), and di-μ-acetato-μ-[1-3:1'-3'-η-(2,2'-biallyl)]dipalladium (148) point to the formation of 2,2'-biallyl. A mononuclear Rh(I) complex containing this ligand was recently isolated (149). Accepting this biallyl formation, then the next step is the insertion to form the trimer ligand in complex IV. Thus the entire reaction paths leading to complexes I, II, and III may be depicted (Scheme 8).

It is now possible to give a reasonable interpretation for the role of phosphorus ligands in this catalysis. The stability of complex IV indicates that phosphorus ligands are effective in stabilizing bisallyl coordination to Ni(0) [IV is the first example of a stable bisallyl phosphine nickel(0) complex]. The inertness is particularly enhanced with a triarylphosphite, which apparently is due to the electron-accepting property stabilizing the zero-valency state. Addition of phosphine to bis(allyl)nickel(0) species generally leads to the formation of biallyl, or to insertion reaction if olefin or diene molecules are present in the system. Thus, the important role of phosphite ligand here is to prevent further insertion.

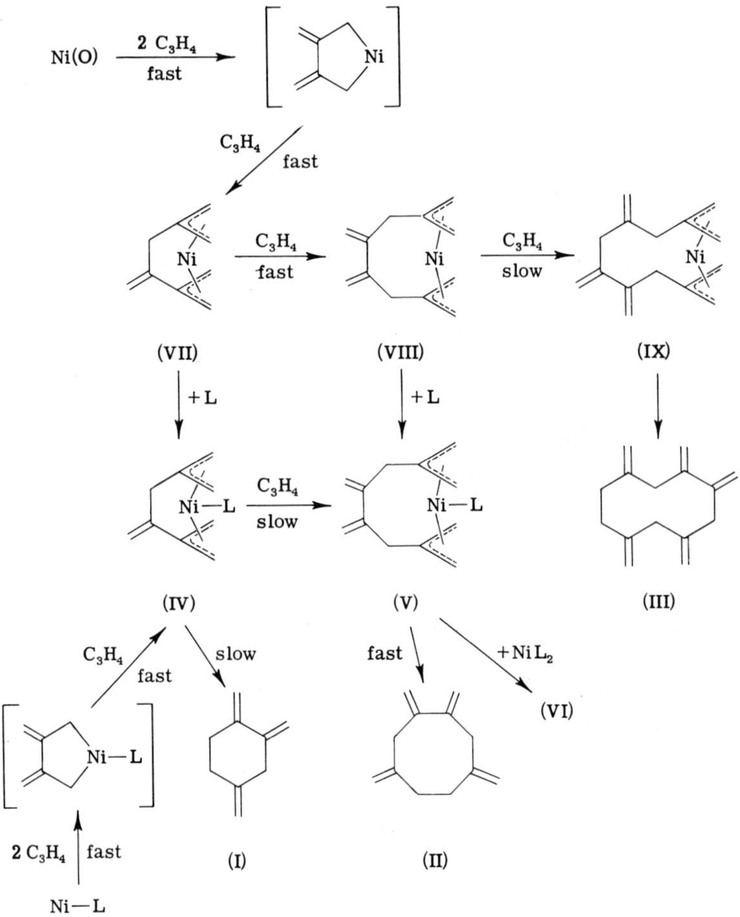

Scheme 8

The monomer insertion was assumed to take place via electrophilic attack of allene at the σ-allyl–metal bond rather than at the π-allyl end. In support of this notion is the preferential formation of Complex V from IV and complex VIII from VII (Scheme 8), indicating that the insertion site is the carbon end of an extended $p\pi$-system in complexes IV and VII, where negative-charge localization is enhanced compared to the isolated allyl system. The ligand effect may also be reasonably interpreted assuming participation of the anionic end of the allyl group in the insertion. It is known that σ-donor ligands stabilize high oxidation states of metals. Here the role of a tertiary phosphine ligand is, contrary to that of electron-withdrawing phosphites, to shift the π–σ allyl equilibrium toward the σ-form, requiring an increase in formal oxidation state of the metal. Electrophilic attack of allene is then facilitated to give higher oligomers. However, the predominant formation of complex II from V and of complex III from VIII suggests that blockage of a coordination site by a ligand L in complex V is effective in raising the potential barrier for allene coordination (hence the insertion). In addition the free energy of activation for thermal decomposition of complex V should be low as the kinetics indicate. Thus, further insertion of allene into V to produce pentamer III becomes a minor reaction path (Table VIII) (*123*) compared to the formation of complex II. The foregoing interpretation is consistent with the observed relative rate, i.e., pentamerization with naked Ni(0) > tetramerization with Ni(0)—PR_3 > trimerization with Ni(0)—$P(OR)_3$.

2. *Rhodium Oligomerization*

The reaction path of allene oligomerization on Rh(I) differs somewhat from that on Ni(0) species. The monomer complex can be isolated. Complex $RhCl(PPh_3)_3$ with a stoichiometric amount of allene gives $RhCl(PPh_3)_2(C_3H_4)$ (X) (*136*). The bromo or iodo analog is prepared from complex X. The structure of the iodo compound has been determined (*129*). In the absence of strongly coordinating substances such as phosphorus ligands, RhCl species take up 5 moles of allene to give the cyclopentamer complex $RhCl(C_{15}H_{20})$ (XIV) (*150*). For example, [RhCl-$(C_2H_4)_2$]$_2$ is a good source for the production of RhCl species. Lower allene oligomers could not be detected in this reaction. Dimer and tetramer complexes have been obtained with Rh(I) having a chelating anion, e.g. acetylacetonate(acac) (*149*). Thus the low-temperature (−78°C) reaction of allene with $Rh(acac)(C_2H_4)_2$ produces unstable $Rh(acac)(C_3H_4)_3$ of unknown structure which gives, upon treatment with pyridine, very stable $Rh(acac)py_2(C_6H_8)$ (XII). The structure of complex XII has been established by an X-ray study. Hence the unstable compound is believed to have a rhodacyclopentane unit (XI). The unstable five-membered ring (XI) is apparently stabilized in complex XII which may be regarded as an octahedral Rh(III) (d^6) complex. The unstable $Rh(acac)(C_3H_4)_3$ is a precursor of the tetramer complex $Rh(acac)(C_{12}H_{16})$ (XIII), which is also directly obtainable from the room-temperature reaction of allene with $Rh(acac)(C_2H_4)_2$ in pentane. The other β-diketonato complex, Rh(dpd)-

($C_{12}H_{16}$) (dpd = 1,3-diphenylpropane-1,3-dionato) was also made. The structure of complex XIII was established by an X-ray study (*151*). The reaction path from the unstable complex containing (XI) to complex XIII remains to be elucidated. The structure of pentamer ligand of XIV, derivable from the tetramer ligand of XIII, is different from the nickel

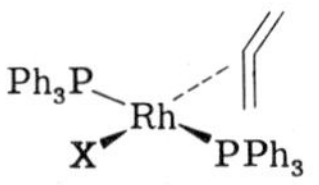

(X) 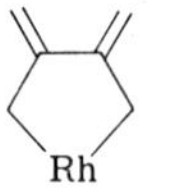   (XI) 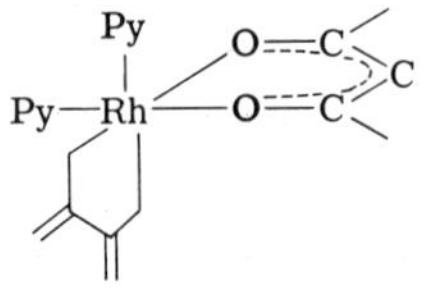   (XII)

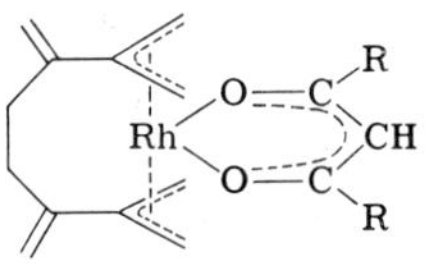

(XIII) 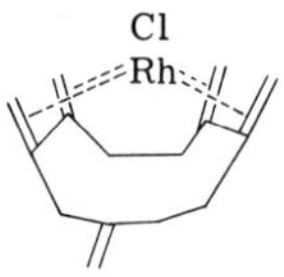   (XIV)

pentamer III. The key step determining the pentamer structure is thus the tetramer stage. Complex XIII is best described as an Rh(III) complex. The Rh(III) ion apparently prefers coordination of the allylic group conjugated with a double bond so that the negative charge localization at the allyl end matches with the high metal oxidation state.

Rhodium(I)–phosphine systems lead to catalytic tetramerization. For example, the system $[RhCl(C_2H_4)_2]_2$ with 1 to 2 moles of PPh_3 is effective in the selective formation of an interesting spiro compound (XV) (*152*) free from other isomers. Although the detailed reaction path is unknown due to the inaccessibility of the intermediate complexes, the formation of (XV) may be visualized from a tetramer complex as follows:

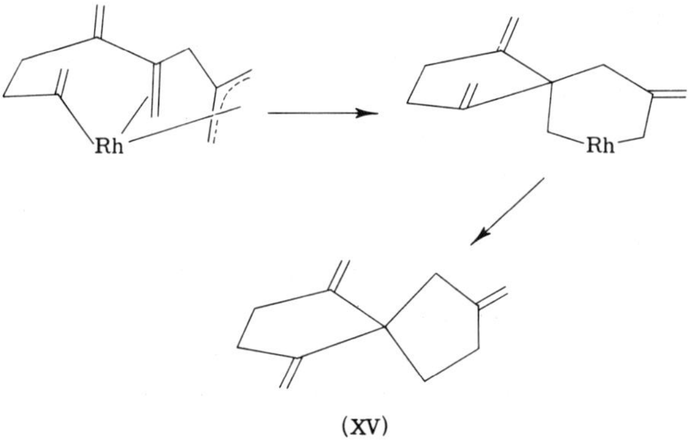

(XV)

REFERENCES

1. Bowden, F. L., and Lever, A. B. P., *Organometal. Chem. Rev.* **3**, 227 (1968).
2. Kemmitt, R. D. W., *MTP Int. Rev. Sci., Inorg. Chem., Ser. 1* **6** (part 2), 227 (1972).
3. Hübel, W., *in* "Organic Syntheses via Metal Carbonyls" (I. Wender and P. Pino, eds.), Vol. 1, p. 273. Wiley (Interscience), New York, 1968.
4. Hartley, F. R., *Chem. Rev.* **69**, 799 (1969); **73**, 163 (1973); *Angew. Chem.* **84**, 657 (1972); "Chemistry of Platinum and Palladium," p. 361. Applied Science Publ., London, 1973.
5. Heck, R. F., "Organotransition Metal Chemistry," pp. 167–200. Academic Press, New York, 1974.
6. Green, M. L. H., *in* "Organometallic Compounds" (G. E. Coates, K. Wade, and M. L. H. Green, eds.), Vol. 2, pp. 288–333. Methuen, London, 1968.
7. Maitlis, P. M., "Organic Chemistry of Palladium," Vol. 1, pp. 110–130; Vol. 2, pp. 31, 47–58. Academic Press, New York, 1971.
8. Jolly, P. W., and Wilke, G., "Organic Chemistry of Nickel," Vol. 1, pp. 305–315. Academic Press, New York, 1974.
9. Pettit, L. D., and Barnes, D. S., *Fortschr. Chem. Forsch.* **28**, 85 (1972).
10. Dickson, R. S., and Fraser, P. J., *Advan. Organometal. Chem.* **12**, 323 (1973).
11. Shaw, B. L., and Stringer, A. J., *Inorg. Chim. Acta Rev.* **7**, 1 (1973).
12. Baker, R., *Chem. Rev.* **73**, 487 (1973).
13. Jolly, P. W., and Wilke, G., "The Organic Chemistry of Nickel," Vol. 2. Academic Press, New York, (1975).
14. Nelson, J. H., Wheelock, K. S., Cusachs, L. C., and Jonassen, H. B., *J. Amer. Chem. Soc.* **91**, 7005 (1969); *Inorg. Chem.* **11**, 422 (1972); Wheelock, K. S., Nelson, J. H., Cusachs, L. C., and Jonassen, H. B., *J. Amer. Chem. Soc.* **92**, 5110 (1970); Nelson, J. H., and Jonassen, H. B., *Coord. Chem. Rev.* **6**, 27 (1971).
15. Bokiy, N. G., Gatilov, Yu. V., Struchkov, Yu. T., and Ustynyuk, N. A., *J. Organometal Chem.* **54**, 213 (1973).
16. Nesmeyanov, A. N., Gusev, A. I., Pasynskii, A. A., Anisimov, K. N., Kolobova, N. E., and Struchkov, Yu. T., *Chem. Commun.* 277 (1969).
17. Bennett, M. A., Robertson, G. B., Whimp, P. O., and Yoshida, T., *J. Amer. Chem. Soc.* **93**, 3797 (1971); Robertson, G. B., and Whimp, P. O., *J. Organometal. Chem.* **32**, C69 (1971).
18. Davies, B. W., and Payne, N. C., *Inorg. Chem.* **13**, 1848 (1974).
19. Glanville, J. O., Stewart, J. M., and Grim, S. O., *J. Organometal. Chem.* **7**, P9 (1967).
20. Panattoni, C., and Graziani, R., *in* "Progress in Coordination Chemistry" (M.Cais, ed.), p. 310. Elsevier, London, 1968.
21. Davies, B. W., Puddephatt, R. J., and Payne, N. C., *Can. J. Chem.* **50**, 2276 (1972).
22. Davies, B. W., and Payne, N. C., *Inorg. Chem.* **13**, 1843 (1974).
23. Davies, G. R., Hewertson, W., Mais, R. H. B., Owston, P. G., and Patel, C. G., *J. Chem. Soc. A* 1873 (1970).
24. McGinnety, J. A., *J. Chem. Soc., Dalton Trans.* 1038 (1974).
25. Jacobson, S., Carty, A. J., Mathieu, M., and Palenik, G. J., *J. Amer. Chem. Soc.* **96**, 4330 (1974).
26. Dickson, R. S., and Ibers, J. A., *J. Organometal. Chem.* **36**, 191 (1972).
27. Nesmeyanov, A. N., Gusev, A. I., Pasynskii, A. A., Anisimov, K. N., Kolobova, N. E., and Struchkov, Yu. T., *Chem. Commun.* 739 (1969).
28. Kirchner, R. M., and Ibers, J. A., *J. Amer. Chem. Soc.* **95**, 1095 (1973).
29. Laine, R. M., Moriarty, R. E., and Bau, R., *J. Amer. Chem. Soc.* **94**, 1402 (1972).
30. Davidson, G., *Organometal. Chem. Rev.* **8**, 342 (1972): see Maitlis (7), Vol. 1, pp. 120–121.

31. Wilke, G., and Herrmann, G., *Angew. Chem.* **74**, 693 (1962).
32. McClure, G. L., and Baddley, W. H., *J. Organometal. Chem.* **27**, 155 (1971).
33. Baddley, W. H., *Inorg. Chem. Acta Rev.* **2**, 1 (1968).
34. Otsuka, S., Yoshida, T., and Tatsuno, Y., *J. Amer. Chem. Soc.* **93**, 6462 (1971).
35. Nakamura, A., and Otsuka, S., *J. Amer. Chem. Soc.* **94**, 1886 (1972).
36. Brintzinger, H. H., and Thomas, J. L., *J. Amer. Chem. Soc.* **96**, 3694 (1974); Thomas, J. L., *ibid.* **95**, 1838 (1973).
37. Tsumura, R., and Hagihara, N., *Bull. Chem. Soc. Jap.* **38**, 861 (1965).
38. Mann, B. E., Shaw, B. L., and Tucker, N. I., *J. Chem. Soc. A* 2667 (1971).
39. Church, M. J., Mays, M. J., Simpson, R. N. F., and Stefanini, F. D. S., *J. Chem. Soc. A* 2909 (1970).
40. Cherwirski, W. J., Johnson, B. F. G., and Lewis, J., *J. Chem. Soc., Dalton Trans.* 1405 (1974).
41. Chisholm, M. H., and Clark, H. C., *Inorg. Chem.* **10**, 2557 (1971); Kong, P. C., and Theophanides, T., *Can. J. Chem.* **45**, 3193 (1967); **48**, 1084 (1970).
42. Green, M., *MTP Int. Rev. Sci., Inorg. Chem., Ser. 1* **6**, 198 (1972); Heck (*5*), p. 168.
43. Schwartz, J., Hart, D. W., and Holden, J. L., *J. Amer. Chem. Soc.* **94**, 9269 (1972).
44. Treichel, P. M., Pitcher, E., and Stone, F. G. A., *Inorg. Chem.* **1**, 511 (1962).
45. Wilford, J. B., and Stone, F. G. A., *Inorg. Chem.* **4**, 93 (1965).
46. Nakamura, A., and Otsuka, S., *J. Amer. Chem. Soc.* **94**, 1886 (1972).
47. Nakamura, A., and Otsuka, S., "Prospects in Organotransition Metal Chemistry" (M. Tsutsui and Y. Ishii, eds.), Plenum Press, New York, 1975; *J. Mol. Catal.* To be published.
48. Clark, H. C., and Tsang, W. S., *J. Amer. Chem. Soc.* **89**, 533 (1967).
49. Clark, H. C., and Mittal, R. K., *Can. J. Chem.* **51**, 1511 (1973).
50. Blackmore, T., Bruce, M. I., Stone, F. G. A., Davis, R. E., and Garza, A., *Chem. Commun.* 852 (1971); *J. Chem. Soc., Dalton Trans.* 106 (1974).
51. Booth, B. L., and Lloyd, A. D., *J. Organometal. Chem.* **35**, 195 (1972).
52. Bruce, M. I., Harbourne, D. A., Waugh, F., and Stone, F. G. A., *J. Chem. Soc.* 895 (1968).
53. Harbourne, D. A., and Stone, F. G. A., *J. Chem. Soc. A* 1765 (1968).
54. Trocha-Grimshaw, M. J., and Henbest, H. B., *Chem. Commun.* 757 (1968).
55. Eaborn, C., Farrell, N., and Pidcock, A., *Chem. Commun.* 766 (1973).
56. Schrauzer, G. N., and Windgassen, R. J., *J. Amer. Chem. Soc.* **89**, 1999 (1967).
57. Yamazaki, H., and Hagihara, N., *J. Organometal. Chem.* **21**, 431 (1970).
58. Booth, B. L., and Hargreaves, G., *J. Chem. Soc. A* 2766 (1969).
59. Dubeck, M., and Schell, R. A., *Inorg. Chem.* **3**, 1757 (1964).
60. Hagihara, N., *et al.*, unpublished result.
61. Booth, B. L., and Hargreaves, G., *J. Organometal. Chem.* **33**, 365 (1971).
62. Baddley, W. H., and Fraser, M. S., *J. Amer. Chem. Soc.* **91**, 3661 (1969); *J. Organometal. Chem.* **36**, 377 (1972).
63. Wailes, P. C., Weigold, H., and Bell, A. P., *J. Organometal. Chem.* **43**, C29 (1972).
64. Baddley, W. H., and Tupper, G. B., *J. Organometal. Chem.* **67**, C16 (1974).
65. Zanella, R., Canziani, F., Ros, R., and Graziani, M., *J. Organometal. Chem.* **67**, 449 (1974).
66. Cullen, W. R., Dowson, D. S., and Styan, G. E., *Can. J. Chem.* **43**, 3365 (1965).
67. Truce, W. E., and Goldhamer, D. L., *J. Amer. Chem. Soc.* **81**, 5795, 5798 (1959).
68. Nakamura, A., and Otsuka, S., *J. Amer. Chem. Soc.* **95**, 7262 (1973).
69. Epiotis, N. D., *J. Amer. Chem. Soc.* **95**, 1191 (1973); Pearson, R. G., *Accounts Chem. Res.* **4**, 125 (1971).

70. Hoogzand, C., and Hübel, W., "Organic Syntheses via Metal Carbonyls" (I. Wender, and P. Pino, eds.), Vol. 1, p. 343. Wiley, New York, 1968.
71. Yamazaki, H., and Hagihara, N., *J. Organometal. Chem.* **7**, P22 (1967).
72. Zeiss, H. H., and Tsutsui, M., *J. Amer. Chem. Soc.* **81**, 6090 (1959).
73. Burt, R., Cooke, M., and Green, M., *J. Chem. Soc. A* 2981 (1970).
74. Sears, C. J., Jr., and Stone, F. G. A., *J. Organometal. Chem.* **11**, 644 (1968).
75. Collman, J. P., and Kang, J. W., *J. Amer. Chem. Soc.* **89**, 844 (1967).
76. Brateman, L. R., Maitlis, P. M., and Dahl, L. F., *J. Amer. Chem. Soc.* **91**, 7292 (1969).
77. Kang, J. W., Childs, R. F., and Maitlis, P. M., *J. Amer. Chem. Soc.* **92**, 721 (1970).
78. Clarke, B., Green, M., and Stone, F. G. A., *J. Chem. Soc. A* 951 (1970).
79. Mague, J. T., and Wilkinson, G., *Inorg. Chem.* **7**, 542 (1968).
80. Collman, J. P., Cawse, J. N., and Kang, J. W., *Inorg. Chem.* **8**, 2574 (1969).
81. Mague, J. T., *Inorg. Chem.* **9**, 1610 (1970); Mague, J. T., Nutt, M. O. and Gause, E. H., *J. Chem. Soc., Dalton Trans.* 2578 (1973).
82. McVey, S., and Maitlis, P. M., *J. Organometal. Chem.* **19**, 169 (1969); Kang, J. W., McVey, S., and Maitlis, P. M., *Can. J. Chem.* **46**, 3189 (1968).
83. Moseley, K., and Maitlis, P. M., *J. Chem. Soc., Dalton Trans.* 169 (1974).
84. Ito, Ts., Hasegawa, S., Takahashi, T., and Ishii, Y., *Chem. Commun.* 629 (1972); Roe, D. M., Calvo, C., Krishnamachari, M., Moseley, K., and Maitlis, P. M., *J. Chem. Soc., Chem. Commun.* 436 (1973).
85. Ashley-Smith, J., Green, M., and Wood, D. C., *J. Chem. Soc. A* 1847 (1970).
86. Collman, J. P., Kang, J. W., Little, W. F., and Sullivan, M. F., *Inorg. Chem.* **7**, 1298 (1968).
87. Singer, H., and Wilkinson, G., *J. Chem. Soc. A* 849 (1968).
88. Kern, R. J., *Chem. Commun.* 706 (1968).
89. Clemens, J., Green, M., Kuo, M. C., Fritchie, C. J., Mague, J. T., and Stone, F. G. A., *Chem. Commun.* 53 (1972).
90. Otsuka, S., Nakamura, A., Tatsuno, Y., and Miki, M., *J. Amer. Chem. Soc.* **94**, 3761 (1972).
91. Wakatsuki, Y., and Yamazaki, H., *Tetrahedron Lett.* 4549 (1974).
92. Blackmore, T., Bruce, M. I., Stone, F. G. A., Davis, R. E., and Gartza, A., *Chem. Commun.* 852 (1971).
93. Chisholm, M. H., and Clark, H. C., *J. Amer. Chem. Soc.* **94**, 1532 (1972); Chisholm, M. H., Clark, H. C., and Hunter., D. H., *Chem. Commun.* 809 (1971).
94. Tsumura, R., and Hagihara, N., *Bull. Chem. Soc. Jap.* **38**, 1901 (1965).
95. King, R. B. and Fronzaglia, A., *Chem. Commun.* 547 (1965).
96. Tate, D. P., Augl, J. M., Ritchey, W. M., Ross, B. L., and Grosselli, J. G., *J. Amer. Chem. Soc.* **86**, 3261.
97. Strohmeier, W. and von Hobe, D., *Z. Naturforsch. B* **19**, 959 (1964).
98. Browning, J., Green, M., Penfold, B. R., Spencer, J. L., and Stone, F. G. A., *J. Chem. Soc., Chem. Commun.* 31 (1973); *J. Chem. Soc., Dalton Trans.* 97 (1974).
99. Baddley, W. H., *Chem. Commun.* 762 (1972).
100. Greatrex, R., Greenwood, N. N., and Pauson, P. L., *J. Organometal. Chem.* **13**, 533 (1968).
101. Kaska, W. C., and Kimball, M. E., *Inorg. Nucl. Chem. Lett.* **4**, 719 (1967).
102. See Heck (*5*), pp. 238–255.
103. Yamazaki, H., Aoki, K., Yamamoto, Y., and Wakatsuki, Y., *Abstr. 22nd Symp. Organometal. Chem. Jap.*, p. 208B (1974); *J. Amer. Chem. Soc.* **97**, 3546 (1975).
104. Hubert, A. J., and Dale, J., *J. Chem. Soc.* 3160 (1965).

105. Schrauzer, G. N., *Advan. Organometal. Chem.* **2**, 1 (1964).
106. Bird, C. W., "Transition Metal Intermediates in Organic Syntheses," pp. 1–29. Academic Press, New York, 1967.
107. Reppe, W., Schlichting, O., Klager, K., and Toepel, T., *Ann. Chem.* **560**, 1 (1948).
108. Schröder, G., "Cyclooctatetraen," pp. 12–15. Verlag Chemie, Weinheim, 1965.
109. Cope, A. C., and Fenton, J. *Amer. Chem. Soc.* **73**, 1195 (1951).
110. Craig, L. E., and Larrabee, L., *J. Amer. Chem. Soc.* **73**, 1191 (1951).
111. Nakamura, A., *Mem. Inst. Sci. Ind. Res., Osaka Univ.* **19**, 81 (1962); *Chem. Abstr.* **59**, 8786 (1963).
112. Meriwether, L. S., Colthup, E. C., Kennerly, G. W., and Reusch, R. N., *J. Org. Chem.* **26**, 5155 (1961); Meriwether, L. S., Leto, M. F., Colthup, E. C., and Kennerly, G. W., *J. Org. Chem.* **27**, 3930 (1962).
113. Brown, C. K., Georgion, D., and Wilkinson, G., *J. Chem. Soc. A* 3120 (1971).
114. Tohda, Y., Sonogashira, K., and Hagihara, N., unpublished result.
115. Tohda, Y., Sonogashira, K., and Hagihara, N., *J. Chem. Soc., Chem. Commun.* 54 (1975).
116. Suzuki, Y., and Takizawa, T., *J. Chem. Soc., Chem. Commun.* 837 (1972).
117. See Bird (*106*), pp. 174–191; Thompson, D. T., and Whyman, R., *in* "Transition Metals in Homogeneous Catalysis" (G. N. Schrauzer, ed.), p. 147. Dekker, New York, 1971.
118. Jautelat, M., and Ley, K., *Synthesis* 593 (1970); Otsuka, S., Nakamura, A., and Yamagata, T., presented at *Symp. Org. Synthetic Chem., 1971.*
119. Yamazaki, H., and Wakatsuki, Y., *Tetrahedron Lett.* 3383 (1973).
120. Bönnemann, H., *Angew. Chem.* **85**, 1024 (1973); Bönnemann, H., Brinkmann, R., and Schenkluhn, H. *Synthesis* 575 (1974).
121. Wakatsuki, Y., Yamazaki, H., and Iwasaki, H., *J. Amer. Chem. Soc.* **95**, 5781 (1973).
122. Otsuka, S., Nakamura, A., and Tani, K., *J. Organometal. Chem.* **14**, P30 (1968).
123. Otsuka, S., Tani, K., and Yamagata, T., *J. Chem. Soc., Dalton Trans.* 2491 (1973).
124. Lichtenberg, D. W., and Wojcicki, A., *J. Amer. Chem. Soc.* **94**, 8271 (1972).
125. Benaim, J., Merour, J. Y., and Roustan, J. L., *Compt. Rend. Acad. Sci., Ser. C* **272**, 789 (1972).
126. Clark, H. C., and Manzer, L. E., *J. Amer. Chem. Soc.* **95**, 3812 (1973); *Inorg. Chem.* **13**, 7996 (1974).
127. Racanelli, P., Pantini, G., Immirzi, A., Allegra, G., and Porri, L., *Chem. Commun.* 361 (1969).
128. Hewitt, T. G., and De Boer, J. J., *J. Chem. Soc. A* 817 (1971).
129. Kashiwagi, T., Yasuoka, N., Kasai, N., and Kakudo, M., *Chem. Commun.* 361 (1969); *Technology Rep. Osaka Univ.* **24**, 355 (1974).
130. Kadonaga, M., Yasuoka, N., and Kasai, N., *Chem. Commun.* 1597 (1971); Kashiwagi, T., Yasuoka, N., Kasai, N., and Kakudo, M., *Technol. Rep., Osaka Univ.* **24**, 1188 (1974).
131. Okamoto, K., Yasuoka, N., and Kasai, N., unpublished.
132. Yasuoka N., Morita, M., Kai, Y., and Kasai, N., *J, Organometal, Chem.* **90**, 111 (1975),
133. Okamoto, K., Kai, Y., Yasuoka, N., and Kasai, N., *J. Organometal. Chem.* **65**, 427 (1974).
134. Mason, R., and Rae, A. I. M., *J, Chem, Soc. A* 1767 (1970).
135. Kashiwagi, T., Yasuoka, N., Ueki, T., Kasai, N., Kakudo, M., Takahashi, S., and Hagihara, N., *Bull. Chem. Soc. Jap.* **41**, 296 (1968).

136. Otsuka, S., Nakamura, A., and Tani, K., *Kogyo Kagaku Zasshi* (*J. Chem. Soc. Jap., Ind. Chem. Sect.*) **70**, 2007 (1967).
137. Vrieze, K., Volger, H. C., Gronert, M., and Praat, A. P., *J. Organometal. Chem.* **16**, P19 (1969).
138. Vrieze, K., Volger, H. C., and Praat, A. P., *J. Organometal. Chem.* **21**, 467 (1970).
139. Otsuka, S., and Tani, K., unpublished.
140. For example see Weinstein, B., and Fenselau, A. H., *J. Chem. Soc.* 368 (1967); *J. Org. Chem.* **32**, 2278, 2988 (1967); Fischer, H., "The Chemistry of Alkenes" (S. Patai, ed.), p. 1025. Interscience, New York, 1964.
141. Dolbier, W. R., Jr., and Dai, S.-H., *J. Amer. Chem. Soc.* **92**, 1774 (1970).
142. Woodward, R. B., and Hoffmann, R., "The Conservation of Orbital Symmetry," pp. 163–166. Academic Press, New York, 1970.
143. Hoover, F. W., and Lindsey, R. V., *J. Org. Chem.* **34**, 3059 (1969).
144. Otsuka, S., Nakamura, A., Ueda, S., and Tani, K., *Chem. Commun.* 863 (1971).
145. Englert, M., Jolly, P. W., and Wilke, G., *Angew. Chem.* **83**, 84 (1971); *ibid.* **84**, 120 (1972).
146. Nakamura, A., *Bull. Chem. Soc., Jap.* **39**, 543 (1966); Nakamura, A., and Hagihara, N., *J. Organometal. Chem.* **3**, 480 (1965).
147. Otsuka, S., Nakamura, A., and Tani, K., *J. Chem. Soc. A* 2248 (1968).
148. Hughes, R. P., and Powell, J., *J. Organometal. Chem.* **20**, P17 (1969).
149. Ingrosso, G., Immirzi, A., and Porri, L., *J. Organometal. Chem.* **60**, C35 (1973).
150. Otsuka, S., Tani, K., and Nakamura, A., *J. Chem. Soc. A* 1404 (1969).
151. Pantini, G., Racanelli, P., Immirzi, A., and Porri, L., *J. Organometal. Chem.* **33**, C17 (1971).
152. Otsuka, S., Nakamura, A., and Minamida, H., *Chem. Commun.* 191 (1969).

High Nuclearity Metal Carbonyl Clusters

P. CHINI, G. LONGONI, and V. G. ALBANO

Istituto di Chimica Generale dell'Università
Milano, Italy

I. Introduction	285
II. Structural Data in the Solid State	286
III. Structural Data in Solution	306
IV. Syntheses	311
V. Methods of Separation	316
VI. Reactivity	317
A. Reduction	319
B. Oxidation	320
C. Ligand Substitution	322
D. Oxidative Addition	322
VII. Iron Derivatives	323
VIII. Ruthenium Derivatives	324
IX. Osmium Derivatives	325
X. Cobalt Derivatives	325
XI. Rhodium Derivatives	327
XII. Iridium Derivatives	332
XIII. Nickel Derivatives	333
XIV. Platinum Derivatives	334
XV. Bonding Theories	336
References	341

I
INTRODUCTION

In the last 5 years at least eight reviews concerning polynuclear metal carbonyls have been published (*16, 18, 34, 41, 86, 87, 91, 108*), and it may appear unlikely that any new material could be added at this time. Nevertheless, we have willingly accepted the invitation of the Editors to present a comprehensive review which, beside the material published up to June 1974, also summarizes the most relevant of our recent results, as yet only published as preliminary notes.

In order to present a fresh view, we have confined ourselves to compounds containing 5 or more metal atoms. The slow rate of publication in this area is mainly due to the number of steps required by this research, i.e., synthesis, crystallization, structural identification, and chemical characterization.

In 1943, Hieber and Lagally reported that the reaction of anhydrous rhodium trichloride with carbon monoxide at 80°C, under pressure, and in the presence of silver and copper as halogen acceptors, gave a black crystalline product which, on the basis of elemental analysis, was formulated as $Rh_4(CO)_{11}$ (75). The exact nature of this compound was established 20 years later by Dahl using three-dimensional X-ray analysis which led to its reformulation as $Rh_6(CO)_{16}$ (53). This discovery can be regarded as the birthday of the chemistry of high nuclearity clusters.

In 1962, Dahl had also structurally characterized $Fe_5(CO)_{15}C$, the first high nuclearity carbide (26). This compound was originally prepared in extremely low yields (0.5%) by the reaction of $Fe_3(CO)_{12}$ with substituted acetylenes, and, probably due to the peculiarity of this synthesis, was considered for some time much more a curiosity rather than being recognized as the precursor of today's large family of carbide–carbonyl clusters.

Finally the hexanuclear dianion $[Co_6(CO)_{15}]^{2-}$ was the first anionic high nuclearity cluster to be isolated (33). Its discovery in 1967 prompted extension of such investigations to other transition metals and originated the present chemistry of the high nuclearity anionic clusters.

More than fifty different examples of high nuclearity carbonyl clusters (HNCC) are presently known, all of which contain Group VIII transition metals (Table I). In post-transition metals the increased separation between the $(n-1)d$ and ns-np orbitals is probably responsible for the low stability of their bonds with the highly π-acidic carbon monoxide ligand; a number of high nuclearity clusters with less π-acidic ligands, such as tertiary phosphines (17, 50) or organic donor groups (51, 72), is known, however.

Approximate calculations on some of the more crowded clusters, such as $[Fe_4(CO)_{13}]^{2-}$, $Fe_5(CO)_{15}C$, and $Ru_6(CO)_{18}H_2$, show that, at the level of the carbon atoms, about 96% of the available surface is occupied (95). This figure seems very high particularly if one takes into consideration that the distribution of the carbonyl groups is not homogeneous. The high nuclearity clusters of the transition metals that precede Group VIII are, therefore, expected to be destabilized by steric crowding, although some carbides and mixed nitrosyl–carbonyl derivatives should be sterically possible.

II

STRUCTURAL DATA IN THE SOLID STATE

The main bond distances found in HNCC are reported in Table II, which has been divided into three sections, corresponding to the three

TABLE I

HIGH NUCLEARITY METAL CARBONYL CLUSTERS

$Fe_5(CO)_{15}C$ (26, 117)	$Co_6(CO)_{16}$ (1, 36)	$[Ni_5(CO)_{12}]^{2-}$ (95)
$[Fe_5(CO)_{14}C]^{2-}$ (79)	$[Co_6(CO)_{15}]^{2-}$ (2, 35)	$[Ni_3Cr_2(CO)_{16}]^{2-}$ (111)
$[Fe_6(CO)_{16}C]^{2-}$ (49, 117)	$[Co_6(CO)_{14}]^{4-}$ (3, 37)	$[Ni_3Mo_2(CO)_{16}]^{2-}$ (111)
	$[Co_4Ni_2(CO)_{14}]^{2-}$ (8, 44)	$[Ni_3W_2(CO)_{16}]^{2-}$ (111)
	$[Co_6(CO)_{15}C]^{2-}$ (9)	$[Ni_4Mo_2(CO)_{14}]^{2-}$ (111)
	$[Co_8(CO)_{18}C]^{2-}$ (13)	$[Ni_6(CO)_{12}]^{2-}$ (28)
		$[Ni_8(CO)_{14}H_2]^{2-}$ (95)
		$[Ni_9(CO)_{18}]^{2-}$ (95)
		$[Ni_{11}(CO)_{20}H_2]^{2-}$ (95)
$Ru_5(CO)_{15}C$ (59)	$Rh_6(CO)_{16}$ (38, 53)	
$Ru_6(CO)_{17}C$ (85, 113)	$Rh_6(CO)_{16-x}(L)_x$ (21, 84)	
$Ru_6(CO)_{16}(L)C$ (85)	$Rh_6(CO)_{14}(diene)$ (92)	
$Ru_6(CO)_{14}(arene)C$ (81, 103)	$Rh_4Co_2(CO)_{16}$ (100)	
$Ru_6(CO)_{18}H_2$ (48)	$[Rh_6(CO)_{15}]^{2-}$ (99)	
	$[Rh_6(CO)_{14}]^{4-}$ (39)	
	$[Rh_6(CO)_{15}X]^{-}$ (6, 45)	
	$[Rh_6(CO)_{14}X_2]^{2-}$ (45)	
	$[Rh_6(CO)_{15}C]^{2-}$ (7, 10)	
	$[Rh_7(CO)_{16}]^{3-}$ (4, 99)	
	$[Rh_7(CO)_{16}X]^{2-}$ (11, 101)	
	$Rh_8(CO)_{19}C$ (9, 12)	
	$[Rh_{12}(CO)_{30}]^{2-}$ (5, 40, 47)	
	$Rh_{12}(CO)_{25}(C_2)$ (13)	
	$[Rh_{13}(CO)_{24}H_{5-n}]^{n-}$ (13)	
	$[Rh_{15}(CO)_{28}(C)_2]^{-}$ (9)	
$Os_5(CO)_{15}C$ (59)	$Ir_6(CO)_{16}$ (96)	$[Pt_6(CO)_{12}]^{2-}$ (29)
$Os_5(CO)_{16}$ (58)	$[Ir_6(CO)_{15}]^{2-}$ (96)	$[Pt_9(CO)_{18}]^{2-}$ (29)
$Os_5(CO)_{15}H_2$ (60)	$[Ir_8(CO)_{22}]^{2-}$ (97)	$[Pt_{12}(CO)_{24}]^{2-}$ (29, 95)
$Os_5(CO)_{16}H_2$ (60)		$[Pt_{15}(CO)_{30}]^{2-}$ (29)
$Os_6(CO)_{18}$ (58, 104)		$[Pt_{18}(CO)_{36}]^{2-}$ (29, 95)
$Os_6(CO)_{18}H_2$ (60)		
$Os_7(CO)_{21}$ (58)		
$Os_7(CO)_{19}(C)H_2$ (60)		
$Os_8(CO)_{23}$ (58)		
$Os_8(CO)_{21}C$ (61)		

transition periods, in order to make a comparison of the data easier. Such comparison should be done with some caution since the data reported are mean values and since there are the usual uncertainties due to discrepancies among sterically equivalent interactions (packing effects, thermal motion, and crystal disorder). Moreover, because the expanded orbitals of the low-valent transition metals can suffer considerable dimensional variation on changing the electron density at the metal atoms, only large differences, or comparison between strictly related species, are meaningful.

TABLE II

BOND DISTANCES IN HIGH NUCLEARITY METAL CARBONYL CLUSTERS

Nuclearity	Formula	Idealized and crystallographic symmetry	d_{M-M} (Å)	d_{M-CO} (Å)			d_{C-O} (Å)			Ref.
				Terminal	Edge bridging	Face bridging	Terminal	Edge bridging	Face bridging	
(a) Iron, cobalt, and nickel										
5	$Fe_5(CO)_{15}C$	C_s	2.64	1.75	—	—	1.17	—	—	26
6	$[Fe_6(CO)_{16}C]^{2-}$	C_s	2.67	1.70	—	—	1.18	—	—	49
6	$[Co_6(CO)_{15}]^{2-}$	C_{3v}	2.51	1.74	1.90	—	1.15	1.17	1.19	2
6	$[Co_6(CO)_{14}]^{4-}$	C_i	2.50	1.70	—	2.00	1.17	—	1.21	3
6	$[Co_4Ni_2(CO)_{14}]^{2-}$	S_6	2.50	1.76	—	2.04	1.13	—	1.15	8
8	$[Co_8(CO)_{18}C]^{2-}$	D_{3d}	2.52	1.72	1.94	2.09	1.16	1.18	—	13
5	$[Ni_{15}(CO)_{12}]^{2-}$	D_2	2.36^a 2.81^b	1.76^a 1.86^b	1.84	—	1.16^a 1.04^b	1.13^a	—	95
5	$[Mo_2Ni_3(CO)_{16}]^{2-}$	C_{3v}	2.34^a 3.10^c	1.89^a	1.91^a	—	1.01^a	1.07^a	—	111
5	$[W_2Ni_3(CO)_{16}]^{2-}$	C_{2v}	2.34^a 3.10^c	1.85^a	1.82^a	—	1.05^a	1.14^a	—	111
6	$[Ni_6(CO)_{12}]^{2-}$	D_{3d}	2.38^d 2.77^e	1.75	1.90	—	1.13	1.17	—	28
9	$[Ni_9(CO)_{18}]^{2-}$	D_3	2.44^d 2.70^e	—	—	—	—	—	—	95

288   P. CHINI, G. LONGONI, AND V. G. ALBANO

(b) Ruthenium and rhodium

6	Ru$_6$(CO)$_{17}$C	C_{2v}	C_2	2.89	—	—	—	—	—	113
6	Ru$_6$(CO)$_{14}$(C$_9$H$_{12}$)C	C_s	C_s	2.88	—	—	—	—	—	103
6	Ru$_6$(CO)$_{18}$H$_2$	D_{3d}	C_i	2.91	1.90	—	1.14	—	—	48
6	Rh$_6$(CO)$_{16}$	T_d	C_2	2.78	1.86	2.17	1.16	—	1.20	53
6	[Rh$_6$(CO)$_{15}$I]$^-$	C_s	—	2.75	1.85	2.17	1.15	—	1.19	6
6	[Rh$_7$(CO)$_{16}$]$^{3-}$	C_{3v}	C_s	2.76	—	—	—	—	—	4
7	[Rh$_7$(CO)$_{16}$I]$^{2-}$	C_s	—	2.77, 2.93	1.82	2.19	1.17	1.14	1.17	11
7	[Rh$_{12}$(CO)$_{30}$]$^{2-}$	C_{2h}	C_i	2.79	1.87	2.19	1.15	1.17	1.19	5
12	[Rh$_{13}$(CO)$_{24}$H$_3$]$^{2-}$	C_s	—	2.80	—	—	—	—	—	13
13	[Rh$_6$(CO)$_{16}$C]$^{2-}$	D_{3h}	C_2	2.79	1.89	—	—	—	—	7
6	Rh$_8$(CO)$_{19}$C	C_1	—	2.81	1.91	2.06	1.13	1.15	1.15	12
8	Rh$_{12}$(CO)$_{25}$(C$_2$)	C_1	—	2.80	1.87	2.08	1.09	1.15	1.20	13
12	[Rh$_{15}$(CO)$_{28}$(C)$_2$]$^-$	C_{2v}	C_2	2.87	1.89	2.02	1.12	1.17	—	9

(c) Osmium and platinum

6	Os$_6$(CO)$_{18}$	C_{2v}	—	2.80	—	—	—	—	104
7	Os$_7$(CO)$_{21}$	C_{3v}	—	2.86	—	—	—	—	61, 104
6	[Pt$_6$(CO)$_{12}$]$^{2-}$	D_{3h}	C_2	2.66^d / 3.04^e	1.77	2.03	—	—	29
9	[Pt$_9$(CO)$_{18}$]$^{2-}$	D_3	C_2	2.66^d / 3.05^e	1.79	2.00	—	—	29
15	[Pt$_{15}$(CO)$_{30}$]$^{2-}$	C_3	—	2.66^d / 3.08^e	1.80	2.03	—	—	29

^{a} Values in the equatorial fragment.
^{b} Values for the apical groups.
^{c} M—Ni values.
^{d} Intratriangular values.
^{e} Intertriangular values.

TABLE III

CALCULATED VALUES FOR THE CAVITY OF REGULAR POLYHEDRA

Polyhedron[a]	Circumradius	Calculated hole (in Å) for $d = 1.4$ Å [b]	Comparison with some covalent radii (Å)
Tetrahedron	$1.225d$	0.315	H = 0.37
Square pyramid	$1.414d$	0.58	
Octahedron	$1.414d$	0.58	
Trigonal prism	$1.525d$	0.735	C = 0.77
Square antiprism	$1.64d$	0.90	
Triangular dodecahedron	$1.701d$	0.98	
Cube	$1.732d$	1.02	Si = 1.17

[a] Regular polyhedron edge = $2d$.
[b] Average value for Rh, the central element of Group VIII.

A peculiarity of the three-dimensional clusters, containing more-or-less regular polyhedra of metal atoms, is the existence of a central cavity whose dimensions, as shown in Table III, are a function of the particular geometry of the polyhedron. The existence of such a hole is confirmed by the formation of a large number of carbide derivatives.

The known structures of pentanuclear HNCC are based both on the square pyramid and on the trigonal bipyramid, and are illustrated in Fig. 1. The square pyramid of iron atoms in $Fe_5(CO)_{15}C$ (26) is essentially regular and corresponds to one-half of an octahedron. By contrast, the trigonal bipyramids found in the bis(triphenylphosphino)iminium (PPN) salts of anions $[Ni_5(CO)_{12}]^{2-}$ (95) and $[M_2Ni_3(CO)_{16}]^{2-}$ (M = Mo, W) (111) are not only elongated along the threefold axis but also show significant differences in the distances between the central $Ni_3(CO)_3(\mu_2\text{-}CO)_3$ triangle, of D_{3h} symmetry and the apical groups. These differences are on average 0.03 Å for apical $Ni(CO)_3$ groups of C_{3v} local symmetry, and 0.1 Å for apical $Mo(CO)_5$ groups of C_{4v} local symmetry. In the central triangle, all the Ni—C distances are comparatively long, and the C—O distances correspondingly short, indicating predominantly dative bonding.

Deformation and other bonding peculiarities are often found in HNCC, and our present ability merely to describe most of these phenomena gives an idea of the present state of the theory.

Preliminary data on the salt $[NEt_4]_2[Mo_2Ni_4(CO)_{14}]$ indicate a structure with D_{2h} symmetry based on an equatorial $Ni_4(CO)_6$ rhombus with two $Mo(CO)_4$ groups placed above and below the plane (111). It is probable that there are only terminal carbonyl groups.

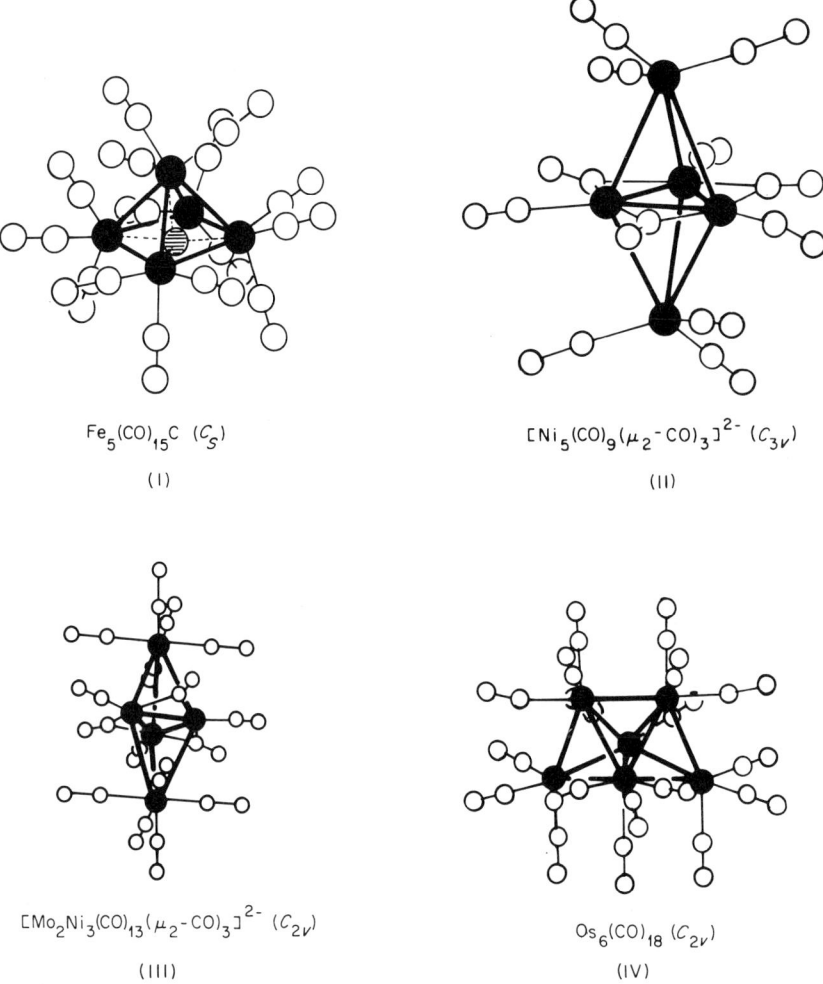

FIG. 1. Schematic molecular structures of the pentanuclear high nuclearity metal carbonyl clusters and of $Os_6(CO)_{18}$.

As shown by the structures so far discussed, and as previously pointed out by King (90), a triangular network of metal atoms is the most common basic unit in transition metal clusters, suggesting that bonding between metal atoms in a triangular network is not confined to the edges but can also occur within the triangle itself. This hypothesis is reasonable because some orbital overlap could still occur at the center of the triangles (1.155 times the metallic radius), whereas there can be little bonding interaction

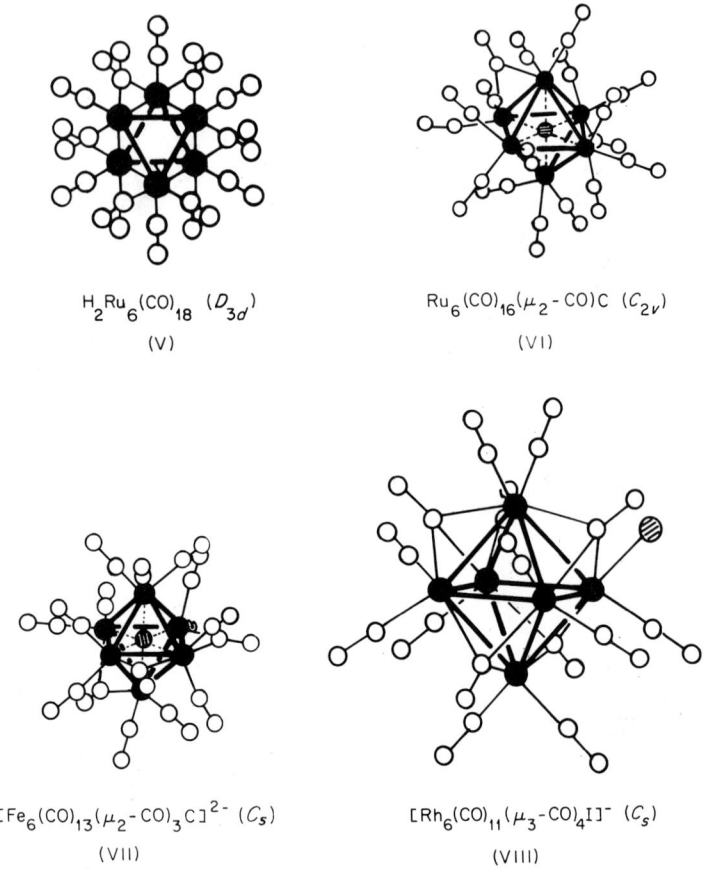

FIG. 2. Schematic molecular structures of some high nuclearity metal carbonyl clusters based on octahedra of metal atoms (section a).

through the center of a square, where the vertex–center distance is 1.414 times the metallic radius.

A good example of the tendency toward triangulated polyhedra is given by the bicapped tetrahedron present in $Os_6(CO)_{18}$ *(104)*. In this cluster (complex IV in Fig. 1), each of the two osmium atoms shared by the basal triangles is directly bonded to 5 other osmium atoms. In spite of theoretical considerations *(90)*, the presence of pentaconnected vertices is common in HNCC, indicating bonds of metallic type and implying considerable delocalization. The Os—Os distances orthogonal to the symmetry axis, between the atoms of greater metallic character (tetra- and pentaconnected vertices), are considerably shorter (2.74 Å as compared to 2.84 Å).

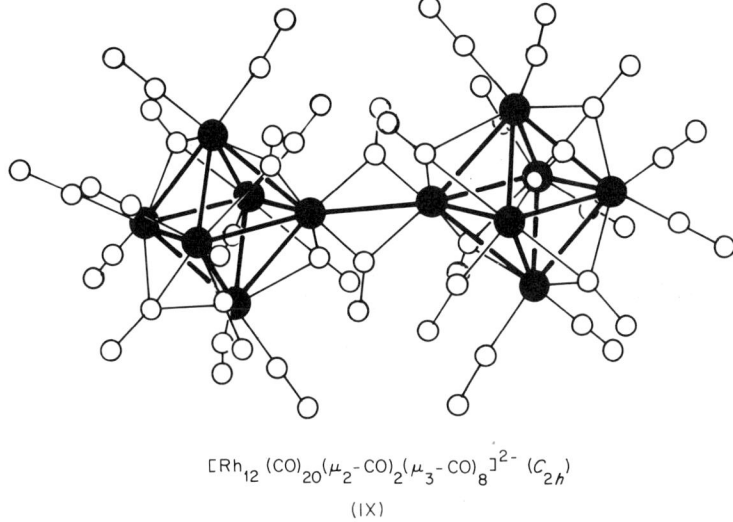

$[Rh_{12}(CO)_{20}(\mu_2\text{-}CO)_2(\mu_3\text{-}CO)_8]^{2-}$ (C_{2h})

(IX)

Fig. 3. Schematic representation of the molecular structure of the dianion $[Rh_{12}(CO)_{30}]^{2-}$.

The octahedron, another triangulated polyhedron, is the most common type in HNCC; about half of the compounds reported in Table I contain octahedra or deformed octahedra of metal atoms. The structures of some octahedral HNCC are reported in Figs. 2–4.

The high symmetry of octahedral $Ru_6(CO)_{18}H_2$ is illustrated in Fig. 2 by structure V (48). Tricoordination toward carbon monoxide is maintained in $Ru_6(CO)_{17}C$, structure VI (113), and in the analogous $Ru_6(CO)_{14}$ (mesitylene) C (103) by formation of a carbonyl bridge. The same tendency is also evident in structure VII, which has been found in the anion $[Fe_6(CO)_{16}C]^{2-}$ (49), although here the bridges are very unsymmetrical and one of the iron atoms is bonded to 4 carbon monoxides. All of the other octahedral clusters bearing sixteen carbonyl groups, and also some of their derivatives, have the much more symmetrical structure VIII. Typical examples of this stereochemistry are $Rh_6(CO)_{16}$ (53) and the anion $[Rh_6(CO)_{15}I]^-$ (6).

Comparison among the octahedral structures VI–XI (Figs. 2–4) shows that the number of CO ligands coordinated to each metal atom increases from three to five. This indicates that the metal–ligand and metal–metal systems of bonds are largely independent, a fact which is of considerable theoretical interest (see Section XV).

The formal coordination numbers of the metal atoms in these clusters

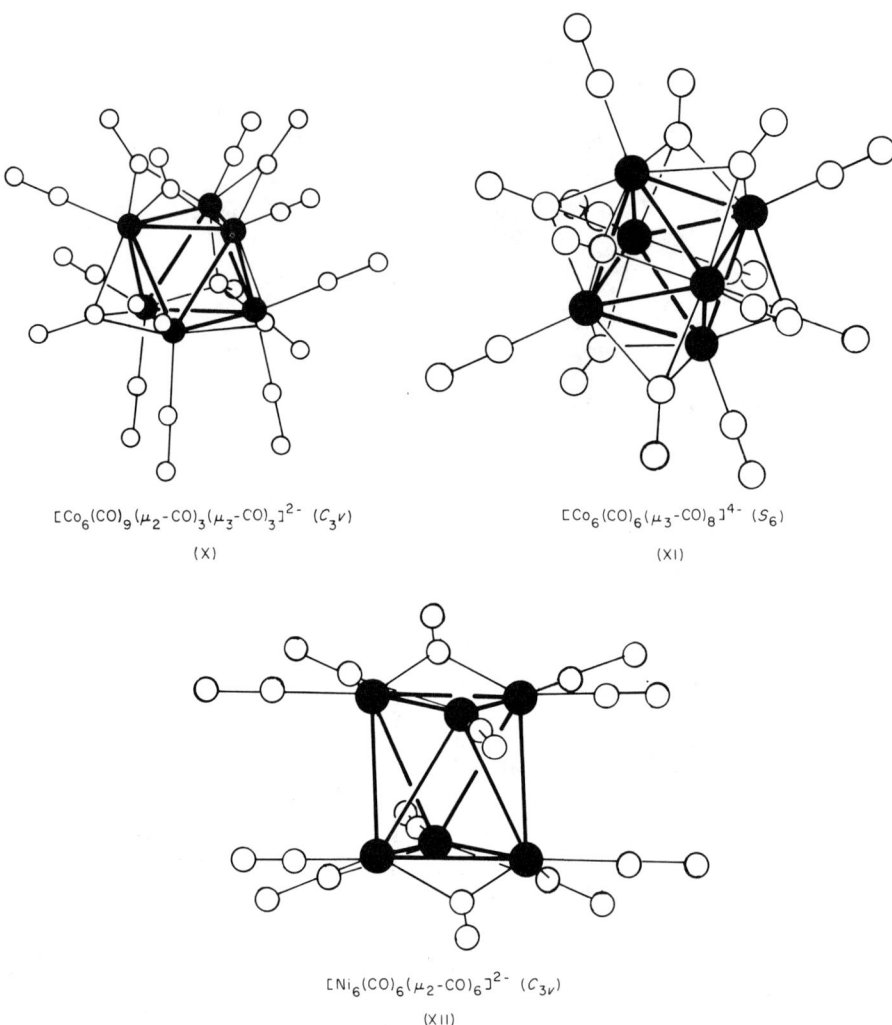

FIG. 4. Schematic molecular structures of some high nuclearity metal carbonyl clusters based on octahedra of metal atoms (section b).

are not only variable but are also unusually high (*34, 106*). However, their stereochemical significance cannot be compared with that usually accepted in simple compounds, because in HNCC part of the bonds are metallic in character and cannot be represented as simple electron pair bonds. The related unsymmetrical distribution of this high number of bonds along the directions of the quaternary axes of the octahedra is readily apparent in structure VIII, where there are two bonds (terminal CO's) pointing out from the cluster and six bonds (two bridging CO's and four metal–metal bonds) pointing toward the cluster. In structure XI the corresponding distribution is 1:8. These unsymmetrical distributions of the bonds obviously require the presence of some counterbalancing electron density in the proper directions.

The bond distances in species $Rh_6(CO)_{16}$ and $[Rh_6(CO)_{15}I]^-$ are very similar, the only exception being the terminal CO group accompanying the iodide ligand. The Rh—C distance is 0.02 Å greater, and C—O length 0.08 Å shorter, than in the remaining metal carbonyl fragments, in agreement with a locally lower backdonation.

Although the dioctahedral anion $[Rh_{12}(CO)_{30}]^{2-}$, structure IX of Fig. 3, exhibits mean values for the bonding interactions similar to those found in the two preceding species, an examination of the individual distances shows considerable deformations (5). For example, perpendicular to the twofold axis joining the two moieties, the equatorial planes of the octahedra are rectangularly deformed with mean edges 2.68 and 2.84 Å. These deformations have been tentatively explained in several ways (5, 40).

The infrared spectra of the solids suggest that $Co_6(CO)_{16}$ and $Ir_6(CO)_{16}$ are isostructural with $Rh_6(CO)_{16}$ (36, 96), and for the former compound this hypothesis is confirmed by the isomorphism of the crystals (1). The same structure is probably present in the dianion $[Rh_6(CO)_{14}(CN)_2]^{2-}$ (45).

Tetracoordination toward carbon monoxide is maintained in the $[Co_6(CO)_{15}]^{2-}$ dianion (complex X in Fig. 4) (2) and, probably, in the homologous anions $[Rh_6(CO)_{15}]^{2-}$ (99) and $[Ir_6(CO)_{15}]^{2-}$ (96).

The progressive lengthening of the Co—C distances (1.74–1.90–2.00 Å) for the sequence, *terminal–edge bridging–face bridging*, and the parallel increase in the C—O values (1.15–1.17–1.19 Å), are in agreement both with the increasing multicenter character and related steric request of the metal-carbonyl interaction, and with the lowering of the C=O stretching absorptions observed in the IR spectra (34).

Structure XI (Fig. 4) is common to the isoelectronic anions of the salts $[NMe_4]_2[Co_4Ni_2(CO)_{14}]$ (8), $K_4[Co_6(CO)_{14}]\cdot 6H_2O$ (3), and to the isomorphous $K_4[Rh_6(CO)_{14}]\cdot 6H_2O$ (39). In this structure, the number of CO ligands coordinated to each metal atom is increased to five. The decrease in negative charge on going from $[Co_6(CO)_{14}]^{4-}$ to $[Co_4Ni_2(CO)_{14}]^{2-}$ and the consequent lowering of back donation is experimentally observed in the lengthening of the M—C and the corresponding shortening of the C—O distances (about 0.05 Å for both values).

Figure 5 shows that, for the octahedral carbonylrhodates, there is, surprisingly, a linear relationship between the fraction of terminal carbonyl

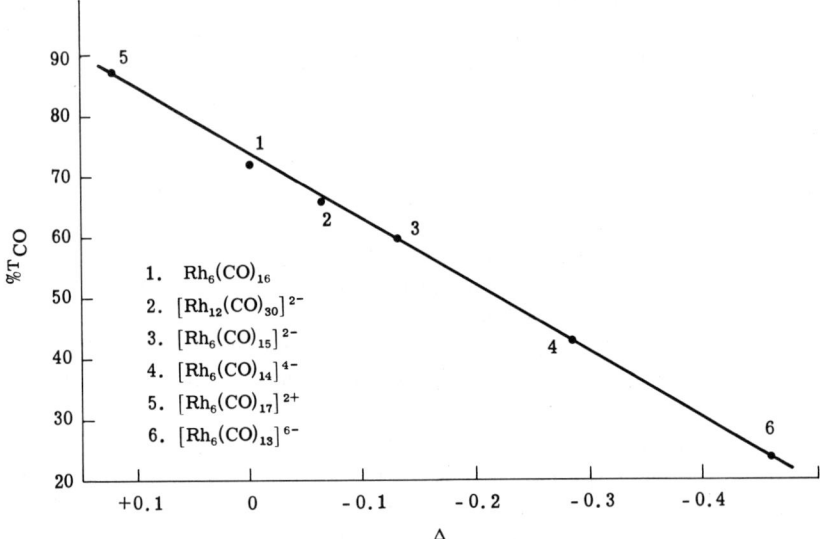

FIG. 5. Relationship between percent terminal CO and average negative charge per carbonyl group.

groups and the average negative charge per carbonyl group (74a). This plot not only proves that dissipation of negative charge is mainly responsible for the interconversion of terminal into bridging carbonyl groups but shows also that no significant deviations due to steric effects are present in this particular series of compounds. The same plot allows limited predictions to be made concerning other members of this family. For example, the hypothetical species $[Rh_6(CO)_{17}]^{2+}$ and $[Rh_6(CO)_{13}]^{6-}$ should have fifteen and three terminal groups, respectively.

A significant and unexpected tendency of the octahedral framework to degenerate into a trigonal antiprism can already be noticed in the distribution of the M—M distances of anions $[Co_6(CO)_{15}]^{2-}$, $[Co_6(CO)_{14}]^{4-}$, and $[Co_4Ni_2(CO)_{14}]^{2-}$; however, the best example of such a deformation is in the dianion $[Ni_6(CO)_{12}]^{2-}$ (structure XII in Fig. 4), which is correctly described as a trigonal antiprism with mean basal and interbasal Ni—Ni distances of 2.38 and 2.77 Å, respectively (28). The decrease in the intratriangular distances can be related to the presence of bridging carbonyl groups spanning all six edges of these faces. In fact, the carbonyl bridge represented as a polycentric bond generates *di per se* a bonding component between the metals (34).

Further twisting of the trigonal antiprism along the ternary axes generates a trigonal prism. Such peculiar packing has been found in the dianions $[Pt_3(CO)_6]_n^{2-}$ ($n = 2, 3, 4, 5$) (29, 95), which contain a repeated trigonal prismatic stacking of metal triangles along the pseudo threefold axis (structures XIII–XV in Fig. 6).

Why the platinum atoms in $[Pt_6(CO)_{12}]^{2-}$ prefer a prismatic arrangement is not clear. The antiprismatic packing found in $[Ni_6(CO)_{12}]^{2-}$ should be the more favorable in both cases as such packing requires less steric in-

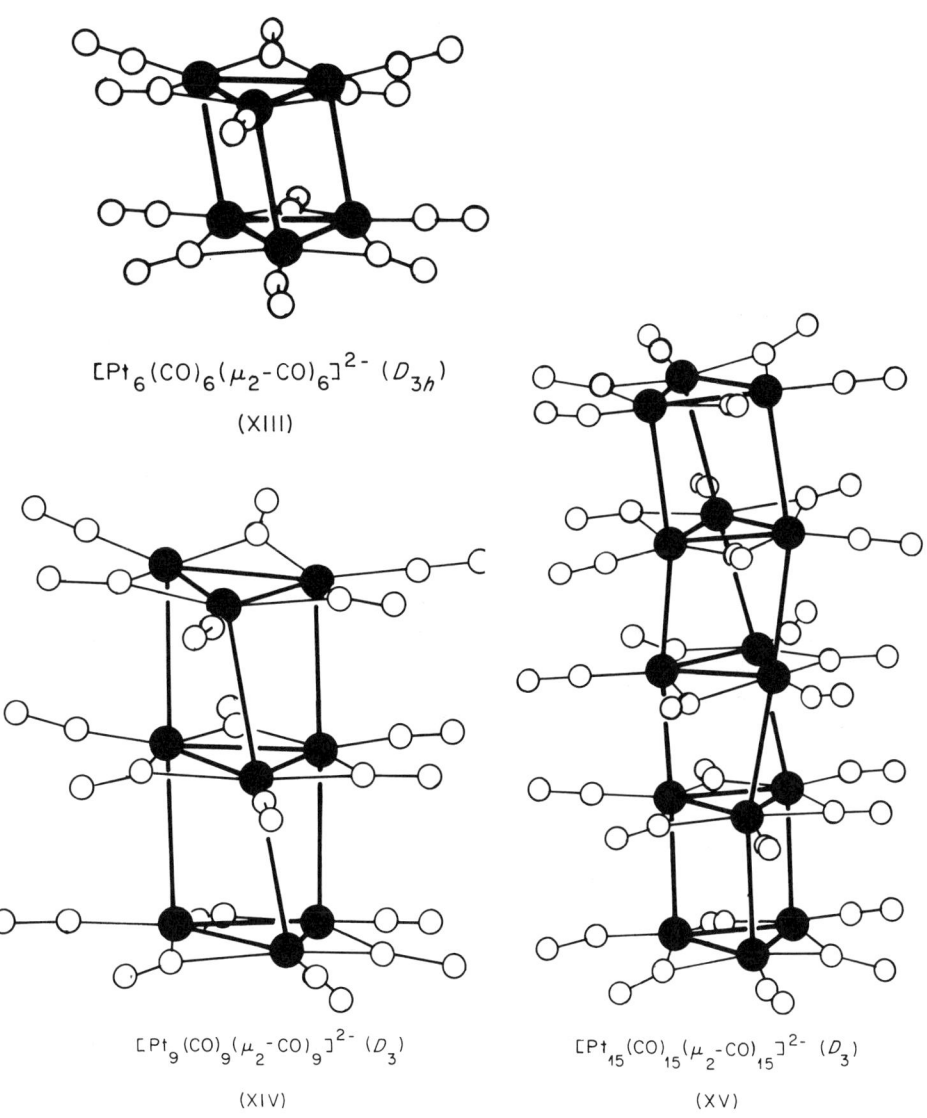

FIG. 6. Schematic representation of the molecular structures of some $[Pt_3(CO)_6]_n^{2-}$ ($n = 2, 3, 4, 5$) anions.

teractions between the two halves of the dianion and formally allows a gain of three M—M bonds. Probably the prismatic geometry results from a total energy minimization in which electronic reasons, such as the reluctance of platinum to use all of its valence orbitals, are the key factors.

In all of these platinum clusters the intratriangular metal–metal bonds (2.66 Å) are supported by carbonyl bridges and are significantly shorter than the intertriangular bonds (3.04–3.08 Å). The observed deviations from a regular prismatic stacking of platinum atoms, which include a translational sliding by ca. 0.51 Å along one edge in $[Pt_6(CO)_{12}]^{2-}$ and a top–bottom helical twisting along the pseudo threefold axis in $[Pt_9(CO)_{18}]^{2-}$ and $[Pt_{15}(CO)_{30}]^{2-}$ (by ca. 26° and 64.1°, respectively), may result from a tendency to minimize nonbonding repulsions, mainly among the carbonyl groups of adjacent layers. The same kind of structure has also been found in the dianion $[Pt_{12}(CO)_{24}]^{2-}$ (95) and is most probable for the dianion $[Pt_{18}(CO)_{36}]^{2-}$.

Dianion $[Ni_9(CO)_{18}]^{2-}$ has a structure very similar to that of the congener $[Pt_9(CO)_{18}]^{2-}$ but in this case the top–bottom helical twisting amounts to a total of ca. 60° (95). The intratriangular distances (2.44 Å) are once again much shorter than the intertriangular ones (2.70 Å).

Only three structures of heptanuclear clusters are known, all containing metal frameworks based on the monocapped octahedron. Figure 7 shows the structures of $[Rh_7(CO)_{16}]^{3-}$ (XVI) and of $[Rh_7(CO)_{16}I]^{2-}$ (XVII) (4, 11). Structure XVI formally derives from that of $[Rh_6(CO)_{15}]^{2-}$, represented by structure X in Fig. 4, by insertion of an $Rh(CO)^-$ unit into the unbridged lower face of the octahedron, whereas structure XVII derives from that of $Rh_6(CO)_{16}$ by insertion of an RhI^{2-} unit into one of the four unbridged faces. In both cases there would be subsequent formation of three edge carbonyl bridges along the new tetrahedral edges.

In the octahedral part of the two clusters, the average Rh—Rh distances are very similar (2.76 versus 2.77 Å). By contrast the average distances along the new tetrahedral edges differ considerably in the two cases (2.76 versus 2.93 Å), in agreement with the presence of one extra electron pair in the substituted dianion. This seems to be another example, as previously pointed out by Dahl (23), of the effect of an excess of electron density in MO's mainly antibonding with respect to the metals.

The capped octahedral structure of $Os_7(CO)_{21}$, which is isoelectronic with $[Rh_7(CO)_{16}]^{3-}$, has only been preliminarly reported (104); all the Os—Os distances are similar (61).

Despite the existence of 86-electron octahedral carbide clusters, such as $Ru_6(CO)_{17}C$, the dianion $[Rh_6(CO)_{15}C]^{2-}$ has 90 valence electrons and a trigonal prismatic array of metal atoms (structure XVIII in Fig. 8) (7). This fact could indicate that the three M—M bonds formally lost in the

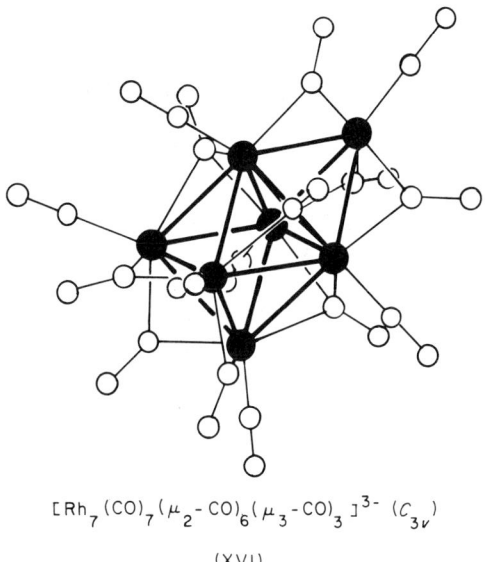

$[Rh_7(CO)_7(\mu_2\text{-}CO)_6(\mu_3\text{-}CO)_3]^{3-}$ (C_{3v})

(XVI)

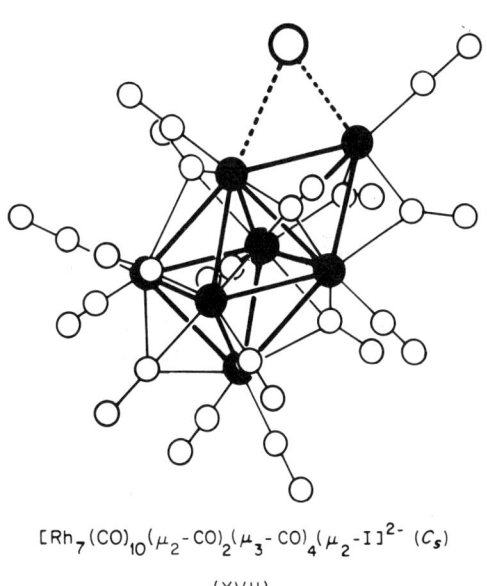

$[Rh_7(CO)_{10}(\mu_2\text{-}CO)_2(\mu_3\text{-}CO)_4(\mu_2\text{-}I)]^{2-}$ (C_s)

(XVII)

FIG. 7. Schematic representation of the molecular structures of the heptanuclear clusters.

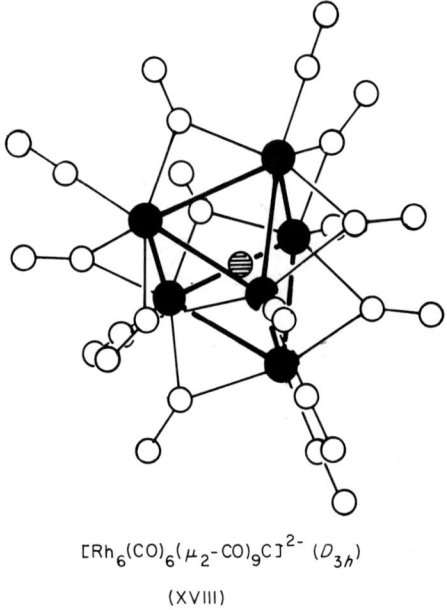

$[Rh_6(CO)_6(\mu_2-CO)_9C]^{2-}$ (D_{3h})

(XVIII)

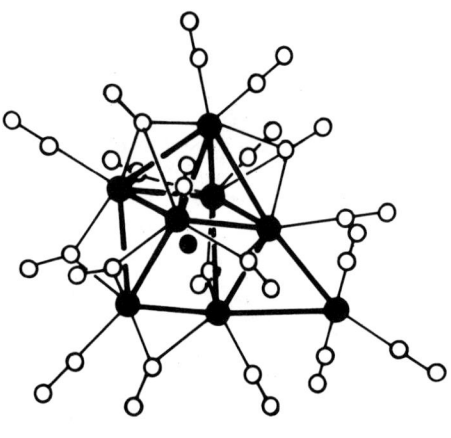

$Rh_8(CO)_{11}(\mu_2-CO)_6(\mu_3-CO)_2C$ (C_1)

(XIX)

FIG. 8. Schematic representation of different molecular structures in carbide high nuclearity metal carbonyl clusters.

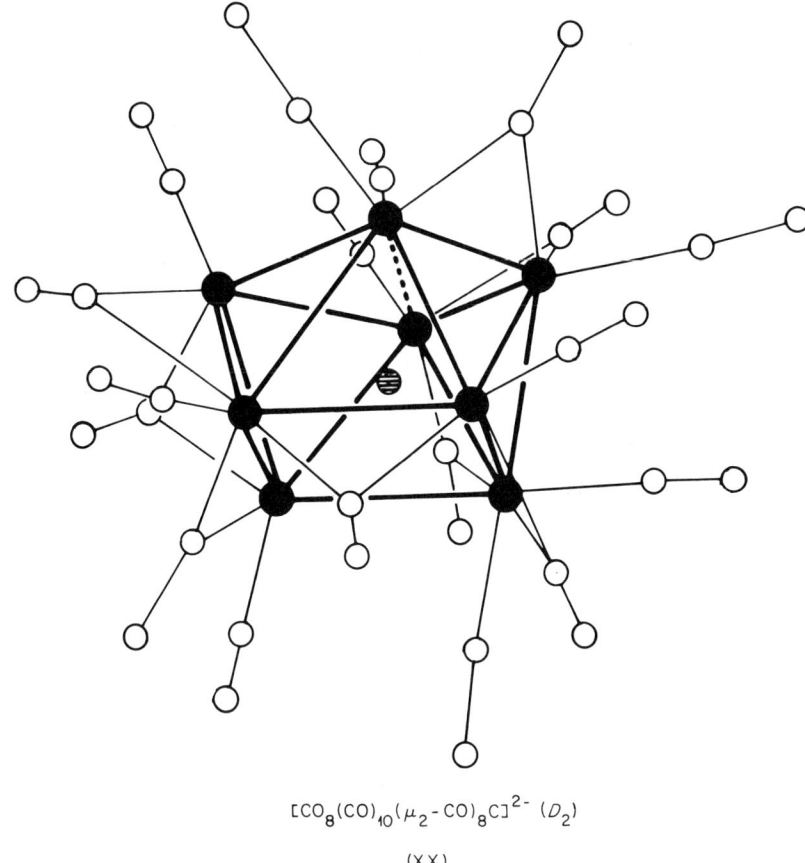

$[Co_8(CO)_{10}(\mu_2\text{-}CO)_8C]^{2-}$ (D_2)

(XX)

FIG. 8. (Continued.)

octahedron–trigonal prism transformation are energetically compensated for by bonding two more carbonyl groups and by allowing more space for the central carbide (see later). Such an interpretation appears reasonable because the $\bar{D}_{Rh-CO}$ (ca. 39 kcal mole^{-1}) (27) is expected to be almost twice as much as the $\bar{D}_{Rh-Rh}$ (ca. 27 kcal mole^{-1}).

The congener $[Co_6(CO)_{15}C]^{2-}$, whose trimethylbenzylammonium (NMe$_3$Bz) salt is isomorphous with the corresponding salt of $[Rh_6(CO)_{15}C]^{2-}$, appears to have the same structure (9).

In the carbide rhodium clusters, the Rh—C distances of the terminal carbonyl groups are significantly longer than those of the noncarbide derivatives, whereas the corresponding C—O distances are shorter (see Table II). This trend indicates a decrease in backbonding to the carbonyl

TABLE IV

EXPERIMENTAL VALUES OF THE CARBIDE CARBON RADIUS IN
HIGH NUCLEARITY METAL CARBONYL CLUSTERS

Polyhedron	Species	Experimental radius from $d_{M-C}-\frac{1}{2}d_{M-M}$ (in Å)	Ref.
Square pyramid	$Fe_5(CO)_{15}C$	1.89–2.64/2 = 0.57	*26*
Octahedron	$[Fe_6(CO)_{16}C]^{2-}$	1.91–2.67/2 = 0.57	*49*
	$Ru_6(CO)_{17}C$	2.05–2.89/2 = 0.61	*113*
	$Ru_6(CO)_{14}(C_9H_{12})C$	2.04–2.88/2 = 0.60	*103*
	$[Rh_{15}(CO)_{28}(C)_2]^-$	2.04–2.87/2 = 0.60	*9*
Half trigonal prism	$Co_3(CO)_9CX$	1.92–2.48/2 = 0.68	*109*
Trigonal prism	$[Rh_6(CO)_{15}C]^{2-}$	2.13–2.79/2 = 0.74	*7*
	$Rh_8(CO)_{19}C$	2.13–2.81/2 = 0.72	*12*
Square antiprism (distorted)	$[Co_8(CO)_{18}C]^{2-}$	1.99–2.52/2 = 0.73	*13*

groups and, in the particular case of $[Rh_6(CO)_{15}C]^{2-}$, is in agreement with the shift to higher frequencies of the carbonyl absorptions in the IR (see Section III). However in this case, the ^{13}C NMR of the carbide carbon atom falls at very low field 264.7 ppm (*10*) (see resonance D in Fig. 13) suggesting that the carbide carbon could be positively polarized.

The M—C carbide distances found in some HNCC are summarized in Table IV. It seems probable that in the smaller octahedral cavities (compare Tables III and IV) the positive charge on the carbide atom will become higher to allow the necessary contraction.

A trigonal prismatic basic unit, containing a carbide atom, is still present in $Rh_8(CO)_{19}C$ (structure XIX in Fig. 8) (*12*). The presence of a rhodium atom in an exceptional bonding situation, spanning just one edge of the original prism, explains the particular reactivity of this compound. Structure XIX does not possess any symmetry element, even considering the metal skeleton alone, and challenges the common opinion that clusters have high symmetry.

Dianion $[Co_8(CO)_{18}C]^{2-}$ is isoelectronic with $Rh_8(CO)_{19}C$ but presents a very different stereochemistry as shown in structure XX (Fig. 8) (*13*). Its geometry can be described as a deformed square antiprism with a carbide atom in the center of the polyhedron. This structure can be derived from that of the bicapped trigonal prism by stretching the common edge of the two capped square faces, as indicated in Fig. 8 by the dotted line. In such a process the loss of an M—M bond formally generates a square face. The idealized symmetry of a tetragonal antiprism is D_{4d}. However, since the cluster is stretched along one of the twofold axes, only D_2 symmetry is retained. One reason for this deformation is that the square antiprismatic cavity is too large to allow sufficient bonding interactions between the central carbon and the cobalt atoms, and the most favorable arrangement leads to the 4 cobalt atoms moving nearer to the central carbon.

The structure of the monoanion $[Rh_{15}(CO)_{28}C_2]^-$, which is one of the biggest known clusters, is reported in Fig. 9 (9). The metal atom framework can be described as a centered tetracapped pentagonal prism in which the bases and two side faces are capped. The pentagonal prism is not regular and the Rh—Rh distances between the two capped side faces are too long to be considered bonding distances (3.33 Å). In agreement with the high average value of all the other Rh—Rh distances (2.87 Å), the chemical reactivity of this compound is high. The polyhedron of $[Rh_{15}(CO)_{28}C_2]^-$ is chemically derived from that of $[Rh_6(CO)_{15}C]^{2-}$, and the presence of two different carbide atoms in octahedral cavities strongly suggests that the two octahedral fragments derive from the condensation and isomerization

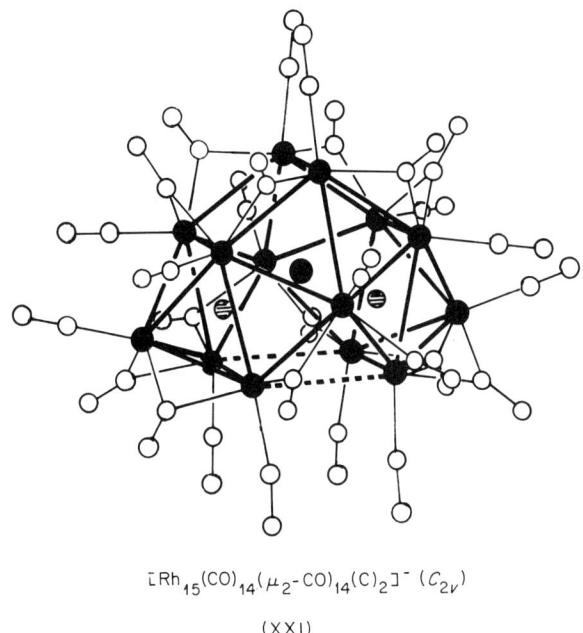

$[Rh_{15}(CO)_{14}(\mu_2\text{-}CO)_{14}(C)_2]^-$ (C_{2v})

(XXI)

FIG. 9. Schematic representation of the molecular structure of the anion $[Rh_{15}(CO)_{28}(C)_2]^-$.

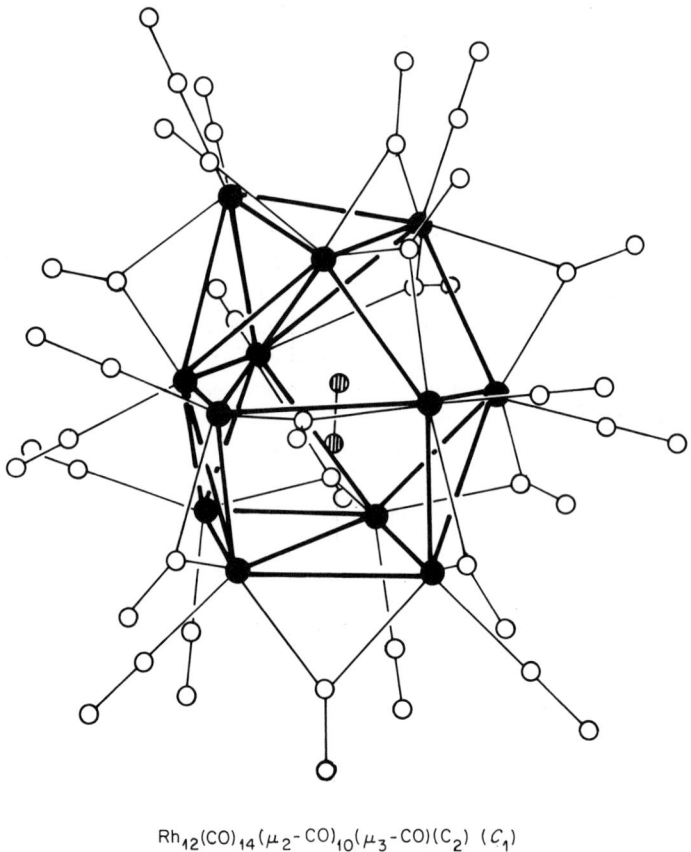

$Rh_{12}(CO)_{14}(\mu_2-CO)_{10}(\mu_3-CO)(C_2)$ (C_1)
(XXII)

FIG. 10. Schematic representation of the molecular structure of $Rh_{12}(CO)_{25}(C_2)$.

of two formerly prismatic units. The central rhodium atom is coordinated to 12 external rhodiums (d = 2.90 Å) and to the 2 carbides and, hence, can be considered to be in a true metallic situation. This special structure is probably due to the additional bonding contribution of the 2 carbide heteroatoms; such an energetic contribution seems able to stabilize cluster geometries that are likely to be unstable in a noncarbide analog.

The structure of $Rh_{12}(CO)_{25}(C_2)$, reported in Fig. 10, is a further example of carbide stabilization (13). This is the most irregular cluster as yet characterized; like $Rh_8(CO)_{19}C$, it has no symmetry element. This cluster is also derived chemically from the oxidation of $[Rh_6(CO)_{15}C]^{2-}$. The 2 central carbide atoms are definitely bonded together (1.47 Å) and lead to the classification of this derivative as an ethanide. The metal atoms, although irregularly bonded, can be considered as distributed on three different layers. Such a layered arrangement is similar to hexagonal packing with 2 atoms missing from the central layer and 1 extra atom in the lower layer (compare Figs. 10 and 11).

The tendency of the finite clusters toward close packing of metals is clearly evident in the structure of the anions $[Rh_{13}(CO)_{24}H_{5-n}]^{n-}$, ($n =$ 2, 3), reported in Fig. 11 (13). The central metal atom is dodecacoordinated, whereas the 12 rhodium atoms on the surface are pentacoordinated toward the other metal atoms and tricoordinated toward carbon monoxide. The average Rh—Rh distance is 2.81 Å without significant differences between internal and surface bonds. The high symmetry of the metal skeleton, D_{3h}, decreases to apparent C_s in the molecule; it has not been possible so far to determine the positions of the hydrogen atoms.

Structure XXIII (Fig. 11) corresponds to the smallest possible unit of

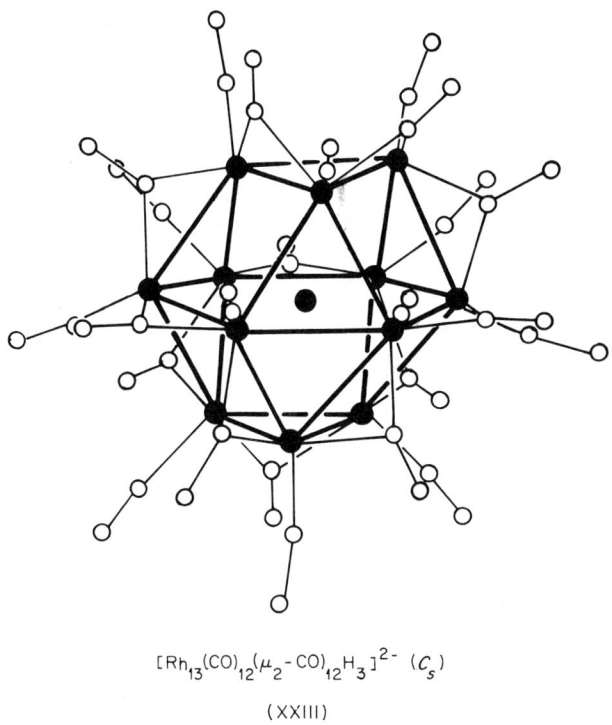

$[Rh_{13}(CO)_{12}(\mu_2\text{-}CO)_{12}H_3]^{2-}$ (C_s)

(XXIII)

FIG. 11. Schematic representation of the molecular structure of the dianion $[Rh_{13}(CO)_{24}H_3]^{2-}$.

a close-packed metal and conclusively demonstrates that the metal skeleton of a cluster can be regarded as a "round surface." Such a relationship between round surfaces and HNCC also provides a formal rationalization of the unsymmetrical distribution of bonds around the metal centers (as, for example, has been pointed out before for octahedral clusters). Formally, this is simply a consequence of the fact that on a surface the distribution of bonds can never be symmetrical relative to the plane of the surface itself. Therefore, important connections between HNCC and surface chemistry are expected, and HNCC could be a reliable model for carbon monoxide absorbed on metallic surfaces.

III

STRUCTURAL DATA IN SOLUTION

A simple technique that provides structural information in solution is infrared spectroscopy, particularly in the carbonyl stretching region (2100–1600 cm^{-1}) (25). Unfortunately this method does not generally provide sufficient information for a complete structural characterization of HNCC. This limitation is mainly due to the predominance of local coupling between carbonyl groups; in other words, local symmetry is mainly responsible for the spectrum. Nevertheless, the infrared spectrum very often allows an unequivocal decision about the presence of bridging carbonyl groups, although further differentiation between edge and face bridging groups is sometimes very difficult. Generally a difference of 150 to 210 cm^{-1} between the main absorptions of terminal and bridging carbonyl groups indicates edge bridges, whereas a difference of 210 to 250 cm^{-1} {but 400 cm^{-1} in [Co$_3$(CO)$_{10}$]$^-$ (66)} indicates face bridges. The assignment is complicated both by the possible presence of asymmetric bridges and by the broadness and consequent low resolution of these absorptions.

The salts of carbonyl metalates are generally soluble only in polar solvents, in which absorptions are broader due to the dipole interactions of the solvent. The spectra are also often considerably dependent on the particular cation and solvent, due to the formation of ion pairs. The largest differences have been observed on going from alkali salts to salts of the large tetrasubstituted ammonium and phosphonium cations; the latter generally provide simpler spectra that indicate minor formation of ion pairs. A similar trend is observed in going from less polar (e.g., THF) to more polar solvents (e.g., CH$_3$CN).

In the carbonylmetalates the main absorption of terminal carbonyl groups is fairly well related to the ratio of the number of metal atoms to

High Nuclearity Metal Carbonyl Clusters

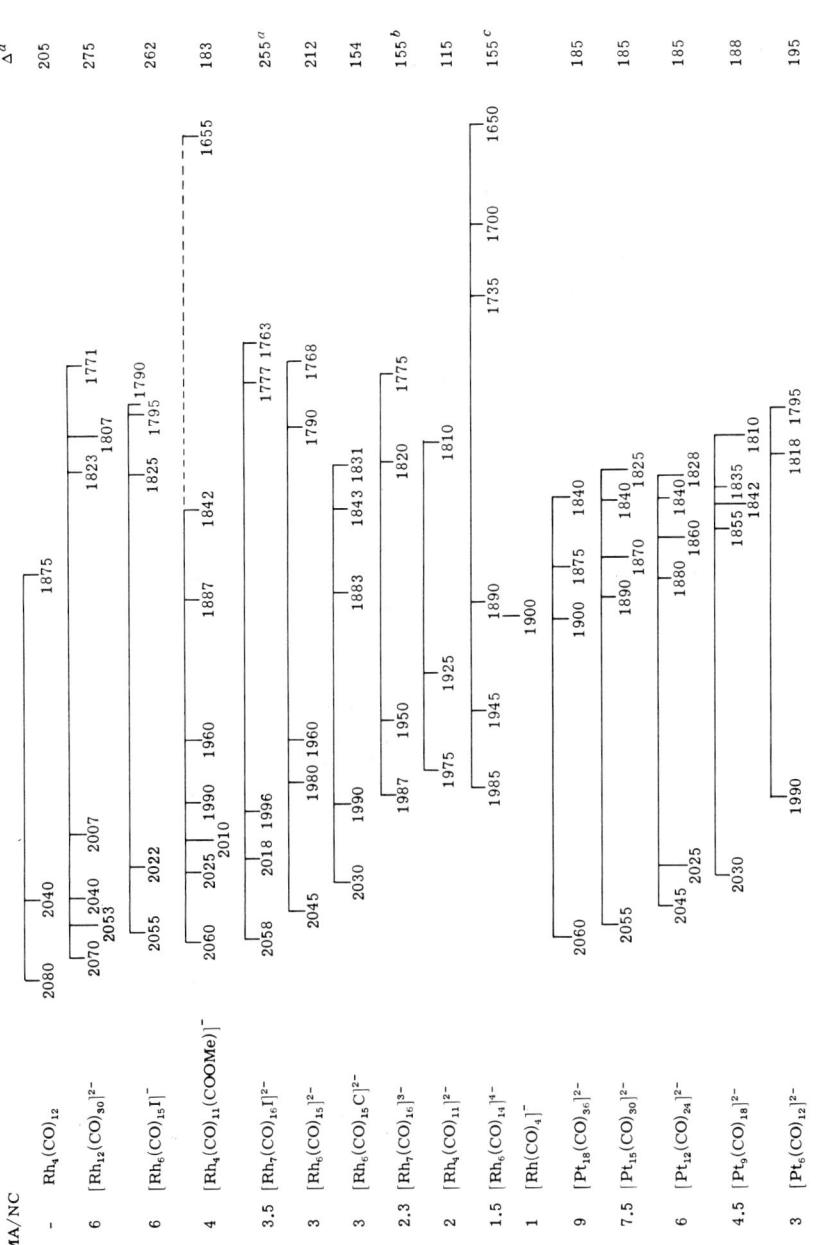

FIG. 12. Infrared spectra of rhodium and platinum polynuclear carbonyl anions in the carbonyl stretching region (tetrasubstituted ammonium or phosphonium salts in THF). MA/NC, ratio of number of metal atoms to the number of negative charges.

[a] CH_3CN. [b] $CH_3O-CH_2-CH_2OH$. [c] CH_3CN, potassium salt. [d] Separation between strongest terminal and bridging absorptions.

the number of negative charges (MA/NC) (*34*), as shown for the carbonyl-rhodates and platinates in Fig. 12. This is a very useful relationship, because it is often possible to interpolate with reasonable accuracy the MA/NC ratio of a new and unknown species just from the IR spectrum. Obviously, the reliability of this information is higher when comparing carbonylmetalates of the same metal, and when both local symmetries and the ratio between different types of carbonyl groups are similar (as with platinum). A surprisingly reasonable agreement is, however, often obtained in less related cases.

Introduction of a carbide atom results in a lowering of the stretching absorptions of the terminal groups (e.g., $[Co_6(CO)_{15}]^{2-}$ 1980 cm^{-1} and $[Co_6(CO)_{15}C]^{2-}$ 1975 cm^{-1}) in agreement with a positive character for this central heteroatom (*10*). Sometimes, however, the opposite shift is observed (compare $[Rh_6(CO)_{15}]^{2-}$ and $[Rh_6(CO)_{15}C]^{2-}$ in Fig. 12).

Very useful information on the actual structure in solution can be obtained from ^{13}C NMR spectra (*118*), particularly when using samples enriched in ^{13}CO (now available commercially). Enrichment can often be accomplished by direct exchange at 25°C and 1 atm. This ^{13}C NMR technique is particularly useful with the salts of polynuclear carbonylmetalates, because their solubilities in polar solvents (such as CD_3COCD_3) are generally sufficient even at low temperatures (-70°C). The use of a low temperature not only generally reduces the fluxional character of these molecules but often results in a marked sharpening of the signals when metal atoms of high nuclear magnetic spin are present (*119*).

A typical ^{13}C NMR spectrum is shown in Fig. 13 for the case of the $[Rh_6(CO)_{15}C]^{2-}$ dianion (*10*), whose structure is shown in Fig. 8. Here

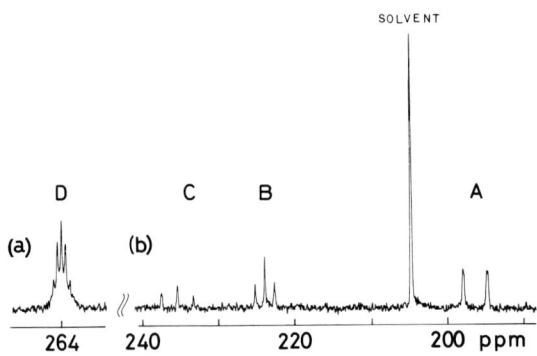

FIG. 13. The ^{13}C NMR spectrum of (a) $[Rh_6(CO)_{15}{}^{13}C]^{2-}$ (*ca.* 90% ^{13}C) and (b) $[Rh_6(^{13}CO)_{15}C]^{2-}$ (*ca.* 92% ^{13}CO) at -70°C in perdeuteroacetone solution in the presence of Cr(acac)$_3$.

the coupling with the rhodium atoms ($I = \frac{1}{2}$) splits the terminal carbonyl resonance into a doublet (A), and the two different sets of bridging carbonyl groups into two triplets (B and C). Similarly, it has also been possible to show that at low temperature anions $[Rh_{12}(CO)_{30}]^{2-}$, $[Rh_7(CO)_{16}]^{3-}$, $[Rh_6(CO)_{15}I]^-$, and $[Rh_6(CO)_{14}]^{4-}$ all maintain the structure found in the solid state (47, 74, 74a).

Generally absorptions at about 180 to 210 ppm [low field from tetramethylsilane (TMS)] are characteristic of terminal carbonyls, whereas those at about 220 to 260 ppm are typical of bridging carbonyl groups.

Figure 14 confirms that the main variable responsible for the ^{13}CO chemical shift toward low-field positions is the net partial negative charge that can be transferred to the carbonyl group (74). This has previously been shown for monomeric carbonyl derivatives (18a).

The main limitation of the method derives from the fluxionality of the carbonyl groups, which, on the other hand, is itself a potential source of basic information.

The low-temperature fluxional behavior of the carbonyl groups can be ascribed to the nucleophilicity of the delocalized skeleton electron density, which can accumulate on particular metal centers giving rise to intermediate carbonyl bridges. Fluxionality is, therefore, expected to increase on increasing the negative charge, or on decreasing the MA/NC ratio. This

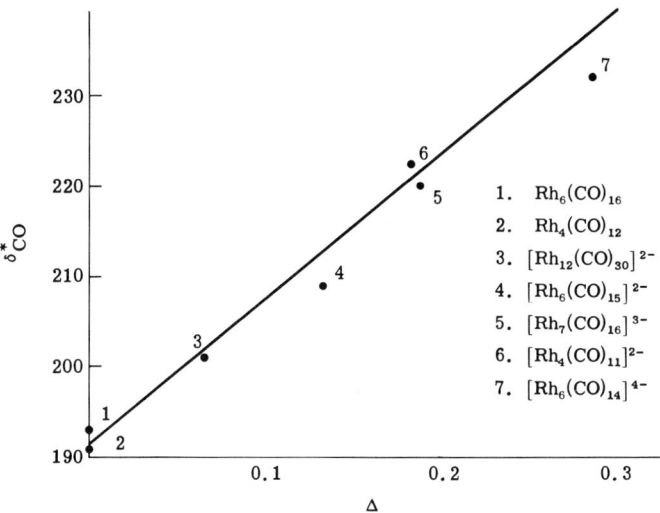

FIG. 14. Relationship between weighted average of ^{13}CO chemical shift (δCO*) and fraction of negative charge per carbonyl group (Δ) in rhodium high nuclearity metal carbonyl clusters.

TABLE V

COMPARISON OF FLUXIONAL BEHAVIOR AND REACTIVITY
OF SOME RHODIUM CARBONYL CLUSTERS

No.	Compound	Temp. (°C)	Behavior	ppm	Relative "reactivity"
1	$Rh_4(CO)_{12}$	+63	Fluxional	189.5 (quintet)[a]	
					1 > 2[b]
2	$Rh_6(CO)_{16}$	+70	Not fluxional	231.5 (quadruplet) 180.1 (doublet)	
3	$[Rh_6(CO)_{15}]^{2-}$	−69	Fluxional	209.2 (septet)	
					3 > 4[c]
4	$[Rh_7(CO)_{16}]^{3-}$	−70	Not fluxional	254 (broad) 229.4, 218.05 (triplets) 206.4, 205.7, 198.2 (doublets)	
5	$[Rh_6(CO)_{15}I]^{-}$	−30	Not fluxional	ca. 244–232 (broad) 187.8, 184.3, 183.4, 181.0 (doublets)	6 > 5[d]
6	$[Rh_7(CO)_{16}I]^{2-}$	−31	Fluxional	218.5 (octet)	—

[a] Data from Cotton et al. (55).
[b] Reactivity toward H_2 + olefins, H_2O, Na_2CO_3 + CH_3OH, H_2SO_4 + CH_3CN.
[c] Reactivity toward OH^- + CO, $[Rh(CO)_4]^-$, $Rh_2(CO)_4I_2$.
[d] Reactivity toward CO and H_2O.

is not the whole story, however, as differences in conformational energies, due to the different geometrical distribution of the carbonyl groups, could also affect such processes. Although only limited information is presently available on the fluxionality of HNCC (74), a low activation energy concerted mechanism, involving formation and breaking of bridges, seems generally reasonable. Several similar mechanisms have already been discussed in the literature for less complicated clusters (55, 64).

An empirical relationship between fluxionality and "reactivity" is presently emerging (Table V) in agreement with the common observation of the ease in changing the original distribution of the ligands and in producing vacancies in the barrier of carbonyl groups.

Although information on the force constants of metal–metal bonds can be obtained from Raman spectra (115), no study of HNCC has yet been reported, probably because these clusters are highly colored and readily decompose in the laser beam. Comparison between several colorless mononuclear carbonyls and the strong colors of HNCC shows that the metal–

metal bond is a chromophore, probably because of the decreased separation between the frontier orbitals. The obvious limit to this separation is apparent in the continuous absorption found in pure metals (41). Attempts to use the electronic spectra of HNCC for identification purposes have given poor results, mainly due to the broad character of these absorptions (95).

IV
SYNTHESES

Condensation reactions, which are at the very heart of the synthetic methods for HNCC, are expected to be subject to basic energetic conditions. The available experimental data show that condensation reactions such as

$$2Co_2(CO)_8 \longrightarrow Co_4(CO)_{12} + 4CO \qquad \Delta H \simeq 33 \text{ kcal} \quad (62) \qquad (1)$$

$$3Rh_4(CO)_{12} \longrightarrow 2Rh_6(CO)_{16} + 4CO \qquad \Delta H \simeq 64 \text{ kcal} \quad (27) \qquad (2)$$

are clearly endothermic. The positive values of ΔH result from the substitution of stronger metal-to-carbon monoxide bonds by weaker metal–metal bonds and agree with the generally favorable effect of a temperature increase on the condensation. Conversely, the opposite process, the addition of carbon monoxide and the resulting degradation of the cluster, is known to be favored at low temperature (99).

Estimation of these enthalpy variations would require some knowledge of the average bond energies $\bar{D}_{M-CO}$ and $\bar{D}_{M-M}$. Recently, an impressive series of standard enthalpies of polynuclear metal carbonyls has been determined (52a), but the problem of dividing these figures into "reasonable" bond energy terms that agree with the known chemical facts is still open. At present we have no direct reliable knowledge of metal–metal bond energies (73) although the limiting low value could be given by

$$\bar{D}_{M-M} = \frac{\Delta H_f^0 M_{(g)}}{6}$$

which is based on the standard atomization energies of metals having close packed structures.

Unfortunately the situation concerning metal-to-carbon monoxide bond energies is worse. The first obvious complication is the occurrence of different types of bonding of the carbon monoxide (terminal, edge, and face bridging), although there is experimental evidence that, as a first approximation and assuming constancy of the metal–metal energies, such a difference could be ignored (107). More difficult to ignore is the expected change in average energies which should take place on substantially changing the carbonyl group-to-metal ratio, owing to the high π-acidity of carbon monoxide. Therefore, the known values of $\bar{D}_{M-CO}$, (30, 52), which refer to low nuclearity species rich in carbonyl groups, are probably significantly smaller ($\sim 10\%$?) than the real $\bar{D}_{M-CO}$ terms of the HNCC.

Experimental data show that the difference $\bar{D}_{M-CO} - \bar{D}_{M-M}$ decreases on descending a subgroup, that is, the $\bar{D}_{M-M}$ term increases more rapidly than $\bar{D}_{M-CO}$. In fact, of the following analogous pairs of compounds, only the first member of each pair is degraded by carbon monoxide (in solution at 25°C and 1 atm):

(a) $Fe_3(CO)_{12}$ — $Ru_3(CO)_{12}$

(b) $Co_4(CO)_{12}$ — $Rh_4(CO)_{12}$

(c) $Co_6(CO)_{16}$ — $Rh_6(CO)_{16}$

(d) $[Co_6(CO)_{15}]^{2-}$ — $[Rh_6(CO)_{15}]^{2-}$

(e) $[Co_6(CO)_{15}C]^{2-}$ — $[Rh_6(CO)_{15}C]^{2-}$

(f) $[Ni_6(CO)_{12}]^{2-}$ — $[Pt_6(CO)_{12}]^{2-}$

(g) $[Ni_9(CO)_{18}]^{2-}$ — $[Pt_9(CO)_{18}]^{2-}$

This difference in behavior reflects different enthalpic conditions, because for each pair one could assume very similar contributions of the ΔS term (although in the a, f, and g pairs there is some approximation since the species are not strictly isostructural). Moreover, the ready exchange between some species, such as $[Rh_6(CO)_{15}]^{2-}$, and ^{13}CO shows that kinetic factors may probably not be responsible for this behavior.

The relative increase in $\bar{D}_{M-M}$ in descending a subgroup is of general synthetic significance, because, whereas synthesis of HNCC of first-row transition metals is generally possible only in the absence of carbon monoxide, in the second and third rows there is less dependence on carbon monoxide pressure. In the first row it is, therefore, necessary to start from preformed carbonyl species and to condense them in a second stage, whereas in the second and third rows reduction to carbonyl species and condensation to HNCC can often be carried out in one step. This direct preparative method is illustrated by the nearly quantitative reactions (95, 102):

$$7K_3RhCl_6 + 48KOH + 28CO \xrightarrow[\text{MeOH}]{25°C, 1\text{ atm}} K_3[Rh_7(CO)_{16}]$$
$$+ 42KCl + 12K_2CO_3 + 24H_2O \quad (3)$$

$$15Na_2PtCl_6 + 84NaOH + 61CO \xrightarrow[\text{MeOH}]{25°C, 1\text{ atm}} Na_2[Pt_{15}(CO)_{30}]$$
$$+ 90NaCl + 22NaHCO_3 + 9CO_2 + 31H_2O \quad (4)$$

In the preparation of the HNCC of rhodium, decomposition by carbon monoxide always becomes a limiting condition when the ratio MA/NC is low. This is shown by the behavior of the hexanuclear clusters (38, 99, 102):

$$2Rh_6(CO)_{16} + 4CO \xrightarrow{60°C, 1\ atm} 3Rh_4(CO)_{12} \quad (5)$$

$$5[Rh_6(CO)_{15}]^{2-} + 17CO \underset{25°C, 1\ atm}{\overset{-70°C, 1\ atm}{\rightleftarrows}} 2[Rh_{12}(CO)_{34}]^{2-} + 6[Rh(CO)_4]^- \quad (6)$$

$$[Rh_6(CO)_{14}]^{4-} + 5CO \xrightarrow{25°C, 1\ atm} [Rh_4(CO)_{11}]^{2-} + 2[Rh(CO)_4]^- \quad (7)$$

Neutral $Rh_6(CO)_{16}$ is perfectly stable under a carbon monoxide atmosphere at 25°C, and the dianion $[Rh_6(CO)_{15}]^{2-}$ is still stable at 25°C (although degraded at $-70°C$), but the tetraanion $[Rh_6(CO)_{14}]^{4-}$ is readily demolished under the same conditions. The interpretation of this behavior is that the term $\bar{D}_{M-CO}$ is very dependent on the MA/NC ratio, due to the related change in back donation. Probably this is a very general effect and most of the HNCC are expected to become increasingly unstable toward carbon monoxide on increasing the reduction of the cluster.

On the basis of mechanism, condensation processes can be divided into two broad categories: (a) reactions induced by coordinatively unsaturated species; and (b) reactions between coordinatively saturated species in different oxidation state (redox condensation).

Among the usual methods for generation of coordinatively unsaturated species (34), only pyrolisis has been applied to the synthesis of HNCC. The main examples are in ruthenium and osmium chemistry (58):

$$Os_3(CO)_{12} \xrightarrow[12\ hr]{195°-200°C} Os_4(CO)_{13} + Os_5(CO)_{16} + Os_6(CO)_{18} +$$
$$Os_7(CO)_{21} + Os_8(CO)_{23} + Os_8(CO)_{21}C \quad (8)$$

Pyrolisis of a $Rh_4(CO)_{12}$ solution at about 60°–80°C represents one of the best methods for obtaining pure $Rh_6(CO)_{16}$, whereas formation of the metals is observed at the high temperatures required for the thermal decomposition of $Co_4(CO)_{12}$ ($\sim$100°C) and $Ir_4(CO)_{12}$ ($\sim$160°C). In all of these cases, photochemical methods (122) could both significantly improve the selectivity of the syntheses and allow the isolation of new HNCC not accessible by thermal routes.

The relevance of redox condensation, which by contrast generally requires very mild conditions, began to emerge slowly after 1965 (15, 110) when Hieber and Shubert reported the first example of such a reaction (78):

$$[Fe_3(CO)_{11}]^{2-} + Fe(CO)_5 \xrightarrow[THF]{25°C} [Fe_4(CO)_{13}]^{2-} + 3CO \quad (9)$$

Here we mention only some recent applications of similar pairs of reagents to the synthesis of HNCC, such as (13, 102)

$$[Rh_6(CO)_{15}]^{2-} + Rh_6(CO)_{16} \xrightarrow[THF]{25°C} [Rh_{12}(CO)_{30}]^{2-} + CO \quad (10)$$

$$2[Co_6(CO)_{15}C]^{2-} + Co_4(CO)_{12} \xrightarrow[Et_2O]{25°C} 2[Co_8(CO)_{18}C]^{2-} + 6CO \quad (11)$$

since more examples are presented later.

Redox condensation is not restricted to the reaction of a carbonylmetalate with a neutral carbonyl, and it is possible to condense the carbonylmetalate with other simple cationic species; for example (35, 102),

$$[Rh(CO)_2(CH_3CN)_2]^+ + [Rh_6(CO)_{14}]^{4-} \xrightarrow[CH_3CN]{25°C} [Rh_7(CO)_{16}]^{3-} + 2CH_3CN \quad (12)$$

$$7[Co(EtOH)_x][Co(CO)_4]_2 \xrightarrow[vacuum]{60°C, EtOH} 3[Co(EtOH)_x][Co_6(CO)_{15}] + 11CO \quad (13)$$

Redox condensation between anionic and cationic species is often highly dependent on the medium, and the use of less basic solvents of high steric requirements, which do not coordinate strongly to the cationic center, may result in a marked improvement in the ease of condensation (35).

The following are further examples of redox condensations, showing the general possibility of such a reaction between species in different oxidation states (99, 101):

$$[Rh_6(CO)_{15}]^{2-} + [Rh(CO)_4]^- \xrightarrow[THF]{25°C} [Rh_7(CO)_{16}]^{3-} + 3CO \quad (14)$$

$$2[Rh_6(CO)_{15}]^{2-} + Rh_2(CO)_4I_2 \xrightarrow[THF]{25°C} 2[Rh_7(CO)_{16}I]^{2-} + 2CO \quad (15)$$

The great utility of redox condensation is due both to the ease with which the reactions take place, probably because of the redox character of the reactions themselves, and to the high number of possible combinations of reagents. This number is much higher than suspected at first glance because it is often unnecessary to start from a preformed pair of reagents. It is sufficient to generate *in situ* a second oxidation state by addition of a suitable reducing or oxidizing agent. Probably most of the syntheses of HNCC, and generally of polynuclear carbonylmetalates (34), in reducing conditions are of this type. A detailed discussion of some redox condensations induced by simple reduction or oxidation is presented for rhodium and nickel in Sections XI and XIII, respectively.

In the HNCC carbides the presence of a central carbide atom is expected to contribute significantly to the bonding energy of the cluster and, therefore, to result in less dependence on the partial pressure of carbon monoxide. Unfortunately, the present stage of our understanding of these synthetic reactions is very primitive.

The most common synthetic method (Fe, Ru, Os) is via pyrolysis of noncarbide metal carbonyl derivatives. It seems probable that the high temperature (140°–260°C) is not only responsible for formation of coordinatively unsaturated species and related condensations but also for the necessary disproportionation of carbon monoxide (58):

$$2CO \dashrightarrow C + CO_2 \quad \Delta G^0_{298} = -28.64 \text{ kcal mole}^{-1} \quad (16)$$

An example of such a reaction is (85)

$$Ru_3(CO)_{12} \xrightarrow[\text{6 hr, H}_2\text{O}]{n\text{-Bu}_2\text{O, 142°C}} \underset{53\%}{Ru} + \underset{10\%}{Ru_4(CO)_{12}H_4} + \underset{3\%}{Ru_4(CO)_{13}H_2} + \underset{30\%}{Ru_6(CO)_{17}C} \quad (17)$$

A different approach, based on the use of carbon halides, has been discovered more recently for cobalt and rhodium HNCC carbides, but its extension to other metals has not yet been successful. For cobalt the reaction is conveniently carried out in two different steps; first, the well-known chloromethynyl derivative, $Co_3(CO)_9CCl$ is prepared (63),

$$5Co_2(CO)_8 + 3CCl_4 \xrightarrow[\text{CCl}_4]{40°C} 2Co_3(CO)_9CCl + 4CoCl_2 + 22CO +$$
$$\tfrac{1}{2}Cl_2C{=}CCl_2 \quad (\text{yield} \sim 90\%) \quad (18)$$

and then the transformation into a carbide is completed via the tetracarbonylcobaltate (-1) anion (9),

$$Co_3(CO)_9CCl + 3Na[Co(CO)_4] \xrightarrow[25°C]{i\text{-Pr}_2\text{O}} Na_2[Co_6(CO)_{15}C] + NaCl + 6CO \quad (19)$$

Otherwise oxidation of the $[Co(CO)_4]^-$ anion to $Co_2(CO)_8$ by the CCl_4 is observed (13).

For rhodium the reaction can be conveniently carried out in one step,

$$6[NMe_3Bz][Rh(CO)_4] + CCl_4 \xrightarrow[25°C]{i\text{-PrOH}} [NMe_3Bz]_2[Rh_6(CO)_{15}C] +$$
$$4[NMe_3Bz]Cl + 9CO \quad (20)$$

but high yields ($\sim 90\%$) are obtained only when the $[Rh_6(CO)_{15}C]^{2-}$ dianion separates as an insoluble salt during the reaction. Using $^{13}C Cl_4$, it has been possible to determine unequivocally the source of the central carbide atom by the presence of a resonance at 264.7 ppm (resonance D in Fig. 13) (10). Resistance of the rhodium HNCC to degradation by carbon monoxide also allows, in this case, a more convenient direct preparation from K_3RhCl_6 and $CHCl_3$ [see Eq. (60)] (7).

Reasonable hopes of extending this method of synthesis to other halides of nontransition elements are presently under experimental verification

(13). A preliminary claim (46) of a related $[Rh_6(CO)_{15}Si]^{2-}$ dianion was consequently proved incorrect by full X-ray and elemental analysis (13). The observed formation of the $[Rh_6(CO)_{15}C]^{2-}$ dianion, from $SiCl_4$ and $[Rh(CO)_4]^-$ anion, can be explained on the basis of the known transformation of $Me_3SiCo(CO)_4$ into $Me_3SiO-CCo_3(CO)_9$ (80). Insertion of a Si—Co bond into a carbonyl group is believed to be due to the high stability of the resulting Si—O bond (80) and may prove to be a serious limitation.

V
METHODS OF SEPARATION

The synthesis of HNCC is rarely very selective; more often a mixture of high molecular weight compounds is obtained and the separation problem becomes crucial.

Separation is carried out using various techniques, depending on the nature of the compounds. For nonionic species the solubilities of the compounds are generally low and similar, and it has been necessary to use either fractionation by continuous extraction with low boiling solvents (36) or thin-layer chromatography (58, 60). The first method has been used with air-sensitive compounds, whereas the second has been applied only to air-stable substances. In both cases, it is possible to separate only limited amounts of compounds, whose characterization is, therefore, carried out using particular techniques such as mass spectroscopy (58–60).

Separation of salts of carbonylmetalates is generally easier, because solubility of the salts of anions of similar molecular weight is mainly dependent on the MA/NC ratio. For instance, at MA/NC of 6 the sodium salts generally separate from aqueous solutions by simple addition of excess NaCl {e.g., $[Rh_{12}(CO)_{30}]^{2-}$ and $[Rh_6(CO)_{15}I]^-$}, whereas at an MA/NC value of 3, precipitation of the potassium salts is possible by addition of potassium ions {e.g., $[Co_6(CO)_{15}]^{2-}$, $[Rh_6(CO)_{15}C]^{2-}$, and $[Ni_6(CO)_{12}]^{2-}$}. At MA/NC of 1.5 the potassium salt can be obtained only from very concentrated solutions of potassium ions {e.g., $[Co_6(CO)_{14}]^{4-}$ and $[Rh_6(CO)_{14}]^{4-}$}, and these anions are often more conveniently precipitated as tetraalkylammonium salts. Great difficulties in separation have been found only in cases when the MA/NC ratios are very close.

Sometimes it is not possible to use an aqueous medium because of the reaction between water and the carbonylmetalate. In these cases, separation can be attempted using salts of large cations, such as $[NBu_4]^+$,

[PPh$_4$]$^+$, and [PPN]$^+$, and organic solvents such as THF and isopropanol (*102*).

The preparation of HNCC in the form of well-shaped crystals suitable for diffractometric analysis is extremely important because of the basic need for this type of characterization. In the last few years, we have used extensively crystallization methods involving slow diffusion of solvents in which the complex is insoluble. Generally, it is sufficient to stratify slowly a lower-density solvent over the solution of the compound (e.g., *i*-PrOH on acetone, *i*-Pr$_2$O on MeOH, toluene on THF); crystallization occurs by simple diffusion for a couple of days. Unfortunately, with salts of carbonylmetalates the ease of formation of crystals is generally unpredictable and dependent on the particular cation. Often it is necessary to try a whole series of cations, which may be very time-consuming, in order to obtain proper crystals.

The elimination of solvent coordinated to the alkali cations may disrupt the original crystal structure during drying of the crystals; in this case, the apparent morphology of the crystals may remain unchanged, but only a powder pattern is observed on X-ray diffraction. Sometimes, similar behavior has been observed with large cations, such as [NMe$_4$]$^+$ and [PPh$_4$]$^+$, possibly due to the original presence of clathrated solvent molecules (*95*).

VI
REACTIVITY

It is well known that the carbon atoms of carbonyl groups are readily attacked by nucleophilic agents (*41, 56*), whereas the oxygen atoms of the same groups are available to electrophilic attack (*93*). It, therefore, seems reasonable to envisage the presence of an alternatively polarized double barrier of negative oxygen and positive carbon atoms.

Unfortunately at the present level of knowledge, it is difficult to say under which conditions the high nuclearity clusters react by an associative or a dissociative mechanism. However, since the number of carbonyl groups bonded to each metal atom is rather small, it is reasonable to assume that the dissociative energies should be high and that an associative mechanism should, therefore, predominate. Associative reactions are expected to take place either indirectly, by prior attack on the external barrier and transfer onto the core of the cluster, when the ligands around the cluster are crowded, or directly on the cluster core when the crowding is irregular.

Figure 15 shows the polyhedra described by the oxygen atoms of some hexanuclear clusters. It is evident that, on increasing the number of car-

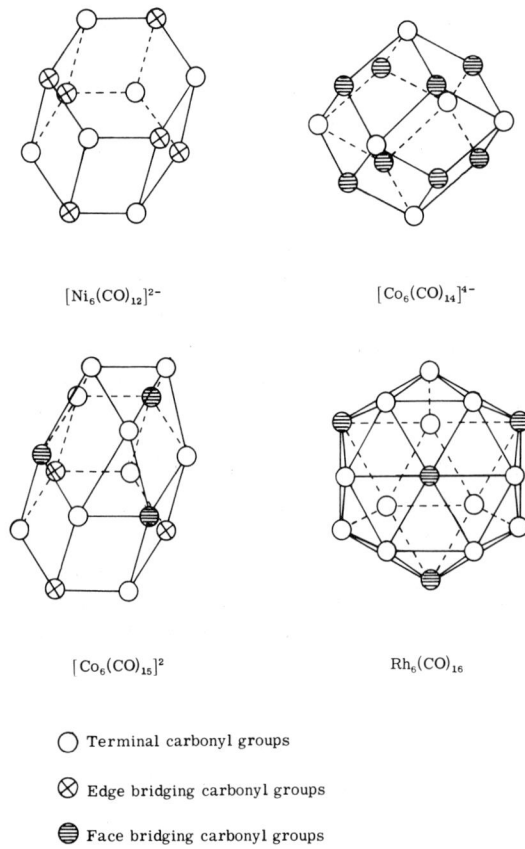

[Ni$_6$(CO)$_{12}$]$^{2-}$ [Co$_6$(CO)$_{14}$]$^{4-}$

[Co$_6$(CO)$_{15}$]$^{2-}$ Rh$_6$(CO)$_{16}$

○ Terminal carbonyl groups

⊗ Edge bridging carbonyl groups

⊜ Face bridging carbonyl groups

FIG. 15. Idealized polyhedra resulting from the distribution of the external oxygen atoms in some hexanuclear clusters.

bonyl groups around the metal core, the packing of the oxygen atoms becomes more and more compact. Whereas the polyhedra of the less crowded species, such as [Ni$_6$(CO)$_{12}$]$^{2-}$ and [Co$_6$(CO)$_{14}$]$^{4-}$, have hexagonal and quadrilateral faces, the more crowded Rh$_6$(CO)$_{16}$ has only triangular ones. The external geometry can greatly affect the reactivity because the presence of hexagonal or quadrilateral faces is expected to facilitate the direct attack of a reagent on the metallic core. However, the absolute significance of these steric considerations is difficult to assess not only due to the lack of knowledge about associative or dissociative character, but also due to the dynamic character of the external barrier of ligands. It has already been pointed out that fluxionality and "reactivity" present similar trends, and it seems probable that dynamic effects controlling the distribution of the ligands can often dominate the properties of the cluster.

Owing to this basic lack of knowledge we have preferred a more formal classification of the reactivity of HNCC based on the product obtained from the reaction itself. Reactions of HNCC have therefore been classified in four main types: reduction, oxidation, substitution, and oxidative addition. Simple addition of neutral ligands, which results in destruction of the cluster, is a common secondary process; it is mainly discussed with substitution.

A. Reduction

Reduction of HNCC is seldom a simple process in which there is substitution of a carbon monoxide ligand, or of a metal–metal bond, with an electron pair. A rare example of this type is (102)

$$[Rh_{12}(CO)_{30}]^{2-} + 2Na \xrightarrow[25°C]{THF} 2[Rh_6(CO)_{15}]^{2-} + 2Na^+ \quad (21)$$

Very often reduction is complicated by the formation of free carbon monoxide and by consequent degradation. For example, the simple reduction of $[Co_6(CO)_{15}]^{2-}$ to $[Co_6(CO)_{14}]^{4-}$,

$$11[Co_6(CO)_{15}]^{2-} + 22Na \longrightarrow 11[Co_6(CO)_{14}]^{4-} + 22Na^+ + 11CO \quad (a)$$

$$2[Co_6(CO)_{14}]^{4-} + 11CO \longrightarrow [Co_6(CO)_{15}]^{2-} + 6[Co(CO)_4]^- \quad (b)$$

$$10[Co_6(CO)_{15}]^{2-} + 22Na \longrightarrow 9[Co_6(CO)_{14}]^{4-} + 6[Co(CO)_4]^- + 22Na^+ \quad (22)$$

takes place with formation of carbon monoxide (a), and this, as shown by separate experiments, adds to the reduced cluster with formation of the $[Co(CO)_4]^-$ anion (b) (37).

A further complication is due to bond redistribution reactions as, for example, in the following sequence:

$$[Pt_9(CO)_{18}]^{2-} + 2Li \longrightarrow [Pt_6(CO)_{12}]^{2-} + [Pt_3(CO)_6]^{2-} + 2Li^+ \quad (a)$$

$$[Pt_3(CO)_6]^{2-} + [Pt_9(CO)_{18}]^{2-} \longrightarrow 2[Pt_6(CO)_{12}]^{2-} \quad (b)$$

$$2[Pt_9(CO)_{18}]^{2-} + 2Li \longrightarrow 3[Pt_6(CO)_{12}]^{2-} + 2Li^+ \quad (23)$$

Similar bond redistribution reactions have been carried out starting from preformed reagents, and have great significance in the chemistry of the oligomeric carbonylplatinates (95).

The formation of free carbon monoxide, and consequent degradation reactions can be avoided using alkali hydroxides or alkoxides as reducing agents. In these cases there is simultaneous oxidation of the carbon monoxide to carbon dioxide. For example, the reaction between $Rh_6(CO)_{16}$ and concentrated aqueous potassium hydroxide takes place according to the stoichiometry (39):

$$Rh_6(CO)_{16} + 8KOH \xrightarrow[H_2O, KOH]{25°C} K_4[Rh_6(CO)_{14}] + 2K_2CO_3 + 4H_2O \qquad (24)$$

The mechanism of this type of reaction, originally elucidated by Kruck (94), is related to the simple addition of the nucleophile observed with alkali alkoxides. In the present case the following reactions, in which the saponification step is also of preparative significance (99, 102), have been observed (45):

$$Rh_6(CO)_{16} + OR^- \xrightarrow[ROH]{25°C} [Rh_6(CO)_{15}(COOR)]^-$$

$$(R = Me, Et) \qquad (25)$$

$$[Rh_6(CO)_{15}(COOR)]^- + 2NaOH \xrightarrow{25°C} [Rh_6(CO)_{15}]^{2-} + Na^+ +$$

$$NaHCO_3 + ROH \qquad (26)$$

Often the carboalkoxy group can be hydrolyzed by simple reaction with water, and the ease of this hydrolysis is higher at lower MA/NC ratio, e.g., in the series $[Rh_6(CO)_{15}(COOMe)]^- < [Rh_4(CO)_{11}(COOMe)]^- < [Rh_6(CO)_{14}(COOMe)_2]^{2-}$. Sometimes, when the carboalkoxy group is not extremely sensitive to hydrolysis, the carboalkoxy anion has been obtained by simple use of anhydrous sodium carbonate suspended in alcohols (99).

Electrochemical reduction has been applied to a number of low nuclearity carbonyl clusters (57, 65), but to date there has been no application to HNCC.

An interesting reducing agent is cobaltocene (37), although its use is somewhat limited by the low solubility of the cobalticinium salts.

B. Oxidation

Simple oxidation, resulting in substitution of negative charges by carbon monoxide or by metal–metal bonds, is not a common process. Complications due to redox condensation reactions, to decomposition from the oxidizing agent, and to degradation by the evolved carbon monoxide, are frequent. For example, in the reaction (36),

$$[Co_6(CO)_{15}]^{2-} \xrightarrow[H_2O]{Hg^{2+}, 25°C} Co_6(CO)_{16} + Co_4(CO)_{12} + Co^{2+} \qquad (27)$$
$$10\%$$

formation of $Co_6(CO)_{16}$ is accompanied by large amounts of $Co_4(CO)_{12}$ and by some cobalt(II) cation. The best results have generally been obtained using stoichiometric amounts of oxidant or by working under conditions in which the reaction product is insoluble and not very sensitive to the excess of oxidizing agent.

Often oxidation reactions are possible via intermediate formation and decomposition of hydride derivatives; for example (96, 117),

$$[Fe_6(CO)_{16}C]^{2-} + 2H^+ \xrightarrow{\text{conc. } H_2SO_4} Fe_5(CO)_{15}C \quad\quad (28)$$

$$[Ir_6(CO)_{15}]^{2-} + 2H^+ \xrightarrow[CH_3COOH]{CO} Ir_6(CO)_{16} + H_2 \quad\quad (29)$$

17%

The relative acid strength in a series of anions $[M_n(CO)_x]^-$ is expected to increase on increasing the number of metal atoms n, because of an increased ability to accommodate the delocalized negative charge. Significant deviations from this simple behavior may be predicted, however, due to the need for new coordination positions for the hydrido groups and to the related change in the overall distribution of ligands. Moreover, as pointed out by Kaesz (87), both protonation and deprotonation are generally slow and show high kinetic isotope effects; this behavior is considered to be characteristic of a tunneling mechanism imposed by the barrier of carbonyl groups. Therefore it is not surprising that formation of hydride derivatives of HNCC often requires reaction with concentrated acids. This behavior should be compared with the result obtained on addition of a stoichiometric amount of dilute sulfuric acid to the barium salt of the $[Rh_{12}(CO)_{30}]^{2-}$ dianion:

$$Ba^{2+} + [Rh_{12}(CO)_{30}]^{2-} + H_2SO_4 \xrightarrow{CH_3OH} 2[H(CH_3OH)_x]^+ +$$

$$[Rh_{12}(CO)_{30}]^{2-} + BaSO_4 \quad (30)$$

In this case the IR spectrum of the dianion remains unchanged, showing that there is no detectable formation of Rh—H bonds but only of hydronium ions (43).

Finally, selective oxidations, in which the oxidant is added to the polynuclear species, are sometimes possible using halogens or pseudohalogens in stoichiometric amounts, for example (45, 101),

$$[Rh_6(CO)_{15}]^{2-} + I_2 \xrightarrow{THF} [Rh_6(CO)_{15}I]^- + I^- \quad\quad (31)$$

C. Ligand Substitution

Simple substitution reactions of carbonyl groups with tertiary phosphines generally occur only with the most robust species; for example (85),

$$Ru_6(CO)_{17}C + L \xrightarrow[\text{reflux}]{n\text{-hexane}} Ru_6(CO)_{16}(L)C + CO \quad (32)$$

(L = PPh_3, $P(p\text{-}FC_6H_4)_3$, $AsPh_3$; yield ~ 60%)

Concurrent degradation is very common, as, for example, in the following reactions (21, 102):

$$Rh_6(CO)_{16} + 12PPh_3 \xrightarrow[C_6H_6]{25°C} 3[Rh(CO)_2(PPh_3)_2]_2 + 4CO \quad (33)$$
$$91\%$$

and (95)

$$[Pt_9(CO)_{18}]^{2-} + 9PPh_3 \xrightarrow[\text{THF}]{25°C} [Pt_6(CO)_{12}]^{2-} + 3Pt(CO)(PPh_3)_3 + 3CO \quad (34)$$

The latter is complicated by the equilibrium between $Pt(CO)(PPh_3)_3$ and $Pt(CO)_2(PPh_3)_2$ (42). Reaction (33) contrasts with the substitution of $Rh_6(CO)_{16}$ at higher temperature (~80°C) which is reported to give products such as $Rh_6(CO)_{10}(PPh_3)_6$ (21, 84). This apparent discrepancy is explained in terms of initial degradation followed by thermal condensation; otherwise some orthometalation of the ligand can be suspected.

A similar contrast between substitution and degradation reactions has been observed with halide ions, and the "noninnocence" of these ions should always be considered. For instance, the facile substitution (45),

$$Rh_6(CO)_{16} + MX \xrightarrow[25°C]{\text{THF}} M[Rh_6(CO)_{15}X] + CO \quad (35)$$

(MX = NaI, NBu_4I, NEt_4Br, NMe_3B_xCl, $AsPh_4Cl$, KCN, KSCN; yields 50–80%)

compares with the degradation (101),

$$4[Rh_{12}(CO)_{30}]^{2-} + 12Cl^- \longrightarrow 7[Rh_6(CO)_{15}]^{2-} + 6[Rh(CO)_2Cl_2]^- + 3CO \quad (36)$$

Degradation by the solvent is also possible, as it is well known from the chemistry of the simple carbonyls. An example of such a reaction in the field of HNCC is (13)

$$Rh_8(CO)_{19}C \xrightleftharpoons[CH_2Cl_2]{CH_3CN} 2[Rh(CO)_2(CH_3CN)_2]^+ + [Rh_6(CO)_{15}C]^{2-} \quad (37)$$

D. Oxidative Addition

Oxidative addition is a particular addition that results in an increase in the formal oxidation state of the metal atoms and that is known to occur readily on metal–metal bonds. Although it is expected to be a common reaction in HNCC chemistry, only a few examples are known, and most of them are related to the dianion $[Rh_{12}(CO)_{30}]^{2-}$ (45, 102):

$$[Rh_{12}(CO)_{30}]^{2-} + X_2 \longrightarrow 2[Rh_6(CO)_{15}X]^- \qquad (X_2 = I_2, H_2) \qquad (38)$$

It is probable that the hydrogen halides also react in a similar way, although in this case the expected hydride monoanion has never been observed (45). At ambient pressure the rapid addition of hydrogen requires about 50°C, and the resulting hydride anion is more readily isolated as an acyl anion after further reaction with ethylene and carbon monoxide (102):

$$[Rh_6(CO)_{15}H]^- + CH_2\!\!=\!\!CH_2 + CO \xrightarrow[THF]{H_2} [Rh_6(CO)_{15}(COEt)]^- \qquad (39)$$

Otherwise there is slow formation of a new series of brown anions of which only $[Rh_{13}(CO)_{24}H_{5-n}]^{n-}$ ($n = 2, 3$) have as yet been characterized (structure XXIII of Fig. 11).

A similar reactivity toward hydrogen has been found with the carbonylplatinates $[Pt_3(CO)_6]_n{}^{2-}$ when n is greater than 3, although only formation of hydronium ions corresponding to a net reduction process can be observed; for example (95),

$$3[Pt_{12}(CO)_{24}]^{2-} + H_2 + 2H_2O \xrightarrow{THF,\ 25°C,\ 1\ atm} 4[Pt_9(CO)_{18}]^{2-} + 2H_3O^+ \qquad (40)$$

VII
IRON DERIVATIVES

The reaction in refluxing diglyme of $Fe(CO)_5$ with a wide variety of carbonylmetalates, such as $[Co(CO)_4]^-$, $[Fe(CO)_4]^{2-}$, $[Mn(CO)_5]^-$, and $[V(CO)_6]^-$, gives the hexanuclear dianion $[Fe_6(CO)_{16}C]^{2-}$ in good yields (49, 117). Under analogous conditions, however, $[CpMo(CO)_3]^-$ reacts with $Fe(CO)_5$ to give the pentanuclear dianion $[Fe_5(CO)_{14}C]^{2-}$ (79), which has not yet been structurally characterized.

The extreme dependence of these reactions on the experimental conditions is well exemplified by the following syntheses:

$$Fe(CO)_5 + [Mn(CO)_5]^- \begin{cases} \xrightarrow{\text{refluxing THF (79)}} [Fe(CO)_4]^{2-} + Mn_2(CO)_{10} & (41) \\ \xrightarrow[\text{5 min reaction}]{\text{refluxing diglyme (14)}} [Fe_2Mn(CO)_{12}]^- & (42) \\ \xrightarrow[\text{1 hr reaction}]{\text{refluxing diglyme (49)}} [Fe_6(CO)_{16}C]^{2-} & (43) \end{cases}$$

The hexanuclear carbide dianion $[Fe_6(CO)_{16}C]^{2-}$ reacts either with strong acids (98% H_2SO_4, 85% H_3PO_4) or with oxidizing agents ($AgBF_4$, Ph_3CPF_6) to give low yields (up to 20%) of neutral $Fe_5(CO)_{15}C$ (*117*). So far all attempts to prepare the hypothetical $Fe_6(CO)_{17}C$ have failed.

Finally, 1-electron oxidation of $[Fe_2Re(CO)_{12}]^-$ with tropilium bromide has been reported to give a neutral mixed-metal cluster formulated as $[Fe_2Re(CO)_{12}]_2$ on the basis of elemental analyses (*67*); however, its IR spectrum, which shows carbonyl absorptions quite similar to those of the starting material, is inconsistent with such a formulation.

VIII

RUTHENIUM DERIVATIVES

Reduction of $Ru_3(CO)_{12}$ depends primarily on the experimental conditions. Reduction with either sodium borohydride or sodium amalgam in THF, or with methanolic potassium hydroxide, followed by acidification, has been reported to give the tetranuclear species $Ru_4(CO)_{13}H_2$ and $Ru_4(CO)_{12}H_4$ (*82*). On the other hand, reduction in THF with carbonylmetalates such as $[Mn(CO)_5]^-$ and $[CpFe(CO)_2]^-$ gives, after acidification, a more complicated mixture, which has been successfully separated into its components by selective extraction. After elimination of some tetranuclear hydride derivatives by dissolution in light petroleum, the sparingly soluble hexanuclear $Ru_6(CO)_{18}H_2$ was isolated by extraction in dichloromethane (*48*). Its parent dianion, $[Ru_6(CO)_{18}]^{2-}$, although predicted theoretically (*120*), has been neither isolated nor observed even in solution.

Ruthenium carbide–carbonyl clusters have been obtained through pyrolysis reactions. Thus, by heating $Ru_4(CO)_{12}H_4$ at 130°C in the presence of ethylene (10–12 atm), trace quantities of $Ru_5(CO)_{15}C$ have been obtained along with higher yields (30%) of $Ru_6(CO)_{17}C$ (*59*). The latter is, however, better synthesized by direct pyrolysis of $Ru_3(CO)_{12}$ in di-*n*-butyl ether (*85*). Heating $Ru_3(CO)_{12}$ in aromatic hydrocarbons, such as benzene, toluene, *m*-xylene, or mesitylene, gives a mixture of $Ru_6(CO)_{17}C$ and of the π-arene derivatives $Ru_6(CO)_{14}(arene)C$, which in the last solvent is the major product (*58, 83*).

Monosubstituted derivatives $Ru_6(CO)_{16}(L)C$ [L = PPh_3, $P(p\text{-}FC_6H_4)_3$, $AsPh_3$] have been obtained by boiling $Ru_6(CO)_{17}C$ in hexane with excess tertiary phosphines or arsines (*85*).

IX
OSMIUM DERIVATIVES

Pyrolysis at 200°C of $Os_3(CO)_{12}$ in a sealed, evacuated tube afforded a mixture of at least seven different carbonyl clusters which could be separated by thin-layer chromatography. In addition to some unreacted $Os_3(CO)_{12}$, the new compounds, $Os_4(CO)_{13}$, $Os_5(CO)_{16}$, $Os_6(CO)_{18}$, $Os_8(CO)_{23}$, and $Os_8(CO)_{21}C$, were identified by mass spectroscopy (58); the last compound was originally formulated as $Os_5(CO)_{15}C_4$ (61). Further pyrolysis of $Os_6(CO)_{18}$ at 255°C gives the pentanuclear carbide derivative, $Os_5(CO)_{15}C$, in 40% yield (59).

In the presence of trace quantities of water, pyrolysis of $Os_3(CO)_{12}$ at 230°C results in a mixture of hydride carbonyl clusters. The new derivatives, $Os_5(CO)_{15}H_2$, $Os_5(CO)_{16}H_2$, $Os_6(CO)_{18}H_2$, and $Os_7(CO)_{19}(C)H_2$, along with the known complexes $Os_3(CO)_{10}H(OH)$, $Os_4(CO)_{13}H_2$, and $Os_4(CO)_{12}H_4$ (60), have been separated by thin-layer chromatography and identified by mass spectroscopy, as before.

The IR spectra of all of these compounds do not show absorptions due to bridging carbonyl groups, and the NMR spectra indicate that the hydrogen atoms are always in bridging positions. Future structural determinations of such an impressive series of polynuclear derivatives will make a considerable contribution to the chemistry of high nuclearity clusters. Until now only the structures of $Os_6(CO)_{18}$ and $Os_7(CO)_{21}$ have been determined (104).

X
COBALT DERIVATIVES

Scheme 1 summarizes the syntheses, and the most significant reactions, of the $[Co_6(CO)_{15}]^{2-}$ dianion, which can be considered the key species in cobalt HNCC chemistry.

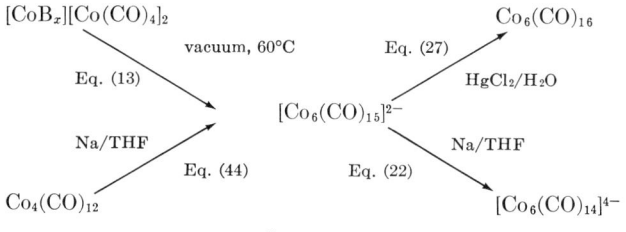

SCHEME 1

Yellow-green $[Co_6(CO)_{15}]^{2-}$ is conveniently synthesized in high yields (80–90%) by merely heating under vacuum an ethanolic solution of $[Co(EtOH)_x][Co(CO)_4]_2$, obtained by the reaction of $Co_2(CO)_8$ with ethanol (35). An alternative route to this could be the reduction of $Co_4(CO)_{12}$ with alkali metals (37),

$$9Co_4(CO)_{12} + 20Na \xrightarrow[\text{THF}]{0°-25°C} 4Na_2[Co_6(CO)_{15}] + 12Na[Co(CO)_4] \quad (44)$$

although this is less convenient both because of the simultaneous formation of a large quantity of tetracarbonylcobaltate and of the further facile reduction to the tetra-anion $[Co_6(CO)_{14}]^{4-}$ according to Eq. (22).

Cobaltocene behaves as a toluene-soluble pseudoalkali metal, and in this nonpolar solvent ionic products precipitate out as soon as they are formed. Thus, the reduction of $Co_4(CO)_{12}$ with cobaltocene in toluene gives an intermediate that analyzes as $[CoCp_2][Co_4(CO)_{10-11}]$, and which is probably dimeric. Unfortunately, its lability has prevented further characterization (37).

Protonation of $[Co_6(CO)_{15}]^{2-}$ at $-70°C$ gives a new unstable species, for which the analytical and spectral data agree with the formula $[Co_6(CO)_{15}H]^-$ (33).

A logical extension of the synthesis of $[Co_6(CO)_{15}]^{2-}$ from $[Co(EtOH)_x]$ $[Co(CO)_4]_2$ has been successfully applied to the preparation of mixed-metal clusters. Thermal decomposition of $[Ni(EtOH)_x][Co(CO)_4]_2$ prepared in situ gives the red hexanuclear dianion $[Ni_2Co_4(CO)_{14}]^{2-}$, through the following redox condensation and redistribution processes (44):

$$2[Ni(EtOH)_x][Co(CO)_4]_2 + 2[Co(CO)_4]^- \longrightarrow 2[NiCo_3(CO)_{11}]^- + 2xEtOH + 2CO \quad (45)$$

$$2[NiCo_3(CO)_{11}]^- \rightleftharpoons [Ni_2Co_4(CO)_{14}]^{2-} + Co_2(CO)_8 \quad (46)$$

The whole process can be represented schematically by

$$2M^{2+} + 6M^- \longrightarrow 2M_4^- \rightleftharpoons M_6^{2-} + M_2 \quad (47)$$

The deep red tetra-anion $[Co_6(CO)_{14}]^{4-}$ is conveniently synthesized by direct reduction of $Co_4(CO)_{12}$ (37):

$$5Co_4(CO)_{12} + 12M \xrightarrow{\text{THF}} 2M_4[Co_6(CO)_{14}] + 8M[Co(CO)_4] \quad (48)$$

$$(M = Li, Na, K)$$

The potassium salt of this anion could be isolated only in 45% yields owing to its high solubility. The corresponding sodium salt has been isolated in two isomeric forms that differ in their solubility in THF. This finding, together with the extreme dependence of the IR spectra of $[Co_6(CO)_{14}]^{4-}$ on the cation and on the solvent, are in agreement with the existence in solution of an easily reversible equilibrium of the type:

$$[Co_6(CO)_{14}]^{4-} \underset{H_2O}{\overset{THF}{\rightleftharpoons}} [Co_6(CO)_{14}]^{4-} \qquad (49)$$

form A   form B

Form A, as confirmed by an X-ray diffraction study (3), has structure XI (see Fig. 4) with six terminal and eight face-bridging CO groups (ν_{CO} = 1640–1680 cm^{-1}), whereas form B should have some edge-bridging carbonyl groups (ν_{CO} = 1710–1760 cm^{-1}) (37). On oxidation both $[Co_6(CO)_{15}]^{2-}$ and $[Co_6(CO)_{14}]^{4-}$ give the neutral hexanuclear cluster $Co_6(CO)_{16}$.

A peculiar metal atom redistribution process was found in the pyrolysis of $Co_2Rh_2(CO)_{12}$, which gives $Co_2Rh_4(CO)_{16}$ (100). This result should apparently require the intermediate formation of an octanuclear species, which rearranges to $Co_2Rh_4(CO)_{16}$ by elimination of $Co_2(CO)_8$.

Finally, cobalt carbide–carbonyl clusters have recently been isolated through a two-step synthesis. First of all, the well-known $Co_3(CO)_9CCl$ is prepared from $Co_2(CO)_8$ and CCl_4, and then the hexanuclear carbide dianion $[Co_6(CO)_{15}C]^{2-}$ is obtained in good yields (9) by further reaction with $Na[Co(CO)_4]$ in diisopropylether [see Eqs. (18) and (19)]. Further redox condensation between $[Co_6(CO)_{15}C]^{2-}$ and $Co_4(CO)_{12}$ [see Eq. (11)] gives the square antiprismatic octanuclear cluster $[Co_8(CO)_{18}C]^{2-}$ (13). Both these carbide derivatives, as well as all of the other cobalt high nuclearity clusters, are sensitive to air and react with carbon monoxide at atmospheric pressure.

XI

RHODIUM DERIVATIVES

Since the time of the original preparation due to Hieber and Lagally (75) the synthesis of $Rh_6(CO)_{16}$ has been considerably improved, and the best methods now available are shown in Scheme 2. This hexanuclear compound is sparingly soluble in organic solvents and its purification is rather difficult. However, thermal decomposition of $Rh_4(CO)_{12}$ in solution affords analytically pure microcrystals (98).

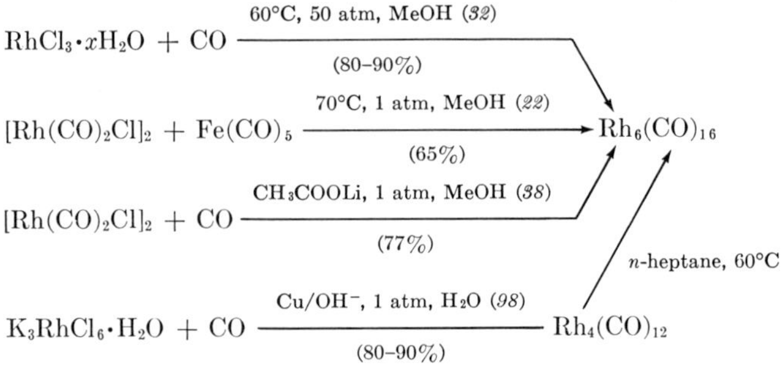

```
                      60°C, 50 atm, MeOH (32)
RhCl₃·xH₂O + CO  ─────────────────────────┐
                      (80-90%)             \
                      70°C, 1 atm, MeOH (22)\
[Rh(CO)₂Cl]₂ + Fe(CO)₅ ───────────────────→ Rh₆(CO)₁₆
                      (65%)                 ↗
                      CH₃COOLi, 1 atm, MeOH (38) /
[Rh(CO)₂Cl]₂ + CO  ───────────────────────/
                      (77%)              /
                                        / n-heptane, 60°C
                      Cu/OH⁻, 1 atm, H₂O (98)
K₃RhCl₆·H₂O + CO  ─────────────────────── Rh₄(CO)₁₂
                      (80-90%)
```

SCHEME 2

In spite of its low solubility, $Rh_6(CO)_{16}$ reacts with a wide variety of neutral ligands to give both substitution and degradation products. In 1970, the syntheses of the species $Rh_6(CO)_7L_9$ (L = PPh_3, $AsPh_3$) and $Rh_6(CO)_{10}(L-L)_3$ (L—L = 1,2-bisdiphenylphosphinoethane) were reported (84). Both compounds result from the reaction of $Rh_6(CO)_{16}$ with excess ligand in refluxing chloroform. However, the first formulation is especially puzzling—it is hard to visualize how nine bulky ligands, such as triphenylphosphine or triphenylarsine, could be accommodated around an octahedral cluster. More recently, other substituted compounds, formulated as $Rh_6(CO)_{10}L_6$ {L = PPh_3, $P(OMe)_3$, and 4-ethyl-2,6,7-trioxa-1-phosphabicyclo[2.2.2]octane (ETPO)} have been isolated by reacting $Rh_6(CO)_{16}$ with excess ligand in benzene at 80°C in a sealed tube (21).

A black, sparingly soluble derivative, tentatively formulated as $Rh_6(CO)_8(PH_3)_8$ has been isolated from the reaction of $[Rh(CO)_2Cl]_2$ with phosphine (89).

Finally, both $Rh_4(CO)_{12}$ and $Rh_6(CO)_{16}$ react with dienes to give the species $Rh_6(CO)_{14}(Diene)$ (Diene = 1,5-cyclooctadiene, norbornadiene, 1,4-cyclohexadiene, and 2,3-dimethyl-1,3-butadiene). The molecular weight of $Rh_6(CO)_{14}$ (1,5-COD) has been confirmed by mass spectroscopy (92).

The reaction of $Rh_6(CO)_{16}$ with a primary amine gives the monoanionic derivative containing a carboamide group, whereas the analogous carboalkoxy derivatives have been obtained by reduction with alkali metal alkoxides in anhydrous alcohol (45):

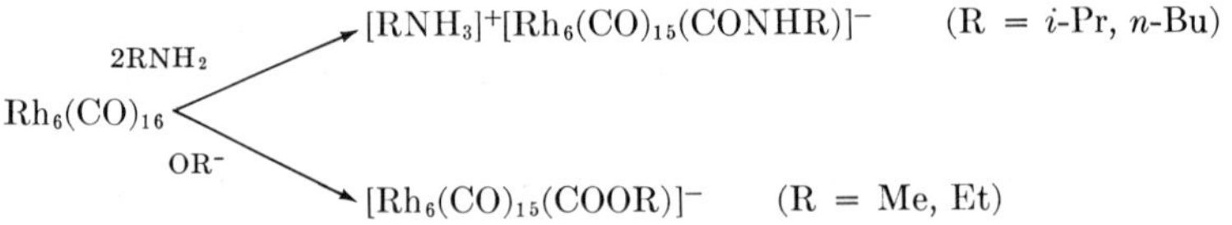

```
              2RNH₂    ↗ [RNH₃]⁺[Rh₆(CO)₁₅(CONHR)]⁻   (R = i-Pr, n-Bu)
Rh₆(CO)₁₆  <                                                                    (50)
              OR⁻      ↘ [Rh₆(CO)₁₅(COOR)]⁻            (R = Me, Et)
```

Halides or pseudohalides give the analogous substitution products $[Rh_6(CO)_{15}X]^-$ (X = Cl, Br, I, CN, SCN) [see Eq. (35)].

In these cases also, initial nucleophilic attack on the carbonyl group seems probable. The least stable of these derivatives is [Rh$_6$(CO)$_{15}$Cl]$^-$ from which the chloride ion can be readily displaced by metathesis with other anions such as SCN$^-$ (45):

$$[Rh_6(CO)_{15}Cl]^- + SCN^- \longrightarrow [Rh_6(CO)_{15}(SCN)]^- + Cl^- \quad (51)$$

The corresponding acyl derivatives, [Rh$_6$(CO)$_{15}$(COR)]$^-$ (R = Et, Pr), were originally obtained as stable solutions of the hydronium salts by reaction of Rh$_4$(CO)$_{12}$ with water in the presence of a mixture of carbon monoxide and an olefin such as ethylene or propylene (43):

$$3Rh_4(CO)_{12} + 2C_2H_4 + 2H_2O \xrightarrow{CH_3COCH_3} 2(H \cdot solv.)^+ +$$
$$2[Rh_6(CO)_{15}(COEt)]^- + 2CO_2 + 2CO \quad (52)$$

It seems probable that their formation derives from olefin addition to the intermediate [Rh$_6$(CO)$_{15}$H]$^-$ [see Eq. (39)].

On increasing the halide concentration, further substitution occurs to give the disubstituted dianions [Rh$_6$(CO)$_{14}$X$_2$]$^{2-}$ (45, 101). Although there is spectroscopic evidence for the presence in solution of the species [Rh$_6$(CO)$_{14}$I$_2$]$^{2-}$, [Rh$_6$(CO)$_{14}$(COOMe)$_2$]$^{2-}$, and [Rh$_6$(CO)$_{14}$(CN)$_2$]$^{2-}$, only the last has been isolated as a pure crystalline product. The stability of the dicyanide derivative probably arises from the backbonding properties of such a substituent. Generally, the disubstituted derivatives are unstable and undergo further transformation to the heptanuclear species [Rh$_7$(CO)$_{16}$X]$^{2-}$ (X = Br, I) (101).

Scheme 3 briefly summarizes the syntheses of these substituted anions:

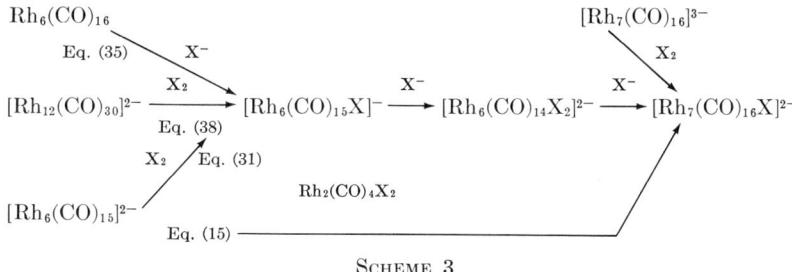

SCHEME 3

The chemistry of unsubstituted polynuclear carbonylrhodates is fascinating, but, although it has been under investigation for several years, it is as yet clarified only in part. The derivatives best characterized are [Rh$_{12}$(CO)$_{30}$]$^{2-}$ and [Rh$_6$(CO)$_{15}$]$^{2-}$, which seem to be the focal points of such a chemistry.

The violet dianion [Rh$_{12}$(CO)$_{30}$]$^{2-}$ is obtained in high yields (80–90%) by reduction under carbon monoxide of [Rh(CO)$_2$Cl]$_2$ or Rh$_4$(CO)$_{12}$ in buffered alkaline solution, such as potassium acetate (40). The reduction proceeds in agreement with the following stoichiometries (38, 40):

$$2Rh_2(CO)_4Cl_2 + 6CO + 2H_2O \longrightarrow Rh_4(CO)_{12} + 2CO_2 + 4HCl \quad (53)$$

$$3Rh_4(CO)_{12} + 2OH^- \longrightarrow [Rh_{12}(CO)_{30}]^{2-} + CO_2 + 5CO + H_2O \quad (54)$$

Equation (54) corresponds to a multistep mechanism in which the key step is the redox condensation

$$[Rh_4(CO)_{11}]^{2-} + 2Rh_4(CO)_{12} \longrightarrow [Rh_{12}(CO)_{\sim 34}]^{2-} + CO \quad (55)$$

This reaction has been thoroughly verified starting from the pure reagents, but no experimental evidence for the formation of the expected octanuclear intermediate could be obtained (102).

The tetranuclear dianion $[Rh_4(CO)_{11}]^{2-}$ has been prepared from $Rh_4(CO)_{12}$ using a sequence of reactions that is believed to be analogous to that involved in the process represented by Eq. (54):

$$Rh_4(CO)_{12} + OCH_3^- \longrightarrow [Rh_4(CO)_{11}(COOCH_3)]^- \quad (56)$$

$$[Rh_4(CO)_{11}(COOCH_3)]^- + OH^- \longrightarrow [Rh_4(CO)_{11}]^{2-} + CO_2 + H_2O \quad (57)$$

In the solid state, the dianion $[Rh_4(CO)_{11}]^{2-}$ possesses a distorted tetrahedral geometry with four terminal and seven edge-bridging carbonyl groups, whereas in solution it is fluxional at $-70°C$ (13, 74).

An important, but not yet understood, reaction of the dianion $[Rh_{12}(CO)_{30}]^{2-}$ is the reversible addition of carbon monoxide. Accurate measurements of the amount of absorbed gas indicate the stoichiometry (40, 102):

$$[Rh_{12}(CO)_{30}]^{2-} + 4CO \underset{\text{vacuum}}{\overset{25°C, 1\ atm,\ THF}{\rightleftharpoons}} [Rh_{12}(CO)_{\sim 34}]^{2-} \quad (58)$$

Although dianion $[Rh_{12}(CO)_{\sim 34}]^{2-}$ has never been isolated in the solid state owing to its lability, it can be considered a definite intermediate. In fact, it has a characteristic and reproducible IR spectrum, and at $-70°C$ it has a peculiar ^{13}C magnetic resonance spectrum consisting of two doublets (191.7 and 208.3 ppm) and a multiplet (247 ppm) (74).

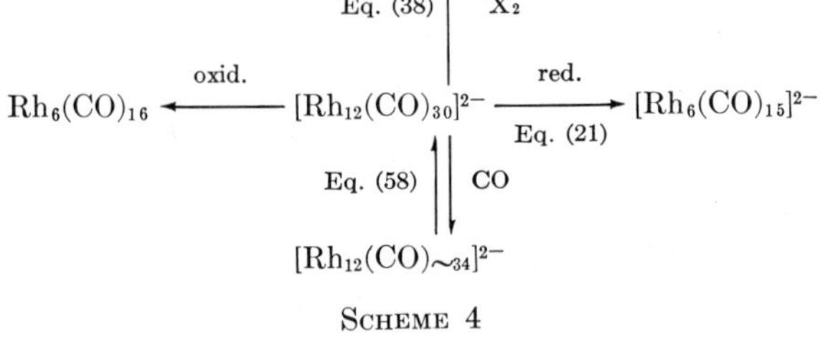

$$
\begin{array}{c}
[Rh_6(CO)_{15}X]^- \\
\uparrow \text{Eq. (38)} \quad \bigg| X_2 \\
Rh_6(CO)_{16} \xleftarrow{\text{oxid.}} [Rh_{12}(CO)_{30}]^{2-} \xrightarrow[\text{Eq. (21)}]{\text{red.}} [Rh_6(CO)_{15}]^{2-} \\
\text{Eq. (58)} \updownarrow \bigg| CO \\
[Rh_{12}(CO)_{\sim 34}]^{2-}
\end{array}
$$

SCHEME 4

As shown in Scheme 4, the chemistry of the violet dianion $[Rh_{12}(CO)_{30}]^{2-}$ is relatively simple: the main feature is the ease of rupture of the Rh—Rh bond connecting the two octahedra. The best synthesis of the green hexanuclear dianion $[Rh_6(CO)_{15}]^{2-}$ involves the reduction of $Rh_6(CO)_{16}$ under nitrogen with a stoichiometric amount of alkali (99):

$$Rh_6(CO)_{16} + 4OH^- \xrightarrow{N_2,\ MeOH} [Rh_6(CO)_{15}]^{2-} + CO_3^{2-} + 2H_2O \quad (59)$$

This should be contrasted with the reduction under carbon monoxide, which is complicated because of concurrent degradation and redox condensation reactions [represented by Eqs. (6) and (14)] and which gives rise to the heptanuclear anion $[Rh_7(CO)_{16}]^{3-}$.

Dianion $[Rh_6(CO)_{15}]^{2-}$ is quite reactive, as shown in Scheme 5. Most of these reactions, as well as the preparations of the $[Rh_7(CO)_{16}]^{3-}$ and $[Rh_6(CO)_{14}]^{4-}$ anions, Eq. (3) and (24), have already been discussed in previous parts of this review. Although the $[Rh_{12}(CO)_{30}]^{2-}$ dianion is moderately air-stable, solutions of the more reduced anions $[Rh_6(CO)_{15}]^{2-}$, $[Rh_7(CO)_{16}]^{3-}$, and $[Rh_6(CO)_{14}]^{4-}$ are extremely air-sensitive.

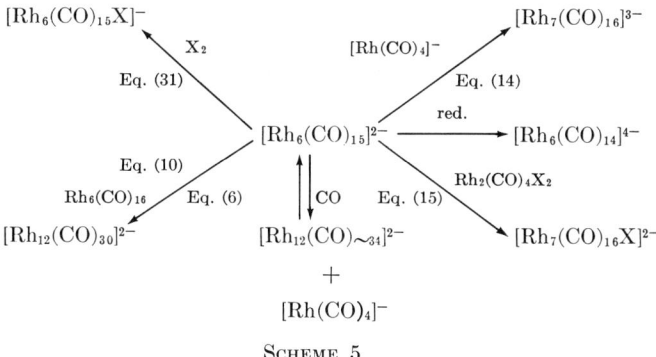

SCHEME 5

Occasionally during the synthesis of the $[Rh_7(CO)_{16}]^{3-}$ trianion, low yields of a yellow anion, which was formulated on the basis of elemental analyses as $[Rh_3(CO)_{10}]^-$, were also obtained (39). Subsequent X-ray structural investigation showed it to be $[Rh_6(CO)_{15}C]^{2-}$ and, furthermore, suggested that its casual formation could be due to the accidental presence in the reaction medium of trace amounts of chloroform (7). As a result, this compound is now readily available in high yields (80–90%) by deliberate addition of small amounts of chloroform to the reaction mixture. The following reaction accounts for the apparent stoichiometry:

$$6[RhCl_6]^{3-} + 23OH^- + 26CO + CHCl_3 \xrightarrow{MeOH} [Rh_6(CO)_{15}C]^{2-} + 39Cl^- + 11CO_2 + 12H_2O \quad (60)$$

The mechanism probably involves a multistep reaction, with initial formation of $[Rh_7(CO)_{16}]^{3-}$, degradation to $[Rh(CO)_4]^-$, and condensation of this anion with the chloroform (10, 13). The $[Rh_6(CO)_{15}C]^{2-}$ dianion is the precursor of a wide series of carbide–carbonyl derivatives. Scheme 6 shows the compounds so far structurally characterized and the necessary conditions for their syntheses (9, 13).

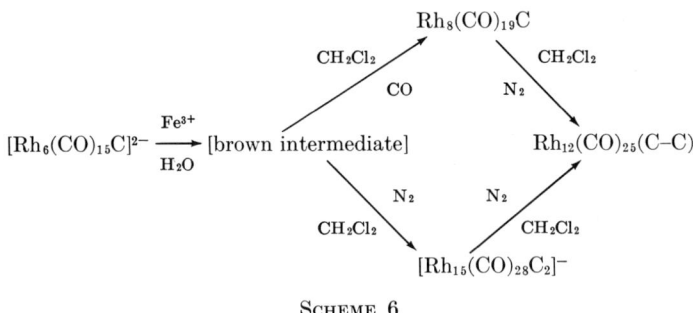

SCHEME 6

Several other species have been isolated in the crystalline state and are presently awaiting definite characterization. Great difficulties arise from the fact that these derivatives are only sparingly soluble in inert solvents and often react with polar solvents [see for instance Eq. (37)]. The octanuclear carbide $Rh_8(CO)_{19}C$ polymerizes THF at room temperature (102).

XII

IRIDIUM DERIVATIVES

Unlike cobalt and rhodium, the chemistry of polynuclear iridium carbonyl derivatives has not been studied in detail (15a). Reduction of $Ir_4(CO)_{12}$ under carbon monoxide with K_2CO_3 in methanol gives the yellow tetranuclear hydride derivative $[Ir_4(CO)_{11}H]^-$, whereas under nitrogen the brown dianion $[Ir_8(CO)_{20}]^{2-}$ has been isolated as a tetraalkylammonium salt (97). It has been suggested that the structure of the dianion could result from the linking of two iridium tetrahedra, although its formulation so far is based only on elemental analyses. Clearly such an interesting compound deserves further chemical and structural characterization.

The reduction of $Ir_4(CO)_{12}$ by sodium metal in THF under carbon monoxide gives hexanuclear $[Ir_6(CO)_{15}]^{2-}$. Its IR spectrum compares well with those of the analogous dianions $[Co_6(CO)_{15}]^{2-}$ and $[Rh_6(CO)_{15}]^{2-}$. As previously shown [see Eq. (29)], dianion $[Ir_6(CO)_{15}]^{2-}$ reacts with acetic acid under carbon monoxide to give red crystals of $Ir_6(CO)_{16}$ (96). This hexanuclear species is much less stable than $Ir_4(CO)_{12}$, and its chemistry has not yet been studied.

XIII
NICKEL DERIVATIVES

There is spectroscopic evidence that the reduction of $Ni(CO)_4$ gives rise to a large number of products, most of which have still to be isolated and characterized. Results obtained so far from a recent reinvestigation of this chemistry (95) suggest that all of the formulations previously reported in the literature are incorrect (76, 77, 116).

Reduction under nitrogen of tetracarbonyl nickel with alkali metals or sodium and lithium amalgams (71) in THF, or with alkali hydroxides in methanol, gives a mixture of the dianions $[Ni_5(CO)_{12}]^{2-}$ and $[Ni_6(CO)_{12}]^{2-}$ (28, 95). The final composition of the reaction mixture greatly depends on the experimental conditions owing to the easily reversed equilibrium:

$$[Ni_5(CO)_{12}]^{2-} + Ni(CO)_4 \rightleftharpoons [Ni_6(CO)_{12}]^{2-} + 4CO \qquad (61)$$

The yellow pentanuclear dianion, $[Ni_5(CO)_{12}]^{2-}$, is rather labile and has been isolated in a pure state only as the bis(triphenylphosphine)iminium salt by crystallization under carbon monoxide in anhydrous solvents. In wet solvents, it reacts readily with carbon monoxide to give a mixture of tetracarbonylnickel and an unstable hydride derivative presently formulated as $[Ni(CO)_3H]^-$ ($\tau = 18.3$), by comparison of its IR spectrum with those of $[Ni(CO)_3X]^-$ (X = Cl, Br, I) (31). This reaction contrasts with that of $[Ni_5(CO)_{12}]^{2-}$ with water under nitrogen when the red dianion $[Ni_6(CO)_{12}]^{2-}$ is formed. The reaction proceeds with formation of tetracarbonylnickel, hydrogen and traces of carbon monoxide and its stoichiometry is believed to be the following:

$$3[Ni_5(CO)_{12}]^{2-} + 2H_2O \longrightarrow 2[Ni_6(CO)_{12}]^{2-} + H_2 + 2OH^- + 3Ni(CO)_4 \qquad (62)$$

The mechanism should involve an easy initial protonation to give a hydride intermediate which then loses hydrogen and condenses to $[Ni_6(CO)_{12}]^{2-}$.

Equation (62) explains why precipitation from the original reaction mixture by addition of an aqueous solution of tetraalkylammonium salts always gives the red hexanuclear dianion $[Ni_6(CO)_{12}]^{2-}$ (yields up to 60%). A comparison of its IR spectrum with that reported in the literature for $[Ni_4(CO)_9]^{2-}$ (76) strongly suggests that the latter should be reformulated as $[Ni_6(CO)_{12}]^{2-}$. Our present knowledge of the reactivity of this hexanuclear dianion is summarized in Scheme 7. Dianion $[Ni_6(CO)_{12}]^{2-}$ reacts slowly under vacuum with $Ni(CO)_4$, to give the dark-red enneanuclear $[Ni_9(CO)_{18}]^{2-}$. The latter may, however, be better synthesized by oxidation of the hexanuclear complex with a stoichiometric amount of $Ni(EtOH)_xCl_2$. Hydrolysis of $[Ni_6(CO)_{12}]^{2-}$ takes place only in an acidic medium and gives, depending on the pH, two hydride derivatives, presently formulated as $[Ni_8(CO)_{14}H_2]^{2-}$ and $[Ni_{11}(CO)_{20}H_2]^{2-}$, but not yet structurally characterized. Such formulations are based on elemental analyses and, in the case of violet $[Ni_{11}(CO)_{20}H_2]^{2-}$, also on a molecular weight calculated from unit cell and density measurements.

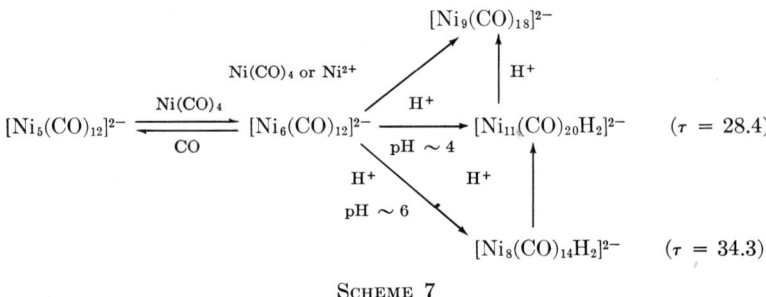

SCHEME 7

It is also worth noting that the ease of hydrolysis decreases in the series: $Ni_5 > Ni_6 > Ni_8 > Ni_{11}$, as expected for a progressively higher delocalization of charge and consequent lowering of the nucleophilicity of the cluster.

Finally mixed-metal clusters with formulas $[M_2Ni_3(CO)_{16}]^{2-}$ (M = Cr, Mo, W) and $[M_2Ni_4(CO)_{14}]^{2-}$ (M = Mo) have been isolated by Ruff by condensation of the corresponding dianions $[M_2(CO)_{10}]^{2-}$ with $Ni(CO)_4$ in refluxing THF (yields 30–68%) (*111*).

XIV

PLATINUM DERIVATIVES

An insoluble compound formulated as $[Pt(CO)_2]_n$ has been obtained by Booth and Chatt both by hydrolysis of $Pt(CO)_2Cl_2$ in benzene and by carbonylation of Na_2PtCl_4 in ethanol (*19, 20*).

More recently, reductive carbonylation of $Na_2PtCl_6 \cdot 6H_2O$ in methanol, in the presence of alkali acetates or hydroxide, has been shown to proceed according to the following reaction scheme (29, 95):

$$[PtCl_6]^{2-} \xrightarrow{OH^-, CO} [Pt(CO)Cl_3]^- \xrightarrow{OH^-, CO} [Pt(CO)_2]_n \text{ or } [Pt_3(CO)_6]_n^{2-}$$
$$(n > 6)$$
$$\xrightarrow{OH^-, CO} [Pt_{18}(CO)_{36}]^{2-} \xrightarrow{OH^-, CO} [Pt_{15}(CO)_{30}]^{2-} \xrightarrow{OH^-, CO}$$

olive-green   yellow-green

$$[Pt_{12}(CO)_{24}]^{2-} \xrightarrow{OH^-, CO} [Pt_9(CO)_{18}]^{2-} \xrightarrow{OH^-, CO} [Pt_6(CO)_{12}]^{2-}$$

blue-green   violet-red   orange-red

During the reduction the first carbonyl product detectable in solution is the well-known anion $[Pt(CO)Cl_3]^-$, which, in the presence of alkali is rapidly transformed into "platinum carbonyl." Although the complete insolubility of this compound makes its characterization difficult, it seems reasonable that its formulation should be changed. Accurate atomic absorption analyses show that the silky precipitate referred to as platinum carbonyl retains small but significant quantities of alkali metal cations, M, even after several washings. The Pt/M ratio is actually near to 15 and suggests that platinum carbonyl could instead be a mixture of carbonylplatinate oligomers with an average nuclearity of about 30.

The so-called platinum carbonyl readily reacts further with alkali, dissolving to give a series of dianions with general formula $[Pt_3(CO)_6]_n^{2-}$ ($n = $ 6, 5, 4, 3, 2). Each of these dianions can be obtained in a pure state by careful control of the initial $OH^-/Pt(IV)$ ratio ($n = 3$, $OH^-/Pt = 8$–9; $n = 4$, $OH^-/Pt = 6$–6.1; $n = 5$, $OH^-/Pt = 5.7$). However, these ratios are quite critical and the reaction must be followed by IR. If the incorrect product is formed, the desired final product can be obtained by adding further small amounts of alkali or of sodium hexachloroplatinate.

Under these experimental conditions, it has proved quite difficult to stop the reduction at $[Pt_{18}(CO)_{36}]^{2-}$. This species has been more conveniently prepared, in the absence of alkali, by oxidation of preformed $[Pt_{15}(CO)_{30}]^{2-}$ with sodium hexachloroplatinate according to the reaction:

$$6[Pt_{15}(CO)_{30}]^{2-} + [PtCl_6]^{2-} \xrightarrow{CO, MeOH} 5[Pt_{18}(CO)_{36}]^{2-} + [Pt(CO)Cl_3]^- + 3Cl^- \quad (63)$$

The orange-red dianion $[Pt_6(CO)_{12}]^{2-}$, which has also been obtained by reductive carbonylation in concentrated methanolic sodium hydroxide solution, is better synthesized by reducing preformed $[Pt_3(CO)_6]_n^{2-}$ ($n = $ 3, 4, or 5) with sodium or lithium metal in THF. Using Na/K alloy, the reduction goes still further to give the unstable pink dianion $[Pt_3(CO)_6]^{2-}$ ($\nu_{CO} = 1950, 1745$ cm^{-1}), as yet not isolated in the solid state.

The most striking peculiarity of all of these derivatives is their ability to de- and reoligomerize by breaking or forming Pt—Pt bonds. In fact all of these clusters are degraded not only by halides and by tertiary phosphines [see Eq. (34)] but also, when $n = 4, 5$, or 6 by molecular hydrogen [see Eq. (40)]. This last reaction can be reversed by oxidation with air and the cycle repeated several times with simple formation of water.

Finally, we should mention that, in the light of these results, which established the first contact points between the chemistry of the carbonyl derivatives of nickel and platinum, some analogous palladium derivatives could be expected. Until now, however, all attempts to prepare them have failed giving only palladium metal under a variety of conditions (*95*).

XV
BONDING THEORIES

The metal frameworks in high nuclearity clusters can be considered as finite parts of close-packed metallic structures stabilized by the external ligands and the negative charges. The apparent simplicity arising from the fact that in a metal cluster there is a finite number of metal atoms is offset by the complication of considering the interactions among the combinations of σ, π, and π^* orbitals of the ·CO groups with suitable combinations of metal orbitals. For example, as pointed out several times (*24, 34*), metal–metal bonds are strongly mixed with the bridging carbonyl groups. Owing to the complexity of these interactions, the number of independent parameters to be used in a sufficiently accurate LCAO-MO calculation is generally so high as to be well out of the range of the most modern computers. Other more advanced calculations, particularly devised for clusters of pure metals (*114*), have not yet been used in this area.

This situation has forced us to compare the experimental data with more-or-less empirical hypotheses, each one of which has its advantages and has enjoyed some popularity. So far the following four different approaches have been proposed: (a) noble gas rule (*112*); (b) topological hypothesis (*88*); (c) relationship between metal clusters and polyboranes and carboranes (*120, 121*); and (d) LCAO-MO treatments limited to the metal atoms (*24, 26, 34, 35, 105, 111*).

The rare gas rule is based on the assumption that bonding interactions are localized, or, more exactly, localizable, along the directions of minimum distance, namely, the edges of the polyhedron; therefore, M—M bonds can be represented by two center–one electron pair bonds. For a metal cluster the rule then takes the form

$$N_3 = 18N_1 - 2N_2 \tag{64}$$

where N_3 is the number of valence shell electrons, given by the number of outer electrons of all of the metal atoms plus the number of electrons donated by the ligands and the eventual negative charge (when a heteroatom is present in the cage, it is considered to donate all of its outer electrons to the cluster). The N_1 is the number of metal atoms, and N_2 is the number of polyhedral edges.

First, it should be noted that, although almost all clusters with 3 or 4 metal atoms have the favored 18 outer electron, inert gas configuration on each metal atom (34), Table VI shows that only a few penta- and hexanuclear clusters and almost none of higher nuclearity obey this rule. Second, the rule requires both an unexplained topological correspondence between the number of polyhedral edges and the number of low-energy metallic molecular orbitals, and a bond order of 1 for each metal–metal bond. We gain, therefore, a picture of metal cluster bonding quite different from that envisaged in an infinite metal structure, even if the M—M bond distances found in such a structure are sometimes shorter by only a few hundredths of an Angström from those found in clusters that obey the rule.

It obviously follows that the noble gas rule is an empirical and formal one and that the electron density effectively concentrated along a polyhedral edge must be less than required by the rule. Nevertheless, deviations are quantitatively small and the rule is often a useful formalism.

The second approach has been proposed by Kettle (88) and is based on an empirical topological correspondence between the sum of edges and triangular faces and the number of bonding skeletal orbitals. This approach tries to take into account, in a qualitative way, the additional possibility of interactions between skeletal and CO orbitals, although in practice it does not provide a reliable picture of the cluster. For example, Kettle found $Rh_6(CO)_{16}$ to be 2 electrons short and foresaw the possibility of synthesizing the anion $[Rh_6(CO)_{16}]^{2-}$, a prediction that has not been fulfilled (99).

Recently, Wade has pointed out a formal analogy between the electronic structures of carboranes and polyboranes and those of metal carbonyl clusters based on the assumption that certain triangulated polyhedra require the same number of skeletal orbitals whether there are BH or CH units as well as transition metal atoms at their apices (120). This assumption is quite reasonable as the synthesis of a large number of polyboranes and carboranes in which a transition metal atom takes the place of a BH skeletal unit may be carried out (70).

According to this approximation, in a metal carbonyl cluster each metal atom, like a BH unit, may effectively contribute only three orbitals to the skeleton, whereas the remaining six are left for bonding with the external ligands and for lone pairs, and are assumed to be completely filled. The

TABLE VI

A COMPARISON OF THE NOBLE GAS RULE WITH WADE'S RULES FOR HIGH NUCLEARITY METAL CARBONYL CLUSTERS

Geometry of the cluster (from X-rays)	Example	Ref.	N_1 = metal atoms	N_2 = edges	N_3 = electrons	Noble gas rule, $18N_1 - 2N_2$	Wade's relationship $14N_1 + x$ [a]	Wade's bond order
Trigonal bipyramid	$[Ni_5(CO)_{12}]^{2-}$	95	5	9	76	72	72	0.87
Square pyramid	$Fe_5(CO)_{15}C$	26	5	8	74	74	74 (nido)	
Bicapped tetrahedron	$Os_6(CO)_{18}$	104	6	12	84	84	84 (capped)	0.50
Bicapped rhombus	$[Mo_2Ni_4(CO)_{14}]^{2-}$	111	6	13	82	82	86	0.78
Trigonal prism	$[Rh_6(CO)_{15}C]^{2-}$	7	6	9	90	90	86	0.58
	$[Pt_6(CO)_{12}]^{2-}$	29	6	9	86	90	86	0.47
Octahedron	$Rh_6(CO)_{16}$	53	6	12	86	84	86	
Capped octahedron	$[Rh_7(CO)_{16}]^{3-}$	4	7	15	98	96	98 (capped)	0.53
	$[Rh_7(CO)_{16}I]^{2-}$	11	7	15	100	96	98 (capped)	
Bicapped trigonal prism	$Rh_8(CO)_{19}C$	12	8	15	114	114	114	0.6
Square antiprism	$[Co_8(CO)_{18}C]^{2-}$	13	8	16	114	112	114	0.56
Condensed trigonal prism	$[Pt_9(CO)_{18}]^{2-}$	29	9	15	128	132	128	0.67
	$[Pt_{12}(CO)_{24}]^{2-}$	95	12	21	170	174	170	0.62
	$[Pt_{15}(CO)_{30}]^{2-}$	29	15	27	212	216	212	0.59
Two connected octahedra	$[Rh_{12}(CO)_{30}]^{2-}$	5	12	25	170	166	170	0.52
Distorted tetracapped cube	$Rh_{12}(CO)_{25}(C-C)$	13	12	27	164	162	170	
Pentagonal tetracapped prism	$Rh_{15}(CO)_{28}(C)_2-$	9	15	31	200	208	198 ($N_1 = 14$)	

[a] Closo, $x = 2$; nido, $x = 4$; arachno, $x = 6$; capped closo, $x = 0$.

analogy with the triangulated polyboranes and carboranes then requires that, in the case of a "closo" cluster containing N_1 metal atoms, only $N_1 + 1$ electron pairs are accommodated in the skeletal bonding orbitals obtained from the $3N_1$ AO's; but accommodation of $N_1 + 2$ or $N_1 + 3$ electron pairs should cause cage opening to "nido" and "arachno" polyhedra. Moreover capping of a triangular face to give a capped "closo" polyhedron (for instance $[Rh_7(CO)_{16}]^{3-}$ of structure XVI in Fig. 7) would require only N_1 skeleton electron pairs (*121a*). Thus, the Wade relationship for simple "closo"-triangulated polyhedra results in the mathematical expression

$$N_3 = 12N_1 + 2N_1 + 2 = 14N_1 + 2 \tag{65}$$

Unfortunately, application of this rule is severely limited by the fact that the correspondence of polyhedra between polyboranes and carboranes with metal clusters is confined to the square pyramid, the trigonal bipyramid, and the octahedron; moreover, the rule fails to account for the 76 electrons of the known trigonal bipyramidal clusters (Table VI).

Indiscriminate application of the rule shows it to be in fairly good agreement with most of the high nuclearity clusters (Table VI). Furthermore, the metal–metal bond orders calculated in this way compare much more reasonably with that expected in the pure metals.

However, after a first glance, it becomes evident that this generalized version of the rule also does not match the experimental data sufficiently well. For example, the bond orders derived from it do not agree with the constant Pt—Pt distance in the series $[Pt_3(CO)_6]_n^{2-}$ ($n = 2, 3, 4, 5$) (*29, 95*); this behavior apparently requires that at least one of the $N_1 + 1$ electron pairs allocated to the metal–metal bonding framework be transferred to a nonbonding situation. Still more relevant is the fact that the generalized form of the rule, being dependent only on the whole number of metal atoms, does not discriminate between different geometries, and, therefore, is of very limited predictive capacity. For instance, with 8 metal atoms and 114 electrons both the bicapped trigonal prism and the square antiprism are known, and with 12 metal atoms and 170 electrons both the condensed trigonal prism and the double octahedron have been found.

The fourth approach, the LCAO-MO treatment, may potentially give the best insight into the metal cluster bonding problem, but its accuracy is limited by the usual assumption of independence between skeletal and ligand orbitals. In favor of this assumption there is, as already discussed, the constant number of valence electrons for the octahedral clusters (86 electrons), which is maintained throughout a concurrent variation in carbon monoxide coordination (from three to five). This behavior shows that, as a first approximation ignoring the significant distorsions of the octahedral core, the separation of MO's at the frontier is independent of both

TABLE VII

EXAMPLES OF THE SIMPLEST LCAO-MO APPROXIMATION IN
HIGH NUCLEARITY METAL CARBONYL CLUSTERS

Compound	AO for bonding with CO's[a]	AO for lone pairs and back donation	AO for skeletal bonding[a]	Ref.
$Rh_6(CO)_{16}$ and $[Co_6(CO)_{15}]^{2-}$	sp^3	$d_{xy}, d_{x^2-y^2}$	d_{xz}, d_{yz}, d_{z^2}	34, 35
$[Co_6(CO)_{14}]^{4-}$	dsp^3	d_{xy} (or $d_{x^2-y^2}$) d_{xz}, d_{yz}, d_{z^2}		24
Skeletal MO's in O_h symmetry: $d_{z^2} = A_{1g} + E_g + T_{1u}$ $d_{(xz, yz)} = T_{1u} + T_{2g} + T_{2u} + T_{1g}$				

[a] These arbitrary divisions have been made in agreement with the stereochemistry of the bonds around each metal center and with the particular choice of Cartesian axes.

the number and type of carbonyl groups. It seems, therefore, reasonable to assume independence between skeleton and ligand MO's if only a first-order justification of the electronic distribution is desired.

In the simplest form of LCAO-MO treatment the metal orbitals are arbitrarily divided into three groups, as shown in Table VII. The $A_{1g}(z^2)$, $T_{2g}(xz, yz)$ and $T_{1u}(xz, yz)$ of Table VII are bonding combinations (54), and accommodate 14 electrons. The remaining 72 electrons are accommodated in the thirty-six other atomic orbitals (or their suitable combinations).

A similar qualitative approach, although with a less rigid separation between the starting atomic orbitals, has been applied by Dahl to $Fe_5(CO)_{15}C$ and $[M_2Ni_3(CO)_{16}]^{2-}$ (M = Cr, Mo, W) (26, 111).

More recently, Mingos has gone further in treating the isolated Co_6 cage of the $[Co_6(CO)_{14}]^{4-}$ anion in a rigorous way (105). On the basis of orbital overlap calculations, the fifty-four cobalt AO's are combined to give thirty-one bonding or weakly antibonding and twenty-three strongly antibonding skeletal orbitals. From the point of view of the isolated Co_6 cage this ordering is both interesting and probably accurate, but, from the $[Co_6(CO)_{14}]^{4-}$ cluster point of view there are similar limitations as before. In fact, metal MO's and ligands MO's are still treated separately and their reciprocal interactions have been estimated only by simple inspection. The final ordering of MO's predicts a frontier separation of about 80,000 cm^{-1}, which is a rather surprising result for a dark red-brown colored cluster.

Finally, we should note that the LCAO-MO approximations profit from the high symmetry of some species and that application to less symmetrical or unsymmetrical clusters such as $Rh_8(CO)_{19}C$ and $Rh_{12}(CO)_{25}(C-C)$ is quite difficult.

Acknowledgments

We wish to acknowledge the contribution of all the co-workers cited in the references, and to mention particularly Dr. S. Martinengo, Prof. L. F. Dahl, and Dr. B. T. Heaton. Direct financial support was obtained from Consiglio Nazionale delle Ricerche, Accademia dei Lincei, the Royal Society, NATO, and the University of Wisconsin.

References

1. Albano, V. G., Chini, P., and Scatturin, V., *J. Chem. Soc. Chem. Commun.* 163 (1968).
2. Albano, V. G., Chini, P., and Scatturin, V., *J. Organometal. Chem.* **15**, 423 (1968).
3. Albano, V. G., Bellon, P. L., Chini, P., and Scatturin, V., *J. Organometal. Chem.* **16**, 461 (1969).
4. Albano, V. G., Bellon, P. L., and Ciani, G. F., *J. Chem. Soc., Chem. Commun.* 1024 (1969).
5. Albano, V. G., and Bellon, P. L., *J. Organometal. Chem.* **19**, 405 (1969).
6. Albano, V. G., Bellon, P. L., and Sansoni, M., *J. Chem. Soc. A* 678 (1971).
7. Albano, V. G., Sansoni, M., Chini, P., and Martinengo, S., *J. Chem. Soc., Dalton Trans.* 651 (1973).
8. Albano, V. G., Ciani, G. and Chini, P., *J. Chem. Soc., Dalton Trans.* 432 (1974).
9. Albano, V. G., Chini, P., Martinengo, S., Sansoni, M., and Strumolo, D., *J. Chem Soc., Chem. Commun.* 299 (1974).
10. Albano, V. G., Chini, P., Martinengo, S., McCaffrey, D. J. A., Strumolo, D., and Heaton, B. T., *J. Amer. Chem. Soc.* **96**, 8106 (1974).
11. Albano, V. G., Ciani, G., Martinengo, S., Chini, P., and Giordano, G., *J. Organometal. Chem.* **88**, 381 (1975).
12. Albano, V. G., Sansoni, M., Chini, P., Martinengo, S., and Strumolo, D., *J. Chem. Soc., Dalton Trans.* 305 (1975).
13. Albano, V. G., Chini, P., Ciani, G., Martinengo, S., Sansoni, M. and Strumolo, D., unpublished results. Albano, V. G., Ceriotti, A., Chini, P., Ciani, G., Martinengo, S., and Anker, M., *J. Chem. Soc., Chem. Commun.* 859 (1975).
14. Anders, U., and Graham, W. A. G., *J. Chem. Soc., Chem. Commun.* 291 (1966).
15. Anders, U., and Graham, W. A. G., *J. Amer. Chem. Soc.* **89**, 539 (1967).
15a. Angoletta, M., Malatesta, L., and Caglio, G., *J. Organometal. Chem.* **96**, 99 (1975).
16. Baird, M. C., *Progr. Inorg. Chem.* **9**, 1 (1968).
17. Bellon, P. L., Manassero, M., and Sansoni, M., *J. Chem. Soc., Dalton Trans.* 2423 (1973) and references therein.
18. Bīryukov, B. P., and Struchov, Y. T., *Russian Chem. Rev.* **39**, 789 (1970).
18a. Bodner, G. M., and Todd, L. J., *Inorg. Chem.* **13**, 1335 (1974).
19. Booth, G., Chatt, J., and Chini, P., *J. Chem. Soc., Chem. Commun.* 639 (1965).
20. Booth, G., and Chatt, J., *J. Chem. Soc. A* 2131 (1969).
21. Booth, B. L., Else, M. J., Fields, R., and Haszeldine, R. N., *J. Organometal. Chem.* **27**, 119 (1971).
22. Booth, B. L., Else, M. J., Fields, R., Goldwhite, H., and Haszeldine, R. N., *J. Organometal. Chem.* **14**, 417 (1968).

23. Boon Keng, T., Hall, M. B., Fenske, R. S., and Dahl, L. F., *J. Organometal. Chem.* **70,** 413 (1974) and references therein.
24. Braterman, P. S., *Structure Bonding (Berlin)* **10,** 57 (1971).
25. Braterman, P. S., "Metal Carbonyl Spectra." Academic Press, New York, 1975.
26. Braye, E. H., Dahl, L. F., Hubel, W., and Wampler, D. L., *J. Amer. Chem. Soc.* **84,** 4633 (1962).
27. Brown, D. L. S., Connor, J. A., and Skinner, H. A., *J. Chem. Soc. Faraday Trans. I* **71** (3), 699 (1975).
28. Calabrese, J. C., Dahl, L. F., Cavalieri, A., Chini, P., Longoni, G., and Martinengo, S., *J. Amer. Chem. Soc.* **96,** 2616 (1974).
29. Calabrese, J. C., Dahl, L. F., Chini, P., Longoni, G., and Martinengo, S., *J. Amer. Chem. Soc.* **96,** 2614 (1974).
30. Cartner, A., Robinson, B., and Gardner, P., *J. Chem. Soc., Chem. Commun.* 317 (1973).
31. Cassar, L., and Foà, M., *Inorg. Nucl. Chem. Lett.* **6,** 291 (1970).
32. Chaston, S. H., and Stone, F. G. A., *J. Chem. Soc. A* 500 (1969).
33. Chini, P., *J. Chem. Soc., Chem. Commun.* 29 (1967).
34. Chini, P., *Inorg. Chim. Acta Rev.* **2,** 31 (1968).
35. Chini, P., and Albano, V. G., *J. Organometal. Chem.* **15,** 433 (1968).
36. Chini, P., *Inorg. Chem.* **8,** 1206 (1969).
37. Chini, P., Albano, V. G., and Martinengo, S., *J. Organometal. Chem.* **16,** 471 (1969).
38. Chini, P., and Martinengo, S., *Inorg. Chim. Acta* **3,** 315 (1969).
39. Chini, P., and Martinengo, S., *J. Chem. Soc., Chem. Commun.* 1092 (1969).
40. Chini, P., and Martinengo, S., *Inorg. Chim. Acta* **3,** 299 (1969).
41. Chini, P., *Pure Appl. Chem.* **23,** 489 (1970).
42. Chini, P., and Longoni, G., *J. Chem. Soc. A* 1542 (1970).
43. Chini, P., Martinengo, S., and Garlaschelli, G., *J. Chem. Soc., Chem. Commun.* 709 (1972).
44. Chini, P., Cavalieri, A., and Martinengo, S., *Coord. Chem. Rev.* **8,** 3 (1972).
45. Chini, P., Martinengo, S., and Giordano, G., *Gazz. Chim. Ital.* **102,** 330 (1972).
46. Chini, P., Martinengo, S., Strumolo, D., Albano, V. G., and Sansoni, M., *Abstr. 165th Meeting Amer. Chem. Soc.,* 57, 1973.
47. Chini, P., Martinengo, S., McCaffrey, D. J. A., and Heaton, B. T., *J. Chem. Soc., Chem. Commun.* 310 (1974).
48. Churchill, M. R., Wormald, J., Knight, J., and Mays, M., *J. Chem. Soc., Chem. Commun.* 458 (1970).
49. Churchill, M. R., and Wormald, J., *J. Chem. Soc., Dalton Trans.* 2410 (1974).
50. Churchill, M. R., Bezman, S. A., Osborn, J. A., and Wormald, J., *Inorg. Chem.* **11,** 1818 (1972).
51. Churchill, M. R., and Bezman, S. A., *Inorg. Chem.* **13,** 1418 (1974).
52. Connor, J. A., Skinner, H. A., and Virmani, Y., *J. Chem. Soc., Faraday Trans. I* 1574 (1972).
52a. Connor, J. A., Skinner, H. A., and Virmani, Y., *Faraday Symp. Chem. Soc.* **8,** 18 (1974).
53. Corey, E. R., Dahl, L. F., and Beck, W., *J. Amer. Chem. Soc.* **85,** 1202 (1963).
54. Cotton, F. A., and Haas, T. E., *Inorg. Chem.* **3,** 10 (1964).
55. Cotton, F. A., Kruckzinski, L., Shapiro, B. L., and Johnson, L. F., *J. Amer. Chem. Soc.* **94,** 6191 (1972).

56. Darensbourgh, D. J., Conder, H. L., Darensbourgh, M. Y., and Hasday, C., *J. Amer. Chem. Soc.* **95**, 5919 (1973).
57. De Beer, J. A., Haines, R. J., Greatrex, R., and vanWyk, J. A., *J. Chem. Soc., Dalton Trans.* 2341 (1973) and references therein.
58. Eady, C. R., Johnson, B. F. G., and Lewis, J., *J. Organometal. Chem.* **37**, C 39 (1972).
59. Eady, C. R., Johnson, B. F. G., Lewis, J., and Matheson, T., *J. Organometal. Chem.* **57**, C82 (1973).
60. Eady, C. R., Johnson, B. F. G., and Lewis, J., *J. Organometal. Chem.* **57**, C84 (1973).
61. Eady, C. R., personal communication.
62. Ercoli, R., and Barbieri-Hermitte, F., *Rend. Acad. Lincei, Cl. Sci. Fis. Nat.* **16**, 249 (1949).
63. Ercoli, R., Santambrogio, E., and Tettamanti-Casagrande, G., *Chim. Ind. (Milano)* **44**, 565 (1962).
64. Evans, J., Johnson, B. F. G., Lewis, J., Norton, J. R., and Cotton, F. A., *J. Chem. Soc., Chem. Commun.* 807 (1973).
65. Ferguson, J. A., and Meyer, T. J., *J. Amer. Chem. Soc.* **94**, 3409 (1972) and references therein.
66. Fieldhouse, S. A., Freeland, B. H., Mann, C. D., and O'Brien, R. J., *J. Chem. Soc., Chem. Commun.* 181 (1970).
67. Flitcroft, N., and Leach, J., *J. Organometal. Chem.* **18**, 367 (1969).
70. Grimes, R. N., *Ann. N. Y. Acad. Sci.* In press.
71. Guerrieri, F., and Foà, M., personal communication.
72. Guss, J. M., Mason, R., Thomas, K. M., van Koten, G., and Noltes, J. G., *J. Organometal. Chem.* **40**, C79 (1972).
73. Haines, L. I. B., and Poe, A. J., *J. Chem. Soc. A* 2826 (1969).
74. Heaton, B. T., Towl, D. C., Chini, P., Fumagalli, A., McCaffrey, D. J. A., and Martinengo, S., *J. Chem. Soc., Chem. Commun.* 523 (1975).
74a. Heaton, B. T., Chini, P., Martinengo, S., McCaffrey, D. J. A., and Fumagalli, A., unpublished results.
75. Hieber, W., and Lagally, H., *Z. Anorg. Allg. Chem.* **251**, 96 (1943).
76. Hieber, W., and Ellermann, J., *Z. Naturforsch. B* **18**, 595 (1963).
77. Hieber, W., Ellermann, J., and Zahn, E., *Z. Naturforsch. B* **18**, 589 (1963).
78. Hieber, W., and Shubert, E. H., *Z. Anorg. Allg. Chem.* **338**, 32 (1965).
79. Hsieh, A. T. T., and Mays, M. J., *J. Organometal. Chem.* **37**, C53 (1972).
80. Ingle, W. M., Preti, G., and MacDiarmid, A., *J. Chem. Soc., Chem. Commun.* 497 (1973).
81. Johnson, B. F. G., Johnston, R. D., and Lewis, J., *J. Chem. Soc., Chem. Commun.* 1057 (1967).
82. Johnson, B. F. G., Johnston, R. D., Lewis, J., Robinson, B. H., and Wilkinson, G., *J. Chem. Soc. A* 2856 (1968).
83. Johnson, B. F. G., Johnston, R. D., and Lewis, J., *J. Chem. Soc. A* 2865 (1968).
84. Johnson, B. F. G., Lewis, J., and Robinson, P. W., *J. Chem. Soc. A* 1100 (1970).
85. Johnson, B. F. G., Lewis, J., and Williams, I. G., *J. Chem. Soc. A* 901 (1970).
86. Johnston, R. D., *Advan. Inorg. Chem. Radiat.* **14**, 471 (1971).
87. Kaesz, H. D., *Chem. Brit.* **9**, 344 (1973).
88. Kettle, S. F. A., *J. Chem. Soc. A* 314 (1967).
89. Klanberg, F., and Muetterties, E. L., *J. Amer. Chem. Soc.* **90**, 3296 (1968).

90. King, R. B., *J. Amer. Chem. Soc.* **94**, 95 (1972).
91. King, R. B., *Progr. Inorg. Chem.* **15**, 288 (1972).
92. Kitamura, T., and Joh, T., *J. Organometal. Chem.* **65**, 235 (1974).
93. Kristoff, J. S., and Shriver, D. F., *Inorg. Chem.* **13**, 499 (1974) and references therein.
94. Kruck, T., Hofler, M., and Noack, M., *Chem. Ber.* **99**, 1153 (1966).
95. Longoni, G., Chini, P., Martinengo, S., Dahl, L. F., Lower, L. and Calabrese, J. C., unpublished results; Longoni, G., Chini, P., Lower, L. D. and Dahl, L. F., *J. Amer. Chem. Soc.* **97**, 5034 (1975).
96. Malatesta, L., Caglio, G., and Angoletta, M., *J. Chem. Soc., Chem. Commun.* 532 (1970).
97. Malatesta, L., and Caglio, G., *J. Chem. Soc., Chem. Commun.* 420 (1967).
98. Martinengo, S., Chini, P., and Giordano, G., *J. Organometal. Chem.* **27**, 389 (1971).
99. Martinengo, S., and Chini, P., *Gazz. Chim. Ital.* **102**, 344 (1972).
100. Martinengo, S., Chini, P., Albano, V. G., Cariati, F., and Salvatori, T., *J. Organometal. Chem.* **59**, 379 (1973).
101. Martinengo, S., Chini, P., Giordano, G., Ceriotti, A., Albano, V. A. G., and Ciani, G., *J. Organometal. Chem.* **88**, 375 (1975).
102. Martinengo, S., Chini, P., Fumagalli, A., and Strumolo, D., unpublished results.
103. Mason, R., and Robinson, W. R., *J. Chem. Soc., Chem. Commun.* 468 (1968).
104. Mason, R., Thomas, K. M., and Mingos, D. M. P., *J. Amer. Chem. Soc.* **95**, 3802 (1973).
105. Mingos, D. M. P., *J. Chem. Soc., Dalton Trans.* 133 (1974).
106. Muetterties, E. L., and Wright, C. M., *Quart. Rev.* **21**, 109 (1967).
107. Noak, K., *J. Organometal. Chem.* **7**, 151 (1967).
108. Penfold, B. R., *Perspect. Struct. Chem.* **2**, 71 (1968).
109. Penfold, B. R., and Robinson, B. H., *Accounts Chem. Res.* **6**, 73 (1973) and references therein.
110. Ruff, J. K., *Inorg. Chem.* **7**, 1818 (1968).
111. Ruff, J. K., White, R. P., and Dahl, L. F., *J. Amer. Chem. Soc.* **93**, 2159 (1971).
112. Sidgwick, N. V., and Bailey, R. W., *Proc. Roy. Soc., Ser. A* **144**, 521 (1934).
113. Sirigu, A., Bianchi, M., and Benedetti, E., *J. Chem. Soc., Chem. Commun.* 596 (1969).
114. Slater, J. C., and Johnson, K. H., *Phys. Rev. B* **5**, 844 (1972).
115. Spiro, G. T., *Progr. Inorg. Chem.* **11**, 1 (1969).
116. Sternberg, H. W., Markby, R., and Wender, I., *J. Amer. Chem. Soc.* **82**, 3638 (1960).
117. Stewart, R. P., Anders, U., and Graham, W. A. G., *J. Organometal. Chem.* **32**, C49 (1971).
118. Todd, L. J., and Wilkinson, J. R., *J. Organometal. Chem.* **77**, 1 (1974).
119. Todd, L. J., and Wilkinson, J. R., *J. Organometal. Chem.* **80**, C31 (1974).
120. Wade, K., *J. Chem. Soc., Chem. Commun.* 792 (1971).
121. Wade, K., *Inorg. Nucl. Chem. Lett.* **8**, 559 (1972) and references therein.
121a. Wade, K., personal communication.
122. Wrighton, M., *Chem. Rev.* **74**, 401 (1974).

Free Radicals in Organometallic Chemistry

M. F. LAPPERT and P. W. LEDNOR*

School of Molecular Sciences
University of Sussex
Brighton, England

I. Introduction	345
II. Metal-Centered Organometallic Radicals	349
A. Main Group Element Compounds	352
B. Transition Metal Compounds	363
III. Other Organometallic Radicals	367
IV. Bimolecular Homolytic Substitution (S_H2) at the Metal Center of an Organometallic Substrate	370
A. Main Group Element Compounds as Substrates	371
B. Transition Metal Compounds as Substrates	373
V. Addition or Elimination Radical Reactions	381
A. Oxidative Addition of Alkyl Halides or Related Reagents and Reductive Elimination	381
B. Metal Alkyl Photolysis or Thermolysis	388
VI. Appendix	390
References	392

I
INTRODUCTION

It is becoming apparent that much organometallic chemistry overlaps with that of free radicals. In this context, an organometallic compound is taken as having at least one metal–carbon bond. The object of this review is to provide a classification of the area, discuss briefly in very general terms the experimental methods, and offer a guide to the relevant literature. We do not claim to cover this comprehensively but aim to give a representative overview, with emphasis on features of interest to the organometallicist rather than the ESR or kinetics specialist. We propose to deal with a relatively small number of systems in detail and point to others more superficially. In keeping with the aim of this volume, we shall draw extensively on examples from our own work; we do this with some temerity, because our direct involvement in this field is of recent origin (since 1971).

We propose the following classification: (i) organometallic paramagnetic compounds (Sections II and III), and (ii) organometallic reactions, in

---

* Present address: Institut für Anorganische Chemie der Universität München, 8 München 2, Meiserstrasse 1, Germany.

which paramagnetic transient species, usually organic free radicals, are implicated in the reaction pathway (Sections IV and V).

Class (i) may be divided into compounds in which the unpaired electron is localized principally on the metal M (I) (see Section II); or remote from the metal (II) (see Section III). The former belong to a wider group, the metal-centered radicals, encompassing not only, for example, the or-

$$\dot{M}L \qquad L'M\sim\sim\sim\sim\cdot$$
$$(I) \qquad\qquad (II)$$

ganometallic $\dot{S}iMe_3$, but also the inorganic $\dot{S}iCl_3$. Corresponding paramagnetic transition metal complexes (I) are rarely named as radicals because they are often stable, and the term "radical" has traditionally had the connotation of a species of low kinetic stability.

Class (ii) is conveniently separated into free radical reactions involving (a) substitution at the metal center, S_H2, or bimolecular homolytic substitutions at M (by analogy with S_N2, *etc.*) (see Section IV), for example,

$$LM-R + \dot{X} \longrightarrow LM-X + \dot{R} \qquad (1)$$

(b) addition or elimination at M (see Section V, A and B), for example,

$$ML + \dot{R} \rightleftharpoons LMR \qquad (2)$$

and (c) a site remote from the metal.

Although the first direct demonstration of the existence of a transient alkyl free radical involved an organometallic system of the type shown in Eq. (2), [Paneth's experiments on the pyrolysis of $PbMe_4$ and the recombination of $\dot{M}e$ with metallic Pb, Zn, Sb, Bi, or Be (*158*)], the topic of free radical organometallic chemistry remained largely neglected except by the gas kineticist until the last decade. This may be demonstrated by reference to the standard general text (1967–1968) on organometallic chemistry (*35, 36*). There is no subject index entry for "radical," "free radical," or "homolysis," although Paneth's and related experiments were discussed, as were also, in free radical terms, the mechanism of thermal decomposition of HgR_2 or of alkyls or acyls of Cu, Ag, and Au, and the addition of R_3Sn-H (via $\dot{S}nR_3$) to olefins or other unsaturated substrates. Even the autoxidation of metal alkyls MR_n was then regarded as polar in nature, although the intermediacy of peroxides, such as $ROOMR_{n-1}$, was recognized. Radical anions, such as $C_{10}H_8^{-\cdot}$, are not within our terms of reference unless they are metal-centered, e.g., the stable $[BAr_3]^{-\cdot}$ (*35*). The existence of numerous transient metal-centered radicals, such as $\dot{H}gMe$, $\dot{P}bMe_3$, or $\dot{S}bMe_2$, was inferred from gas kinetic data, e.g., on thermolysis of $HgMe_2$, $PbMe_3H$, or $SbMe_3$ (*74, 75*).

Electron spin resonance spectroscopy has made a considerable impact on the study of organometallic free radical chemistry. Four significant developments may be identified: (i) the introduction of steady-state photolysis techniques for the detection and identification of transient radicals; (ii) various methods for determining the rates of radical reactions in solution, by measuring relative radical concentrations at different times using competition with a radical reaction of known rate; (iii) the use of spin traps, such as Bu^{t}NO, to intercept a transient radical Ṙ as the kinetically more stable spin-trapped derivative, such as the nitroxide (III); and (iv) the attachment of a stable organic free radical, usually a nitroxide, a spin label, that can interact with a reaction center, as in structure IV for coenzyme B$_{12}$-controlled enzymatic isomerizations (193). A different mag-

netic resonance technique, which has some application in organometallic free radical chemistry, employs ^{1}H or, in principle, ^{13}C NMR spectroscopy and the observation of enhanced absorption or emission as a consequence of chemically induced dynamic nuclear polarization. Proper interpretation of CIDNP (Chemically induced dynamic nuclear polarization) spectra allows the detection of *radical pair* processes.

Other experimental methods, diagnostic for establishing radical pathways, are broadly kinetic or stereochemical in origin, being based on regio- or stereoselectivity. The effect of the addition of a known radical initiator, such as azobisisobutyronitrile, AIBN (V), or of a radical inhibitor, such as galvinoxyl (VI), upon the rate of a reaction may be informative. Similarly, the composition of a copolymer obtained by addition of two monomers, such as PhCH=CH$_2$ and CH$_2$=C(Me)CO$_2$Me, to the reaction mixture may implicate free radicals as initiators for the copolymerization since the copolymer composition is related to the relative propagation rates for each homopolymerization, and these rates are distinct and different for $\dot{R}$, R^+, or R^- initiation. In such experiments, as with spin trapping, great care must be taken to establish by control experiments that the added chemicals do not alter the course of the reaction that takes place in the absence of such addenda.

The participation of a specific free radical in a particular reaction may be inferred from product analysis: for instance, many alkyl radicals $\dot{R}$ undergo competing processes of radical combination and disproportionation, e.g., for $R = Et$,

$$2C_2\dot{H}_5 \begin{array}{c} \text{(a)} \\ \longrightarrow \\ \text{(b)} \\ \longrightarrow \end{array} \begin{array}{l} C_4H_{10} \\ \\ C_2H_6 + C_2H_4 \end{array} \qquad (3)$$

and the ratio of appropriate constants (k_a/k_b) for gas phase reactions of many radicals is known (*122*).

Trivalent, neutral, carbon-centered radicals are generally either planar or slightly pyramidal (unless there are highly electronegative substituents, as in the pyramidal $F_3\dot{C}$) with a low inversion barrier. Consequently, most radical reactions, in which the alkyl reactant is optically active with chirality at the α-carbon, lead to racemic products because free radicals have relatively long lifetimes compared with their relaxation with respect to nearest neighbors. Whereas the formation of a racemic product is consistent with a radical mechanism, it is not a necessary precondition; the possibility of stereoselectivity via radical reactions has long been recognized, especially in the context of geminate combinations of "caged" radical pairs which are among the fastest of chemical reactions (*131a*). A further caution relates to the temptation to generalize from an observation on an optically active compound to a whole class of reaction.

In the context of organometallic chemistry, optically stable chiral centers other than carbon are at this time found only for silicon, germanium, and the Group V elements. The corresponding trivalent radicals are pyramidal and, hence, a radical reaction involving such a metal-centered radical is expected to afford the products having the same configuration as the reactant; this is rarely of diagnostic value. However, radical VII readily racemizes (*175*).

$$\begin{array}{c} Ph_3Si \\ \diagdown \\ \underset{\diagup \ \diagdown}{\overset{\bullet}{Si}} \\ Ph \quad Me \end{array}$$

(VII)

Evidence for the participation of a free radical in an organometallic reaction, using one or more of the above methods, may still leave two major questions. The first is whether the radical pathway is a principal or a side reaction. Rather than discuss this generally (organometallic reactions are, after all, exceedingly diverse and complex), it is suggested that as many as practicable of the experimental methods be brought to bear on the study of a particular reaction. The second is whether the mechanism of the reaction is a chain or a nonchain process. This distinction, in an organometallic context, may not readily be made and requires detailed kinetic data.

The major secondary sources of literature are listed in Table I. The single most important item is the two-volume compendium edited by Kochi (126). It will be noticed that the role of organic free radicals in transition metal chemistry, except for Co(II) derivatives, is a rather new area and constitutes a major theme of our article (Sections III–V). With regard to metal-centered radicals (Section II), we concentrate particularly on those that have considerable kinetic stability, e.g., $\dot{\mathrm{Sn}}[\mathrm{CH}(\mathrm{SiMe}_3)_2]_3$, another recent development (48).

II
METAL-CENTERED ORGANOMETALLIC RADICALS

The study of metal-centered radicals has advanced significantly during the 1970s. Much of the earlier work was concerned with long-lived species, particularly radical anions, or with the interpretation of kinetic data from which the existence of various organometallic radicals was inferred (74, 75, 131a). The emphasis has shifted to the characterization of such transient species by ESR spectroscopy, and to the study of their structure, stability, and reactivity. Initial experiments were with the main group (Group IV) elements but have been extended to boron and the other Group III and V elements [phosphorus (17) including the phosphoranyl radicals (VIII),

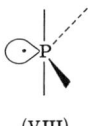

(VIII)

as well as oxygen- (128) and sulfur- (123) centered radicals are outside our scope]. Much attention has been given to the synthesis of stable transi-

TABLE I

REVIEWS RELEVANT TO ORGANOMETALLIC FREE RADICAL CHEMISTRY

Section of this review	Subject/title[a]	References
II	Organotin radical chemistry	135
II	Characterization of organometallic compounds by ESR— mainly concerned with line widths	174
II	Reactivity, selectivity, and polar effects in hydrogenation transfer reactions, including hydrogen abstraction from Ph_3Si—H by $\dot{P}h$	170
II	Atom transfer and substitution reactions, including $\dot{S}nR_3/R'Hal$ and $\dot{R}/R'_3Sn$—H reactions	163
II	Thermochemistry of free radicals, including $\dot{S}iR_3$	155
II	The shape of inorganic paramagnetic compounds, principally from ESR parameters	6
II	Organometallic electrochemistry (this topic is not discussed extensively here)	55
II	ESR of transition organometallic complexes (this topic is not discussed extensively here)	79, 80, 84
II/III	Organometallic radicals of B, Al, Ga, In, Tl, Si, Ge, Sn, and Pb	74
II/III	Organometallic radicals of As, Sb, and Bi	75
II/III	Organometallic radicals of Si, Ge, Sn, and Pb and their reactions	74, 110, 135
II/III	Free radical rearrangements, including $Si_2Me_6 \rightarrow Me_3SiSiMe_2\dot{C}H_2 \rightarrow Me_3SiCH_2\dot{S}iMe_2$	191
II/III	Addition of radicals, $\dot{M}R_3$ (M = Si, Ge, or Sn) to multiple bonds	1
II/III	Structure of free radicals by ESR spectroscopy, including spin trapping (no direct organometallic content)	71
II/III/IV	Organometallic radicals, principally of main group elements, literature survey 1971 to mid-1972	102
II/III/IV	Organometallic radicals, principally of main group elements; literature survey 1972–1973	103
II/III/IV	Organometallic radicals and reactions, excluding transition elements and radical ions; literature survey 1970–1971	111
II/III/IV	Organometallic radicals and reactions, excluding transition elements and radical ions; literature survey October, 1971, to December, 1973	112
[II/IV	Organophosphorus radicals (this topic is not within our scope)	17
II/IV	Rate constants for free radical reactions in solutions, including $\dot{M}R_3$ (M = Si, Ge, or Sn)	107
II/IV/V	ESR studies in nonaqueous solutions of organometallic complexes, particularly Si, Ge, Sn, and Pb	131

Free Radicals

TABLE I—*Continued*

Section of this review	Subject/title[a]	References
II/V	The homolytic pathway for thermal or photochemical decomposition of metal hydrocarbyls MR_n (R = alkyl or aryl); and stable paramagnetic species MR_n	*48, 49*
III/V	Application of magnetic resonance techniques, including the use of a spin label	*193*
IV	S_H2 reactions at some main group metal centers	*50, 51, 108, 151*
IV	S_H2 reactions at boron centers	*85*
IV	Electron transfer reactions of free radicals and metal complexes, including reactions of $\dot{R}$ with M^{n+}, e.g., oxidation by Cu^{2+} or reduction by Cr^{2+} (considered to be outside our scope unless M^{n+} represents an organometallic compound), and reactions of RMgX with transition metal complexes	*127, 129*
IV/V	Autoxidation of organometallic compounds of main group elements	*2*
IV/V	Synthetic aspects of free radical reactions of organoboranes	*23*
V	The kinetics and mechanism of metal alkyl decomposition; a particularly detailed account of HgR_2	*164*
V	CIDNP, e.g., RHal/LiR' or R'MgX systems	*186*
V	Electron transfer reactions of organic anions, including photolysis of LiAr and RHal/LiR' reactions	*76*
V	Spin trapping	*113*
V	Oxidative addition reactions of coordinatively unsaturated, low oxidation state, transition metal complexes, including established radical reactions especially of Co(II)	*89*
V	Organocobalt chemistry, including one-electron transfer reactions, especially of Co(II)	*59*
V	Organometallic transition metal complexes as initiators of free radical polymerization	*8*

[a] Entries in this column are intended to indicate the scope of the review in so far as it is relevant to organometallic chemistry.

tion metal (paramagnetic) alkyls, essentially by judicious choice of alkyl ligands. These, by their bulk and/or by the absence of β- or α-hydrogen, e.g., Me_3SiCH_2, $Ph\bar{C}H_2$, or 1-norbornyl⁻, may insure that normally facile decomposition pathways, such as β or α elimination, become energetically unfavorable (*48, 49*). Additionally, photolysis (and thermolysis) of metal alkyls has been examined critically: homolysis is an uncommon route for metal alkyl decomposition (see Section V, B). Electrochemical techniques have been used in organometallic chemistry; organometallic derivatives of main group elements tend to extrude carbanions or radicals, e.g., $\dot{H}gR$ or $\dot{M}R_3$ from HgRCl or $(MR_3)_2$ (M = Si, Ge, Sn, or Pb), respectively (55).

A. Main Group Element Compounds

Because of the existence of the excellent recent reviews listed in Table I (102, 103, 110–112, 171), it would be inappropriate to deal with this topic extensively. We confine ourselves, therefore, to drawing attention to a selection of such radicals in Table II [section (a)], providing an introduction to the literature on Group IV element-centered radicals [excluding their reactions (110, 171)], in order to place in context our studies on the unusually stable species $\dot{M}[CH(SiMe_3)_2]_3$ (M = Si, Ge, or Sn) (44, 46).

1. Transient Group IV Element-Centered Radicals

a. *Formation of Radicals.* It is advantageous to examine radicals in solution since the ESR spectra are usually isotropic, which results in an average g value, average hyperfine coupling constants, and narrow lines. Assignment is normally much easier than for radicals in the solid state in which anisotropy is generally found. The disadvantage is the high rate constant for combination (typically 10^9 liters/mole/second) that occurs in solution, making it difficult to achieve a steady-state concentration of radicals sufficient for direct detection by ESR.

In situ UV irradiation of a sample containing $(Bu^tO)_2$ in the cavity of an ESR spectrometer has proved valuable for obtaining Group IVB radicals from the parent hydrides,

$$(Bu^tO)_2 \xrightarrow{h\nu} 2Bu^t\dot{O} \qquad (4)$$

$$Bu^t\dot{O} + R_3MH \longrightarrow \dot{M}R_3 + Bu^tOH \qquad (5)$$

This led to the first observation in solution of radicals centered on Si (1969) (16, 130), Ge (1969) (16), and Sn (1972) (188). Hyperfine coupling to the central atom was only observed for Si. A Pb-centered radical has not yet been detected in solution, although the existence of such species has been inferred from ESR results. For example, photolysis of a solution of Pb_2Me_6 in the cavity of a spectrometer gave a lead mirror but no signal attributable to $\dot{P}bMe_3$ (weak $\dot{M}e$ was observed). In the presence of an

TABLE II
Representative Organo Main Group Element Radicals Characterized by ESR

Compounds	References	Compounds	References
(a) Metal-centered radicals		$\dot{O}N(SiMe_2R)_2$	102, 103
$[BAr_3]^{\bar{\cdot}}$	35	$O\dot{N}(SiMe_2Bu^t)(GeMe_3)$	103
$[AlMe_3]^{\bar{\cdot}}$	102	$O\dot{N}(GeMe_3)_2$	103
$\dot{S}iR_3$	102	$\dot{O}SiMe_3$	103
$\dot{S}iMe_n(SiMe_3)_{3-n}$	102	(c) Radicals centered β to a metal	
$\dot{S}i[CH(SiMe_3)_2]_3$	44	$\dot{C}R^1R^2OBR_2^3$	102
$\dot{S}iMe_2F$	102	$\dot{C}(CF_3)OSiEt_3(Me)$	102
$\dot{S}iMe_nCl_{3-n}$	102	$\dot{C}Ph_2OSiPh_3$	102
$[PhC=CSiMe_3]^{\bar{\cdot}}$	102	$\dot{C}H_2CH_2MR_3$ (M = Si, Ge, Sn, or Pb)	102
$[SiXX'X''X''']^{\bar{\cdot}}$	103	$\dot{O}OMMe_3$ (M = Si, Ge, Sn, or Pb)	102
$\dot{G}eR_3$	103	$\dot{C}Bu_2^tCH_2SiX_3$	103
$\dot{G}e[CH(SiMe_3)_2]_3$	44	$\dot{C}Ar_2OSnMe_3$	103
$\dot{S}nR_3$	103	$\dot{C}H_2CH_2AsR_2$	103
$\dot{S}n[CH(SiMe_3)_2]_3$	46	$\dot{O}OAsPh_3(OBu^t)$	103
$\dot{P}bR_3$	102, 103	(d) Radicals centered γ to a metal, or still more remote	
$\dot{A}sR_2$	102		
$[o\text{-}C_6H_4(AsMe_2)_2]^{\ddagger}$	103	$\overline{\dot{C}H_2CH_2CH_2OSiMe_3}$	103
$[AsR_3]^{\ddagger}$	102	$\overline{\dot{C}H_2\dot{C}HCHCH_2SnR_3}$	102
$\dot{A}sMe_3SBu^t$	103		
$\dot{A}sPh_3OBu^t$	102		
$[HgAr_2]^{\bar{\cdot}}$	58	(M = Si, Ge, or Sn)	
(b) Radicals centered α to a metal			
$\dot{C}H_2Li(LiMe_3)_3$	103	$\overline{\dot{C}H_2\dot{C}HCHCHMR_3}$ (M = Si, Ge, or Sn)	103
$\dot{C}H_2MgMe$	103	$[SiR_2Ph_2]^{\bar{\cdot}}$	118
$\dot{C}H_2BMe_2$	102	$[SiMe_2(C_6H_4Ph\text{-}p)_2]^{\bar{\cdot}}$	43
$\dot{C}H_2AlMe_2$	102		
$\dot{C}(SiMe_3)_3$	103		
$\dot{C}H_2MMe_3$ (M = Si, Ge, Sn, or Pb)	102		
$\dot{C}Me_2MMe_3$ (M = Si, Ge, Sn, or Pb)	102		
$\dot{C}H_2MMe_2MMe_3$ (M = Si, Ge, Sn, or Pb)	102		

alkyl halide RX the lead mirror was not formed, but a white precipitate, Me₃PbX, separated and the spectrum of Ṙ was obtained (*42*). The problem of radical recombination is overcome if the radicals are isolated from one another by an inert matrix. This usually leads to anisotropic spectra, but the adamantane technique, in which the trapped radicals are free to rotate and thereby give isotropic spectra, has been successfully applied to the study of $\dot{\text{Sn}}(\text{Me})_n\text{Cl}_{3-n}$ (n = 0, 1, 2, or 3) and $\dot{\text{GeMe}}_3$ (*143*). Other solid state techniques that have been used to generate Group IV radicals include γ irradiation for $\dot{\text{GeMe}}_3$ using GeMe₄ in a matrix (*143*), and a procedure involving a rotating cryostat, for $\dot{\text{SnMe}}_3$ and $\dot{\text{PbMe}}_3$ from ClMMe₃ and Na (*13*). Transient cationic species, e.g., [PbMe₄]‡, are accessible by one-electron oxidation, e.g., from PbMe₄–[IrCl₆]²⁻ (*129*).

b. *Structure of Radicals.* Whereas the methyl radical $\dot{\text{CH}}_3$ is planar, there are now much data suggesting that all other Group IV radicals show varying degrees of deviation from planarity. The main line of evidence uses the hyperfine coupling of the unpaired electron to those isotopes of the central atom that possess nonzero nuclear spin (²⁹Si, ⁷³Ge, ¹¹⁷Sn, and ¹¹⁹Sn). For a planar radical, the unpaired electron occupies a pure p_z orbital which has a node at the central atom, whereas for a pyramidal radical the odd electron is in a hybrid orbital. The magnitude of the isotropic coupling depends on the amount of s character in the orbital containing the unpaired electron. Consequently, the more pyramidal the radical, the more s character in the orbital of the odd electron and, hence, the greater is the hyperfine coupling. In general the greater the difference in electronegativity between the central atom and the atoms bonded to it, the more pyramidal the radical. For example, in the series $\dot{\text{SiMe}}_n(\text{SiMe}_3)_{3-n}$, as methyl groups are replaced by less electronegative SiMe₃ groups there is a trend toward the more planar structures (*41*). From the $a(\text{M})$ data of Table III and on $\dot{\text{MH}}_3$, $\dot{\text{MMe}}_3$, or $\dot{\text{MCl}}_3$, trends are summarized in Scheme 1.

c. *Stability of Radicals.* The trityl radical, $\dot{\text{CPh}}_3$, is perhaps the best-known stable radical of Group IV and was long thought to exist in solution equilibrium with its symmetrical dimer, hexaphenylethane. It has now been established (*138*), however, that there is the following equilibrium:

$$2 \text{ Ph}_3\dot{\text{C}} \rightleftharpoons \text{Ph}_3\text{C}-\underset{}{\overset{H}{\bigcirc}}=\text{C}\underset{\text{Ph}}{\overset{\text{Ph}}{\diagup}} \qquad (6)$$

Dimerization is presumably prevented by crowding around the central carbon. More recent results, including those of Sections II, A, 2 and III, demonstrate the importance of steric hindrance to dimerization in con-

$\dot{C}H_3$, $\dot{C}Me_3$ $\qquad\qquad\qquad\qquad\qquad\qquad\qquad$ $\dot{C}Cl_3$

$$\left.\begin{array}{llll}\dot{S}iH_3 & \dot{S}iMe_3 & \dot{S}iR_3{}^a & \\ \dot{G}eH_3 & \dot{G}eMe_3 & \dot{G}eR_3{}^a & \dot{G}e(NR'_2)_3{}^a \quad \dot{G}eCl_3 \\ \dot{S}nH_3 & \dot{S}nMe_3 & \dot{S}nR_3{}^a & \dot{S}n(NR'_2)_3{}^a \quad \dot{S}nCl_3 \\ & \dot{P}bMe_3 & & \end{array}\right\}$$

$\underbrace{\qquad\qquad\qquad\qquad\qquad}_{\text{Similar geometries}}$

$\underbrace{\qquad\qquad\qquad\qquad\qquad\qquad\qquad\qquad\qquad\qquad}_{\text{All show marked deviations from planarity}}$

Approximately $\qquad$ Increasingly pyramidal $\qquad$ Approximately
planar $\qquad\qquad\qquad\qquad\qquad\qquad\qquad\qquad\qquad$ tetrahedral

[a] Electronegativity acts in opposition to a steric effect; however the former is clearly dominant despite the considerable bulk of R and R'.

SCHEME 1. Variations in geometry around the central metal M in some Group IV radicals $\dot{M}X_3$ [X = H, Cl, R, or NR'_2; R = $(Me_3Si)_2CH$; R' = Me_3Si]

tributing to radical stability. Other factors, such as the nonavailability of disproportionation pathways and the possibility of delocalization into silicon d orbitals, may contribute to stability.

2. Stable Group IV Element-Centered Radicals

This work (44, 46, 142) originated in an attempt to establish whether the interesting compound SnR_2 [R = $(Me_3Si)_2CH$] (47) (formally analogous to a carbene) exists in a singlet or triplet ground state. No evidence was found from ESR spectra of liquid or frozen ($-110°$*) solutions for the latter, but a weak signal near $g = 2$ was detected. It had quartet structure, suggesting $\dot{S}nR_3$, since the three equivalent α protons of $\dot{S}n[CH(SiMe_3)_2]_3$ would give rise to a 1:3:3:1 quartet. Unambiguous assignment, through detection of ^{117}Sn and ^{119}Sn satellites, was not then possible due to the low concentration of the species. The presence of the radical was initially attributed to reaction of SnR_2 with traces of oxygen, but later work showed that irradiation of the solution with UV or visible light caused a dramatic

* Temperatures are all Centigrade unless otherwise noted.

TABLE III

ELECTRON SPIN RESONANCE PARAMETERS FOR THE STABLE GROUP IV TRIS(ALKYL) AND TRIS(AMIDO) RADICALS[a]

Radical[b]	Synthesis				g	$a(H)$[e] or $a(N)$[e]	$a(M)$[e,f,g]	Stability at 20° in solution[h]
	Reactants[c]	Radiation[d]	Solvent					
$\dot{S}iR_3$	Si_2Cl_6/LiR	UV	C_6H_6		2.0027	0.48	19.3	$t_{1/2} \sim 10$ min at 30°
$\dot{S}iR_3$	Si_2Cl_6/LiR	UV	n-C_6H_{14}		2.0028	0.48	19.3	Unchanged after 4 months
$\dot{G}eR_3$	$GeCl_2 \cdot$diox/LiR	i	C_6H_6		2.0078	0.38	9.2	
$\dot{S}nR_3$	$SnCl_2$/LiR [j]	UV or vis.	C_6H_6		2.0094[m]	0.21	{169.8 (^{117}Sn) 177.6 (^{119}Sn)}	$t_{1/2} \sim 1$ year
$\dot{G}e(NR_2')_3$	$GeCl_2 \cdot$diox/$LiNR_2'$ [k]	UV	n-C_6H_{14}		1.9991	1.06	17.1	$t_{1/2} > 5$ months
$\dot{S}n(NR_2')_3$	$SnCl_2$/$LiNR_2'$ [l]	UV	n-C_6H_{14}		1.9912	1.09	{317.6 (^{117}Sn) 342.6 (^{119}Sn)}	$t_{1/2} \sim 3$ months
$\dot{G}e(NBu^tR')_3$[n]	$Ge(NBu^tR')_2$[o]	UV	n-C_6H_{14}		1.9998	1.29	17.3	$t_{1/2} \sim 5$ min[p]
$\dot{S}n(NBu^tR')_3$[n]	$Sn(NBu^tR')_2$[o]	UV	n-C_6H_{14}		1.9928	1.27		$t_{1/2} \sim 5$ min[p]
$\dot{S}nR_3$	$SnR_2 + 6Sn(NR_2')_2$[q]	UV	n-C_6H_{14}		2.0094	0.21	{169.8 (^{117}Sn) 177.6 (^{119}Sn)}	

a Data from Refs. *44*, *46*, *142*.
b R = (Me₃Si)₂CH; R′ = Me₃Si.
c R = (Me₃Si)₂CH, R′ = Me₃Si.
d Under Ar or a vacuum.
e In mT.
f ²⁹Si; ⁷⁵Ge, ¹¹⁷Sn, or ¹¹⁹Sn.
g Corrected for second-order shift (Eq. (7)).
h Based on ESR signal strength of a light-protected, sealed sample.
i None required.
j Or SnR₂.
k Or Ge(NR′₂)₂.
l Or Sn(NR′₂)₂.
m g_1-2.016; g_{11}-1.994.
n Reference *142*.
o Reference *93*.
p The reason for the lower stability of these radicals compared with $\dot{M}(NR'_2)_3$ is uncertain; from g and a(Ge) values on the Ge radicals, bond angles are probably similar; the generation of $\dot{M}(NBu^tR')_3$ was not as clean as of $\dot{M}(NR'_2)_3$, and other paramagnetic species were sometimes detected, possibly $\dot{N}Bu^tR'$.
q This was an attempt at a crossover experiment, possibly relevant to the mechanism of photolysis, Eqs. (8)–(11); a trace of $\dot{S}n(NR'_2)_3$ was also present, but not detected with 1:1 SnR₂:Sn(NR′₂)₂, nor any mixed alkyl-amido-radical (*142*). Irradiation of Sn(NR′₂)₂ with Sn(C₅H₅-η)₂ yielded Sn(NR′₂)(C₅H₅-η), but no paramagnetic species. Radical formation was also not detected by UV irradiation of Sn(C₅H₅-η)₂, SnI₂, SnCl(NR′₂), or Zn(NR′₂)₂.

increase in the signal strength. Neither type of irradiation caused deposition of a tin mirror, and it was also demonstrated that (i) heat did not cause any increase in signal strength (irradiation of a sample without concomitant cooling raises the temperature considerably), and (ii) a sample prepared in the dark gave no ESR signal, but irradiation generated the radical. The intensity of the signal obtained on irradiation allowed identification of satellite peaks due to ^{117}Sn and ^{119}Sn, confirming formulation of the radical as $\dot{\text{S}}$nR$_3$ (*46*).

Extension to the related $\dot{\text{G}}$eR$_3$ and to isoelectronic amides, $\dot{\text{M}}$(NR$_2'$)$_3$ (M = Ge or Sn; R' = Me$_3$Si) was carried out by reacting the metal(II) chloride with the appropriate lithium reagent and irradiating a solution of the product or, alternatively, GeR$_2$ (*93*) or M(NR$_2'$)$_2$ (*92*). For the Si-centered radical, $\dot{\text{S}}$iR$_3$ (see Fig. 1), a different route was required since suitable Si(II) species are unknown, except as short-lived intermediates. Compound Si$_2$Cl$_6$ was reacted with LiR with the view to forming R$_3$SiSiX, (X = R$_3$, Cl$_3$, R$_2$Cl, or RCl$_2$) which would then be expected to fragment readily to $\dot{\text{S}}$iR$_3$. This radical was obtained from Si$_2$Cl$_6$ and LiR, followed by irradiation, but the compound isolated from the reaction, (SiCl$_2$R)$_2$, suggested a different mechanism for radical formation, perhaps via R$_2$SiClSiCl$_3$. *In situ* UV irradiation of a solution of PbR$_2$ (*47*) at 20° gave a complex spectrum containing lines attributable to (Me$_3$Si)$_2$$\dot{\text{C}}$H, other paramagnetic species, and a lead mirror. [Assignment was confirmed by generating the same radical from the low-temperature (−40°) irradiation of (Bu^{t}O)$_2$ and (Me$_3$Si)$_2$CH$_2$: doublet of multiplets, $a(\alpha$-H$)$ = 1.89 mT, $a(\gamma$-H$)$ = 0.037 mT.] Irradiation of a solution of Pb(NR$_2'$)$_2$ (*92*) with visible or UV light at 20°, or UV at −40°, gave no signals. Some decomposition of the sample appeared to occur.

The main feature of the ESR spectra (e.g., Fig. 1) of these metal-centered radicals is a multiplet arising from the coupling of the unpaired electron to three equivalent protons (quartet) or three equivalent nitrogen nuclei (septet). For dilute solutions of the amido radicals, the septets showed further structure, attributed to partially resolved proton coupling. Under conditions of higher gain, satellite lines from those isotopes of the central atom that possess nonzero spin were observed (nucleus, percent abundance, nuclear spin: ^{29}Si, 4.7, $\frac{1}{2}$; ^{73}Ge, 7.6, $\frac{9}{2}$; ^{117}Sn, 7.7, $\frac{1}{2}$; and ^{119}Sn, 8.7, $\frac{1}{2}$). The low abundance of these isotopes makes detection of the satellite lines difficult, but the intensity of these was increased by using high microwave power (e.g., 50 mW) and high modulation amplitude (e.g., 0.5 mT). (For C-centered radicals, values such as these can lead to saturation or loss in resolution, respectively.) For $\dot{\text{S}}$nR$_3$, the satellite lines were very broad but could be sharpened by an increase in temperature. The width of the lines

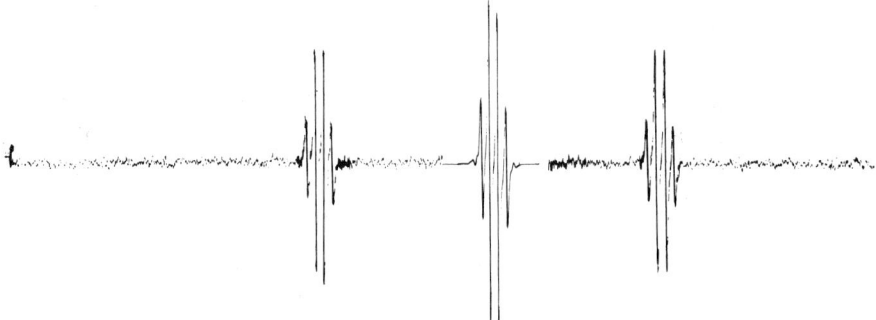

FIG. 1. The ESR spectrum of Ṡi[CH(SiMe₃)₂]₃ in C₆H₆ at 30°.

is attributed to incomplete averaging of the anisotropic contribution to the g and hyperfine tensors, caused by slow tumbling of the radical. Raising the temperature increases the rate of tumbling so that the anisotropy is averaged to zero and the spectrum becomes isotropic.

Measurement of the g values [relative to polycrystalline diphenylpicrylhydrazyl (DPPH)] and the α proton [$a(H)$] or nitrogen [$a(N)$] couplings, was straightforward, but determination of the central atom hyperfine coupling was not, and requires further comment.

It is seen from the spectra (e.g., Fig. 1) that the satellite lines are not symmetrical about the central multiplet; the satellites are shifted downfield, but not by equal amounts. This is a second-order effect and results from the breakdown of the "high-field approximation" in the theory of coupling constants. This approximation assumes no coupling between the spin of the electron and the spin of the nucleus, so that when the former is reversed, the latter remains unchanged. Under conditions of low magnetic field or large hyperfine coupling, this is no longer true and results in a nonlinear divergence of energy levels as the field increases and, hence, an unequal separation of lines in the spectrum. (More rigorous explanations of this effect are given in Refs. 84 and 6.) Corrected values of the coupling constants are obtained from the observed spacings by application of the Breit-Rabi equation (for $I = \frac{1}{2}$),

$$a(M) = \frac{2H_0(H_0 - H_k)}{2H_0 - H_k} \qquad a(M) = \frac{2H_0(H_l - H_0)}{2H_0 - H_l} \qquad (7)$$

where $a(M)$ is the corrected coupling constant, H_0 the field position of the central line, and H_k or H_l the field position of the low- or high-field satellite, respectively (104). Since each nucleus (²⁹Si, ¹¹⁷Sn, and ¹¹⁹Sn) gives rise to a pair of lines (at H_k and H_l), two values of the couplings are obtained in each case, which serves as an internal check. For the germanium-centered radicals, the Breit-Rabi equation was used in the form of Ref. *168*. Computer analysis provides the position of the satellite lines using a trial value of the coupling constant and the observed field position of the central multiplet. The value of the coupling constant is varied until the calculated line positions agree with the measured ones. The results are included in Table III.

Stability measurements on the radicals $\dot{M}X_3$ were made using samples in sealed tubes, protected from light and stored at ambient temperature. Spectra were recorded periodically, and stabilities estimated from the decrease in signal strength (see Table III). The radical $\dot{S}iR_3$ decayed in benzene with a half-life of about 10 minutes at ca. 30°. The decay curve was measured using the spectrometer to plot peak height against time. The value of $t_{1/2}$ remained constant over five half-lives, thereby showing that the radical decayed with first-order or pseudo-first-order kinetics. The decay of $\dot{S}iR_3$ appeared to be reversible in benzene but not in hexane. Irradiation of the sample in C_6H_6 caused formation of the radical up to a constant maximum intensity; shutting off the light led to complete decay. This cycle of formation–decay was repeated several times but does not prove unambiguously that the radical decayed reversibly.

Electron spin resonance data obtained on $\dot{M}R_3$ and $\dot{M}(NR'_2)_3$ are listed in Table III. For comparison, results on $\dot{M}Me_3$ are: $\dot{S}iMe_3$, solution, g_{iso} = 2.0031, $a(H)$ = 0.634 mT, $a(M)$ = 18.3 mT (*15*); $\dot{G}eMe_3$, matrix, g_{iso} = 2.0101, $a(H)$ = 0.53 mT, $a(M)$ = 8.47 mT (*143*); $\dot{S}nMe_3$, matrix, g_{iso} = 2.0163, $a(H)$ = 0.25 mT, $a(M)$ = 153.0 (^{117}Sn) and 161.1 (^{119}Sn) mT (*13*); $\dot{P}bMe_3$, matrix, g_{iso} = 2.0389, $a(M)$ = 185.0 mT (*13*).

The mechanism for $\dot{M}X_3$ formation from photolysis of MX_2 [MX_2 = GeR_2, SnR_2, $Ge(NR'_2)_2$, or $Sn(NR'_2)_2$] may follow either of two routes (shown for $MX_2 = SnR_2$) (*46*),

$$SnR_2 \xrightarrow{h\nu} \dot{R} + \dot{S}nR \quad (8)$$

$$R + SnR_2 \longrightarrow \dot{S}nR_3 \quad (9)$$

or

$$SnR_2 \xrightarrow{h\nu} (SnR_2)^* \quad (10)$$

$$(SnR_2)^* + SnR_2 \longrightarrow \dot{S}nR_3 + \dot{S}nR \quad (11)$$

In the first mechanism, an M—X bond is homolyzed and the resultant radical $\dot{X}$ is trapped by another molecule of MX_2. [Unsuccessful attempts to provide evidence for this proposition were experiments (i)–(iii); the radical precursors $(Bu^tON)_2$ and AIBN, as well as the inhibitor galvinoxyl, reacted with SnR_2 at 20° to give diamagnetic solutions; whereas, in experiment (iv), Ph$_3$Ċ did not react (142).] Alternatively, ṀX$_3$ is formed from a bimolecular reaction between an excited state of MX$_2$ (possibly triplet) and a ground state MX$_2$. Both mechanisms require formation of a M(I) species: since no metal develops on photolysis, ṀX must react with solvent or form a soluble diamagnetic oligomer.

It is possible that ṠiR$_3$ is formed according to the following reactions:

$$Si_2Cl_6 + 6LiR \longrightarrow \underset{\text{isolated}}{R-\underset{\underset{Cl}{|}}{\overset{\overset{Cl}{|}}{Si}}-\underset{\underset{Cl}{|}}{\overset{\overset{Cl}{|}}{Si}}-R} + \underset{\text{postulated}}{R-\underset{\underset{R}{|}}{\overset{\overset{Cl}{|}}{Si}}-\underset{\underset{Cl}{|}}{\overset{\overset{Cl}{|}}{Si}}-Cl} \text{ (+ other products?)} \quad (12)$$

$$R-\underset{\underset{R}{|}}{\overset{\overset{Cl}{|}}{Si}}-\underset{\underset{Cl}{|}}{\overset{\overset{Cl}{|}}{Si}}-Cl \xrightarrow{h\nu} SiR_2 + SiCl_4 \quad (13)$$

$$2SiR_2 \xrightarrow{h\nu} \dot{S}iR_3 + \frac{1}{n}(SiR)_n \quad (14)$$

Supporting evidence for Eq. (13) is provided by (109)

$$Me_2Si\underset{}{\overset{(SiMe_2)_4}{\diagup\diagdown}}SiMe_2 \xrightarrow{h\nu} SiMe_2 + (SiMe_2)_5 \quad (15)$$

The postulated SiMe$_2$ was trapped as an insertion product, but Si$_2$Me$_6$ was not isolated (which might have been expected if the silylene itself photolyzed to ṠiMe$_3$). *In situ* irradiation of (SiMe$_2$)$_6$ (toluene, $-60°$) did not lead to ESR detection of ṠiMe$_3$ (142). However, for SiR$_2$ the bulky R groups may stabilize the silylene sufficiently for photolysis to ṠiR$_3$ to be favored over alternative reactions.

There is precedent for such redox reactions in transition metal chemistry, e.g. (3),

$$(TiX_3)_2 \longrightarrow TiX_4 + \frac{1}{n}(TiX_2)_n \longrightarrow X = NMe_2$$

although photochemical disproportions are less common (25) (but see Section II, B).

The g values for ṀR$_3$ increase down the group (Table III), the same trend as is found for ṀMe$_3$ and ṀH$_3$. It is attributed to the parallel increase in the M-spin-orbit coupling constant, because, in general, Δg is proportional to $\epsilon/\Delta E$, where Δg is the difference between the measured g and the free-spin value of 2.0023, ϵ is the spin-orbit coupling constant, and ΔE is the difference between ground-state and excited-state energies. However, $g[\dot{\text{Ge}}(\text{NR}'_2)_3] > g[\dot{\text{Sn}}(\text{NR}'_2)_3]$ (Table III); cf., $g(\dot{\text{GeCl}}_3) > g(\dot{\text{SnCl}}_3)$. These findings may be due to delocalization of the unpaired electron into ligand π-type orbitals (142, 143).

Anisotropy in g values was only found for $\dot{\text{Sn}}R_3$ (Table III). The g values of the amide radicals are less than those of corresponding alkyls, consistent with the former having more pyramidal structures; bending of an $\dot{\text{M}}X_3$ radical mixes excited states into the wave function for the unpaired electron and, thus, lowers the g value.

The methine splittings $a(\text{H})$ for the radicals $\dot{\text{M}}[\text{CH}(\text{SiMe}_3)_2]_3$ are close to those for $\dot{\text{M}}(\text{CH}_3)_3$. The most notable feature is the large difference between $\dot{\text{C}}(\text{CH}_3)_3$ [$a(\text{H}) = 2.25$ mT] and the analogous Si, Ge, and Sn species [$a(\text{H}) = 0.634, 0.53,$ and 0.275 mT, respectively; see Table III]. This difference is attributed to (i) the more pyramidal geometry of the Si, Ge, Sn, (and Pb) radicals; (ii) the reluctance of the heavier elements to form multiple bonds; and (iii) the greater size of the heavier elements compared with C. All three factors reduce hyperconjugative coupling, which is believed to be the main cause of such splittings in simple alkyl radicals (6). In this process there is a contribution to the bonding from a structure in which the unpaired electron couples with one of the electrons in a C—H bonding orbital,

$$\underset{\text{M}\text{—}\text{C}}{\overset{\text{H}}{\cdot\;\cdot}} \longleftrightarrow \text{M}=\text{C} \quad \text{H}\cdot \tag{16}$$

The extreme longevity of radicals $\dot{\text{M}}R_3$ and $\dot{\text{M}}(\text{NR}'_2)_3$ (Table III) must be mainly due to steric hindrance to dimerization (44, 46). The low values of the M—H bond strengths [$D(\text{M}\text{—}\text{H})$: C, 104; Si, 81; Ge, 73; Sn, 70 kcal mole^{-1}] (110) do not favor H abstraction from the C—H bonds of the solvent, but the shorter lifetimes of the $\dot{\text{Sn}}(\text{NR}'_2)_3$ and $\dot{\text{Ge}}(\text{NR}'_2)_3$ radicals in hexane compared to $\dot{\text{Sn}}R_3$ and $\dot{\text{Ge}}R_3$ in benzene may reflect the ease of H abstraction from the two solvents. A third factor, for the tris(alkyl) radicals is the low probability of disproportionation, such as that shown for a carbon-centered radical,

$$2 \; \underset{\text{CH}_3}{\overset{\text{H}_3\text{C}\diagdown \overset{\cdot}{\text{C}} \diagup \text{CH}_3}{|}} \longrightarrow (\text{CH}_3)_3\text{CH} + \underset{\text{H}_3\text{C}\diagup ^{\text{C}}\diagdown \text{CH}_3}{\overset{\text{CH}_2}{\|}} \tag{17}$$

because stable, double-bonded compounds of the heavier Group IV elements are unknown.

The first-order decay of radical $\dot{\text{S}}\text{iR}_3$ in benzene implies reaction with solvent or an intramolecular rearrangement, such as

$$\dot{\text{Si}}[\text{CH}(\text{SiMe}_3)_2]_3 \longrightarrow \begin{array}{c} \text{H}_2\text{C}\text{-----}\text{H} \\ | \quad\quad \vdots \\ \text{Me}_2\text{Si}\text{....}\dot{\text{Si}}[\text{CH}(\text{SiMe}_3)_2]_2 \\ \text{H}\diagdown_\text{C}\diagup\text{SiMe}_3 \end{array} \tag{18}$$

$$\downarrow \text{H}$$

$$\dot{\text{C}}\text{H}_2-\text{Si}(\text{Me}_2)-\text{C}(\text{H})(\text{SiMe}_3)-\dot{\text{Si}}[\text{CH}(\text{SiMe}_3)_2]_2 \longrightarrow \text{Nonradical products}$$

It is interesting that the stability, unlike the geometry, of these radicals $\dot{\text{M}}\text{X}_3$ is so sensitive to steric effects. Bulky groups X thus hinder the formation of a four-coordinate $\text{X}_3\text{M}-\text{MX}_3$ or $\text{X}_3\text{M}-\text{H}$ and favor the three-coordinate $\dot{\text{M}}\text{X}_3$ (*44*). A similar effect has been noted for transition metal MR_3 and $\text{M}(\text{NR}'_2)_3$ complexes (*48*).

B. Transition Metal Compounds

Many paramagnetic organo-transition metal complexes are stable under ambient conditions. Relevant ligands include CO, R$^-$, olefin, η-C$_5$H$_5^-$, or η-arene, and compounds may be neutral, e.g., [V(CO)$_6$], [Cr(CH$_2$SiMe$_3$)$_4$], [Cr{CH(SiMe$_3$)$_2$}$_3$], and TaCl$_2$(C$_5$H$_5$-η)$_2$], anionic, e.g., [Os$_3$(CO)$_{12}$]$^-$, [Cr(CH$_2$SiMe$_3$)$_4$]$^-$, and [TiCl$_2$(C$_5$H$_5$-η)$_2$]$^-$, or cationic, e.g., [Cr(ArH-η)$_2$]$^+$. These owe their stability, in part, to electron delocalization for π-bonded ligands; whereas for the alkyls, it is due to steric effects (effectively making the metal coordinatively saturated) and the use of ligands that do not allow normal decomposition pathways to be accessible (e.g., β elimination from M—R if R$^-$ has no β-hydrogen). Various recent reviews are available (*48, 49, 55, 79, 80, 84*), but a representative series of compounds is included in Table IV. Experimental results are concerned with preparative methods (e.g., alkali-metal or electrochemical reduction in donor solvents for radical anions), and bulk magnetic susceptibility, ESR measurements, or electronic spectra. These have, in general, been unexceptional. For example, many of the alkyls of first-row transition metals have magnetic moments close to spin-only values, and ESR spectra are consistent with electronic structure and geometry, e.g., tetrahedral and trigonal for the local Cr environment of CrR$_4$ (*19, 153, 185*) or CrR$_3$ (*10*); the tetra-1-norbornyls of Fe(IV) and Co(IV) are low spin with $\mu_{\text{eff}} = 0$ and 2.0 BM, respectively (*19*).

TABLE IV

Representative Organo Transition metallic Paramagnetic Species

	Compounds	References		Compounds	References
d^1 Ti(III):	ca. 30 papers concerning (η-C$_5$H$_5$)– complexes	66, 84, 96, 121, 140	d^2 V(III):	[V(C$_5$H$_5$-η)$_2$X] (X = Cl, Ph, SR, or Tol)	66, 84
	[Ti(C$_5$H$_5$-η)(C$_8$H$_8$-η)]	181		[V{CH(SiMe$_3$)$_2$}$_3$]	10
	[Ti{CH(SiMe$_3$)$_2$}$_3$]	10		[CrR$_4$] (R = Me, Bu^t, Me$_3$SiCH, Me$_3$CCH$_2$, Ph$_3$CCH, PhMe$_2$CCH$_2$ or 1-norbornyl)	19, 153, 185
	[Ti(C$_5$H$_5$)$_2$CHR$_2$], R = Me$_3$Si or Ph	10, 166	Cr(IV):		
Zr(III):	[Zr(C$_5$H$_5$-η)$_2$(PPh$_2$)$_2$]–	84	W(IV):	[W(CH$_2$Ph)$_4$]	49
V(IV):	about 25 papers concerning (η-C$_5$H$_5$)– complexes	{7, 66, 84, 177, 179	d^3 V(II):	complexes containing (η-C$_5$H$_5$)–	67, 84
	[V(CH$_2$Ph)$_4$]	106			
	[V(CH$_2$SiMe$_3$)$_4$]	153		[Cr(CH$_2$SiMe$_3$)$_4$]–	121
	[V(1-norbornyl)$_4$]	19	Cr(III):	[CrR$_3$] [R = (Me$_3$Si)$_2$CH; 2,2,3-Me$_3$-1-norbornyl, Me$_2$P(CH$_2$)$_2$; CH$_2$(PPh$_2$C$_6$H$_4$); o-R$_2$NCH$_2$C$_6$H$_4$]	10, 19, 49
	[V(C$_8$H$_8$-η)$_2$]	181			
	[V(C$_5$H$_5$-η)$_2$Cl$_2$(AlCl$_3$)]	65			
	[V(C$_5$H$_5$-η)Cl$_3$]	180			
Nb(IV):	[Nb(C$_5$H$_5$-η)$_2$X$_2$]	63, 173, 177	Mo(III):	complexes containing (η-C$_5$H$_5$)–	84
	[Nb(C$_5$H$_5$-η)$_2$R$_2$]	63	Mn(IV):	[Mn(1-norbornyl)$_4$]	19
Ta(IV):	[Ta(C$_5$H$_5$-η)$_2$Cl$_2$]	6a, 63	d^4 Fe(IV):	[Fe(1-norbornyl)$_4$]	19
	[Ta(C$_5$H$_5$-η)$_2$R$_2$]	63			
Mo(V):	[Mo(C$_5$H$_5$-η)$_2$X$_2$]$^+$ (X = Cl or Br)	83, 84	d^5 V(0):	[V(CO)$_6$, V(Arene-η)$_2$]	84, 97
W(V):	[W(C$_5$H$_5$-η)$_2$X$_2$]$^+$	83, 84		[V(C$_5$H$_5$-η)(C$_7$H$_7$-η)]	166
	[W(C$_6$F$_5$)$_6$]–	125	Nb(0):	[Nb(C$_5$H$_5$-η)(C$_7$H$_7$-η)]	157

	Compound	Ref.
M(I):	(M = Cr, Mo, or W)	84, 97, 62
	[M(arene-η)$_2$]$^+$,	
	[M(C$_5$H$_5$-η)(CO)$_3$],	
	[M(CO)$_4$(PMe$_3$)]$^-$	
Cr(I):	[Cr(C$_5$H$_5$-η)(C$_6$H$_6$-η)],	62
	[Cr(C$_5$H$_5$-η)(C$_7$H$_7$-η)]$^-$	8
	[Cr(C$_5$H$_5$-η)(C$_8$H$_8$-η)]$^-$	129
	[Cr(CH$_2$Ph)L]$^+$	105
Mo(I):	[Mo(C$_5$H$_5$-η)(CO)$_3$(NOAr)]	105
	[Mo(CO)$_4$(diacetylanil)]$^-$	56
Mn(II):	[Mn(C$_5$H$_4$Me-η)(CO)$_2$-	145
	{C(OMe)(1-ferrocenyl)}]$^+$	105a
	[Mn(THF)$_6$]$^{2+}$	165
Fe(III):	ferrocenium salts	11
	[Fe(olefin)]$^{3+}$	
Co(IV):	[Co(1-norbornyl)$_4$]	19
	[Fe(C$_5$H$_4$C$_6$H$_4$NO$_2$-p-η)$_2$]$^-$	45
	[Fe(C$_5$H$_5$-η){C$_5$H$_4$CO(COR)}]$^-$	146
	and ferrocenebenzo-	
	semiquinones	
Co(II):	[Fe(C$_5$H$_5$-η)(CO)$_2$(NOAr)]	105
	[Co(diphos)R(H)]	95
	(IV)	193
d^7 Cr$^-$(I):	[Cr(C$_{10}$H$_8$)$_2$]$^-$	97
Mn(0)/Mn(I):	[Mn$_2$(CO)$_{10}$]$^-$	4
Mn(0):	[Mn(CO)$_5$Br]$^-$,	4
	[Mn(CO)$_3$(PPh$_3$)$_2$]	154
	[Mn(CO)$_5$O$_2$], [Mn(CO)$_5$]	69, 101
	[Mn(CO)$_4$(P)(NOAr)]	105
	(P = CO or PPhMe$_2$)	
Tc(0):	[Tc(CO)$_x$]	116
Re(0):	[Re(CO)$_3$(P)$_2$] (P = PPh$_3$ or	73, 152
	PMe$_2$Ph)	
	[Re(CO)$_5$(NOAr)]	105
Fe(I):	[Fe(C$_5$H$_5$-η){C$_5$H$_4$NO(Bu^t)-η}]	72
	[Fe(C$_5$H$_5$-η){C$_5$H$_4$CO(Ph)}]$^-$	61
d^9 Cr$^-$(III):	[Cr(C$_{10}$H$_8$)$_2$]$^{3-}$	97
Co(0):	[Co$_3$(CO)$_9$CX]$^-$ (X = R, Hal, or SiMe$_3$)	147
	[Co(CO)$_4$O$_2$], [Co(CO)$_4$]	69, 120
	[Co(CO)$_3$P(OEt)$_3$(NOAr)]	105
	[Co(CO)(NO)phen]$^-$	57
	[Co$_3$(CO)$_9$C]	172
Fe$^-$(I):	[Fe(C$_6$H$_8$)(CO)$_2$(PPh$_3$)]$^-$	178
	[Fe(CO)$_5$]$^-$	159
Fe(0)/Fe$^-$(I):	[Fe$_2$(CO)$_9$]$^-$, [Fe$_3$(CO)$_{12}$]$^-$,	159
	[Fe$_3$(CO)$_{11}$P(OPh)$_3$]$^-$,	
	[RuFe$_2$(CO)$_{12}$]$^-$,	
	[Fe$_2$(CO)$_8$Pt{P(OPh)$_3$}$_2$]$^-$	
Ru(0)/Ru(I):	[Ru$_3$(CO)$_{12}$]$^-$, [Ru$_6$(CO)$_1$C]$^-$	159
Os(0)/Os(I):	[Os$_3$(CO)$_{12}$]$^-$	159
Ir(0)/Ir(I):	[Ir$_4$(CO)$_{12}$]$^-$	159
Ni(I):	[M(PR$_3$)$_3$ or $_4$(E$^-$)]	64
Pt(I):	{[E = C$_2$(CN)$_4$, Cl$_2$(CN)$_2$-	64
	benzoquinone, or chloranil;	
	M = Ni or Pt}	
Cu(II):	[Cu(olefin)]$^{2+}$	11
Ag(II):	[Ag(olefin)]$^{2+}$	77

Numerous transient paramagnetic compounds are known and some of these are also shown in Table IV; there is an overlap with Section V, and Table IV does not duplicate material which is more conveniently treated later. The distinction is arbitrary, but we shall defer consideration of transient transition metal-centered radicals, e.g., Pt(I), if their formation is primarily of interest in connection with an organometallic mechanistic study, e.g., the oxidative addition of an alkyl halide to a Pt(0) substrate. The designation of metal oxidation state in Table IV is somewhat formal; in many cases it might be more appropriate to describe a complex as derived from a paramagnetic ligand, such as a nitroxide or ketyl.

The saga of neutral paramagnetic metal carbonyl transient species is somewhat confused. Claims to having obtained [Co(CO)$_4$] (*120*) or [Mn(CO)$_5$] (*70*) by sublimation of the dimer onto a cold (77 K) finger have been shown to be in error for the latter. In the absence of dioxygen, the Mn condensate is diamagnetic, but in the presence of O$_2$ the radicals were trapped and identified (ESR) as [Co(CO)$_4$O$_2$] or [Mn(CO)$_5$O$_2$], respectively (*69*). Similarly, photolysis of metal–metal bonded dimers in the presence of nitrosodurene in chloroform at −30° yielded the spin-trapped metallonitroxides (ESR) [ML$_n${NO(Ar)}] [ML = Mn(CO)$_5$, Re(CO)$_5$, Mn(CO)$_4$(PPhMe$_2$), Co(CO)$_3$P(OEt)$_3$, Fe(C$_5$H$_5$-η)(CO)$_2$, or Mo(C$_5$H$_5$-η)(CO)$_3$] (*105*) (see Sections III and V, B). Radical [Re(CO)$_5$] has been obtained by photolysis (*194*) and by co-condensing rhenium vapor and CO in an inert matrix and IR analysis (*100*). {Several other metal carbonyl paramagnetic species have been obtained by matrix isolation techniques, e.g., [Mn(CO)$_5$] (*101*).} Irradiation at 350 nm of a degassed THF solution of [Mn$_2$(CO)$_{10}$] produced an ESR signal (*88*); this almost certainly is due to a Mn(II) species (*105, 105a*) rather than to [Mn(CO)$_5$]. Kinetic studies of CO displacement by tertiary phosphine or phosphite from [Mn(CO)$_4$(PPh$_3$)]$_2$, [Mn(CO)$_4${P(OPh)$_3$}]$_2$, [Re(CO)$_4$(PPh$_3$)]$_2$, or [Ru(CO)$_4$SiMe$_3$]$_2$ point to a small steady-state concentration of the corresponding monomer during the course of these reactions (*68*). γ-Irradiation of [Mn$_2$(CO)$_{10}$] provided the radical anion [Mn$_2$(CO)$_{10}$]$^-$ in which the unpaired electron was believed to be principally in the Mn—Mn σ* orbital (*4*). Similarly, [Mn(CO)$_5$Br]$^-$ resulted from the neutral precursor, and the Mn—Br σ* orbital was implicated. A series of stable ($t_{1/2} \sim$ several hours) polynuclear metal carbonyl radical anions has been obtained by alkali-metal and/or electrochemical reduction of the neutral parent, in O$_2$-free ether solvents (*159*), which revert to the starting material by quenching with MeI. Low-field ESR signals were assigned to the nonbridged species, believed to be in temperature-dependent equilibrium with the carbonyl-bridged isomer.

III
OTHER ORGANOMETALLIC RADICALS

Representative lists of radicals in which the unpaired electron is mainly localized remote from the metal are included in Tables III and IV. The distinction can readily be made experimentally by ESR provided that the metal has an isotope with a nonzero nuclear spin. For transition metal species, when π-bonded ligands are often implicated, or radical ions, the demarcation may become somewhat blurred; for this reason Table IV is not subdivided according to whether a particular compound is best formulated as of type I or II (see Section I). The reviews cited in Section II are also relevant for type II radicals.

Metal alkyl or alkoxy radicals in which the unpaired electron is localized on carbon, α or β to the metal, are known for Li, Mg, B, Al, Si, Ge, Sn, Pb, and As (for a selection, see Table II) but not at present for a transition metal. Methods of generating such species are by γ-irradiation of the solid at low temperature or by hydrogen abstraction, for example by $Ph_2\dot{C}-\dot{O}$ or $Bu^t\dot{O}$ radicals, or the addition of a metal-centered radical (e.g., by photolysis of the metal hydride with $Bu^t_2O_2$) to an unsaturated substrate; the following reactions are some examples:

$$(LiMe)_4 + Bu^t\dot{O} \xrightarrow{(33)} \dot{C}H_2Li(LiMe)_3 \quad (19)$$

$$BMe_3 \xrightarrow[(144)]{\gamma- \text{ at } 77 \text{ K}} \dot{C}H_2BMe_2 \quad (20)$$

$$BR^3_3 + R^1R^2\dot{C}-\dot{O} \xrightarrow{(52)} \dot{C}R^1R^2OBR^3_2 + \dot{R}^3 \quad (21)$$

$$\dot{S}iEt_3 + R^1R^2CO \xrightarrow{(20)} \dot{C}R^1R^2OSiEt_3 \quad (22)$$

$$\dot{G}eBu^n_3 + C_2H_4 \xrightarrow[(134)]{-148°} \dot{C}H_2CH_2GeBu^n_3 \equiv \quad (23)$$

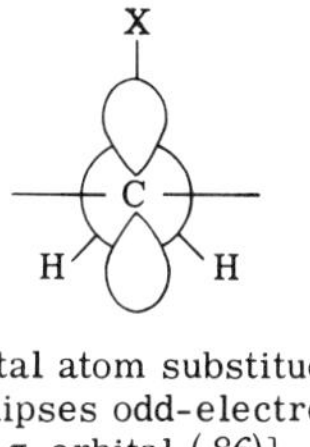

[metal atom substituent eclipses odd-electron π-orbital (86)]

$$\dot{S}nR_3 + C_4H_6 \xrightarrow{(117)} \quad (24)$$

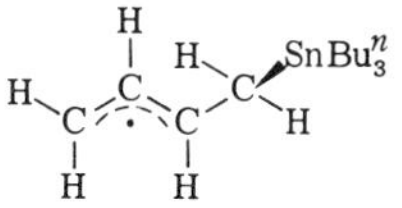

Reactions such as that shown in Eq. (19) are unusual. However, another example is the formation of the stable $\dot{C}(SiMe_3)_3$ from $(Me_3Si)_3CH$ and Bu^tO (*12, 150*) (this decays by a first-order hydrogen abstraction from the silane). More typically, Bu^tO attacks the metal center, displacing the radical $\dot{R}$; this is an example of an S_H2 reaction discussed in Section IV [but see XVI (*104a*)]. The choice of a particular pathway is governed by the activation energies for the competing processes, and relevant bond strengths provide a useful guide.

The ESR spectra of arylsilane radical anions show that the spin density is associated essentially with the aromatic rings. At low temperature, the unpaired electron in $[Ph(SiR_2)_2Ph]^{\overline{\cdot}}$ is delocalized over one of the phenyl rings but, at higher temperatures, over both (*118*). In species such as $[Me_2Si\text{-}(C_6H_4\text{·}Ph\text{-}p)_2]^{\overline{\cdot}}$, the Me_2Si moiety bridges essentially independent π systems (*43*); in the neutral analogs, the triplet electrons are similarly distributed.

Nitroxides derived from organometallic Group IV elements have been obtained by reactions such as (*189*)

$$RMe_2SiNHOSiMe_2R \xrightarrow[\text{2. electrolytic or } O_2 \text{ oxidation}]{\text{1. } -H^+(LiR);} RMe_2Si\ddot{N}OSiMe_2R \rightleftharpoons (RMe_2Si)_2N\bar{O} \quad (25)$$

$$\downarrow -1e^-$$

$$(RMe_2Si_2)\dot{N}O$$

They show low values of $a(N)$, indicating localization of the unpaired electron on the oxygen and strong $\ddot{N}\text{—}M$ (Si > Ge) π bonding, consistent also with higher g values than in $R_2\ddot{N}O$.

Transition metal paramagnetic complexes that warrant further comment are the radical anions derived from metallocene- (especially ferrocene-) ketyls and related species, such as compounds IX (*61*), X (*45*), or XI (*146*), and metallonitroxides. From the ESR spectra of compounds such as IX–XI, generated from alkali–metal and the neutral precursor, there is some evidence for participation of metal orbitals in stabilizing the lone electron, and the analogy has been made with the unusually stable α-ferrocenylcarbonium ions. A similar picture emerges from an ESR and X-ray (!) study of the ferrocenylnitroxide (XII) (*72*), obtained by autoxidation of the corresponding hydroxylamine, which was, in turn, prepared from the ferrocenyl–Grignard reagent and Bu^tNO.

The metallonitroxides, such as compound XIII, were obtained by spin-

(XII)   (XIII)

trapping the metal-centered precursors [which, however, were not detected by ESR in the absence of spin trap (see Section II,B)] and identified by ESR (Table V and Fig. 2) (105). The similarity of $a(N)$ in all the species and the g factors suggest that little spin density is associated with the metal.

The nitroxide (IV) (see Section I) was used as a spin label in connection with a study relevant to vitamin B_{12} coenzyme, which is of type $Co(III)$—R (R = 5′-deoxyadenosyl) and thus an analog of IV, but with a different axial group R (193). Compound (IV), which we may abbreviate $Co(III)$—R′, was prepared from a $Co(I)$ precursor B_{12s} and R′Br, and was found to bind to the enzyme ethanolamine–ammonia–lyase (in this system ethanolamine is converted to acetaldehyde). By ESR it was concluded that the active site is relatively close to the enzyme surface (see also Section V,B).

Another type of complex relevant here is that formed from a stable organic radical and a metal complex. An example of the former is a nitroxide,

TABLE V
Electron Spin Resonance Parameters for Metallonitroxides[a,b]

Compound ML in $ArN(ML)O$	$a(N)$ (mT)	$a(M)$ (mT)	$a(P)$ (mT)	g
$[Mn(CO)_5]$	1.59	0.86 (^{55}Mn)[c]	—	2.006
$[Re(CO)_5]$	1.47	4.09 ($^{185,187}Re$)[c]	—	2.01
$[Mn(CO)_4PPhMe_2]$	1.59	0.86 (^{55}Mn)[c]	0.99	2.006
$[Co(CO)_3P(OEt)_3]$	1.79	1.39 (^{59}Co)[d]	0.45	2.007
$[Fe(C_5H_5\text{-}\eta)(CO)_2]$	1.75	—	—	2.005
$[Mo(C_5H_5\text{-}\eta)(CO)_3]$	1.45	0.40 ($^{95,97}Mo$)[c]	—	2.005

[a] Data from Ref. 105.
[b] In $CHCl_3$ or THF solution at ca. $-30°$.
[c] $I = 5/2$.
[d] $I = 9/2$.

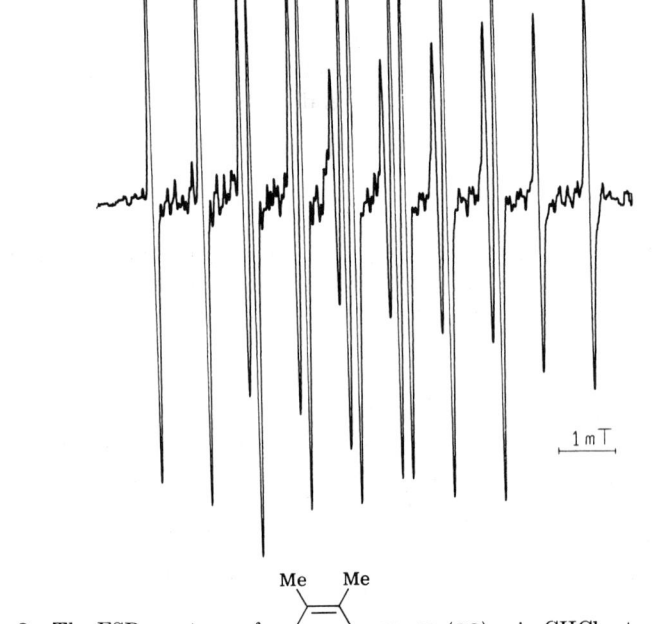

FIG. 2. The ESR spectrum of (2,3,5,6-tetramethylphenyl)–N(O)–Mn(CO)$_5$ in CHCl$_3$ at ca. −30°.

such as Bu$_2^t$NO, which complexes Lewis acids such as SnCl$_4$, TiCl$_4$, or Bu$_2^t$Sn-Cr(CO)$_5$ (*37*), or a nitroxide related to compound IV which can bind to Co (Section V,B) (*193*).

IV

BIMOLECULAR HOMOLYTIC SUBSTITUTION (S_H2) AT THE METAL CENTER OF AN ORGANOMETALLIC SUBSTRATE

Bimolecular heterolytic substitution (S_N2 and S_E2) at a saturated center has been extensively investigated during the formative years of physical–organic chemistry, and the S_E2 mechanism has an important role in the context of organometallic chemistry. The S_H2 mechanism involving a metal center M is defined by Eq. (1), and has been established for main group elements M since 1966. The well-known free radical process of atom (H or Hal) abstraction is an example of the S_H2 mechanism at a terminal atom, e.g., Eq. (19). Relevant reviews, confined to main group metal centers, are in Refs. *23, 50, 51, 102, 103, 108, 111, 112, 131, 151*, and this aspect is therefore not discussed in detail (Section IV,A). We shall limit ourselves to a brief introduction to S_H2 processes, without indication of primary sources [but reference should also be made to later papers by Davies *et al.* (*53*) and S_H2 reactions at Si and Ge (*99*), and Mg (*32*)] (Section IV,A), followed by a discussion of the situation involving a transition metal center M, invoking observations made since 1973 (Section IV,B).

A. Main Group Element Compounds as Substrates

Some indication of the scope of S_H2 reactions is gleaned by considering the types of attacking radicals $\dot{X}$ in Eq. (1), which may be $R\dot{O}$, $R\dot{S}$, $R_2\dot{N}$, $RCO\dot{N}R$, $R\dot{S}O_2$, $\dot{R}$, or Hal. Most of the work involving alkoxyl radicals has used $Bu^t\dot{O}$. This may be generated photochemically from $Bu^t_2O_2$, which is particularly convenient for ESR studies. For example, in 1969, several groups reported that the *in situ* photolysis of a static solution containing $Bu^t_2O_2$ and a substrate gave rise to radicals derived from the latter (*131*). High-intensity UV light is focused onto the solution, causing the formation of $Bu^t\dot{O}$ radicals from symmetrical cleavage of the peroxide. The $Bu^t\dot{O}$ radicals are not detected, due to extreme width of the signal. The two most common reactions are hydrogen atom abstraction, e.g., Eq. (19), or an S_H2 displacement, i.e.,

$$Bu^t\dot{O} + MR_n \longrightarrow Bu^tOMR_{n-1} + \dot{R} \qquad (26)$$

The sensitivity of the technique is indicated by the detection of natural abundance ^{13}C satellites in the spectrum of $\dot{C}H_3$, derived from BMe_3 and $(Bu^tO)_2$ (*131*). The principal limitations of the technique are the solubility of the organometallic sample, its photochemical lability, and the efficiency of the light-generating and focusing apparatus. An interesting example is in the $Bu^t\dot{O}/ClMgPr^i$ system when the S_H2 process (to give $\dot{P}r^i$) was accompanied by H abstraction [to yield (ESR) $\dot{C}Me_2MgCl$] (*32*).

Di-*t*-butyl hyponitrite is a suitable thermal source of $Bu^t\dot{O}$ and nitrogen ($t_{1/2}$ = 29 minutes at 65°) (*124*):

$$(Bu^tON)_2 \longrightarrow 2Bu^t\dot{O} + N_2 \qquad (27)$$

It has been used in the ESR detection of alkyl radicals, as in Eq. (26), when the substrate is photosensitive. As a thermal source of $Bu^t\dot{O}$, $(Bu^tON)_2$ is usually more convenient than Bu^tOOBu^t ($t_{1/2}$ = 1 hour at 150°), or the more labile and somewhat dangerous di-*t*-butyl peroxyoxalate (calculated $t_{1/2}$ = 6.8 minutes at 60°).

A chain reaction, involving S_H2 attack by $Bu^t\dot{O}$, is illustrated by the action of t-butyl hypochlorite on a metal alkyl,

$$Bu^t\dot{O} + MR_n \longrightarrow Bu^tOMR_{n-1} + \dot{R} \qquad (28)$$

$$\dot{R} + Bu^tOCl \longrightarrow RCl + Bu^t\dot{O} \qquad (29)$$

This process has been demonstrated for $MR_n = BR_3$ or R_nSnCl_{4-n} ($n = $ 1–3). A synthetically useful S_H2 reaction is found in the conjugative addition of organoboranes to enones:

$$\dot{R} + \overset{\diagdown}{\underset{\diagup}{C}}=C-C=O \longrightarrow R-\overset{|}{C}-\overset{|}{C}=\overset{|}{C}-\dot{O} \qquad (30)$$

$$R-\overset{|}{C}-\overset{|}{C}=\overset{|}{C}-\dot{O} + BR_3 \longrightarrow R-\overset{|}{C}-\overset{|}{C}=\overset{|}{C}-OBR_2 + \dot{R} \qquad (31)$$

The triplet ketones, $R_2\dot{C}-\dot{O}$, are a special case of alkoxyl radicals. Thus UV irradiation of acetone and tri-n-butylborane in the cavity of an ESR spectrometer led to the observation of two radicals, $\dot{C}Me_2OBBu_2^n$, Eq. (21), and $\dot{B}u^n$. On a preparative scale the coupled products as well as $Bu^nMe_2COBBu_2^n$ were isolated.

Both chain and nonchain reactions involving an S_H2 attack are known for thiyl radicals. An example of the latter is the photolysis of Bu^tSSBu^t in the presence of an organoborane or Bu^tMgCl, when an alkyl radical is displaced by $Bu^t\dot{S}$ and can be detected by ESR. The former, known for B, Sb, and Bi, are illustrated by

$$R'\dot{S} + BR_3 \longrightarrow R'SBR_2 + \dot{R} \qquad (32)$$

$$\dot{R} + R'SH \longrightarrow RH + R'\dot{S} \qquad (33)$$

Dimethylaminyl radicals may be produced photolytically from $Me_2N \cdot N{=}N \cdot Me_2$, and the S_H2 displaced $\dot{R}$ has been observed for BR_3 and SbR_3; the evolution of nitrogen can interfere with the recording of satisfactory ESR spectra. A chain reaction has been observed between Me_2NCl and BR_3, cf. Eqs. (28) and (29), but a competing nonradical process yielding Me_2NR and $ClBR_2$ is also found. N-Halogenosuccinimides undergo a chain S_H2 reaction with SnR_4, in contrast to $Me_2\dot{N}$ (where there is no reaction) or $Bu^t\dot{O}$ (where there is hydrogen atom abstraction).

Another radical chain process is shown in the following propagation sequence:

$$Ph\dot{S}O_2 + BBu_3^n \xrightarrow{70°} PhSO_2BBu_2^n + \dot{B}u^n \qquad (34)$$

$$\dot{B}u^n + PhSO_2Br \longrightarrow Bu^nBr + Ph\dot{S}O_2 \qquad (35)$$

It has been suggested (50) that the widespread insertion reactions of sulfur dioxide into metal–carbon bonds may also involve an S_H2 reaction,

$$R\dot{S}O_2 + MR_n \longrightarrow RSO_2MR_{n-1} + \dot{R} \qquad (36)$$

$$\dot{R} + SO_2 \longrightarrow R\dot{S}O_2 \qquad (37)$$

a sequence similar to that found for the autoxidation of organoboranes.

There are isolated reports of S_H2 reactions involving attack by C-centered radicals. These are typically gas-phase reactions using the photolysis of acetone (sometimes the per-deutero or per-fluoro analog) as a source of $\dot{M}e$, which gives $\dot{C}D_3$ with $Hg(CD_3)_2$; examples involving B and Sn have also been reported.

Reactions involving halogen atoms are poorly defined, due to competing polar processes from the parent X_2, or to alternative reactions by the halogen atom (e.g., hydrogen atom abstraction). However, the reaction of a dialkylmercury with a dihalogen in a nonpolar solvent can be accelerated by light or peroxide, and retarded by oxygen, so that the chain sequence,

$$\dot{X} + HgR_2 \longrightarrow XHgR + \dot{R} \qquad (38)$$

$$\dot{R} + X_2 \longrightarrow RX + \dot{X} \qquad (39)$$

may be operative.

The reaction between Si_2R_6 and $(BrCH_2)_2$, yielding $BrSiR_3$ and $CH_2{=}CH_2$, is inhibited by galvinoxyl but initiated by Bz_2O_2 and, hence, appears to be the first example of S_H2 at Si in solution (99, 171); Ge_2Me_6 behaves similarly.

The S_H2 reaction can, in principle, be stepwise or synchronous; if the former, an intermediate (XIV) may be detectable by ESR,

$$\dot{X} + MR_n \longrightarrow X\dot{M}R_n \longrightarrow XMR_{n-1} + \dot{R} \qquad (40)$$
$$\text{(XIV)}$$

but has so far only been found in the reaction of $Bu^t\dot{O}$ with trimethylphosphine or with phosphites, to give a phosphoranyl radical (VIII).

Many of these S_H2 processes are fast, the fastest approaching the diffusion-controlled limit, e.g., for $Bu^t\dot{O}/BBu_3^n$, $k = 3 \times 10^7$ M^{-1} second^{-1} at 30°. The factors favoring an S_H2 reaction are the presence of energetically low-lying M acceptor orbitals, the formation of a strong bond relative to the bond being broken (e.g., $B-C \longrightarrow B-O$), and the absence of steric effects that hinder access of $\dot{X}$ to M.

B. Transition Metal Compounds as Substrates

At the start of our initial work, published in 1973–4 (29, 142), there was no definite example of an S_H2 reaction at a transition metal. We shall, therefore, describe in detail S_H2 displacement $(-\dot{R})$ reactions of $Bu^t\dot{O}$ (nonchain) and PhṠ (chain) and cis-[PtR$_2$(PR$_3'$)$_2$]. Our entry into this area was accidental and arose from an attempt to generate a radical with an unpaired electron in an alkyl chain and not at a transition metal center [(II), which are still unknown (see Section III)]. We chose to attempt to abstract a hydrogen atom from a methyl group bound to a metal, using Bu^{t}Ȯ. The metal compound selected was a Pt(II) methyl because platinum(II) alkyls combine ease of synthesis with stability to air and moisture; the nuclear spin of platinum might aid ESR identification of Pt—ĊH$_2$; and the methyl group should present a unique site toward H abstraction. However, three other modes of attack by Bu^{t}Ȯ are in principle, also possible (XV). Actually, we have mainly observed S$_H$2 at Pt(II) but, in one case probably at silyl-substituted-Me (*29, 104a*).

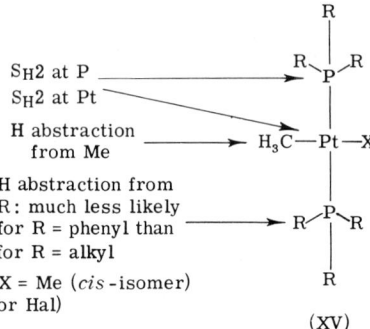

(XV)

Initial experiments with [PtBrMe(PPh$_3$)$_2$] were hampered by its low solubility. For cis-[PtMe$_2$(PEt$_3$)$_2$] the possibility of competing H abstraction from Et existed but did not occur. Instead, ESR experiments involving *in situ* irradiation of this compound in Bu$_2^t$O$_2$, suggested an S$_H$2 reaction at platinum. To confirm and extend this evidence, a number of other platinum(II) alkyl complexes were synthesized and the following types of experiment carried out:

1. Direct observation of the displaced alkyl radical, using Bu^{t}Ȯ radicals generated photochemically from (Bu^{t}O)$_2$.
2. Spin trapping of the displaced alkyl radical, using Bu^{t}Ȯ radicals generated thermally from (Bu^{t}ON)$_2$.
3. Attempts to prepare a complex containing the Pt—OBu^t group by an S$_H$2 reaction.
4. Attempts to find an S$_H$2 chain reaction between a platinum alkyl and a thiol or disulfide.

As for experiment 1, the ESR signals obtained were weak and lasted only a few minutes, making detection difficult. The reasons for this were probably low solubility of the organometallic, competing reactions (see structure XV), and efficiency of the apparatus (increasing the light intensity, say from 250 W to 1 kW, leads to stronger signals, but of shorter duration). Low solubility means rapid sample depletion and a correspondingly short-lived signal. Even the strongest signal obtained from cis-[PtMe$_2$(PEt$_3$)$_2$] decayed considerably over the time taken to record a spectrum. Furthermore, it is not possible significantly to reduce the temperature without extensive precipitation. In general, the lower the temperature the better, because the reactivity of the radicals is reduced, and the population difference between the spin states is increased with consequent increase in the intensity of absorption. [A reduction in temperature is not always beneficial—it can lead to broadened lines, particularly for large radicals, due to slower tumbling (see Section II,A,2).] The best conditions were usually found by dissolving the sample in Bu$_2^t$O$_2$ and irradiating at $-45°$ (the pure peroxide freezes at $-40°$ and the presence of other material reduces this to ca. $-50°$, but, by using a solvent such as cyclopropane, much lower temperatures could be achieved). The signal obtained by irradiating a mixture (ca. 1:1) of dimethoxymethane and (Bu^{t}O)$_2$ at $-45°$ (which is then homogeneous and liquid) was of good strength and lasted for about 1 hour; the spectrum is due to CH$_3$OĊHOCH$_3$ (doublet of septets) and ĊH$_2$OCH$_2$OCH$_3$ (triplet of triplets) and the intensity of these lines was used as a standard, with the optical apparatus adjusted to maximize the signals. In the work with Pt(II) complexes, conditions of solvent and temperature depended mainly on the phosphine. Deoxygenated Bu$_2^t$O$_2$ was used either neat or as a solution in benzene and kept under argon or nitrogen. Solutions of the Pt(II) complex were normally saturated at room temperature but a small amount of solid deposited by cooling did not interfere with the signal. The ESR results are summarized in Scheme 2.

The bis(trialkylphosphine)Pt(II) complexes were studied to obtain better solubility than with cis-[PtMe$_2$(PPh$_3$)$_2$]. Complex cis-[PtEt$_2$(PPr$_3^n$)$_2$] was used (Ėt detected) in attempting to eliminate the possibility that the source of Ṁe formed from cis-[PtMe$_2$(PEt$_3$)$_2$] was β scission of Bu^{t}Ȯ catalyzed by Pt(II); and cis-[PtMe$_2$(PBu$_3^n$)$_2$] was employed in attempting to obtain kinetic data with a highly soluble complex and monitoring the relative decay of Ṁe and Ċ$_5$H$_9$ when using cyclopentane as solvent (finally, this was not pursued because the concentration of Ṁ in pure Bu$_2^t$O$_2$ at $-40°$ was low). With cis-[Pt(CH$_2$SiMe$_3$)$_2$(PMe$_2$Ph)$_2$] (38) in Bu$_2^t$O$_2$, irradiation at $-20°$ gave a strong spectrum ($t_{1/2}$ ca. 5 seconds at $-20°$) (142) consisting of four multiplets, each seemingly a 1:4:6:4:1 quintet which agreed reasonably with a computer-simulated spectrum of the platinum-coordinated methylene radical (XVI), using $a(P) = 5.6, a(Pt) = 14.9, a(H) = 1.6$ mT, and $g = 2.021$ (104a). Attack,

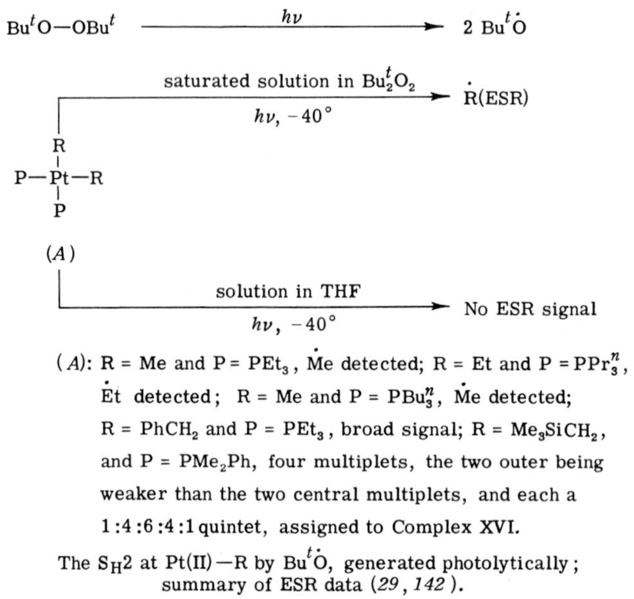

$$Bu^tO-OBu^t \xrightarrow{h\nu} 2\ Bu^t\dot{O}$$

$$\begin{array}{c} R \\ | \\ P-Pt-R \\ | \\ P \end{array}$$

(A)

$$\xrightarrow[h\nu,\ -40°]{\text{saturated solution in } Bu_2^tO_2} \dot{R}(ESR)$$

$$\xrightarrow[h\nu,\ -40°]{\text{solution in THF}} \text{No ESR signal}$$

(A): R = Me and P = PEt$_3$, Me detected; R = Et and P = PPr$_3^n$, Et detected; R = Me and P = PBu$_3^n$, Me detected; R = PhCH$_2$ and P = PEt$_3$, broad signal; R = Me$_3$SiCH$_2$, and P = PMe$_2$Ph, four multiplets, the two outer being weaker than the two central multiplets, and each a 1:4:6:4:1 quintet, assigned to Complex XVI.

The S$_H$2 at Pt(II)—R by Bu$^t\dot{O}$, generated photolytically; summary of ESR data (29, 142).

SCHEME 2. The S$_H$2 at Pt(II)-R by Bu$^t\dot{O}$, generated photolytically; summary of ESR data (29, 142).

in this instance, at Si rather than at Pt may be due to steric hindrance at the metal center. However, this is a tentative assignment.

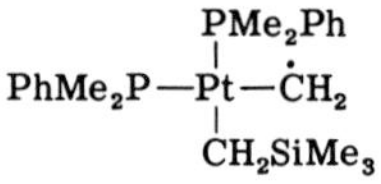

$$\begin{array}{c} PMe_2Ph \\ | \\ PhMe_2P-Pt-\dot{C}H_2 \\ | \\ CH_2SiMe_3 \end{array}$$

(XVI)

Spin trapping is an ESR technique for identifying free radicals in reactions in which their steady-state concentration is inadequate for direct detection. The spin trap, commonly a nitroso compound or a nitrone, and in this work (29, 142) Bu^{t}NO or nitrosodurene, when added to the reaction mixture couples with free radicals to generate the paramagnetic spin adduct of much greater stability. The concentration of the latter, therefore, builds up and is detected by ESR. Bu^{t}NO and 2,3,5,6-Me$_4$C$_6$HNO (ArNO) have no α-H and, hence, in the spin adduct Bu^t(R)$\dot{N}$O or Ar(R)$\dot{N}$O there is no hyperfine coupling from Bu^t or Ar, so that $\dot{R}$ gives a distinctive coupling. The spectrum then observed consists of a 1:1:1 triplet (due to ^{14}N, 99.6% abundance, $I = 1$) further split by magnetic nuclei that are one or two bonds removed from nitrogen. For example, for R = Me, Et, or Me_2CH the pattern observed is a triplet (1:1:1) of quartets (1:3:3:1), a triplet (1:1:1) of triplets (1:2:1), or a triplet (1:1:1) of doublets (1:1), respectively. A disadvantage of Bu^tNO is that it is photochemically (visible) more labile than ArNO and in this way or by heating gives $Bu_2^t\dot{N}O$ [on the other hand, it is adequately soluble in water for use in aqueous media (see Section V,B)]. Our spin-trapping experiments involved mixing $[PtR_2P_2]$, Bu^tNO or ArNO, and $(Bu^tON)_2$ (as a thermal source of $Bu^t\dot{O}$) in benzene solution in an ESR tube, and warming to 40° in the cavity of the spectrometer. For Bu^tNO, this temperature was a compromise between minimal decay of Bu^tNO and maximal production of $Bu^t\dot{O}$. (For the Bu^tNO experiments, the three solid components were introduced in a foil-wrapped ESR tube under argon, and deoxygenated benzene was added by syringe; for ArNO, solutions of known concentration of each component in C_6H_6 were introduced into the tube by means of a calibrated pipette. The foil was removed immediately before placing the tube in the cavity.) Spectra usually grew in intensity over a few minutes at 40° and for Bu^tNO were recorded when the signal for $Bu^t(R)\dot{N}O$ was at a maximum with respect to that for $Bu_2^t\dot{N}O$. For the nitrosodurene (ArNO) experiments, equal volumes (0.1 ml each) of solutions in C_6H_6 of the Pt compound (10^{-1} M), $(Bu^tON)_2$ (10^{-1} M), and ArNO (10^{-2} M) were employed. The results are summarized in Scheme 3, together with details of the necessary control experiments. A typical spectrum, with assignments, is shown in Fig. 3.

The ESR results summarized in Schemes 2 and 3 provide evidence for the reactions,

$$Bu_2^tO_2 \xrightarrow{h\nu} 2Bu^t\dot{O} \quad (41)$$

$$Bu^t\dot{O} + [PtR_2(PR_3')_2] \longrightarrow [PtR(OBu^t)(PR_3')_2] + \dot{R} \quad (42)$$

with the latter representing an S_H2 process at Pt(II).

Confirmation of this result was sought by attempting to synthesize a Pt(II)-t-butoxide, as in Eq. (42), in various ways: photolytically from $Bu_2^tO_2$ using $trans$-$[PtMe(Br)(PPh_3)_2]$ or cis-$[PtMe_2(PEt_3)_2]$; thermally from $(Bu^tON)_2$ and cis-$[PtMe_2(PR_3)_2]$ (R = Et or Ph); or by an attempted chain reaction using Bu^tOCl and cis-$[PtMe_2(PR_3)_2]$ [cf. Eqs. (28) and (29)]. These were unsuccessful. However, compounds containing Pt(II)—O bonds are rare and largely limited to complexes having chelating ligands (14), which may achieve kinetic stability by imposing a conformation on the alkoxide that is unfavorable to normal facile decomposition pathways, such as β elimination. Clearly further work is required using a system suitable for monitoring preparative-scale experiments with ESR studies.

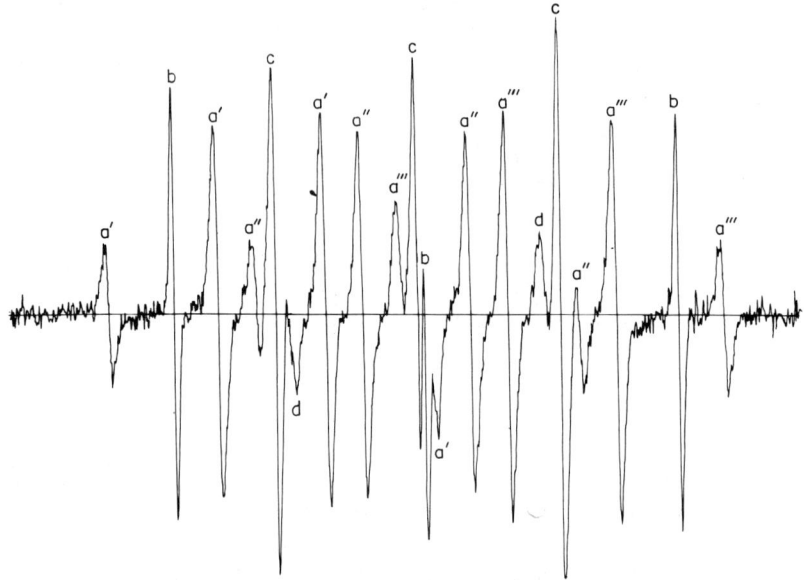

FIG. 3. The ESR spectrum generated from cis-[PtMe$_2$(PBu$_3^n$)$_2$], (Bu^{t}ON)$_2$, and Bu^{t}NO in C$_6$H$_6$ at 40° [a = Bu^t($\dot{\text{N}}$O)Me, b = Bu^t($\dot{\text{N}}$O)OBu^t, c = Bu$_2^t$$\dot{\text{N}}$O].

When attention was turned to S$_H$2 attack at Pt(II) involving thiyl radicals, a different situation was encountered (29, 142), as summarized in Scheme 4 and by the following chain mechanism:

$$\text{Initiation} \longrightarrow \text{Ph}\dot{\text{S}} \tag{43}$$

Propagation
$$\begin{cases} \text{Ph}\dot{\text{S}} + [\text{Pt}(\text{CH}_2\text{SiMe}_3)_2(\text{PMe}_2\text{Ph})_2] \longrightarrow \\ \qquad [\text{Pt}(\text{CH}_2\text{SiMe}_3)\text{SPh}(\text{PMe}_2\text{Ph})_2] + \text{Me}_3\text{Si}\dot{\text{C}}\text{H}_2 \quad (44) \\ \qquad\qquad\qquad (\text{XVII}) \\ \text{Me}_3\text{Si}\dot{\text{C}}\text{H}_2 + \text{Ph}_2\text{S}_2 \longrightarrow \text{Me}_3\text{SiCH}_2\text{SPh} + \text{Ph}\dot{\text{S}} \quad (45) \\ \text{Ph}\dot{\text{S}} + (\text{XVII}) \longrightarrow [\text{Pt}(\text{SPh})_2(\text{PMe}_2\text{Ph})_2] + \text{Me}_3\text{Si}\dot{\text{C}}\text{H}_2 \quad (46) \\ \text{Me}_3\text{Si}\dot{\text{C}}\text{H}_2 + \text{Ph}_2\text{S}_2 \longrightarrow \text{Me}_3\text{SiCH}_2\text{SPh} + \text{Ph}\dot{\text{S}} \quad (47) \end{cases}$$

This mechanism was established by identifying the products of the reaction from [Pt(CH$_2$SiMe$_3$)$_2$(PMe$_2$Ph)$_2$] and Ph$_2$S$_2$ at 60° in C$_6$H$_6$ as trans-[Pt(SPh)$_2$(PMe$_2$Ph)$_2$] and Me$_3$SiCH$_2$SPh and showing that the reaction only took place under similar conditions at a reasonable rate in the presence of (Bu^{t}ON)$_2$ as a free-radical initiator; additionally, in the cis-[PtMe$_2$(PEt$_3$)$_2$]/Ph$_2$S$_2$ system, in the absence of initiator but with nitrosodurene as spin trap, a weak signal of the spin adduct Ar(Me)$\dot{\text{N}}$O was

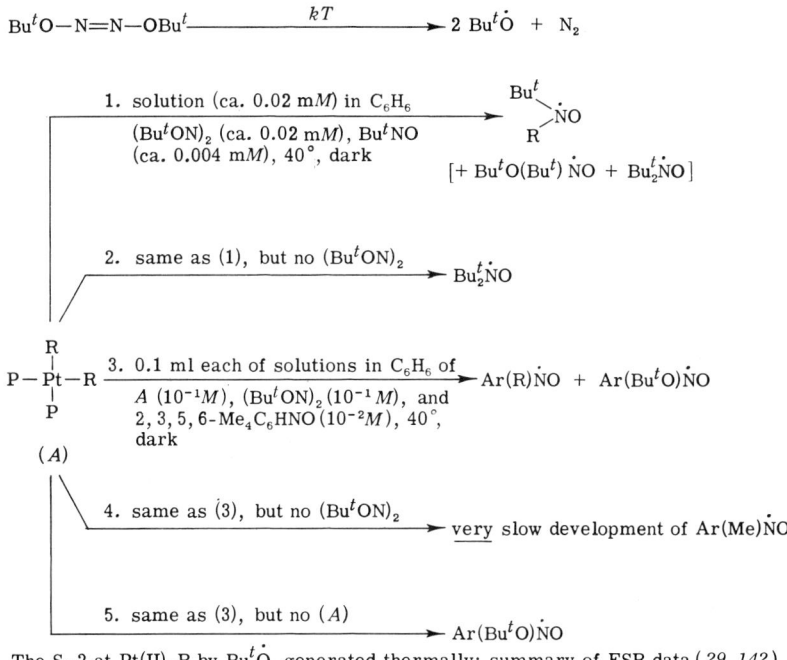

The S_H2 at Pt(II)-R by $Bu^t\dot{O}$, generated thermally; summary of ESR data (29, 142)

SCHEME 3. The S_H2 at Pt(II)-R by $Bu^t\dot{O}$, generated thermally; summary of ESR spin-trapping data (29, 142).

detected (87). Nuclear magnetic resonance experiments on the latter system indicated that rearrangement takes place during the first Me displacement, generating trans-[PtMe(SPh)(PEt$_3$)$_2$] (142), but such an intermediate was not detected for the former.

The initiation may involve $Bu^t\dot{O}$, $\dot{M}e$ (from β scission), or $Me_3Si\dot{C}H_2$ from S_H2 at Pt(II) (formation of compound XVI takes place at ca. 100° lower). An alternative mechanism may include oxidative addition to form [Pt(CH$_2$SiMe$_3$)$_2$(SPh)$_2$(PMe$_2$Ph)$_2$] [this may involve a radical chain mechanism having a five-coordinate Pt(III) intermediate] and subsequent reductive elimination {cf., oxidative addition of p-TolSO$_2$Br to cis-[PtMe$_2$-(PMe$_2$Ph)$_2$] (see Section V,A) in which $\dot{M}e$ is detected, suggestive of S_H2 as a minor pathway (87)}.

Benzenethiol gave methane and [Au(SPh)L], cis-[AuMe$_2$(SPh)L], and trans-[PtMe(SPh)L$_2$] or trans-[Pt(SPh)$_2$L$_2$], with respectively, [AuMeL], [AuMe$_3$L], and cis-[PtMe$_2$L$_2$] (L = PMe$_3$, PMe$_2$Ph, PMePh$_2$,

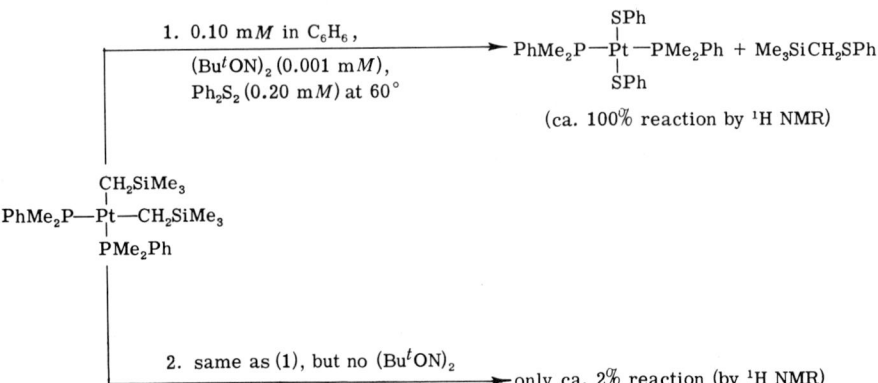

SCHEME 4. The S_H2 at Pt(II)-R by PhṠ, generated thermally (29, 142)
Additional control experiments: (i) [Pt(PPh$_3$)$_3$] + Bu^{t}NO, (ii) PPh$_3$ + Bu^{t}NO, (iii) PPh$_3$ + Bu^{t}NO + MeI.
Additional variations: O$_2$, solvent, concentrations.

or PPh$_3$) in CH$_2$Cl$_2$ at ca. 20°, and the Au(I) and Pt(II) reactions were formulated as involving a free-radical chain mechanism (90). The evidence was based on the observation (NMR) of an induction period, inhibition by galvinoxyl (1%), acceleration by AIBN, and ESR detection of Bu^t(Me)ṄO when the reaction was carried out in the presence of Bu^{t}NO, which has little effect on the rate. Two mechanisms were considered, both involving initial attack by PhṠ at the metal center,

$$\text{PhṠ} + \text{MMe} \xrightarrow{\text{PhSH}} [\text{ṀMe(SPh)}] \longrightarrow \text{MSPh} + \text{Ṁe} \quad (48)$$
$$\xrightarrow{\text{PhSH}} \text{MMe(H)SPh} \longrightarrow \text{CH}_4 + \text{MSPh} \quad (49)$$

We have independently examined a related Pt(II) system, with cis-[PtMe$_2$(PR$_3$)$_2$] (R = Et or Bu^n) (142) and note that galvinoxyl readily reacts with PhSH; AIBN is not expected to be an efficient initiator under the conditions used, $t_{1/2}$ ca. 20 hours at 60° (184) (our reaction is very fast at 20° without initiator); and the lack of effect of Bu^{t}NO on rate may be due to an insufficient amount present as the active monomer (Bu^{t}NO exists as dimer in the solid state).

The autoxidation of TiR'_4, ZrR''_4, $(MoR'''_3)_2$, or $[W(CH_2Ph)_3]_2$ ($R' = Me_3SiCH_2$, $R'' = Me_3SiCH_2$ or $PhCH_2$, and $R''' = Me_3SiCH_2$ or Me_3CCH_2) was studied in hydrocarbon solution at 20° or −74° (*22*). The amount of oxygen consumed (0.5 mole per M—C bond) indicated $M(OR)_4$ as the end-product but only a trace of metal peroxide MOOC was detected. Galvinoxyl reacts with some of the alkyls and, therefore, its use as an inhibitor is mechanistically dubious; however, phenothiazine (and other inhibitors) were free from this complication and retarded the reactions. The analogy with autoxidation of main group element alkyls, such as BR_3, is clear, for which the following propagation sequence is established (*50*):

$$\dot{R} + O_2 \longrightarrow RO\dot{O} \quad (50)$$

$$RO\dot{O} + BR_3 \longrightarrow ROOBR_2 + \dot{R} \quad (51)$$

V
ADDITION OR ELIMINATION RADICAL REACTIONS

We are here concerned with various organometallic reactions for which there is evidence that organic free radicals are implicated in the reaction pathway. Many of these are formally two-electron oxidative additions or their retrogressions, the reductive eliminations (Section V,A). We shall focus attention on systems in which transition metal Group VIII complexes are involved.

A. Oxidative Addition of Alkyl Halides or Related Reagents and Reductive Elimination

The reaction between a low-valent Group VIII metal complex and an alkyl halide belongs to the class known as oxidative addition and has attracted much study and controversy as to the mechanism. Recent evidence suggests free radicals as intermediates in many cases. The oxidative-addition reaction is of widespread occurrence and importance in transition metal chemistry, due in part to its use in synthesis and to its implication in many catalytic systems. In one of its forms it is described by

$$[LM] + A{-}B \longrightarrow [LM(A)B] \quad (52)$$

Related reactions involve addition of AB without fragmentation (e.g., $AB = O_2$) or with fragmentation into three parts [e.g., $AB = Me_2NCHCl_2$ (*30*)]. The metal complex [LM] may be cationic, anionic, or neutral, and the addendum A—B also covers a wide range, including H_2, O_2, RX, R_3SiH, or RHgX. (Reviews covering both mechanistic and preparative aspects of the reaction are in Refs. *24, 39, 40, 54, 59, 81, 82, 89, 119, 167, 182.*) We shall not duplicate this material, but aim to present a brief account of the main developments in the mechanistic work and then to discuss recent evidence implicating radicals.

1. *Mechanistic Ideas Prior to 1972*

The addition of an alkyl halide to the d^7 complex $[Co(CN)_5]^{3-}$ results in a one-electron oxidation of the metal and proceeds by homolytic abstraction of halogen. For a methyl or benzyl halide, an organocobalt product is formed,

$$[Co(CN)_5]^{3-} + RX \longrightarrow [Co(X)(CN)_5]^{3-} + \dot{R} \qquad (53)$$

$$[Co(CN)_5]^{3-} + \dot{R} \longrightarrow [Co(R)(CN)_5]^{3-} \qquad (54)$$

whereas for other halides olefins may be formed in a competing reaction, e.g.,

$$[Co(CN)_5]^{3-} + \dot{C}_2\dot{H}_5 \longrightarrow [Co(H)(CN)_5]^{3-} + C_2H_4 \qquad (55)$$

These and other additions to d^7 complexes are reviewed in Refs. *59* and *89*. Most of the mechanistic work on oxidative addition to d^8 complexes has been concerned with Ir(I) and in particular with Vaska's compound, *trans*-$[Ir(CO)(Cl)(PPh_3)_2]$, or its analogs, e.g.,

$$[Ir(CO)(Cl)(PPh_3)_2] + MeI \longrightarrow [Ir(CO)(Me)(Cl)(I)(PPh_3)_2] \qquad (56)$$

An early kinetic study (*34*; see also *177a*) revealed features similar to that of the Menschutkin reaction ($R_3N + R'X \longrightarrow R_3R'N^+X^-$), and it was suggested that the Ir(I) complex behaved as a nucleophile in an S_N2 type-mechanism,

$$LM + MeI \longrightarrow \left[LM^{\delta+} \cdots \overset{H\ \ H}{\underset{H}{C}} \cdots I^{\delta-} \right] \longrightarrow LM\overset{Me}{\underset{I}{\diagdown}} \qquad (57)$$

$LM = [Ir(CO)(Cl)(PPh_3)_2]$

Later work showed that no incorporation of added anions occurred during the addition of MeI to *trans*-$[Ir(CO)(Cl)P_2]$ (P = PPh_3 or $PMePh_2$) and that solid Ir(I) complexes added gaseous MeI or MeBr, suggestive of a nonionic mechanism (*161*); a molecular process, involving the transition state (XVIII) was proposed. Evidence was presented for retention of configuration at C_1 of optically active $MeCHBrCO_2Et$, accompanying the

$$\left[\mathrm{LM} \begin{smallmatrix} \nearrow \mathrm{Me} \\ | \\ \searrow \mathrm{X} \end{smallmatrix} \right]$$

(XVIII)

addition (*161*). However, objections to this [later experimentally confirmed (*137*)] and also to a claim (*136*) for inversion [later withdrawn (*21*)] have been published (*114*). A kinetic study of alkyl halide addition to Ir(I) has favored a polarized or unsymmetrical, three-center transition state (*183*). Kinetic studies have shown that addition to the d^{10} [Pt(PPh)$_n$] (n = 3 or 4) or to [Pt(C$_2$H$_4$)(PPh$_3$)$_2$] are first-order in both addendum and metal complex; the latter may initially lose ethylene or phosphine (*89, 162*).

2. *Recent Developments*

In 1972, a radical chain mechanism was suggested for addition of PhCHFCH$_2$Br (or deutero analogs) to *trans*-[Ir(CO)Cl(PMe$_3$)$_2$] to give [IrBr(CO)Cl(CH$_2$CHFPh)(PMe$_3$)$_2$] (*21*). [A radical nonchain mechanism had been accepted for one-electron oxidation of Co(II) (*89*).] The evidence for radicals was (i) initiation by O$_2$, AIBN, or Bz$_2$O$_2$; (ii) retardation by duroquinone or hydroquinone (radical inhibitors); (iii) loss of a specific stereochemistry at C$_1$ (deduced from NMR spectra), [the first positive demonstration of racemization during oxidative addition involving Ni, Pd, or Pt was in the reaction of (+) RCHBrCO$_2$Et (R = Me or Ph) to Pd(Bu^{t}NC)$_2$ (*156*)]; and (iv) a reactivity order in which the rate of addition of halide was increased by its having electronegative substituents. The following mechanism was proposed, by analogy with additions to Cr(II) and Co(II) (cf. Refs. *54* and *89*):

$$\dot{Q} + [\mathrm{Ir(I)}] \longrightarrow [\mathrm{Ir(II)-Q}] \qquad (58)$$

$$[\mathrm{Ir(II)-Q}] + \mathrm{RBr} \longrightarrow [\mathrm{BrIr(III)-Q}] + \dot{R} \qquad (59)$$

$$[\mathrm{Ir(I)}] + \dot{R} \longrightarrow [\mathrm{Ir(II)-R}] \qquad (60)$$

$$[\mathrm{Ir(II)-R}] + \mathrm{RBr} \longrightarrow [\mathrm{BrIr(III)-R}] + \dot{R} \qquad (61)$$

Equations (58) and (59) represent initiation by an unknown $\dot{Q}$, and Eqs. (60) and (61) the propagation sequence.

This work was extended as follows (*133*), (*137*). The addition of optically active MeCHBrCO$_2$Et either to [Ir(CO)ClP$_2$] (P = PMe$_3$, PMe$_2$Ph, or PMePh$_2$) or to Pt(PPh$_3$)$_3$ (*137*), or the chloride to [M(PEt$_3$)$_3$] (M = Pd or Pt) (*133*) gave the racemic adduct [Ir(Br)(CO)(Cl)(CHMeCO$_2$Et)-P$_2$], *trans*-[Pt(Br)(CHMeCO$_2$Et)(PPh$_3$)$_2$], or *trans*-[M(Cl)(CHMeCO$_2$-

Et)(PEt$_3$)$_2$], respectively. Inhibition by galvinoxyl was found, which was taken as evidence for a chain mechanism. The oxidative addition reaction may be a complex composite. For example, [Pt(PEt$_3$)$_3$] and Bu^{n}Br in PhMe at 25° reacted (*133*) as follows

$$[\text{Pt}(\text{PEt}_3)_3] + \text{Bu}^n\text{Br} \xrightarrow[\text{2 hours}]{\text{PhMe, 25°}} \begin{cases} \textit{trans-}[\text{Pt}(\text{Br})\text{Bu}^n(\text{PEt}_3)_2] \\ 95\% \\ \textit{trans-}[\text{Pt}(\text{Br})\text{H}(\text{PEt}_3)_2] \\ 4\% \\ \textit{trans-}[\text{PtBr}_2(\text{PEt}_3)_2] \\ 1\% \end{cases} \quad (62)$$

during the first 2 hours; further reaction afforded the dibromide at the expense of [Pt(Br)Bu^n(PEt$_3$)$_2$], whereas the hydride increased to a maximum and then decreased, by reacting with Bu^{n}Br to yield the dibromide. For a *sec*-alkyl bromide, the hydride was the major product in the absence of excess RBr. With neopentyl bromide in toluene, *trans*-[Pt(Br)CH$_2$Ph-(PEt$_3$)$_2$] was obtained in high yield, whereas with BrCH$_2$(CH$_2$)$_3$CH=CH$_2$, the ratio of cycloalkyl– to *n*-alkenyl–Pt(II) complex was 3:1. Evidence for a radical mechanism is accordingly considerable, but indication supporting a *chain mechanism was limited to inhibition by galvinoxyl or duroquinone. Since these scavengers react with the Pt(0) complex,* caution is required in interpreting the observation. We are not convinced that the demonstration of inhibition, even without this caveat, would inevitably support a chain, rather than a nonchain, mechanism; for this to be more persuasive a zeroth-order dependence on inhibitor would be helpful.

The next major line of evidence for radicals as intermediates in related reactions came from our laboratory, where it was found that the ESR technique of spin trapping provided for the first time spectroscopic evidence for the intermediacy of free radicals (see Scheme 5) (*139*). When the addition of CH$_3$I, CD$_3$I, EtI, PhCH$_2$Br, or Ph$_2$CHBr to [Pt(PPh$_3$)$_3$] was carried out in the presence of Bu^{t}NO, strong signals for the corresponding nitroxide Bu^t(R)ṄO were obtained; analogous experiments with 2,3,5,6-Me$_4$C$_6$HNO as the spin trap have yielded similar results (*142*). *Control experiments* showed that the signals were not derived from either of the reactants, nor from either of the products (metal complex and phosphonium salt), nor from the reaction of alkyl halide with phosphine. Furthermore, [Pt(PPh$_3$)$_3$] with MeI in the presence of Bu^{t}NO still gave [PtMe(I)(PPh$_3$)$_2$] as the major product. Reaction of [Pt(PPh$_3$)$_3$] with benzhydryl bromide proceeded according to Eq. 63 below and with Ph$_3$CCl gave [PtCl$_2$(PPh$_3$)$_2$] and Ph$_3$Ċ; the latter is stable and was identified by ESR without a spin trap. {An attempt to observe the moderately stable

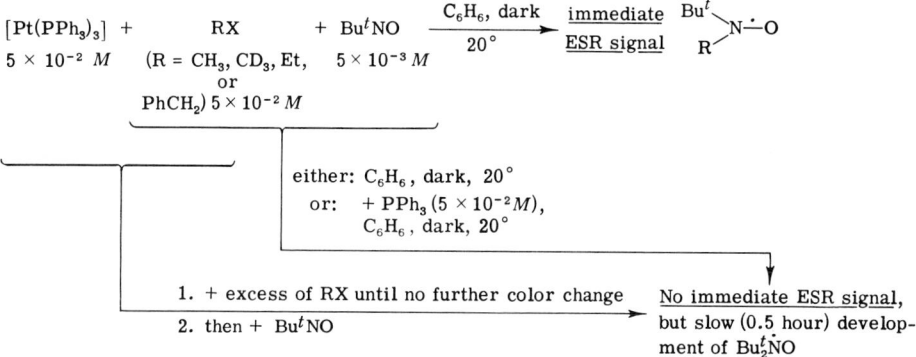

SCHEME 5. Summary of ESR spin-trapping experiments (similarly for 2,3,5,6-Me$_4$C$_6$HNO) for the [Pt(PPh$_3$)$_3$]–RX reaction (*139, 142*)

allyl radical directly in the reaction of [Pt(PPh$_3$)$_3$]/C$_3$H$_5$Br in dilute PhMe solution at low temperature (when PhMe is rather viscous) failed (*142*).}

$$[\text{Pt}(\text{PPh}_3)_3] + 2\text{Ph}_2\text{CHBr} \longrightarrow [\text{PtBr}_2(\text{PPh}_3)_2] + (\text{Ph}_2\text{CH})_2 + \text{PPh}_3 \quad (63)$$

The mechanism we proposed is nonchain (*139*):

$$[\text{Pt}(\text{PPh}_3)_3] \longrightarrow [\text{Pt}(\text{PPh}_3)_2] + \text{PPh}_3 \quad (64)$$

$$[\text{Pt}(\text{PPh}_3)_2] + \text{RX} \underset{}{\overset{\text{slow}}{\rightleftharpoons}} [\text{PtX}(\text{PPh}_3)_2] + \dot{\text{R}} \quad (65)$$
$$\text{(XIX)}$$

$$[\text{PtX}(\text{PPh}_3)_2] + \dot{\text{R}} \overset{\text{fast}}{\rightleftharpoons} trans\text{-}[\text{PtX}(\text{R})(\text{PPh}_3)_2] \quad (66)$$

This mechanism readily accounts for formation of [PtX$_2$P$_2$] or [PtX(H)P$_2$], compounds often found in these reactions, e.g., Eq. (62), by H or X abstraction by the Pt(I) complex (XIX), and is consistent with the second-order kinetics (*89, 162*). Similar spin-trapping experiments have been carried out for the addition of PhCH$_2$Cl or PhMeCHBr to [Pd(PPh$_3$)$_4$] (*192*). Nitroxides were generated, and their most likely origin was the metal product, since the combination [Pd(Cl)(CH$_2$Ph)(PPh$_3$)$_2$] and Bu^{t}NO generated a signal for PhCH$_2$(Bu^t)ṄO. However, we had previously reported that the analogous experiment with [Pt(PPh$_3$)$_3$], PhCH$_2$Br (or other RX), and Bu^{t}NO does *not* give an ESR signal (*139*); this difference between the Pd and Pt systems is not surprising in view of the greater lability of Pd—C than Pt—C bonds (94). Thus, cis-[PtMe$_2$-(PEt$_3$)$_2$] may be distilled at 85°/10^{-4} mmHg (31), but cis-[PdMe$_2$(PEt$_3$)$_2$] decomposes completely at 100°. Palladium(II) alkyl complexes with Ph$_3$P as a ligand are even less stable,e .g., cis-[PdMe$_2$(PPh$_3$)$_2$] decomposed at 35° to 40° in solution (26).[1] Other oxidative addition reactions investigated by the spin-trapping technique using nitrosodurene were on (i) [Co(II)(dmg)$_2$PPh$_3$]/CH$_2$=CHCH$_2$Br (dmg = dimethylglyoximato) and (ii) cis-[PtMe$_2$(PMe$_2$Ph)$_2$]/p-TolSO$_2$Br {to give [Pt(IV)BrMe$_2$-(SO$_2$Tol-p)(PMe$_2$Ph)$_2$] in (ii)}. Compounds Ar(Me)ṄO and Ar(p-TolSO$_2$)ṄO were observed at +30°, but only Ar(Me)ṄO possibly from S$_H$2 at −30°, which lost intensity at the expense of Ar(p-TolSO$_2$)ṄO at +30° but reappeared with concomitant collapse of the sulfonyl nitroxide (perhaps from reductive elimination) at +40°, and control experiments ruled out alternative sources of these nitroxides (87). In the [Pt(PPh$_3$)$_4$]-CCl$_4$ system (see below), a weak signal [about 1% of the intensity of Ar(Me)ṄO from Pt(0)/MeI] due to Ar(Cl$_3$C)ṄO was detected (142). In the [RhCl(PPh$_3$)$_3$]-MeI or trans-[Ir(CO)Cl(PPh$_3$)$_2$]-MeI systems, Ar(Me)ṄO was not observed after 1 hour (142).

The proponents of the chain mechanism have revised their proposal (132): they now believe that for [Pt(PEt$_3$)$_3$] as substrate, a radical pair is formed, as in Eq. (65) (139) which may collapse to the adduct [PtX(R)-(PEt$_3$)$_2$] as in Eq. (66) (139), but the adduct may alternatively (e.g., for PhCH$_2$Cl but not PhCH$_2$Br) be produced by an S$_N$2 displacement of X by Pt(0) from RX (cf. Ref. 141). For a "reactive" halide (e.g., an α-Br–ester, PhCH$_2$Br, or sec-RI), the dihalide [PtX$_2$(PEt$_3$)$_2$] is no longer thought (132) to arise via [PtX(H)(PEt$_3$)$_2$] (itself formed by a radical-chain process) but by halogen abstraction from [Pt$^{(I)}$(R)(PEt$_2$)] (139), and this concept was supported by CIDNP data (132) on the coproduct (e.g., ^{1}H NMR enhancements in the resonances for CH$_2$=CHMe and Pr^{i}I in the Pt(0)/Pr^{i}I system). These CIDNP effects were observed only when the dihalide was rapidly produced in the early stages of the reaction, e.g., for Pr^{i}I but not Pr^{i}Br. For an α-chloroester, the original chain mechanism, in which Eqs. (63) and (64) (139) provide the initiation, is still held (132) to proceed, cf. Eqs. (60) and (61) (21).

Evidence for inversion at C$_1$ in the addition of optically active PhCHDCl (192) or PhMeCHBr (141) to [Pd(PPh$_3$)$_4$] has been cited in support of an S$_N$2 mechanism. However, racemic adducts have been obtained in the

---

[1] An alternative, but considered (192) a less likely possibility, was that Bu^{t}NO induced radical decomposition; the case of an organosilver compound was cited as precedent (190). However, we note that in that work a *nitroxide and not a spin trap* was used.

addition of MeCHBrCO$_2$Et or PhCHBrCO$_2$Et (this rules out the possibility of a racemate being formed via a $\sigma-\pi$ rearrangement) to [Pd(CNBu^t)$_2$] *(156)* and of PhCHClCF$_3$ to [Pd(PPh$_3$)$_4$] *(141)*, for which the reaction (S$_N$2 by Pd(0) at C$_1$ of Pd(II)) was proposed as a possible alternative to a free radical process:

$$(L_n)Pd(0) + \underset{\underset{Ph}{H}}{\overset{F_3C}{C}}-Pd(L_n)Cl \rightleftharpoons \left[(L_n)Pd-\underset{\underset{Ph}{H}}{\overset{CF_3}{C}} \right]^+ [Pd(L_n)Cl]^- \tag{67}$$

$$Cl(L_n)Pd-\underset{\underset{Ph}{H}}{\overset{CF_3}{C}} + Pd(0)(L_n)$$

The secondary alkyl halides EtMeCHBr, EtMeCHI, PhMeCHBr, or MeCHBrCO$_2$Et and [PtP$_n$] (P = PPh$_3$, PMe$_2$Ph, or PEt$_3$; n = 3 or 4) gave halide complexes [PtX$_2$P$_2$], rather than a 1:1 adduct *(160)*.

Although one-electron changes have only recently been recognized in the context of oxidative-addition reactions, the ability of a metal, or complex, to abstract halogen homolytically appears to be widespread. Thus, reactions of Na with RX in the gas phase *(187)*, of Mg with RX (CIDNP experiments) *(18)*, of Ag(0) or Ag(I) with N-chloramines *(60)*, and the transition metal-catalyzed addition of CCl$_4$ to olefins *(148)*, all proceed, at least in part, by halogen abstraction. Additionally, the combination of a low-valent transition metal complex and an organic halide has been extensively investigated as a free radical-initiating system for vinyl–monomer polymerization *(8)*. The initiating step was considered to be electron transfer to the halide, followed in some cases by a choice of pathways *(9a)*, such as

$$RX + e^- \begin{cases} \text{(a)} \rightarrow \dot{R} + X^- \\ \text{(b)} \rightarrow R^- + \dot{X} \end{cases} \tag{68}$$

For [Pt(PPh$_3$)$_4$] and CCl$_4$, reaction (a) was a minor pathway (cf. spin-trapping, above) *(9b)*; i.e., only a small amount of $\dot{C}Cl_3$ was formed *(9b)* with a molecular mechanism as the major route. Further support for electron transfer from d^{10} species was provided by identifying the organic radical anion obtained by mixing a Ni(0) or Pt(0) complex, such as [Ni(PEt$_3$)$_4$], with an electron acceptor, such as C$_2$(CN)$_4$ *(64)*.

Finally, it is noted that polar effects often found in oxidative additions (usually such that increasing the electron density at the metal, or decreasing it at C$_1$ of a halide, facilitates the reaction) are not inconsistent with radical reactions. Similar features have been found and rationalized in the homolytic abstraction of halogen by tin-centered radicals (*149*).

It is likely that many reductive eliminations also proceed by radical pathways, but few studies have been reported. The reaction of $[PtR_2L_2]$ (R = Me, CD_3, or Et, and L_2 = bipy; or R = Me and L_2 = phen or 1,5-cyclooctadiene) with diethyl fumarate or maleate in the dark at room temperature in the presence of Bu^tNO led to the detection (ESR) of complex XX, which was interpreted in terms of the following sequence (*91*):

$$[PtR_2L_2] + \{CH(CO_2Et)\}_2 \longrightarrow [PtR_2L_2\{CH(CO_2Et)\}_2] \longrightarrow$$
$$[PtL_2\{CH(CO_2Et)\}_2] + 2\dot{R} \quad (69)$$
$$2RCH(CO_2Et)\dot{C}H(CO_2Et) \longleftarrow cis\text{- or } trans\text{-}\{CH(CO_2Et)\}_2$$
(XX)

However, a Pt(I) species could be an intermediate. The insertion reaction of $trans$-$[Pt(X)MeL_2]$ (X = Cl, Br, or I; and L = PMe_2Ph) with $RC\equiv CCO_2Me$ in $CHCl_3$ afforded $trans$-$[PtXL_2\{C(CO_2Me)=C(Cl)R\}]$ in the presence of Bz_2O_2 as a radical initiator (R = CO_2Me) or by addition of HCl (R = CO_2Me, Ph, Me, or H) (*5*); the proposed mechanism involved initial formation of the 1:1 adduct with the acetylene, followed by nucleophilic attack by HCl generated by a radical process.

The homolysis of a metal alkyl may be regarded as a reductive elimination; however, it is convenient to consider such reactions separately (Section V,B). Electron-transfer mechanisms for organometallic intermediates in catalytic reactions have been reviewed (*129*); examples are in the formation of transient RCu(I) or RCr(III) in oxidation (by Cu(II)) or reduction (by Cr(II)) of $\dot{R}$, and in the role of Fe in the Kharasch-Grignard reaction (e.g., Fe catalysis of disproportionation of $EtMgBr + EtBr \longrightarrow C_2H_6 + C_2H_4$, via $Fe(I) + RBr \longrightarrow Fe(II)Br + \dot{R}$).

B. Metal Alkyl Photolysis or Thermolysis

It was formerly assumed that homolysis is a common mode of decomposition for metal alkyls. In fact it is rather rare (*49*). A detailed survey of the topic is available elsewhere (*49*). For main group element alkyls, the clearest case is the thermolysis of Hg(II) or Pb(IV) compounds, and other examples are for alkyls of Zn, Cd, Si, Ge, Sn, and methyls of Ga, In, Tl, As, Sb, and Bi (also a minor pathway for B and Al). In the transition metal series, data are available for Mn(I), Ni(II), Pt(IV), Cu(I), and Ag(I), with homolysis playing a minor role for Ti(IV) and Zr(IV). Evidence is based on stereochemical arguments, the isolation of products appropriate for radical reactions (e.g., methylcyclopentane from M—$CH_2CH_2CH_2CH=CHMe$ compounds), the effect of radical transfer agents (e.g., amine) or inhibitors, deuterium-labeling, and CIDNP or ESR experiments.

There seems to be only one example of direct ESR observation of an organic radical derived from photolysis of a metal alkyl, the stable $\dot{C}(SiMe_3)_3$ from $[Hg\{C(SiMe_3)_3\}_2]$ (12). However, spin trapping by nitrosodurene (ArNO) has been used, for UV irradiation of $[Mn(CO)_5\text{-}CH_2Ph]$, when compound XIII (Sections II,B and III) and $Ar(PhCH_2)\dot{N}O$ (105) were detected (ESR). The acylpentacarbonylmanganese(I) compounds $[Mn(CO)_5(COR)]$ (R = $PhCH_2$ or Ph_2CH) underwent similar homolysis, although the spin adduct formed was generally derived from $\dot{R}$ rather than $\dot{R}CO$. Other examples of spin trapping the photolysis products of metal alkyls relate to Sn(IV), Pb(IV), and Hg(II) (cf. Ref. 113). Homolysis of M—R bonds may well play a role in a number of catalytic processes (129).

The photolysis of Co(III) alkyls has attracted considerable attention (59), because homolysis of such a bond may well be implicated in enzymatic isomerization reactions catalyzed by vitamin B_{12} coenzyme (see Refs. 27, 98, 115a, 193). Early evidence was based on characterization by UV or ESR of the d^7 Co(II) corrinoid. However, recently the organic fragment has been spin-trapped and identified by ESR for the case of (i) the coenzyme (5'-deoxyadenosylcobalamin) (10^{-3} M) with Bu^tNO (10^{-2} M) and photolysis in water at 50° to yield $Ar(5'\text{-deoxyadenosyl})\dot{N}O$ (XXI)

(XXI)

[Fig. 4; $a(N)$, 1.64 mT; $a(H^1)$, 1.41 mT; $a(H^2)$, 0.81 mT; and $a(H^3)$, 0.06 mT] and Co(II) ($g \sim 2.2$); and (ii) ethylcobalamin, which, under similar conditions but at 20°, gave $Bu^t(Et)\dot{N}O$ [$a(N)$ 1.71 mT and $a(H)$ 1.13 mT] and Co(II) ($g \sim 2.2$) (115). The anaerobic photolysis of alkylcobaloxime–pyridine adducts has been studied by ESR [Co(II) identification] and shown to be a homolytic process for Pr^i, Bu^i, n-C_5H_{11}, or cyclo-C_6H_{11}, but for Me or $PhCH_2$ it is a one-electron transfer process in-

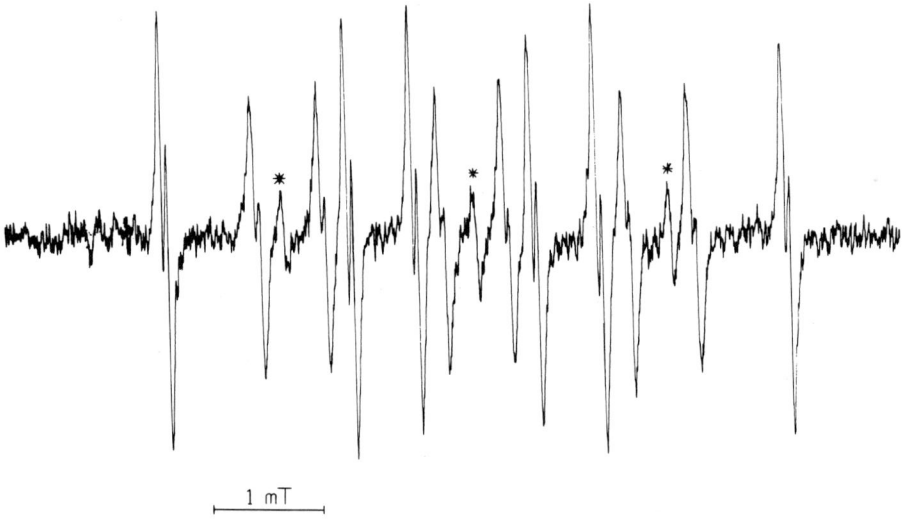

FIG. 4. The ESR spectrum of Bu^t(5'-deoxyadenosyl)ṄO generated by photolysis of vitamin B$_{12}$-coenzyme in the presence of Bu^{t}NO in H$_2$O at 50°.

volving the equatorial ligands or solvent (78). Spin-labeled 5'-deoxyadenosylcobinamide, shown schematically in XXII, has been used as a co-

R—|Co(III)|—O—Ṅ(Me$_2$)(Me$_2$)

(XXII)

factor for the enzyme ethanolamine–ammonia–lyase and the ESR spectrum followed during catalysis (193). The ESR signal disappeared upon adding the substrate ethanolamine, which indicates that the Co(III)—R bond may well cleave homolytically upon adding the substrate.

VI

APPENDIX

Brief reference is made here to some papers which have appeared since the submission of the manuscript in February 1975; this section was added at the proof stage, in December 1975.

Section II,A: The photolysis of hydrides of XX'X"Ge-H in presence of Bu$_2^t$O$_2$ has been reported (*171a*); phenyl-substituted radicals may be more planar than alkyl analogs, and there is significant delocalization of the unpaired electron into the aromatic ring. Further examples of bulky stable Group IV metal-centered radicals have been found: $\dot{M}$(C$_6$H$_2$Me$_3$-2,4,6)$_3$ (M = Ge or Sn), $\dot{M}$[N(GeMe$_3$)$_2$]$_3$ (M = Ge or Sn), and Sn[N-(GeEt$_3$)$_2$]$_3$ (*86c*); the amido-metal compounds were obtained by photolysis of corresponding M[N(GeR$_3$)$_2$]$_2$ derivatives [cf. Eqs. (8)–(11)], but this method failed to yield radicals from Ge[N(GeEt$_3$)$_2$]$_2$ or M[N(GePh$_3$)$_2$]$_2$ (M = Ge or Sn), possibly for steric reasons (*86c*). A new high-yield method has been discovered (*86a*) for the fomation of stable metal-centered radicals from a halide and an electron-rich olefin, e.g., (*86c*),

$$2\text{ClGe}(\text{C}_6\text{H}_2\text{Me}_3\text{-}2,4,6)_3 + (\text{Me}_2\text{N})_2\text{C}=\text{C}(\text{NMe}_2)_2 \xrightarrow[h\nu]{\text{C}_6\text{H}_6} [\text{C}_2(\text{NMe}_2)_4]^{2+}(\text{Cl}^-)_2$$
$$+ \dot{\text{G}}\text{e}(\text{C}_6\text{H}_2\text{Me}_3\text{-}2,4,6)_3$$

Section II,B: Paramagnetic organometallic actinide and lanthanide complexes were omitted from the previous discussion and Table IV. However, reference may be made to reviews (*49, 86d, 94a, 146a, 146b*); compounds include [M(C$_5$H$_5$-η)$_2$X] (e.g., X = Cl, OPh, or OAc; M = a lanthanide or U), [M(C$_5$H$_5$-η)$_3$] (M = a lanthanide, U, Pu, Am, Cm, Bk, or Cf), [M(C$_5$H$_5$-η)$_2$] (M = Eu, Sm, or Yb), [M(C$_5$H$_5$-η)Cl$_2$·3THF] (M = Sm, Eu, Gd, Dy, Ho, or Yb), [M(C$_8$H$_8$-η), liqd. NH$_3$] (M = Eu or Yb), [U(C$_5$H$_5$-η)$_4$], [M(C$_8$H$_8$-η)$_2$] (M = Th, U, or Np), [M(C$_5$H$_5$-η)$_3$X] (X = OR, BH$_4$, R, or Cl; M = Th, U, or Np). Further interesting species are the alkyl-bridged XXIII (M = Yb, Er, Ho, or Dy) (*98b*) and XXIV (M = Yb, Er, Ho, Dy, or Gd) (*98c*) and [Yb{CH(SiMe$_3$)$_2$}$_3$·2THF] (*98a*); other lanthanide complexes [M(C$_5$H$_5$-η)$_2$R]$_n$ have been described (*64a*).

$$(\eta\text{-C}_5\text{H}_5)_2\text{M}\underset{\text{Me}}{\overset{\text{Me}}{\diamond}}\text{M}(\text{C}_5\text{H}_5\text{-}\eta)_2 \qquad (\eta\text{-C}_5\text{H}_5)_2\text{M}\underset{\text{Me}}{\overset{\text{Me}}{\diamond}}\text{AlMe}_2$$

(XXIII)   (XXIV)

A further contribution has been made to the controversy relating to the nature of the long-lived paramagnetic species obtained by photolysis of [Mn$_2$(CO)$_{10}$] in THF, suggesting this to be [Mn(CO)$_4$] (*135a*); however, detailed ESR line-shape analysis, IR detection of [Mn(CO)$_5$]$^-$, and analogy with the Fe(CO)$_5$/base system supports the [Mn(THF)$_6$]$^{2+}$[Mn(CO)$_5$]$_2^-$ assignment (*105a, 105b*). Further photolysis studies refer to [Mo(CO)$_3$(C$_5$H$_5$-η)$_2$] (*105c, 194b*), [Mn$_2$(CO)$_{10}$] and [Re$_2$(CO)$_{10}$] and related derivatives (*194a*). Paramagnetic intermediates were proposed in kinetic schemes relating to ligand (phosphine or phosphite) substitution in [Mn(CO)$_4$(PPh$_3$)]$_2$ (*67a*) or [ReH(CO)$_5$] (*25a*). A Co(IV) complex resulted from electrochemical or Ce(IV) oxidation of a Co(III)-R cobaloxime species (*89a*).

Sections IV,A and IV,B: A CIDNP study of S$_H$2 reactions has been reported (*116a*) for [M(L)R]/(PhCO)$_2$O$_2$ systems, by heating an appropriate mixture {[M(L)R = SnEt$_4$, SnMe$_3$Cl, SnBu$_3^n$Br, PbEt$_4$, PbMe$_3$Cl, AuMe(PPh$_3$), or *cis*-PtMe$_2$L$_2'$} in an NMR tube and monitoring the PhCO$_2$R, PhR, or C$_2$H$_6$ (from R = Et) signals; the reactions afforded a metal benzoate, CO$_2$, and products from $\dot{\text{P}}$h or $\dot{\text{R}}$.

Section V,A: Halide exchange, e.g., MeI/CH$_2$Cl$_2$ at 100° under 150 lb/sq. inch of Ar or CO, was catalyzed by [RuCl$_2$(PPh$_3$)$_3$], [RhCl(PPh$_3$)$_3$], [Rh(CO)Cl(PPh$_3$)$_2$], or [Ir(CO)Cl(PPh$_3$)$_2$], and this was ascribed to sequential oxidative-additions and reductive eliminations (*144a*); but another study attributes catalysis to halide ion formed by quarternization of free PPh$_3$ (*72b*). Halide ion catalysis was observed in the addition of alkyl halides to some Rh(I) substrates, e.g., for [Rh(CO)Cl(Q)$_2$]/MeI with [Bu$_4^n$N]$^+$I$^-$ for Q = As, Sb, (*72a*) but not P. Alkyl or aryl halides R″X oxidatively add to (a) SnR$_2$ or (b) Sn(NR$_2'$)$_2$ [R = (Me$_3$Si)$_2$CH, R′ = Me$_3$Si] to yield SnR$_2$(R″)X or Sn(NR$_2'$)$_2$(R″)X (*86b*). These reactions are radical in character as shown by (i) the spin trapping and ESR characterization of Ar(R″)$\dot{\text{N}}$O (Ar = 2,3,5,6-Me$_4$C$_6$H) for (a), (ii) the ESR detection of a tin-centered radical, probably $\dot{\text{S}}$nR$_2$(X) for (a) and (iii) the obtaining of a racemic product in (a) when R″X = (+)-*n*-C$_6$H$_{13}$(Me)CHCl, and (iv) the catalytic effect of a trace of EtBr in the oxidative addition of PhBr, for (a) or (b).

Section V,B: A further ESR paper has appeared on the photolysis of a Co(III)-R cobaloxime (*78a*) and similar observations have been made on the corresponding vitamin B$_{12}$ coenzyme in propane-1,3-diol (*28*). Photolysis of a dimethylcobalt(III) chelate gave CH$_4$ and a methylcobalt(II) complex which was believed to disproportionate to a Co(I) and a Co(III) species (*191a*).

ACKNOWLEDGMENTS

We thank Messrs. T. L. Hall and J. J. MacQuitty, and Drs. M. J. S. Gynane, A. Hudson, R. A. Jackson, and B. K. Nicholson for useful discussions.

REFERENCES

1. Abell, P. I., *in* "Free Radicals" (J. K. Kochi, ed.), Vol. 2, Chapter 13. Wiley (Interscience), New York, 1973.
2. Alexandrov, Yu. A., *J. Organometal. Chem.* **55**, 1 (1973).

3. Alyea, E. C., Bradley, D. C., Lappert, M. F., and Sanger, A. R., *J. Chem. Soc., Chem. Commun.* 1064 (1969).
4. Anderson, O. P., and Symons, M. C. R., *J. Chem. Soc., Chem. Commun.* 1020 (1972).
5. Appleton, T. G., Chisholm, M. H., and Clark, H. C., *J. Amer. Chem. Soc.* **94,** 8912 (1972); Appleton, T. G., Chisholm, M. H., Clark, H. C., and Yasufuku, K., *ibid.* **96,** 6600 (1974).
6. Ayscough, P. B., "Electron Spin Resonance in Chemistry." Methuen, London, 1967; Atkins, P. W., and Symons, M. C. R., "The Structure of Inorganic Radicals." Elsevier, Amsterdam, 1967.
6a. Baallen, A. van, Groenenboom, C. J., and Liefde Meijer, H. J. de, *J. Organometal. Chem.* **74,** 245 (1974).
7. Bakalik, D. P., and Hayes, R. G., *Inorg. Chem.* **11,** 1734 (1972).
8. Bamford, C. H., *in* "Reactivity, Mechanism, and Structure in Polymer Chemistry" (A. D. Jenkins and A. Ledwith, eds.), Vol. 3, p. 52. Wiley (Interscience), New York, 1974.
9. (a) Bamford, C. H., Eastmond, G. C., and Whittle, D., *J. Organometal. Chem.* **17,** P33 (1969); (b) Bamford, C. H., Eastmond, G. C., and Hargreaves, K., *Trans. Farad. Soc.* **64,** 175 (1968).
10. Barker, G. K., and Lappert, M. F., *J. Organometal. Chem.* **76,** C45 (1974).
11. Bartelink, H. J. M., Ostendorf, H. K., Roest, B. C., and Schepers, H. A. J., *J. Chem. Soc., Chem. Commun.* 878 (1971).
12. Bassindale, A. R., Bowles, A. J., Cook, M. A., Eaborn, C., Hudson, A., Jackson, R. A., and Jukes, A. E., *J. Chem. Soc., Chem. Commun.* 559 (1970).
13. Bennett, J. E., and Howard, J. A., *Chem. Phys. Lett.* **15,** 322 (1972).
14. Bennett, M. A., Robertson, G. B., Whimp, P. O., and Yoshida, T., *J. Amer. Chem. Soc.* **95,** 3028 (1973).
15. Bennett, S. W., Eaborn, C., Hudson, A., Jackson, R. A., and Root, K. D. J., *J. Chem. Soc., A* 348 (1970).
16. Bennett, S. W., Eaborn, C., Hudson, A., Hussain, M. A., and Jackson, R. A., *J. Organometal. Chem.* **16,** P36 (1969).
17. Bentrude, W. G., *in* "Free Radicals" (J. K. Kochi, ed.), Vol. 2, Chapter 22. Wiley (Interscience), New York, 1973.
18. Bodewitz, H. W. H. J., Blomberg, C., and Bickelhaupt, F., *Tetrahedron* **29,** 719 (1973).
19. Bower, B. K., and Tennent, H. G., *J. Amer. Chem. Soc.* **94,** 2512 (1972).
20. Bowles, A. J., Hudson, A., and Jackson, R. A., *J. Chem. Soc., B* 1947 (1971).
21. Bradley, J. S., Connor, D. E., Dolphin, D., Labinger, J. A., and Osborn, J. A., *J. Amer. Chem. Soc.* **94,** 4043 (1972).
22. Brindley, P. B., and Hodgson, J. C., *J. Organometal. Chem.* **65,** 57 (1974).
23. Brown, H. C., and Midland, M. M., *Angew. Chem., Int. Ed. Engl.* **11,** 692 (1972).
24. Burgess, J., *Inorg. React. Mech.* **1,** 311 (1971).
25. Burkett, A. R., Meyer, T. J., and Whitten, D. G., *J. Organometal. Chem.* **67,** 67 (1974).
25a. Byers, B. H., and Brown, T. L., *J. Amer. Chem. Soc.* **97,** 947 (1975).
26. Calvin, G., and Coates, G. E., *J. Chem. Soc.* 2008 (1960).
27. Cardin, D. J., Joblin, K. N., Johnson, A. W., Lang, G., and Lappert, M. F., *Biochim. Biophys. Acta* **371,** 44 (1974).
28. Cardin, D. J., Joblin, K. N., and Lowe, D. J., unpublished work (1975).
29. Cardin, D. J., Lappert, M. F., and Lednor, P. W., *J. Chem. Soc., Chem. Commun.* 350 (1973).

30. Çetinkaya, B., Lappert, M. F., McLaughlin, G. M., and Turner, K., *J. Chem. Soc., Dalton Trans.* 1591 (1974).
31. Chatt, J., and Shaw, B. L., *J. Chem. Soc.* 705 (1959).
32. Chen, K. S., Batlioni, J.-P., and Kochi, J. K., *J. Amer. Soc.* **95,** 4439 (1973).
33. Chen, K. S., Bertini, F., and Kochi, J. K., *J. Amer. Chem. Soc.* **95,** 1340 (1973).
34. Chock, P. B., and Halpern, J., *J. Amer. Chem. Soc.* **88,** 3511 (1966).
35. Coates, G. E., Green, M. L. H., and Wade, K., "Organometallic Compounds," 3rd ed., Vol. 1: "The Main Group Elements" by G. E. Coates and K. Wade. Methuen, London, 1967.
36. Coates, G. E., Green, M. L. H., and Wade, K., "Organometallic Compounds," 3rd ed., Vol. 2: "The Transition Elements" by M. L. H. Green. Methuen, London, 1968.
37. Cohen, A. H., and Hoffman, B. M., *Inorg. Chem.* **13,** 1484 (1974).
38. Collier, M. R., Eaborn, C., Jovanović, B., Lappert, M. F., Manojlović-Muir, Lj., Muir, K. W., and Truelock, M. M., *J. Chem. Soc., Chem. Commun.* 613 (1972).
39. Collman, J. P., *Accounts Chem. Res.* **1,** 136 (1968).
40. Collman, J. P., and Roper, W. R., *Advan. Organometal. Chem.* **7,** 53 (1968).
41. Cooper, J., Hudson, A., and Jackson, R. A., *Mol. Phys.* **23,** 209 (1972).
42. Cooper, J., Hudson, A., and Jackson, R. A., *J. Chem. Soc., Perkin Trans. II* 1056, 1933 (1973).
43. Correa-Duran, F., Allred, A. L., and Lloyd, R. J., *J. Organometal. Chem.* **49,** 365, 373 (1973).
44. Cotton, J. D., Cundy, C. S., Harris, D. H., Hudson, A., Lappert, M. F., and Lednor, P. W., *J. Chem. Soc., Chem. Commun.* 651 (1974).
45. Danen, W. C., and West, T. C., *Tetrahedron Lett.* 219 (1970).
46. Davidson, P. J., Hudson, A., Lappert, M. F., and Lednor, P. W., *J. Chem. Soc., Chem. Commun.* 829 (1973).
47. Davidson, P. J., and Lappert, M. F., *J. Chem. Soc., Chem. Commun.* 317 (1973).
48. Davidson, P. J., Lappert, M. F., and Pearce, R., *Accounts Chem. Res.* **7,** 209 (1974).
49. Davidson, P. J., Lappert, M. F., and Pearce, R., *Chem. Rev.* **76,** (1976)., in press.
50. Davies, A. G., and Roberts, B. P., *Accounts Chem. Res.* **5,** 387 (1972).
51. Davies, A. G., and Roberts, B. P., in "Free Radicals" (J. K. Kochi, ed.), Vol. 1, Chapter 10. Wiley (Interscience), New York, 1973.
52. Davies, A. G., Roberts, B. P., and Scaiano, J. C., *J. Chem. Soc., B* 2171 (1971).
53. Davies, A. G., and Scaiano, J. C., *J. Chem. Soc., Perkin Trans. II* 1777 (1973); and earlier papers in this series.
54. Deeming, A. J., *MTP Int. Rev. Sci., Inorg. Chem., Ser.* 1 **9,** 117 (1972).
55. Dessy, R. E., and Bares, L. A., *Accounts Chem.* **5,** 415 (1972).
56. Dessy, R. E., Charkoudian, J. C., Abeles, T. P., and Rheingold, A. L., *J. Amer. Chem. Soc.* **92,** 3947 (1970).
57. Dessy, R. E., Charkoudian, J. C., and Rheingold, A. L., *J. Amer. Chem. Soc.* **94,** 738 (1972).
58. Dessy, R. E., Kleiner, M., and Cohen, S. C., *J. Amer. Chem. Soc.* **91,** 6800 (1969).
59. Dodd, D., and Johnson, M. D., *J. Organometal. Chem.* **52,** 1 (1973).
60. Edwards, O. E., Paskovich, D. H., and Reddoch, A. H., *Can. J. Chem.* **51,** 978 (1973).
61. Elschenbroich, C., and Cais, M., *J. Organometal. Chem.* **18,** 135 (1969); Bigam, G., Hooz, J., Linke, S., McLung, R. E. D., Mosher, M. W., and Tanner, D. D., *Can. J. Chem.* **50,** 1825 (1972); Gogan, N. J., Chu, C. K., and Gray, G. W., *J. Organometal. Chem.* **51,** 323 (1973).

62. Elschenbroich, C., and Grerson, F., *J. Organometal. Chem.* **49,** 445 (1973); Elschenbroich, C. H., Grerson, F., and Stohler, F., *J. Amer. Chem. Soc.* **95,** 6956 (1973); Karthe, W., and Kleinwachter, W., *Z. Phys. Chem. (Leipzig)* **247,** 241 (1971).
62a. Elson, I. H., and Kochi, J. K., *J. Amer. Chem. Soc.* **97,** 1262 (1975).
63. Elson, I. H., Kochi, J. K., Klabunde, U., Manzer, L. E., Parshall, G. W., and Tebbe, F. N., *J. Amer. Chem. Soc.* **96,** 7374 (1974).
64. Elson, I. H., Morrell, D. G., and Kochi, J. K., *J. Organometal. Chem.* **84,** C7 (1975).
64a. Ely, N. M., and Tsutsui, M., *Inorg. Chem.* **14,** 2680 (1975).
65. Evans, A. G., Evans, J. C., and Moon, E. H., *J. Chem. Soc., Dalton Trans.* 2390 (1974).
66. Fachinetti, G., and Floriani, C., *J. Chem. Soc., Dalton Trans.* 2433 (1974).
67. Fachinetti, G., Fochi, G., and Floriani, C., *J. Organometal. Chem.* **57,** C51 (1973).
67a. Fawcett, J. P., Jackson, R. A., and Poë, A. J., *J. Chem. Soc., Chem. Commun.* 733 (1975).
68. Fawcett, J. P., Poe, A. J., and Twigg, M. V., *J. Organometal. Chem.* **51,** C17 (1973).
69. Fieldhouse, S. A., Fulham, B. W., Neilson, G. W., and Symons, M. C. R., *J. Chem. Soc., Dalton, Trans.* 567 (1974).
70. Fischer, E. O., Offhaus, E., Muller, J., and Nothe, D., *Chem. Ber.* **105,** 3027 (1972).
71. Fischer, H., in "Free Radicals" (J. K. Kochi, ed.), Vol. 2, Chapter 19. Wiley (Interscience), New York, 1973.
72. Forrester, A. R., Hepburn, S. P., Dunlop, R. S., and Mills, H. H., *J. Chem. Soc., Chem. Commun.* 698 (1969).
72a. Foster, D., *J. Amer. Chem. Soc.* **97,** 951 (1975).
72b. Foster, D., *J. Chem. Soc., Chem. Commun.* 917 (1975).
73. Freni, M., Giusto, D., and Romiti, P., *J. Inorg. Chem. Nucl. Chem.* **29,** 761 (1967).
74. Friswell, N. J., and Gowenlock, B. G., *Advan. Free Radical Chem.* **1,** 39 (1967).
75. Friswell, N. J., and Gowenlock, B. G., *Advan. Free Radical Chem.* **2,** 1 (1968).
76. Garst, J. F., in "Free Radicals" (J. K. Kochi, ed.), Vol. 1, Chapter 9. Wiley (Interscience), New York, 1973.
77. Gee, D. R., and Wan, J. K. S., *Can. J. Chem.* **49,** 160 (1971).
78. Giannotti, C., and Bolton, J. R., *J. Organometal. Chem.* **80,** 379 (1974).
78a. Giannotti, C., Merle, G., and Bolton, J. R., *J. Organometal. Chem.* **99** 145 (1975).
79. Gibson, J. F., *Electron Spin Reson.* **1,** 156 (1973).
80. Gibson, J. F., *Electron Spin Reson.* **2,** 111 (1974).
81. Green, M., *Organometal. Chem.* **1,** 431 (1972).
82. Green, M., *Organometal. Chem.* **2,** 491 (1973).
83. Green, M. L. H., and Lindsell, W. E., *J. Chem. Soc., A* 2215 (1969).
84. Goodman, B. A., and Raynor, J. B., *Advan. Inorg. Chem. Radiochem.* **13,** 136 (1970).
85. Griller, D., *J. Magn. Reson.* **6,** 402 (1972).
86. Griller, D., and Ingold, C. K., *J. Amer. Soc.* **95,** 6459 (1973).
86a. Gynane, M. J. S., and Lappert, M. F., unpublished work (1975).
86b. Gynane, M. J. S., Lappert, M. F., Miles, S. J., and Power, P. P., *J. Chem. Soc., Chem. Commun.* in press (1976).
86c. Gynane, M. J. S., Lappert, M. F., Power, P. P., Rivière, P., and Rivière-Baudet, M., unpublished work (1975).
86d. Gyslin, H., and Tsutsui, M., *Advan. Organometal. Chem.* **9,** 361 (1970).
87. Hall, T. L., and Lappert, M. F., unpublished work (1975).
88. Hallock, S. A., and Wojcicki, A., *J. Organometal. Chem.* **54,** C27 (1973).
89. Halpern, J., *Accounts Chem. Res.* **3,** 386 (1970).

89a. Halpern, J., Chan, M. S., Hanson, J., Roche, T. S., and Topich, J. A., *J. Amer. Chem. Soc.* **97**, 1606 (1975).
90. Hargreaves, N. G., Johnson, A., Puddephatt, R. J., and Sutcliffe, L. H., *J. Organometal. Chem.* **69**, C21 (1974); Johnson A., and Puddephatt, R. J., *J. Chem. Soc., Dalton, Trans.* 115 (1975).
91. Hargreaves, N. G., Puddephatt, R. J., Sutcliffe, L. H., and Thompson, P. J., *J. Chem. Soc., Chem. Commun.* 861 (1973).
92. Harris, D. H., and Lappert, M. F., *J. Chem. Soc., Chem. Commun.* 895 (1974).
93. Harris, D. H., and Lappert, M. F., unpublished work (1974).
94. Hartley, F. R., "The Chemistry of Platinum and Palladium." Applied Science Publ. London, 1973.
94a. Hayes, R. G., and Thomas, J. L., *Organometal. Chem. Rev.* **7A**, 361 (1971).
95. Henrici-Olivé, G., and Olivé, S., *J. Chem. Soc., Chem. Commun.* 1482 (1969).
96. Henrici-Olivé, G., and Olivé, S., *J. Organometal. Chem.* **23**, 155 (1970).
97. Henrici-Olivé, G., and Olivé, S., *J. Amer. Chem. Soc.* **92**, 4831 (1970).
98. Hill, H. A. O., *in* "Inorganic Biochemistry" (G. I. Eichhorn, ed.), Chapter 30. Elsevier, Amsterdam, 1973.
98a. Holton, J., and Lappert, M. F., unpublished work (1975).
98b. Holton, J., Lappert, M. F., Ballard, D. G. H., Pearce, R., Atwood, J. L., and Hunter, W. E., *J. Chem. Soc., Chem. Commun.* in press (1976).
98c. Holton, J., Lappert, M. F., Scollary, G. R., Ballard, D. G. H., Pearce, R., Atwood, J. L., and Hunter, W. E., *J. Chem. Soc., Chem. Commun.* (1976).
99. Hosomi, A., and Sakurai, H., *J. Amer. Chem. Soc.* **94**, 1384 (1972).
100. Huber, H., Kündig, E. P., and Ozin, G. A., *J. Amer. Chem. Soc.* **96**, 5585 (1974).
101. Huber, H., Kündig, E. P., Ozin, G. A., and Poe, A. J., *J. Amer. Chem. Soc.* **97**, 308 (1975).
102. Hudson, A., *Electron Spin Reson.* **1**, 253 (1973).
103. Hudson, A., *Electron Spin Reson.* **2**, 270 (1974).
104. Hudson, A., personal communication (1973).
104a. Hudson, A., Lappert, M. F., Lednor, P. W., MacQuitty, J. J., and Nicholson, B. K., unpublished work (1975).
105. Hudson, A., Lappert, M. F., Lednor, P. W., and Nicholson, B. K., *J. Chem. Soc., Chem. Commun.* 966 (1974).
105a. Hudson, A., Lappert, M. F., and Nicholson, B. K., *J. Organometal. Chem.* **92**, C11 (1975).
105b. Hudson, A., Lappert, M. F., Nicholson, B. K., and MacQuitty, J. J., unpublished work (1975).
105c. Hughey, J. L., Bock, C. R., and Meyer, T. J., *J. Amer. Chem. Soc.* **97**, 4440 (1975).
106. Ibekwe, S. D., and Myatt, J., *J. Organometal. Chem.* **31**, C65 (1971).
107. Ingold, K. U., *in* "Free Radicals" (J. K. Kochi, ed.), Vol. 1, Chapter 2. Wiley (Interscience), New York, 1973.
108. Ingold, K. U., and Roberts, B. P., "Free Radical Substitution Reactions." Wiley (Interscience), New York, 1971.
109. Ishikawa, M., and Kumada, M., *J. Chem. Soc., Chem. Commun.* 612 (1970).
110. Jackson, R. A., *Advan. Free Radical Chem.* **3**, 231 (1969); *Chem. Soc. Spec. Publ.* **24**, 295 (1970).
111. Jackson, R. A., *MTP Int. Rev. Sci., Org. Chem., Ser.* 1 **10**, 205 (1973).
112. Jackson, R. A., *MTP Int. Rev. Sci., Org. Chem., Ser.* 2 **10**, 205 (1975).
113. Janzen, E. G., *Accounts Chem. Res.* **4**, 31 (1971).

114. Jensen, F. R., and Knickel, B., *J. Amer. Chem. Soc.* **93,** 6339 (1971).
115. Joblin, K. N., Johnson, A. W., Lappert, M. F., and Nicholson, B. K., *J. Chem. Soc., Chem. Commun.* 441 (1975).
115a. Joblin, K. N., Johnson, A. W., Lappert, M. F., Hollaway, M. R., and White, H. A., *FEBS Lett.* **53,** 193 (1975); and references cited therein.
116. Jong, I. G. de, and Wiles, D. R., *Inorg. Chem.* **12,** 2519 (1973).
116a. Kaptein, R., Leuwen, P. W. N. M. van, and Huis, R., *J. Chem. Soc., Chem. Commun.* 568 (1975).
117. Kawamura, T., and Kochi, J. K., *J. Organometal. Chem.* **30,** C8 (1971).
118. Kean, E. S., Fisher, K., and West, R., *J. Amer. Chem. Soc.* **94,** 3246 (1972).
119. Kemmitt, R. D. W., and Burgess, J., *Inorg. React. Mech.* **2,** 350 (1972).
120. Keller, H. J., and Waversik, H., *Z. Naturforsch. B* **20,** 938 (1965).
121. Kenworthy, J. G., Myatt, J., and Symons, M. C. R., *J. Chem. Soc., A* 1020 (1971).
122. Kerr, J. A., *in* "Free Radicals" (J. K. Kochi, ed.), Vol. 1, Chapter 1. Wiley (Interscience), New York, 1973.
123. Kice, J. L., *in* "Free Radicals" (J. K. Kochi, ed.), Vol. 2, Chapter 24. Wiley (Interscience), New York, 1973.
124. Kiefer, H., and Traylor, T. G., *Tetrahedron Lett.* 6163 (1966).
125. Kinsella, E., Smith, V. B., and Massey, A. G., *J. Organometal. Chem.* **34,** 181 (1972).
126. Kochi, J. K., ed., "Free Radicals," Vols. 1 and 2. Wiley (Interscience), New York, 1973.
127. Kochi, J. K., *in* "Free Radicals" (J. K. Kochi, ed.), Vol. 1, Chapter 11. Wiley (Interscience), New York, 1973.
128. Kochi, J. K., *in* "Free Radicals" (J. K. Kochi, ed.), Vol. 2, Chapter 23. Wiley (Interscience), New York, 1973.
129. Kochi, J. K., *Accounts Chem. Res.* **7,** 351 (1974).
130. Kochi, J. K., and Krusic, P. J., *J. Amer. Chem. Soc.* **91,** 3938 (1969).
131. Kochi, J. K., and Krusic, P. J., *Chem. Soc. Spec. Publ.* **24,** 147 (1970).
131a. Koenig, T., and Fischer, H., *in* "Free Radicals" (J. K. Kochi, ed.), Vol. 1, Chapter 4. Wiley (Interscience), New York, 1973.
132. Kramer, A. V., and Osborn, J. A., *J. Amer. Chem. Soc.* **96,** 7832 (1974).
133. Kramer, A. V., Labinger, J. A., Bradley, J. S., and Osborn, J. A., *J. Amer. Chem. Soc.* **96,** 7145 (1974).
134. Krusic, P. J., and Kochi, J. K., *J. Amer. Chem. Soc.* **93,** 846 (1971); Krusic, P. J., Meakin, P., and Jesson, J. P., *J. Phys. Chem.* **75,** 3438 (1971).
134a. Krusic, P. J., Stocklosa, H., Manzer, L. E., and Meakin, P., *J. Amer. Chem. Soc.* **97,** 667 (1975).
135. Kuivila, H. G., *Accounts Chem. Res.* **1,** 299 (1968).
135a. Kwan, C. L., and Kochi, J. K., *J. Organometal. Chem.* **101,** C9 (1975).
136. Labinger, J. A., Braus, R. J., Dolphin, D., and Osborn, J. A., *J. Chem. Soc., Chem. Commun.* 612 (1970).
137. Labinger, J. A., Kramer, A. V., and Osborn, J. A., *J. Amer. Chem. Soc.* **95,** 7908 (1973).
138. Lankamp, H., Nauta, W. T., and Maclean, C., *Tetrahedron Lett.* 249 (1968).
139. Lappert, M. F., and Lednor, P. W., *J. Chem. Soc., Chem. Commun.* 948 (1973).
140. Lappert, M. F., and Sanger, A. R., *J. Chem. Soc., A* **847,** 1314 (1971).
141. Lau, K. S. Y., Fries, R. W., and Stille, J. K., *J. Amer. Chem. Soc.* **96,** 4983 (1974).
142. Lednor, P. W., D. Phil. Thesis, Sussex, 1974.

143. Lloyd, R. V., and Rogers, M. T., *J. Amer. Chem. Soc.* **95,** 2459 (1973).
144. Lyons, A. R., and Symons, M. C. R., *J. Chem. Soc. Faraday Trans. II* **68,** 502 (1972).
144a. Lyons, J. E., *J. Chem. Soc., Chem. Commun.* 418 (1975).
145. McCleverty, J. A., Orchard, D. G., Connor, J. A., Jones, E. M., Lloyd, J. P., and Rose, P. D., *J. Organometal. Chem.* **30,** C75 (1971).
146. McDonnell, J. J., and Pochopien, D. J., *J. Org. Chem.* **36,** 2092 (1971); **37,** 4064 (1972).
146a. Marks, T. J., *J. Organometal. Chem.* **79,** 181 (1974).
146b. Marks, T. J., *J. Organometal. Chem.* **95,** 301 (1975).
147. Matheson, T. W., Peake, B. M., Robinson, B. H., Simpson, J., and Watson, D. J., *J. Chem. Soc., Chem. Commun.* 894 (1973).
148. Matsumoto, H., Nakano, T., and Nagai, Y., *Tetrahedron Lett.* 5147 (1973).
149. Menapace, L. W., and Kuivila, H. G., *J. Amer. Chem. Soc.* **86,** 3047 (1964).
150. Mendenhall, G. D., and Ingold, K. U., *J. Amer. Chem. Soc.* **95,** 3422 (1973).
151. Milovskaya, E. B., *Russ. Chem. Rev.* **42,** 384 (1973).
152. Moelwyn-Hughes, J. T., Garner, A. W. B., and Gordon, N., *J. Organometal. Chem.* **26,** 373 (1971).
153. Mowat, W., Shortland, A., Yagupsky, G., Hill, N. J., Yagupsky, M., and Wilkinson, G., *J. Chem. Soc., Dalton Trans.* 533 (1972); Mowat, W., Shortland, A. J., Hill, N. J., and Wilkinson, G., *J. Chem. Soc., Dalton Trans.* 770 (1973).
154. Nakayama, H., *Bull. Chem. Soc. Jap.* **43,** 2057 (1970).
155. O'Neal, H. E., and Benson, S. W., *in* "Free Radicals" (J. K. Kochi, ed.), Vol. 2, Chapter 17. Wiley (Interscience), New York, 1973.
156. Otsuka, S., Nakamura, A., Yoshida, T., Naruto, M., and Ataka, K., *J. Amer. Chem. Soc.* **95,** 3180 (1973).
157. Oven, H. O. van, Groenenboom, C. J., and Liefde Meijer, H. J. de, *J. Organometal. Chem.* **81,** 379 (1974).
158. Paneth, F. A., and Hofeditz, W., *Ber.* **62,** 1335 (1929).
159. Peake, B. M., Robinson, B. H., Simpson, J., and Watson, D. J., *J. Chem. Soc., Chem. Commun.* 945 (1974).
160. Pearson, R. G., Louw, W., and Rajaram, J., *Inorg. Chim. Acta* **9,** 251 (1974).
161. Pearson, R. G., and Muir, W. R., *J. Amer. Chem. Soc.* **92,** 5519 (1970).
162. Pearson, R. G., and Rajaram, J., *Inorg. Chem.* **13,** 246 (1974).
163. Poutsma, M. L., *in* "Free Radicals" (J. K. Kochi, ed.), Vol. 2, Chapter 14. Wiley (Interscience), New York, 1973.
164. Price, S. J. W., *in* "Comprehensive Chemical Kinetics," (C. H. Bamford, and C. F. H. Tipper, eds.), Vol. 4, p. 197. Elsevier, Amsterdam.
165. See, e.g., Prins, R., and Kortbeek, A. G. T. G., *J. Organometal. Chem.* **33,** C33 (1971); Horsefield, A., and Wassermann, A., *J. Chem. Soc., Dalton Trans.* 187 (1972); Cowan, D. O., Candela, G. A., and Kaufman, F., *J. Amer. Chem. Soc.* **93,** 3889 (1971).
166. Rettig, M. F., Stout, C. D., Klug, A., and Farnham, P., *J. Amer. Chem. Soc.* **92,** 5100 (1970).
167. Robinson, S. D., *Organometal. Chem.* **3,** 376 (1975).
168. Roncin, J., Debuyst, R., *J. Chem. Phys.* **51,** 577 (1969).
169. Rowe, M. D., Gale, R., and McCaffery, A. J., *Chem. Phys. Lett.* **21,** 360 (1973).
170. Russell, G. A., *in* "Free Radicals" (J. K. Kochi, ed.), Vol. 1, Chapter 7. Wiley (Interscience), New York, 1973.

171. Sakurai, H., in "Free Radicals" (J. K. Kochi, ed.), Vol. 2, Chapter 25. Wiley (Interscience), New York, 1973.
171a. Sakurai, H., Mochida, K., and Kira, M., *J. Amer. Ch m. Soc.* **97,** 929 (1975).
172. Seyferth, D., and Hallgren, J. E., *J. Organometal. Chem.* **49,** C41 (1973).
173. Siegert, F. W., and Liefde Meijer, H. J. de, *J. Organometal. Chem.* **23,** 177 (1970).
174. Smentowski, F. J., *Chem. Anal.* **26,** 481 (1971).
175. Sommer, L. H., and Ulland, L. A., *J. Org. Chem.* **37,** 3878 (1972).
176. Starkie, H. C., and Symons, M. C. R., *J. Chem. Soc., Dalton Trans.* 731 (1974).
177. Stewart, C. P., and Porte, A. L., *J. Chem. Soc., Dalton Trans.* 722 (1973).
177a. Stieger, H., and Kelm, H., *J. Phys. Chem.* **77,** 290 (1973).
178. Symons, M. C. R., and Anderson, O. P., *Inorg. Chem.* **12,** 1932 (1973).
179. Teuben, J. H., and Liefde Meijer, H. J. de, *J. Organometal. Chem.* **17,** 87 (1969).
180. Thiele, K.-H., and Wagner, S., *J. Organometal. Chem.* **20,** P25 (1969).
181. Thomas, J. L., and Hayes, R. G., *Inorg. Chem.* **11,** 348 (1972).
182. Ugo, R., *Coord. Chem. Rev.* **3,** 319 (1968).
183. Ugo, R., Pasini, A., Fusi, A., and Cenini, S., *J. Amer. Chem. Soc.* **94,** 7364 (1972).
184. Walling, C., and Huyser, E. S., *Org. React.* **13,** 91 (1963).
185. Ward, G. A., Kruse, W., Bower, B. K., and Chien, J. C. W., *J. Organometal. Chem.* **42,** C43 (1972); Kruse, W., *ibid.* **42,** C39 (1972); Gramlich, V., and Pfefferkorn, K., *ibid.* **61,** 247 (1973).
186. Ward, H. R., in "Free Radicals" (J. K. Kochi, ed.), Vol. 1, Chapter 6. Wiley (Interscience), New York, 1973.
187. Warhurst, E., *Quart. Rev.* **5,** 44 (1951).
188. Watts, G. B., and Ingold, K. U., *J. Amer. Chem. Soc.* **94,** 491 (1972).
189. West, R., and Boudjouk, P., *J. Amer. Chem. Soc.* **95,** 3983 (1973).
190. Whitesides, G. M., Bergbreiter, D. E., and Kendall, P. E., *J. Amer. Chem. Soc.* **96,** 2806 (1974).
191. Wilt, J. W., in "Free Radicals" (J. K. Kochi, ed.), Vol. 1, Chapter 8. Wiley (Interscience), New York, 1973.
191a. Witman, M. W., and Weber, J. H., *Inorg. Nucl. Chem. Lett.* **11,** 591 (1975).
192. Wong, P. K., Lau, S. K. Y., and Stille, J. K., *J. Amer. Chem. Soc.* **96,** 5956 (1974).
193. Wood, J. M., and Brown, D. G., *Struct. Bonding* **11,** 47 (1972).
194. Wrighton, M., and Bredesen, D., *J. Organometal. Chem.* **50,** C35 (1973).
194a. Wrighton, M. S., and Ginley, D. S., *J. Amer. Chem. Soc.* **97,** 2065 (1975).
194b. Wrighton, M. S., and Ginley, D. S., *J. Amer. Chem. Soc.* **97,** 4246 (1975).

Subject Index

A

Acetylene complexes, 55–57, 245–265
  catalytic reactions, 261–265
    activation of acetylene on coordination, 261–262
    cocyclization with isocyanides, 263–265
    cyclo-oligomerization, 262
    linear oligomerization, 262–263
  electron-deficient species, 258–259
  in homogeneous catalysis, 245–265
  via insertion reactions, 251–261
    geometry of transition state, 255
    mechanism, 253–255
    NMR studies, 251
    stereochemistry of product, 251–253
  metalococyclization reactions, 260–261
    with carbon monoxide, 260
    with isocyanides, 260
  metalocyclization reactions, 256–260
    formation of metalocycloheptatrienes, 259–260
    formation of metalocyclopentadienes, 256–257
  structure and bonding, 246–251
    MO scheme, 246
    nature of interacting orbitals, 246–247
    twisting of $C \equiv C$ bond, 56
    variation of $C \equiv C$ bond length, 56, 247
    variation of $C \equiv C$—C bond angle, 247
    thermal stability and extent of back-bonding, 248
    variation in stretching frequencies of metal–acetylene unit, 248–251
      effect of other ligands, 250
Acetylenedicobalt hexacarbonyl, 138–140
  analogy with alkylidynetricobaltnonacarbonyls, 139–140

Acetylene trimerization, catalysis by alkylidynetricobaltnonacarbonyls, 137
Acylation reactions, with tricobaltcarbon decacarbonyl cation, 111–119
Alkyl compounds
  of antimony (V), 232–236
  of arsenic (V), 229–231
  indium halides, reaction with triorganostibine sulfide, 196
  of lithium, in reduction of $ClCCo_3(CO)_9$, 103
  of niobium, 237–238
  of phosphorus (V), 209–224
  of tantalum, 238–239
  of tin, reaction with tricobaltcarbon decacarbonyl cation, 115
  of zinc, in alkylation of tricobaltcarbon decacarbonyl cation, 115
Alkylidene trialkylarsoranes, 224–228
Alkylidene trialkylphosphoranes, 209–214
Alkylidynetricobalt nonacarbonyl complexes, 97–114
  analogy with acetylenedicobalt hexacarbonyls, 138–140
  as catalysts, 137–138
  carbon-functional derivatives, 97–110
    esters, 110–112
  catalytic hydrogenation of unsaturated derivatives, 97–110
  decomposition reactions, 135–138
    to acetylenes, 135–136
    to acetylene complexes, 135–136
    with methoxide ion, 136–137
    oxidation by ceric ion, 135–136
  Hammett σ-constant for $CCo_3(CO)_9$ group, 128
  infra-red spectra of ketone derivatives, 123
  mechanism of formation, 101
  mechanism of reactions, 140–141

preparation of alcohol derivatives, 119
  using triethylsilane, 121–122
reduction of ketone derivatives, 119–122
  by triethylsilane and trifluoroacetic acid, 120
reduction by organolithium reagents, 103
stable carbonium ions from, 119–134
steric hindrance in, 99, 111
structure, 99
synthesis, 100–110
  from acetylenedicobalt hexacarbonyls, 100
  via addition to olefinic bonds, 107–108
  using arylmercury compounds, 103–106
  from cobalt carbonyl and halocarbons, 101–102
  effects of hydrolysis, 101
  via Friedel-Crafts reactions, 106–107
  using Group III halides, 109–110
  via α-haloalkylmercurials, 103–106
  via radical reactions, 107–108
  via substitution at apical carbon atom, 103
tricobaltcarbon decacarbonyl cation from, 110–119
Allene, oligomerization reactions, 270
Allene complexes, cyclo-oligomerization reactions, 270–278
  catalysis by Ni(O) complexes, 271–277
  catalysis by rhodium complexes, 277–278
  kinetic studies, 273
  role of phosphorus ligands, 275–277
dissociation of, 269–270
effect of coordination on allene bond lengths, 267
in homogeneous catalysis, 265–278
^{1}H NMR studies, 267
relative reactivities of Ni, Pd, and Pt complexes, 269–270
X-ray structural data, 265–267
Aluminum halides, reactions
  with carbene complexes, 27
  with ClCCo$_3$(CO)$_9$, 116–119
  with cobalt carbonyl, 110
Antimony
  organometallic chemistry, 187–204

ylides, 231–232
  aryl derivatives, 231
Antimony (V) alkyls, 232–236
Arsenic ylides, 224–228
  effect of silylation on stability, 228–229
  NMR spectra, 226–227
  photoelectron spectra, 228
  reactions, 225
  structure, 226
  synthesis, 224–225
  thermal stability, 226
Arsenic (V) pentamethyl, 229–231
Azobene
  π-backbonding ability, 57
  MO calculations on, 57
  Ni(O) complexes, 57
Azobisisobutyronitrile, as radical initiator, 347, 380

B

Bent rehybridization theory, application to methyltin halides, 71
Bimolecular homolytic substitution reactions involving free radicals, 370–381
  with main group compound substrates, 371–373
  with transition metal complex substrates, 373–381
Boron halides, reactions
  with carbene complexes, 21–27
  with ClCCo$_3$(CO)$_9$, 118–119
  with cobalt carbonyl, 109
Boron hydrides, 145–150
  structural relationship with high nuclearity metal carbonyl clusters, 337–339

C

Carbene complexes, 2–20, 24–28
  acidity of α-carbon atom of alkoxyalkylcarbenes, 13
  addition-rearrangement reactions, 13
    with hydrogen halides, 13
  bonding, 4–6
  IR spectral studies, 5–6
  liberation of carbene ligand, 14–21
    by acid, 14–16
    by O, S, or Se, 17

Subject Index

by pyridine, 16
thermally, 16
trapping of ligand, 15
^{13}C NMR studies, 6, 133
in peptide synthesis, 11–12
positive charge on α-carbon atom, 14
reactions
  with acids, 14–16
  addition at carbene carbon, 9
  with amines, 11–12
  carbene displacement, 10
  carbonyl substitution, 9–10
  with electrophilic carbenes, 20–21
  with phosphines, 9–10
  with pyridine, 16
  substitution at carbene carbon, 11–12
  with vinyl ethers, 17–18
  with N-vinyl-2-pyrrolidones, 18–19
relationship with ylide complexes, 240
substitution of hydrogen at α-carbon atom, 13–14
deuterium exchange, 13
synthesis, 3–4, 6–8
  from alcohols and isocyanide complexes, 7
  via carbene transfer, 7
  via cleavage of electron-rich olefins, 8
  from 1,1-dichloro-2,3-diphenyl-2-cyclopropene, 6
  via organolithium reagents, 3–4
X-ray structural studies, 4–5, 7
Carbonium ions
  nonacarbonyltricobaltcarbon substituted, 119–134
  stability of ferrocenylmethyl, 131
Carboranes
  degradation reactions, 147
  metal complexes, 145–186
  polyhedral rearrangements, 146, 149
Carbyne complexes
  cleavage of carbyne ligand, 28–29
  ^{13}C NMR, 22
    chemical shifts of carbyne carbon atom, 133
  reactions, 28–29
  synthesis, 21–28
    via aluminum halides, 27
    via boron halides, 21, 24–26
    via gallium halides, 27
  X-ray structural studies, 22–23

Catalysis
  by alkylidynetricobalt nonacarbonyls, 137–138
    acetylene trimerization, 137
    olefin polymerization, 137–138
  of cocyclization of acetylenes and isocyanides, 263–265
  of cyclo-oligomerization of allenes, 270–278
    by Ni(O) complexes, 271–277
    by rhodium complexes, 277–278
  of oligomerization of acetylenes, 262–263
  by rhodacarborane complex, 183
Chemically induced dynamic nuclear polarization (CIDNP) spectra, and organometallic radicals, 347, 392
Cobalt
  alkylidyne nonacarbonyl clusters, 97–144
  high nuclearity carbonyl clusters, 287–288, 325–327
    carbido-derivatives, 327
    IR spectra, 326–327
    reactions, 325–327
    structural data, 288
    synthesis, 325–327
Cyclo-oligomerization
  of acetylenes, 262
  of allenes, 270–278
Cyclopropanes
  bonding, 36
  from carbene complexes and ethylvinyl ether, 17

D

Dewar-Chatt-Duncanson model
  for allene-metal bonding, 267
  for olefin-metal bonding, 35
Diazene-transition metal complexes, structures, 57
Dihalogenocarbenes, reaction with carbenecarbonyl complexes, 20–21
Dimethyltin dihalides, 71–72, 84–90
  electron diffraction studies, 71
  molecular complexes, 84–90
    with triorganostibine sulfides, 195–196
  Mössbauer studies, 72

NQR studies, 72
X-ray diffraction studies, 71–72
Dipole moments, of methyltin halides, 68, 76

E

Electron diffraction studies, 67, 69, 71, 72
Electron spectroscopy for chemical analysis (ESCA)
 measurement of electron density in platinum complexes, 44
Electron spin resonance (ESR), of organometallic radicals, 346–391

F

Free radicals, in organometallic chemistry, 345–392
Friedel-Crafts reaction
 acylation
  of a cobaltacarborane, 178
  of $H_2C=CHCCo_3(CO)_9$, 125
 in synthesis of alkylidynetricobalt nonacarbonyls, 106–107

G

Gallium halides, reaction with carbene complexes, 27
Galvinoxyl, as radical inhibitor, 347, 380
Grignard reagents in synthesis,
 of alkylidynetricobalt nonacarbonyls, 103–104
 of tertiary stibines, 197
Group V elements, penta-alkyls and alkylidenetrialkyls, 205–243

H

Hammett σ-constant, for $CCo_3(CO)_9$ substituent, 128
High nuclearity metal carbonyl clusters, 285–344
 bonding, 336–341
  analogy with polyboranes, 337–339
  application of noble gas rule, 336
  LCAO-MO theories, 339–341
  topological theories, 337
 cobalt, 325–327
 iridium, 332–333
 iron, 323–324
 nickel, 333–334
 osmium, 325
 platinum, 334–336
 reactions, 317–323
  effect of core enclosure by carbonyls, 317–318
  ligand substitution, 322
  oxidation, 320–321
  oxidative addition, 322–323
  reduction, 319–320
 rhodium, 327–332
 ruthenium, 324
 separation, 316–317
 solid state structures, 286–306
  carbido-carbon radius, 302
  of carbido-species, 300–302
  effect of charge on bond lengths, 295
  of heptanuclear species, 299
  metal coordination numbers, 293–295
  occurrence of triangular metal arrays, 291–292
  octahedron-trigonal prism transformation, 298–301
  tendency toward close packing of metals, 305–306
 solution state structures, 306–311
  ^{13}C NMR studies, 308–310
  IR studies, 306–308
 synthesis, 311–316
  bond energy considerations, 311–313
  of carbido-complexes, 314–315
  by condensation reactions, 311–313
  dependence on CO pressure, 312–313
  by pyrolysis, 313–315
  by redox condensation, 313–314
Homogeneous catalysis involving
 acetylene complexes, 245–265
 allene complexes, 265–278
Hückel calculations, on phosphorus ylides, 212
Hydroformylation, catalyzed by rhodacarborane, 183
Hydrosilylation, catalyzed by rhodacarborane, 183

I

Imines, transition metal complexes, 58
Infra-red studies
 on acetylene complexes, 248–251
 C=C bond length versus $\nu_{C=C}$ in olefin complexes, 39, 44

on high nuclearity metal carbonyl clusters, 306–308
on methyltin halides, 64–65, 68–75
molecular complexes, 79–91
Sn—C bond frequencies, 64–65
structural studies, 64–65
variation of ν_{CN} in isocyanide complexes, 45–46
Insertion reactions
of acetylenes, 251–261
mechanisms of reactions, 253–255
into metal-acetylene bonds, 251
into metal-σ-carbon bonds, 251
into metal-chlorine bonds, 251
into metal-hydrogen bonds, 251
metalococyclization, 260–261
metalocyclization, 256–260
stereochemistry of products, 251–253
or iridium, into B—H bonds, 181–182
of ylides, into silacyclobutanes, 215
Iridium, high nuclearity carbonyl clusters, 287
structural data, 288
synthesis, 332–333
Iron
carbene complexes, 4, 7
high nuclearity carbonyl clusters, 287
reactions, 324
structural studies, 288, 290, 293
synthesis, 323–324
Isocyanide complexes
carbene formation with alcohols, 7
variation of ν_{CN} with metal oxidation state, 44–45, 248–251
Isomerization, of carbenecarbonyl complexes, 9–11

K

Kinetic studies
oligomerization of allenes by nickel complexes, 273
isomerization of carbenecarbonyl complexes, 10–11
Ketone-metal complexes, 57–58
structure of nickel compound, 57

M

Mercurials, in synthesis of alkylidynetricobalt nonacarbonyls, 103, 105

Metallocarboranes, 145–186
with eleven vertices, 171–175
bimetallic species, 173–175
monometallic species, 171–173
polyhedral rearrangement, 175
reactions, 173–175
structures, 171, 174
synthesis, 171–175
with fourteen vertices, 171
geometry and number of polyhedral vertices, 148–149
in homogeneous catalysis, 182–183
alkene isomerization, 183
deuterium exchange, 183
with nine vertices, 178–180
bimetallic species, 180
NMR spectra, 179
structures, 178–179
synthesis, 178–180
oxidative-addition to B—H bonds, 180–182
in deuteration, 181–182
by iridium complexes, 181–182
relative reactivity of BH groups, 182
stabilization of high oxidation states, 156
synthesis, 150–155
from nido-carborane anions, 150–151
via polyhedral contraction, 152–153
via polyhedral expansion, 151–152
via polyhedral subrogation, 153
by thermal metal-transfer, 153–154
with ten vertices, 175–178
bimetallic species, 177–178
mixed sandwich complexes, 175
polyhedral rearrangements, 175–176
reactions, 178
synthesis, 175–178
trimetallic species, 178
with thirteen vertices, 167–171
polyhedral contraction, 169
polyhedral expansion, 169
polyhedral subrogation, 168–169
rearrangement reactions, 167–168
synthesis, 167–171
with twelve vertices, 155–167
bimetallic species, 166–167
mixed ligand complexes, 163–166
monometallic species, 155–163
polyhedral rearrangement, 158–159

reactions, 159–161
structures, 157
synthesis, 155–159, 161–167
Methyltin halides, 63–96
molecular complexes, 76–92
spectral studies, 64–67
structural studies, 68–76
Methyltin trihalides, 72–76
electron diffraction studies, 72
molecular complexes, 90–91
NMR data, 74–75
NQR data, 74
structures, 72–76
Molecular orbital calculations
on azobenzene, 57
on boron hydride derivatives, 147
for coordinated acetylenes, 246–247
on high nuclearity metal carbonyl clusters, 339–341
on phosphorus ylides, 212
Mössbauer studies
of methyltin halides, 66–67, 69, 71–72, 74
molecular complexes, 85–87, 89, 91

N

Nickel
azobenzene complex, 57
carbene complexes, 4
high nuclearity carbonyl clusters, 287
reactions, 333–334
structural data, 288, 290, 287–298
synthesis, 333–334
Niobium, penta-alkyls, 207, 237–238
Nitrogen ylides, 207–209
reactions, 207
synthesis, 207
Noble gas rule, and high nuclearity metal carbonyl clusters, 336–338
Nonacarbonyltricobaltcarbon-substituted carbonium ions, 119–134
consequences of charge distribution, 123–128
fluxional behavior, 134
σ-π hyperconjugation, 133
NMR spectra, 129–134
reactions at carbon, 122–124
structure, 129, 134
synthesis,
from alcohols and HPF_6, 122

via protonation of vinyl derivatives, 125
as weak electrophiles, 122–123
^{13}C Nuclear magnetic resonance studies
arsenic ylides, 226
carbene complexes, 14
carbyne complexes, 25
high nuclearity metal carbonyl clusters, 308–310
chemical shift and charge on CO, 309
fluxional behavior and reactivity, 309–310
nonacarbonyltricobaltcarbon carbonium ions, 130–134
penta-alkylphosphoranes, 216
phosphorus ylides, 212–213
^1H Nuclear magnetic resonance studies,
arsenic ylides, 226
methyltin halides, 65–66, 69, 71–72, 75–76
dependence of $J(^{119}Sn—C—^1H)$ on solvent, 65–66
methyltin halide complexes, 77, 79–81, 83–4, 86–91
nonacarbonyltricobaltcarbon carbonium ions, 130–134
olefin complexes, 44
organoantimony compounds, 189–191
penta-alkylphosphoranes, 216
phosphorus ylides, 212
^{31}P Nuclear magnetic resonance studies
penta-alkylphosphoranes, 216
phosphorus ylides, 212
Nuclear quadrupole resonance (NQR) studies, methyltin halides, 66, 72
molecular complexes, 79, 81, 83, 87, 89–91

O

Olefin hydrogenation, catalyzed by rhodacarborane, 183
Olefin isomerization, catalyzed by rhodacarborane, 183
Olefin polymerization, catalyzed by alkylidynetricobaltnonacarbonyls, 137–138
Olefin transition-metal complexes
bonding, 34–37
geometry, 53–55
five coordinate metal, 55

four coordinate metal, 54
three coordinate metal, 53–54
nonplanarity of bound olefin, 48–51
structural studies, 37–55
twist of olefin, 51–53
  orientation of π-substituents, 52
  pointing of trans-ligand, 52–53
variation of C=C bond length, 38–46
  correlation with chemical shift of olefinic protons, 44
  correlation with metal ionization potential, 44
  correlation with $\nu_{C=C}$, 39, 44
variation of M—C bond length, 46–48
  with olefin substituent, 46–47
  with trans-ligand-metal bond length, 47–48
Organoantimony compounds
  hexacoordinate species, 188–191
    cis-trans isomerism, 188–189
    IR studies, 189–191
    NMR studies, 189–191
    synthesis, 188–189
    X-ray studies, 189–190
  pentavalent species, 188
  tertiary stibines, 197–202
  triorganostibine sulfides, 192–197
Organometallic radicals, 345–392
  actinide species, 391
  generation, in homolytic substitution reactions, 370–381
    of main group species, 371–373
    of platinum derivatives, 373–381
  of Group IV elements, 352–363
    ESR spectra, 356–362
    mechanism of formation, 360–361
    stable species, 355–363
    structure, 354–355
    transient species, 352–355
    via UV irradiation, 352–354, 358
  lanthanide species, 391
  in metal-alkyl photolysis, 389–390
  in metal-alkyl thermolysis, 388–389
  metal-centered species, 346, 349–366
  not centered on metal, 367–370
    ESR spectra, 367–369
    metallocenes, 368
    nitroxide derivatives, 368–369
  in oxidative-addition reactions, 383–388
    ESR evidence, 384

in reductive-elimination reactions, 388
transition-metal compounds, 363–366
  carbonyl species, 366
  ESR spectra, 363
  by photolysis of dimers, 366
Organotin compounds, 63–96
  reaction with triorganostibine sulfides, 195–197
Osmium, high nuclearity metal carbonyl clusters, 287
  IR studies, 325
  structural data, 288, 292
  synthesis, 325
Oxidative addition
  of alkyl halides to Group VIII metal complexes, 381–388
    mechanism, 382–388
  to B—H bonds, 180–182
  to B—H—B bridges, 183
Oxymercuration, of alkylidynetricobalt-nonacarbonyls, 125

P

Penta-alkylphosphoranes, 214–224
  cyclic species, 215
    via insertion of ylides into silacyclobutanes, 215
  as intermediates, 217–224
    in ylide cleavage of 1,3-disilacyclobutanes, 218–219
    in ylide cleavage of monosilacyclobutanes, 219–224
  polycyclic species, 215–217
    spectral studies, 216
    synthesis, 216
    thermal stability, 216–217
Pentaalkylstiboranes, 232–236
  NMR spectra, 234
  reactions, 234–236
    with Bronsted acids, 235
    with Lewis acids, 235
    with oxidizing agents, 235–236
  synthesis, 232–233
    of alkenyl species, 233
    of methyl compound, 232
    via organometallics, 232–233
    of trimethylsilylmethyls, 233
  thermal decomposition, 233–234
  vibrational spectra, 234

Pentaarylphosphoranes, 214
Pentamethylarsorane, 229–231
  NMR spectra, 230
  reactions, 230
  structure, 230
  synthesis, 229–230
  thermal decomposition, 231
Peptide synthesis, via carbene complexes, 11–12
Phosphorus(V) alkyls, 214–224
Phosphorus ylides,
  bonding, 212–214
    d-orbital participation, 212
  dipole moments, 212
  effect of silylation on stability, 228–229
  NMR data, 212
  photoelectron spectra, 214
  properties, 210–211
  reactions, 210
    with silacyclobutanes, 218–220
  structures, 211
  synthesis, 209–210
  vibrational spectra, 211
Phosphorylide complexes, from carbene complexes and phosphines, 9–11
  irradiation, 9–10
Photoelectron spectra
  of arsenic ylides, 228–229
  of methyltin halides, 67
  of phosphorus ylides, 214
Platinum
  carbene complexes, 7
  high nuclearity carbonyl clusters, 287
    reactions, 335–336
    structures, 288, 296–298
    synthesis, 334–338
  radicals, from homolytic substitution reactions, 373–383

R

Radical anion, of $ClCCo_3(CO)_9$, 103
Radical reactions, rate determination by ESR studies, 347
Raman studies
  on high nuclearity metal carbonyl clusters, 310–311
  on methyltin halides, 64–65
Reaction mechanisms, in organocobalt cluster chemistry, 140

Reductive elimination reactions, involving radicals, 388
Rhodium, high nuclearity carbonyl clusters, 287
  anionic species, 329–332
  reactions, 328–332
  structural studies, 288, 295–296, 298, 302–305
  synthesis, 328–332
Ruthenium, high nuclearity carbonyl clusters, 287
  structural studies, 288, 293
  synthesis, 324

S

Spin traps, 347
  use in S_H2 reactions at Pt(II), 376–378
Stability constants, of methyltin halide complexes, 81–82

T

Tantalum
  pentaalkyls, 238–239
  ylide complexes, 207, 238–239
    mechanism of formation, 239
    X-ray structural studies, 239
Tertiary stibines, 197–202
  asymmetric species, 198–200
    rate of pyramidal inversion, 199–200
  cleavage of Sb-phenyl bonds, 197–198
  quaternization, 200
    resolution of asymmetric salts, 200
  reactions,
    of allyl compounds with $CpFe(CO)_2Cl$, 202
    with metal carbonyls, 200–202
  synthesis, 197–198
Tin, methylhalides and their molecular complexes, 63–96
Topology, and high nuclearity metal carbonyl clusters, 337
Transition metal complexes of unsaturated molecules, 33–61
Tricobaltcarbon decacarbonyl cation, 110–119
  electrophilic nature, 115
  mechanism of formation, 116–119
  reactions
    acylation, 111–119

with alkyltin compounds, 115–116
with triethylsilane, 116
synthesis
via aluminum chloride, 115
from $(CO)_9Co_3CCO_2R$, 111–112
of hexafluorophosphate salt, 112
Trimethyltin halides, 68–71
fluoride, 68–69
structural studies, 68–69
molecular complexes, 77–83
spectral studies, 68–71
Triorganostibine sulfides, 192–197
nature of Sb—S bond, 193–194
π-bonding, 194
spectral studies, 193–194
reactions, 195–197

U

UV spectra, of triorganostibine sulfides, 193–194

V

Vanadium, ylides, 236–237
Vitamin B_{12} coenzyme
ESR studies of active site, 369, 389–390, 392
homolysis of Co—alkyl bond, 389–390

X

X-Ray crystallographic studies
acetylene complexes, 55–57, 247
alkylidynetricobalt nonacarbonyls, 99, 110
alkyltin halides, 67–69, 71–72, 78, 83–86, 89
allene complexes, 265–267, 271
criteria for accuracy, 37–38
diazene complexes, 57
high nuclearity metal carbonyl clusters, 286–306, 327
imine complexes, 58
ketone complexes, 57–58
metallocarboranes, 148, 155–156, 168, 174, 178
olefin complexes, 37–55
phosphorus ylides, 211
six-coordinate methylantimony compound, 189–190
tantalum-ylide complex, 239

Y

Ylides
ammonium, 208
antimony, 231–232
arsenic, 224–229
donor properties, 206
nitrogen, 207–209
phosphorus, 209–214
tantalum, 238–239
transition-metal, relationship with carbene complexes, 240
vanadium, 236–237

Z

Zeise's salt, 37

Cumulative List of Contributors

Abel, E. W., **5**, 1; **8**, 117
Aguilo, A., **5**, 321
Albano, V. G., **14**, 285
Armitage, D. A., **5**, 1
Atwell, W. H., **4**, 1
Bennett, M. A., **4**, 353
Birmingham, J., **2**, 365
Brook, A. G., **7**, 95
Brown, H. C., **11**, 1
Brown, T. L., **3**, 365
Bruce, M. I., **6**, 273; **10**, 273; **11**, 447; **12**, 379
Cais, M., **8**, 211
Callahan, K. P., **14**, 145
Cartledge, F. K., **4**, 1
Chalk, A. J., **6**, 119
Chatt, J., **12**, 1
Chini, P., **14**, 285
Churchill, M. R., **5**, 93
Coates, G. E., **9**, 195
Collman, J. P., **7**, 53
Corey, J. Y., **13**, 139
Coutts, R. S. P., **9**, 135
Coyle, T. D., **10**, 237
Craig, P. J., **11**, 331
Cullen, W. R., **4**, 145
Cundy, C. S., **11**, 253
de Boer, E., **2**, 115
Dessy, R. E., **4**, 267
Dickson, R. S., **12**, 323
Emerson, G. F., **1**, 1
Ernst, C. R., **10**, 79
Fischer, E. O., **14**, 1
Fraser, P. J., **12**, 323
Fritz, H. P., **1**, 239
Furukawa, J., **12**, 83
Fuson, R. C., **1**, 221
Gilman, H., **1**, 89; **4**, 1; **7**, 1
Green, M. L. H., **2**, 325
Griffith, W. P., **7**, 211
Gubin, S. P., **10**, 347
Gysling, H., **9**, 361
Harrod, J. F., **6**, 119
Hawthorne, M. F., **14**, 145
Heck, R. F., **4**, 243
Heimbach, P., **8**, 29

Henry, P. M., **13**, 363
Hieber, W., **8**, 1
Ibers, J. A., **14**, 33
Ittel, S. A., **14**, 33
Jolly, P. W., **8**, 29
Jukes, A. E., **12**, 215
Kaesz, H. D., **3**, 1
Kawabata, N., **12**, 83
Kettle, S. F. A., **10**, 199
Kilner, M., **10**, 115
King, R. B., **2**, 157
Kingston, B. M., **11**, 253
Kitching, W., **4**, 267
Köster, R., **2**, 257
Kühlein, K., **7**, 241
Kuivila, H. G., **1**, 47
Kumada, M., **6**, 19
Lappert, M. F., **5**, 225; **9**, 397; **11**, 25; **14**, 345
Lednor, P. W., **14**, 345
Longoni, G., **14**, 285
Luijten, J. G. A., **3**, 397
Lupin, M. S., **8**, 211
McKillop, A., **11**, 147
Maddox, M. L., **3**, 1
Maitlis, P. M., **4**, 95
Mann, B. E., **12**, 135
Manuel, T. A., **3**, 181
Mason, R., **5**, 93
Matsumura, Y., **14**, 187
Moedritzer, K., **6**, 171
Morgan, G. L., **9**, 195
Mrowca, J. J., **7**, 157
Nagy, P. L. I., **2**, 325
Nakamura, A., **14**, 245
Nesmeyanov, A. N., **10**, 1
Neumann, W. P., **7**, 241
Okawara, R., **5**, 137; **14**, 187
Oliver, J. P., **8**, 167
Onak, T., **3**, 263
Otsuka, S., **14**, 245
Parshall, G. W., **7**, 157
Paul, I., **10**, 199
Petrosyan, V. S., **14**, 63
Pettit, R., **1**, 1
Poland, J. S., **9**, 397

Pratt, J. M., **11**, 331
Prokai, B., **5**, 225
Reutov, O. A., **14**, 63
Rijkens, F., **3**, 397
Ritter, J. J., **10**, 237
Rochow, E. G., **9**, 1
Roper, W. R., **7**, 53
Roundhill, D. M., **13**, 273
Rubezhov, A. Z., **10**, 347
Schmidbaur, H., **9**, 259; **14**, 205
Schrauzer, G. N., **2**, 1
Schwebke, G. L., **1**, 89
Seyferth, D., **14**, 97
Silverthorn, W. E., **13**, 47
Skinner, H. A., **2**, 49
Slocum, D. W., **10**, 79
Smith, J. D., **13**, 453
Stafford, S. L., **3**, 1
Stone, F. G. A., **1**, 143

Tamao, K., **6**, 19
Taylor, E. C., **11**, 147
Thayer, J. S., **5**, 169; **13**, 1
Todd, L. J., **8**, 87
Treichel, P. M., **1**, 143; **11**, 21
Tsutsui, M., **9**, 361
Tyfield, S. P., **8**, 117
van der Kerk, G. J. M., **3**, 397
Wada, M., **5**, 137
Walton, D. R. M., **13**, 453
Wailes, P. C., **9**, 135
West, R., **5**, 169
Wiles, D. R., **11**, 207
Wilke, G., **8**, 29
Wojcicki, A., **11**, 87; **12**, 31
Yashina, N. S., **14**, 63
Ziegler, K., **6**, 1
Zuckerman, J. J., **9**, 21

Cumulative List of Titles

Acetylene and Allene Complexes: Their Implication in Homogeneous Catalysis, **14**, 245
Alkali Metal Derivatives of Metal Carbonyls, **2**, 157
Alkyl and Aryl Derivatives of Transition Metals, **7**, 157
Alkylcobalt and Acylcobalt Tetracarbonyls, **4**, 243
Allyl Metal Complexes, **2**, 235
π-Allylnickel Intermediates in Organic Synthesis, **8**, 29
Applications of ^{119m}Sn Mössbauer Spectroscopy to the Study of Organotin Compounds, **9**, 21
Arene Transition Metal Chemistry, **13**, 47
Boranes in Organic Chemistry, **11**, 1
Carbene and Carbyne Complexes, On the Way to, **14**, 1
Carboranes and Organoboranes, **3**, 263
Catalysis by Cobalt Carbonyls, **6**, 119
Catenated Organic Compounds of the Group IV Elements, **4**, 1
Chemistry of Carbon-Functional Alkylidynetricobalt Nonacarbonyl Cluster Complexes, **14**, 97
^{13}C NMR Chemical Shifts and Coupling Constants of Organometallic Compounds, **12**, 135
Compounds Derived from Alkynes and Carbonyl Complexes of Cobalt, **12**, 323
Conjugate Addition of Grignard Reagents to Aromatic Systems, **1**, 221
Coordination of Unsaturated Molecules to Transition Metals, **14**, 33
Cyclobutadiene Metal Complexes, **4**, 95
Cyclopentadienyl Metal Compounds, **2**, 365
Diene-Iron Carbonyl Complexes, **1**, 1
Electronic Effects in Metallocenes and Certain Related Systems, **10**, 79
Electronic Structure of Alkali Metal Adducts of Aromatic Hydrocarbons, **2**, 115
Fast Exchange Reactions of Group I, II, and III Organometallic Compounds, **8**, 167
Fluorocarbon Derivatives of Metals, **1**, 143
Free Radicals in Organometallic Chemistry, **14**, 345
Heterocyclic Organoboranes, **2**, 257
α-Heterodiazoalkanes and the Reactions of Diazoalkanes with Derivatives of Metals and Metalloids, **9**, 397
High Nuclearity Metal Carbonyl Clusters, **14**, 285
Infrared Intensities of Metal Carbonyl Stretching Vibrations, **10**, 199
Infrared and Raman Studies of π-Complexes, **1**, 239
Insertion Reactions of Compounds of Metals and Metalloids, **5**, 225
Insertion Reactions of Transition Metal-Carbon σ-Bonded Compounds I. Carbon Monoxide Insertion, **11**, 87
Insertion Reactions of Transition Metal-Carbon σ-Bonded Compounds II. Sulfur Dioxide and Other Molecules, **12**, 31
Isoelectronic Species in the Organophosphorus, Organosilicon, and Organoaluminum Series, **9**, 259
Keto Derivatives of Group IV Organometalloids, **7**, 95
Lewis Base-Metal Carbonyl Complexes, **3**, 181
Ligand Substitution in Transition Metal π-Complexes, **10**, 347

Literature of Organo-Transition Metal Chemistry 1950–1970, **10,** 273
Literature of Organo-Transition Metal Chemistry 1971, **11,** 447
Literature of Organo-Transition Metal Chemistry, 1972, **12,** 379
Mass Spectra of Metallocenes and Related Compounds, **8,** 211
Mass Spectra of Organometallic Compounds, **6,** 273
Metal Carbonyl Cations, **8,** 117
Metal Carbonyls, Forty Years of Research, **8,** 1
Metal π-Complexes formed by Seven- and Eight-Membered Carbocyclic Compounds, **4,** 353
Metallocarboranes, Ten Years of, **14,** 145
Methyltin Halides and Their Molecular Complexes, **14,** 63
Nitrogen Groups in Metal Carbonyl and Related Complexes, **10,** 115
Nitrosyls, **7,** 211
Nuclear Magnetic Resonance Spectra of Organometallic Compounds, **3,** 1
Of Time and Carbon-Metal Bonds, **9,** 1
Olefin Oxidation with Palladium Catalyst, **5,** 321
Organic and Hydride Chemistry of Transition Metals, **12,** 1
Organic Chemistry of Copper, **12,** 215
Organic Chemistry of Lead, **7,** 241
Organic Complexes of Lower-Valent Titanium, **9,** 135
Organic Substituted Cyclosilanes, **1,** 89
Organoantimony Chemistry, Recent Advances in, **14,** 187
Organoarsenic Chemistry, **4,** 145
Organoberyllium Compounds, **9,** 195
Organolanthanides and Organoactinides, **9,** 361
Organometallic Aspects of Diboron Chemistry, **10,** 237
Organometallic Benzheterocycles, **13,** 139
Organometallic Chemistry: A Historical Perspective, **13,** 1
Organometallic Chemistry: A Forty Years' Stroll, **6,** 1
Organometallic Chemistry, My Way, **10,** 1
Organometallic Chemistry of Nickel, **2,** 1
Organometallic Chemistry of the Main Group Elements—A Guide to the Literature, **13,** 453
Organometallic Chemistry, Some Personal Notes, **7,** 1
Organometallic Complexes with Silicon-Transition Metal or Silicon-Carbon-Transition Metal Bonds, **11,** 253
Organometallic Nitrogen Compounds of Germanium, Tin, and Lead, **3,** 397
Organometallic Pseudohalides, **5,** 169
Organometallic Reaction Mechanisms, **4,** 267
Organometallic Reactions Involving Hydro-Nickel, -Palladium, and -Platinum Complexes, **13,** 273
Organopolysilanes, **6,** 19
Organosulphur Compounds of Silicon, Germanium, Tin, and Lead, **5,** 1
Organothallium Chemistry, Recent Advances, **11,** 147
Organotin Hydrides, Reactions with Organic Compounds, **1,** 47
Organozinc Compounds in Synthesis, **12,** 83
Oxidative-Addition Reactions of d^8 Complexes, **7,** 53
Palladium-Catalyzed Organic Reactions, **13,** 363
Pentaalkyls and Alkylidene Trialkyls of the Group V Elements, **14,** 205
Preparation and Reactions of Organocobalt(III) Complexes, **11,** 331

Redistribution Equilibria of Organometallic Compounds, **6,** 171
Radiochemistry of Organometallic Compounds, **11,** 207
Strengths of Metal-to-Carbon Bonds, **2,** 49
Structural Aspects of Organotin Chemistry, **5,** 137
Structural Chemistry of Organo-Transition Metal Complexes, **5,** 93
Structures of Organolithium Compounds, **3,** 365
Transition Metal-Carborane Complexes, **8,** 87
Transition Metal-Isocyanide Complexes, **11,** 21